Introducing the experts who bring you

Electronic Media...

John E. Craft
Frederic A. Leigh
Donald G. Godfrey

John E. Craft, Ph.D.

Director of Graduate Studies, Walter Cronkite School of Journalism and Telecommunication, Arizona State University

Frederic A. Leigh, Ed.D.

Associate Director and Clinical Professor, Walter Cronkite School of Journalism and Telecommunication, Arizona State University

Donald G. Godfrey, Ph.D.

Past President of Broadcast Education Association; Instructor and Researcher at the Walter Cronkite School of Journalism and Telecommunication

Welcome to John Craft, Frederic Leigh, and Donald Godfrey's new text, *Electronic Media*. Inside, you'll find a cogent, savvy discussion of broadcasting essentials, driven by the author team's many years of active involvement in both the field and the classroom. Craft, Leigh, and Godfrey's keen insights into students' needs *and* the industry's new directions allow them to produce a text that not only provides your students with the most current introduction to the field, but also gives them the background and skills that will form the foundation for their entire careers within the industry.

Turn the page to find out more about this groundbreaking new text!

PREVIEW

Craft, Leigh, and Godfrey's *Electronic Media*...

The first text for broadcasting students of the new millennium!

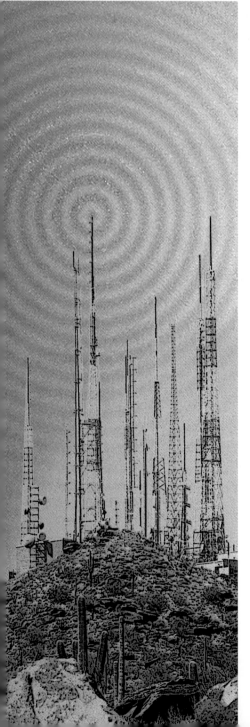

New technology, changes to regulatory policy, increased competition, unprecedented business mergers —these elements have, unquestionably, shaken up the broadcasting industry as we once knew it. For your students to become savvy media professionals and consumers in this current climate, they'll need to be able to examine, comprehend, and analyze these developments in light of the industry's past, present, and future.

Enter Craft, Leigh, and Godfrey's new text, *Electronic Media*— the text that expertly balances discussion of new developments with ample coverage of the industry's history, traditions, and fundamental concepts. Their broad overview of the electronic media, which progresses from a theoretical to a professional perspective, helps students increase their understanding of the challenges presented by the increasing levels of competition between new technologies and the traditional arenas of radio, television, and cable.

Once your students have grasped the foundational information and knowledge laid out in this text, they'll be better prepared to develop the skills and talents needed for their potential roles in the broadcasting industry. Whether they get involved on the air or behind the scenes—as professionals or analysts—your students will be sure to have a solid footing in an ever evolving world.

PREVIEW

Up-to-the-minute discussion of telecommunications and its impact on individuals, education, and business

...content that you'll find only in this text!

Thanks to their expertise in the field, the authors are able to seamlessly incorporate themes of new technology, business, global information, and change throughout the text, without disrupting their focus on the key concepts necessary to your students' introduction to the world of broadcasting.

In addition to thorough coverage of new technology in Part III, Electronic Media Technology, the authors integrate discussion of critical developments within the field in appropriate sections throughout the text. As a result, influential, contemporary issues such as digitization and the rapid evolution of delivery systems used in broadcast journalism are explored in appropriate detail—while kept within the context of the fundamental principles of broadcast communications.

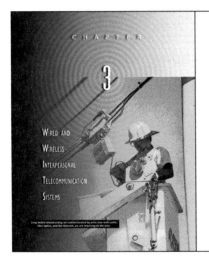

Chapter 3, **Wired and Wireless Interpersonal Telecommunication Systems,** covers the evolution of interpersonal technologies and the successful integration of telecommunications with computers.

Chapter 12, **Corporate and Instructional Electronic Media,** focuses on the increasing use of corporate video applications (a growing area of entry for students seeking jobs in the broadcasting industry), as well as instructional electronic media, such as distance learning.

PREVIEW

Organization, writing style, and features built on a belief in *user-friendliness*

In creating *Electronic Media*, the authors took great care to ensure that it would meet your needs for thorough coverage and logical organization, while meeting your students' needs for clearly written, engaging, and succinct descriptions of the content. To accomplish both goals, the authors incorporate numerous organizational and pedagogical features that enable you and your students to make this text an effective teaching and learning tool.

Unique four-part organization

The text's unique four-part organization successfully illustrates the connections among concepts related to the industry's foundations, business aspects, rapidly evolving technology, and societal consequences, allowing your students to witness the evolution of the industry *and* prepare themselves for the changes ahead. This clear-cut organization also helps you use the material *as you need it*—each topically organized part stands independently, allowing you the flexibility to adapt the text to your syllabus.

Clear, student-friendly writing, structure, and layout

The authors' writing style—praised by reviewers as crisp and highly readable—imparts information in a conversational yet authoritative tone that makes their descriptions and analysis of electronic technology clear and comprehensible. (See Chapter 13, Audio and Video Systems, for examples of their student-friendly tone.) Plus, they've made sure that the structure and layout of the book conform to the needs of today's students, many of whom are used to contemporary television's fast-paced, succinct "sound bites."

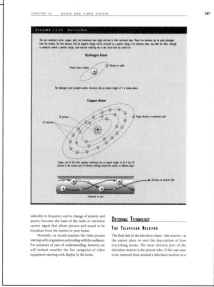

PREVIEW

Effective pedagogical elements

The authors have incorporated numerous features, designed to make the text's concepts relevant to today's students and easy to review and retain. These features include:

- *Boxes* that contain colorful biographies, history, quotations, legal issues, and other useful bits of information that highlight important people, concepts, and events from the life span of the industry (see examples on the facing page)
- A rich *art program,* comprising photos and diagrams that illustrate and demonstrate concepts and also represent the excitement of today's fast-paced media environment
- *Chapter summaries* that synthesize material and provide a broad perspective on electronic media
- *References and Suggested Readings* for each chapter that provide students with direction to additional information about the chapter topic
- A *Glossary* that provides definitions for the many key terms that have become the jargon of the electronic media industry

Let your students take skillful advantage of online resources with the text's

Thorough integration of the Web and *InfoTrac*® *College Edition!*

Web Watch

This reviewer-praised feature, found at the end of each chapter, presents a list of pertinent URLs that direct students to additional avenues of research or noteworthy points of interest on the Internet.

InfoTrac College Edition Exercises

These chapter-ending questions urge students to use *InfoTrac College Edition,* the online library featuring numerous broadcasting-related journals, to seek out chapter-related information that broadens their understanding of the material. (See the following page of this Preview for more information about *InfoTrac College Edition*).

RESOURCES

An EXCLUSIVE learning tool, available only from Wadsworth/Thomson Learning!

InfoTrac® College Edition

The latest news and research articles online—updated daily and spanning four years! *InfoTrac College Edition* is packaged **FREE** with every copy of this text to give you and your students four months of free access to an easy-to-use online database of reliable, full-length articles (not abstracts) from hundreds of top academic journals and popular sources. Ideal for launching lectures and igniting discussions, *InfoTrac College Edition* opens whole new worlds of information and research for students. Best of all, you and your students can take advantage of the *InfoTrac College Edition* exercises integrated throughout the text, which can further enhance your students' understanding of key issues in broadcasting!

Journals available on *InfoTrac College Edition* include...

Broadcasting & Cable	*Journal of Broadcasting & Electronic Media*	*U.S. News & World Report*
Communication Quarterly		*USA Today Magazine*
Communication World	*Multimedia Publisher*	*Variety*
EMedia Professional	*Newsweek*	*Video Age International*
Film Quarterly	*Public Broadcasting Report*	*Video Magazine*
Historical Journal of Film, Radio, and Television	*Television Digest*	*Worldwide Telecom*
	Time	...and many, many more!

Exclusive to Wadsworth/Thomson Learning. Available to North American college and university students only. Journals subject to change.

Also available...

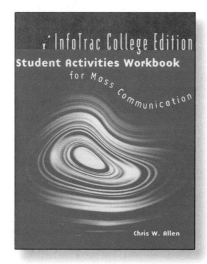

InfoTrac College Edition Student Activities Workbook for Mass Communication, 1.0

Chris W. Allen, Ph.D.
University of Nebraska, Omaha   0-534-56087-3

To enhance your students' experience with *InfoTrac College Edition*, choose this handy workbook. It features extensive individual and group activities that let students use *InfoTrac College Edition* to explore mass communication topics. It also includes guidelines for faculty and students on maximizing this resource. To bundle this workbook with *Electronic Media* and *InfoTrac College Edition*, please use ISBN 0-534-71295-9.

RESOURCES

Wadsworth's Communication Café Web site...

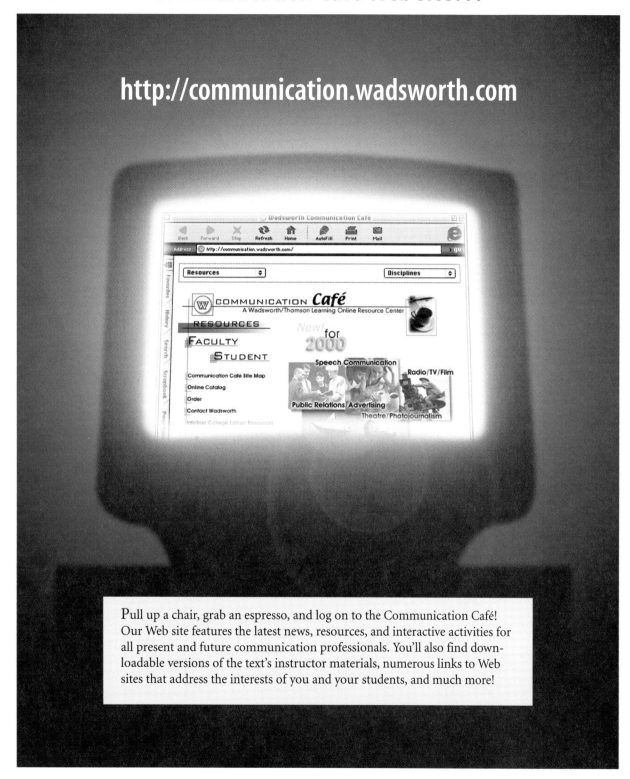

Pull up a chair, grab an espresso, and log on to the Communication Café! Our Web site features the latest news, resources, and interactive activities for all present and future communication professionals. You'll also find downloadable versions of the text's instructor materials, numerous links to Web sites that address the interests of you and your students, and much more!

RESOURCES

Exclusive teaching and learning tools that let you...

Bring the power of multimedia into your classroom!

CNN Today Videos

Mass Communication
- Volume I: 0-534-54813-X
- Volume II: 0-534-54818-0
- Volume III: 0-534-54823-7

Coming soon:
Electronic Media, Volume I

Launch your lectures with riveting footage from CNN, the world's leading 24-hour global news television network. *CNN Today Videos* allow you to integrate the newsgathering and programming power of CNN into the classroom to show students the relevance of course topics to their everyday lives. Organized by topics covered in a typical course, these videos are divided into short segments—perfect for introducing key concepts. A Wadsworth/Thomson Learning exclusive.

The Wadsworth Communication Video Library

Qualified adopters can make a selection from our library of videos, featuring a diverse range of topics covered in communication courses. Contact your Wadsworth/Thomson Learning representative for a detailed policy and list of titles. Videos are complimentary to qualified adopters based on adoption size.

PowerPoint® CD-ROM

Marc van Horne
Arizona State University

0-534-19571-7

This text-specific software is designed to work with the PowerPoint presentation program. It includes text and images that help you enhance your classroom presentations and discussions with material that illustrates important concepts in the text.

Also available...

Instructor's Manual

Marianne Barrett, Ph.D.
Arizona State University   0-534-19567-9

This manual includes sample syllabi, chapter outlines, chapter summaries, suggested lecture and discussion topics, classroom exercises, transparency masters, suggested assignments, and a test bank.

ExamView®

Cross-platform for Windows®
and Macintosh®   0-534-19570-9

Create, deliver, and customize tests and study guides with this easy-to-use assessment and tutorial system. *ExamView* offers "wizards" that guide you step-by-step through the process of creating tests, while its unique "WYSIWYG" capability allows you to see the test exactly as it will print or display online. You can build tests of up to 250 questions using up to 12 question types. Using *ExamView*'s complete word processing capabilities, you can also enter an unlimited number of new questions or edit existing questions.

ExamView® and ExamView Pro® are registered trademarks of FSCreations, Inc. PowerPoint and Windows are registered trademarks of Microsoft Corporation used herein under license. Macintosh and Power Macintosh are registered trademarks of Apple Computer, Inc., used herein under license.

www.wadsworth.com

wadsworth.com is the World Wide Web site for Wadsworth and is your direct source to dozens of online resources.

At *wadsworth.com* you can find out about supplements, demonstration software, and student resources. You can also send e-mail to many of our authors and preview new publications and exciting new technologies.

wadsworth.com
Changing the way the world learns®

FROM THE WADSWORTH SERIES IN PRODUCTION

Albarran, Alan B., *Management of Electronic Media*

Alten, Stanley, *Audio in Media*, 5th Ed.

Armer, Alan, *Writing the Screenplay*, 2nd Ed.

Craft, John, Frederic Leigh, and Donald G. Godfrey, *Electronic Media*

Eastman, Susan Tyler, and Douglas A. Ferguson, *Broadcast/Cable Programming: Strategies and Practices*, 5th Ed.

Gross, Lynne S., and Larry W. Ward, *Electronic Moviemaking*, 4th Ed.

Hausman, Carl, Philip Benoit, and Lewis B. O'Donnell, *Modern Radio Production*, 5th Ed.

Hausman, Carl, Lewis B. O'Donnell, and Philip Benoit, *Announcing: Broadcast Communicating Today*, 4th Ed.

Hilliard, Robert L., *Writing for Television and Radio*, 7th Ed.

Kaminsky, Stuart, *American Film Genres*, 2nd Ed.

Kaminsky, Stuart, and Jeffrey Mahan, *American Television Genres*

Mamer, Bruce, *Film Production Technique: Creating the Accomplished Image*, 2nd Ed.

MacDonald, J. Fred, *One Nation Under Television: The Rise and Decline of Network TV*

MacDonald, J. Fred, *Blacks and White TV: African Americans in Television Since 1948*, 2nd Ed.

Meeske, Milan D., *Copywriting for the Electronic Media*, 3rd Ed.

Morley, John, *Scriptwriting for High-Impact Videos: Imaginative Approaches to Delivering Factual Information*

Viera, Dave, *Lighting for Film and Electronic Cinematography*

Zettl, Herbert, *Sight Sound Motion*, 3rd Ed.

Zettl, Herbert, *Television Production Handbook*, 7th Ed.

Zettl, Herbert, *Television Production Workbook*, 7th Ed.

Zettl, Herbert, *Video Basics*, 3rd Ed.

Zettl, Herbert, *Zettl's VideoLab 2.1*

ELECTRONIC MEDIA

John E. Craft

Frederic A. Leigh

Donald G. Godfrey

Arizona State University

Australia • Canada • Mexico • Singapore
Spain • United Kingdom • United States

Radio-TV-Film Editor: Karen Austin
Executive Editor: Deirdre Cavanaugh
Publisher: Clark Baxter
Executive Marketing Manager: Stacey Purviance
Project Editor: Cathy Linberg
Print Buyer: Barbara Britton
Permissions Editor: Bob Kauser
Production Service: Gary Palmatier, Ideas to Images

Photo Researcher: Roberta Broyer
Copy Editor: Elizabeth von Radics
Illustrator: Robaire Ream, Ideas to Images
Cover Designer: Gary Palmatier
Cover Images: Corbis, Index Stock
Cover Printer: Phoenix Color
Compositor: Ideas to Images
Printer: Courier, Kendallville

COPYRIGHT © 2001 Wadsworth, a division of Thomson Learning, Inc. Thomson Learning™ is a trademark used herein under license.

ALL RIGHTS RESERVED. No part of this work covered by the copyright hereon may be reproduced or used in any form or by any means—graphic, electronic, or mechanical, including photocopying, recording, taping, Web distribution, or information storage and retrieval systems—without the written permission of the publisher.

Printed in the United States of America

1 2 3 4 5 6 7 04 03 02 01 00

For permission to use material from this text, contact us by
Web: http://www.thomsonrights.com
Fax: 1-800-730-2215
Phone: 1-800-730-2214

ExamView® and ExamView Pro® are registered trademarks of FSCreations, Inc. Windows is a registered trademark of Microsoft Corporation used herein under license. Macintosh and Power Macintosh are registered trademarks of Apple Computer, Inc., used herein under license.

Library of Congress Cataloging-in-Publication Data
Craft, John.
    Electronic media / John Craft, Frederic Leigh, Donald G. Godfrey.
      p. cm.
    Includes bibliographical references and index.
    ISBN 0-534-19566-0
    1. Broadcasting—United States. 2. Cable television—United States. I. Leigh, Frederic. II. Godfrey, Donald G. III. Title.
PN1990.6.U5 C73 2000
384.54′0973—dc21

00-020640

For more information, contact
Wadsworth/Thomson Learning
10 Davis Drive
Belmont, CA 94002-3098
USA

For information about our products, contact us:
Thomson Learning Academic Resource Center
1-800-423-0563
http://www.wadsworth.com

International Headquarters
Thomson Learning
International Division
290 Harbor Drive, 2nd Floor
Stamford, CT 06902-7477
USA

UK/Europe/Middle East/South Africa
Thomson Learning
Berkshire House
168-173 High Holborn
London WC1V 7AA
United Kingdom

Asia
Thomson Learning
60 Albert Street, #15-01
Albert Complex
Singapore 189969

Canada
Nelson Thomson Learning
1120 Birchmount Road
Toronto, Ontario M1K 5G4
Canada

Dedication

The authors would like to thank their spouses

Elizabeth Craft
Helen Leigh
Christina Godfrey

for their support during the preparation of this book.

Brief Contents

Introduction: Convergence and the Electronic Media 2

Part I Electronic Media History, Foundations, and the Law
1 Foundations of Electronic Media 12
2 Electronic Media: World War II to the Telecommunication Revolution 38
3 Wired and Wireless Interpersonal Telecommunication Systems 58
4 Foundations of Electronic Media Regulation 78
5 Operational Regulations for the Electronic Media 98

Part II The Business of Electronic Media: Systems, Sales, and Programming
6 Patterns of Media Organization and Ownership 118
7 Commercial and Economic Structures 148
8 Radio Programming 166
9 Broadcast Television Programming 182
10 News and Information Programming 206
11 Audience Analysis and Marketing 232
12 Corporate and Instructional Electronic Media 254

Part III Electronic Media Technology
13 Audio and Video Systems 270
14 Distribution Systems 302

Part IV Electronic Media and Society
15 The Global Village 330
16 Effects and Influences of the Electronic Media 348

Detailed Contents

Preface xvii

About the Authors xxi

Introduction: Convergence and the Electronic Media 2

The Electronic Media Today 3
The Mass Communication Model 5
Media Functions 6
   The Convergence of Telephone and Television 7
   Multimedia 8

Future Media Businesses 10
   Information and Entertainment 10
Preparing for the Future 10
   InfoTrac College Edition Exercises 10
   Web Watch 11

Part I Electronic Media History, Foundations, and the Law

1 Foundations of Electronic Media 12

From Wire to Wireless to Wire 13
The Industrial Age: 1850–1900 13
The Telegraph 13
Broadcasting's Prehistory 14
   Maxwell and Hertz 14
   Marconi: Inventor and Entrepreneur 14
      Precedents: Electronic Pioneers 15
   Fessenden: Voice Experiments 15
   De Forest: The Father of Radio 15
   Corporate Players Emerge 16
Maritime Radio: The Era of American Expansion 16
   Radio Technology and Sea Power 16
   The *Titanic:* A Catalyst 16
      Precedents: The Titanic 17
   The Radio Act of 1912 17
      Evolution of a Myth 17
   World War I 18
      The End of Patent Rivalry 18
The Roaring Twenties: Anything Goes 18

Radio After World War I 19
   Corporate Stations 20
   KDKA: "First in the United States" 21
   Popularizing the New Medium 21
   WEAF: AT&T Advances 21
      Toll Broadcasting 21
      Herbert Hoover on Radio Advertising 22
      Advertising Controversy, Yesterday and Today 23
      The First Network Broadcast 23
      Precedents: The Roaring Twenties 24
   WJZ: RCA's Key Station 24
   The Birth of NBC 24
      Sarnoff 26
   The Birth of CBS 26
      Precedents: The Network Contract 26
   William S. Paley Creates Networks 27
   The Mutual Broadcasting System 27
   Programming 28
The Thirties: A Need for Escapism 28
   Herbert Hoover on Radio Programming 28
   Replacing the Hearth 29

    The Thirties' Favorite Recreation 29
        War of the Worlds 29
        Programming Matures 30
Radio News Grows Up 31
    The Press-Radio War 31
        Precedents: The Depression and World War II 31
Radio During World War II 32
    Armstrong and FM 33

        Armstrong's FM Demonstration 33
Television: RCA Versus Farnsworth 34
        Precedents: FM 34
    Wartime Transition 35
        Precedents: Television 35
    SUMMARY 35
    INFOTRAC COLLEGE EDITION EXERCISES 36
    WEB WATCH 37

2 Electronic Media: World War II to the Telecommunication Revolution 38

Radio Growth and Transition 39
    Pioneers of Top 40 Radio Format 40
    Top 40 41
    Payola 41
    FM Emerges 41
    FM Growth 42
        Music Formats Defined 43
Television: Growth and Transition 43
The Freeze Is Lifted 44
The Golden Age of Television: The Fifties 44
        The First TV Situation Comedy 45
        Gunsmoke: From Radio to Television 46
    The Move to Hollywood 46
    The Game Show Scandal 46

Television Grows Up: The Sixties 47
    Government Regulation and Public Pressure 48
    The Rise of Syndication 48
The Noncommercial Alternative 49
    Transition to Public Broadcasting 50
PBS into the Marketplace 51
New Competition: Cable TV 52
The Video Marketplace: The Seventies 54
    Cable Expands 55
Other Technologies Emerge 55
    SUMMARY 55
    INFOTRAC COLLEGE EDITION EXERCISES 56
    WEB WATCH 57

3 Wired and Wireless Interpersonal Telecommunication Systems 58

Visual Communication Systems 59
The Telegraph 59
    The Invention 59
        The Pony Express 60
    The Expanding Business 60
        Samuel F. B. Morse 61
    Early Competition 61
        Cyrus W. Field 62
    The Industry Peaks 62
The Telephone 63
    The Invention 63
        Alexander Graham Bell 64
        Theodore N. Vail 65
    The Expanding Business 65
    Early Competition 66
    AT&T Moves into Mass Media 67
    The World's Largest Corporation 67

    Satellite Communications 68
    The Beginning of the End for Ma Bell 69
        William McGowan 70
    The Breakup of AT&T 70
        The Original Baby Bells 71
    The Outcome: Bell Today 71
Converging Systems: Telcos and Cable 71
Data Communication Networks 72
    Telecommunication Networks 73
    The Internet 73
Wireless Telecommunication Networks 74
    Cellular Telephone 74
    VSAT 75
    SUMMARY 75
    INFOTRAC COLLEGE EDITION EXERCISES 76
    WEB WATCH 77

4 Foundations of Electronic Media Regulation   78

Key Issues 79
  The Airwaves as a Natural Resource 79
    The Business of the Nation Was Business 79
    Clarence C. Dill 80
  Censorship 80
    Herbert C. Hoover 81
  Corporate Monopoly 82
    Wallace H. White Jr. 83
The 1927 Radio Act 83
  Chaos on the Air 84
  Precedents: Balanced Programming 85
    Radio Act of 1927 Declares "Public Interest" Primary 86
The Communications Act of 1934 86
  The Great Lakes Opinion 86
Loosening the Chains 87

Programming and the Law 87
  The Blue Book 88
  The 1960 Program Policy Statement 88
    Program Policy Statement of 1960 89
  The Fairness Doctrine 89
    The Fairness Doctrine 90
  Political Campaign Law: Section 315 91
    Sarnoff on Equal Time Law 92
  New Technology, Deregulation, and the Marketplace 92
  The Telecommunications Act of 1996 93
    Summary of the Telecommunications Act of 1996 94
  SUMMARY 95
  INFOTRAC COLLEGE EDITION EXERCISES 96
  WEB WATCH 97

5 Operational Regulations for the Electronic Media   98

The Organization of the FCC 99
  Rule Making and Enforcement Process 101
Statutory and Related Applied Law 103
  The First Amendment 103
  The Sixth Amendment 103
FCC Rules and Regulations 104
  Sections 315 and 312-7a: Questions and Obligations 104
    "If any licensee shall permit..." 104
    "legally qualified candidate..." 104
    "use..." 104
    Political Broadcast Law 105
    "afford equal opportunities..." 106
    "licensee shall have no power of censorship..." 106
    "charges..." 106
  The Fairness Doctrine 106
  Indecent, Obscene, and Profane Material 107
  News Distortion 108
  Hypoing 109
  False or Deceptive Advertising 109
  Multiple Ownership 109
  Equal Employment Opportunity Regulations 109
    FCC Rules on Contests 109

  Children's Television Advertising 110
  Professional Advertising 110
  International Advertising 110
  Public Files 111
  Station Identification 111
  Sponsored Programs 111
  Recordings 111
  Payola and Plugola 111
  Fraudulent Billing 112
  Advertising Tobacco Products 112
Cable and New Technology Regulation 112
  Copyright Issues 112
  "Must Carry" and Retransmission Consent 113
  New-Technology Issues 113
    AM Band Expansion and Stereo 113
    High-Definition Television 113
    Media-Related Federal Laws 114
    The Telephone Companies 114
  SUMMARY 115
  INFOTRAC COLLEGE EDITION EXERCISES 116
  WEB WATCH 117

Part II The Business of Electronic Media: Systems, Sales, and Programming

6 Patterns of Media Organization and Ownership 118

Ownership of Radio and Television Stations 119
   The Broadcast Station: The Retailer 119
   Obtaining a Broadcast License 120
   Ownership Limits 121

Network Owners 122
   Group Owners 124
   Cross-ownership 125

Local Owners 125

Station Organization 126

Ownership Patterns in Cable Television 127
   Early Locally Owned Systems 127
   The Multisystem Cable Owner 128
   Vertical Integration 129
      Media Megamergers 130
   Multinational Ambitions 131

Cable System Organizational Structure 131
   Media Merger Chronology 131

Ancillary Businesses in Broadcasting 132
   The Networks: Then 133
   The Networks: Now 133
   Network/Affiliate Relationships 134
      Affiliate Contracts 135
   The Future of the Networks 135
   Satellite-Delivered Cable Networks 136
      Cable Network and Cable System Relationships 138
   Content Providers 138
      Television Production Companies 139
      Radio Production and Distribution 139
      Syndication 139
   The Consultants 140
   Unions and Guilds 142
   Music Licensing 143
   Professional Associations 143

Business Infrastructure 144
   SUMMARY 145
   INFOTRAC COLLEGE EDITION EXERCISES 146
   WEB WATCH 147

7 Commercial and Economic Structures 148

Traditional Economics 149
   Market Size 150
   Coverage Area 150
   Competition 150
   Audience Demographics 150

The Advertiser as Customer 151

The Sales Department 152
   Selling Tools of the Broadcast Station 154

Other Sales Strategies 156

Advertising Agencies 157

Station Representative Companies 158

The Audience as Customer 158

Programming for the Customer 159

Promotion of the Program 160

Cable Economics 161

Revenue Structure of the New Media 163
   SUMMARY 164
   INFOTRAC COLLEGE EDITION EXERCISES 165
   WEB WATCH 165

8 Radio Programming — 166

Contemporary Radio Emerges 167
Radio Dayparts 167
The Radio Clock 168
Contemporary Radio Formats 169
    Adult Contemporary 169
    Top 40/Contemporary-Hit Radio 169
    AOR/Classic Rock 170
    Country 170
    Urban Contemporary 170
    Easy Listening 170
    Other Music Formats 171
        News, News/Talk Radio Shares Jump 172
    Information: News/Talk 173
Local Programming 173
    Who Dominates Talk Radio? 174
Sources of Programming 174
    Diversified Radio Network: ABC 176
Radio Programming Issues 176
    Indecency 176
    Consolidation 177
New Technology and the Future 178
    SUMMARY 179
    INFOTRAC COLLEGE EDITION EXERCISES 180
    WEB WATCH 181

9 Broadcast Television Programming — 182

The Programming Department 183
Television Audiences 185
Programming on the Product Shelf 185
Television Dayparts 186
    Programming Appeals 187
    Program Types 187
    Program Sources 187
    Scheduling Strategies 188
        ABC 2000—Millennium Coverage 189
Network Programming 192
    Game Show Frenzy 193
The Business of Syndication 193
Programming Costs 196
Cable Programming 197
    Must Carry 198
    Superstations 199
    Pay Networks 199
    Basic Cable 199
        A Short History of HBO 200
        A Short History of CNN 201
    Local-Origination and Access Channels 202
        Local Origination 202
        Access Channels 202
    SUMMARY 203
    INFOTRAC COLLEGE EDITION EXERCISES 204
    WEB WATCH 205

10 News and Information Programming — 206

Defining News 207
    Network News 207
        Edward R. Murrow 208
        The Television Documentary 209
    Local News 210
    Deregulation and Local News 210
    Technology: Reshaping the News 211
The Newsroom 212
    The News Director 212
    Producers and Assignment Editors 212
    Reporters and Field Producers 214
    Anchors and Talent 214
    Support Staff 215
Sources: AP and UPI 215
    Specialized Wire and Computer Services 216
    Network Feeds 216
    News Bureaus 216
        DIALOG 217

Satellite News Gathering 217
Cable News Network 218
Video News Releases 219
News Inserts 220
News Programming Strategies 220
Content Treatment 221
Promotions 221
Leads—In and Out 222
Consultant Research 222
The Consultant's Report 223
Talent 223
Critical Trends in News and Information Media 224
Edward R. Murrow: A Broadcaster Talks to His Colleagues 225
John Chancellor 226
SUMMARY 229
INFOTRAC COLLEGE EDITION EXERCISES 230
WEB WATCH 231

11 AUDIENCE ANALYSIS AND MARKETING 232

Understanding Research 233
Early Research Companies 234
Defining Market 235
The Research Sample 236
Data Gathering 237
Diaries 237
Telephone Surveys 238
Electronic Meters 238
Personal Interviews 238
In-house and Consultant Research 239
Data Analysis: Ratings and Shares 239
Research Reports 242
Radio: Arbitron 242
Television: Nielsen 244
Applying the Research 244
Radio 244
Television 246
Sales and Marketing Applications 246
Criticism of Audience Research 250
SUMMARY 251
INFOTRAC COLLEGE EDITION EXERCISES 252
WEB WATCH 253

12 CORPORATE AND INSTRUCTIONAL ELECTRONIC MEDIA 254

Corporate Video 255
Corporate Programming Goals 256
Internal Communications 258
Orientation 258
Training 258
Information 259
Motivation 260
Documentation 260
External Communications 261
Sales Information 261
Product Demonstrations 261
Public Relations 262
News Releases 262
Corporate Video Departments 262
The Future of Corporate Video 263
Distance Learning 263
History of Instructional Television 263
Distance-Learning Technologies 266
Distance Learning Today 266
SUMMARY 268
INFOTRAC COLLEGE EDITION EXERCISES 268
WEB WATCH 269

Part III Electronic Media Technology

13 Audio and Video Systems — 270

Tools of the Trade 271
Properties of Sound 272
   Frequency and Amplitude 273
The Sound Path 274
   Signal Generation 275
   Signal Processing 278
   Audio Signal Storage Equipment 279
   Analog Versus Digital Transduction 281
   Signal Transmission or Distribution 282
   Signal Reception, Decoding, and Presentation 282
   Digital Broadcasting 284
The Television Sound Path 284
Properties of Vision 285
Decoding Technology 287
   The Television Receiver 287

New Display Technologies 289
Encoding Technology 291
   The Video Camera 291
   Computer-Generated Video 293
Signal Manipulation 293
   The Video Production Switcher 293
   Monitoring Systems 294
   Processing Amplifiers 295
   Edit Controllers 296
Storage 297
   Videotape Recording 297
   Digital Recording 299
      Summary 299
      InfoTrac College Edition Exercises 300
      Web Watch 301

14 Distribution Systems — 302

Broadcast Communication Systems 304
   The Electromagnetic Spectrum 304
      Frequency Allocation 307
   The Carrier Wave 308
      The Antenna 309
   AM Radio 311
      Properties of AM 311
   FM Radio 311
      Properties of FM 312
      Frequency Allocations 312
      FM Stereo and Subsidiary Services 312
   Broadcast Television 313
      Properties of Television Frequencies 313
      Stereo and Vertical-Blanking Services 315
   Other Broadcast Technologies 315
      Low-Power Television 315
      Microwave 316
   Satellites 316
      Uplinks and Downlinks 318

Wired Communication Systems 320
   Network Standards 320
   Video Compression 320
   Cable Television Systems 321
      Star Versus Tree Designs 323
   Fiber-optic Distribution Systems 324
Physical Distribution 325
   Home Audio: The Phonograph 325
   Magnetic Tape 325
   Optical Audio 325
   Home Video 326
   Optical Video 326
Digitizing the Technology 327
      Summary 328
      InfoTrac College Edition Exercises 328
      Web Watch 329

Part IV Electronic Media and Society

15 The Global Village — 330

A Framework for Regulation   331
International Telecommunication Agreements   332
International Broadcasting   333
National Broadcast Systems   335
Comparative Domestic Broadcast Systems   336
   Canada   336
   Mexico   337
   Great Britain   337
   Germany   338
   Turkey   339
   Russia   339
   Saudi Arabia   340
   India   341
   Japan   342
   People's Republic of China   343
      Hong Kong   343
      Republic of China   343
Worldwide Media Distribution   344
   Summary   345
   InfoTrac College Edition Exercises   346
   Web Watch   347

16 Effects and Influences of the Electronic Media — 348

Effects and Influences of the Media   349
Effects Theories   349
   Violence Theories   350
Media Stereotypes   351
Children and Media Effects   352
Politics and Media Effects   354
Advertising and the Media   355
   Summary   356
   InfoTrac College Edition Exercises   356
   Web Watch   357

Epilogue   358

Glossary   359

References and Suggested Readings   374

Index   384

Preface

Purpose

Nearly every institution of higher education that has a mass communication curriculum teaches an introductory course in broadcasting. This class is usually designed to provide the fundamentals of the business of electronic mass media from history to regulation and from technology to programming. It is often the first class that a student takes in the subject, and the information and knowledge gained become the foundation on which a career can be built. The introductory class is perhaps the most important in the student's academic preparation, because it is where the student's first attitudes concerning the study of broadcasting are formulated.

The first forty years of broadcast instruction have been marked with little change except for the inclusion into the curriculum of television, management, and ethics. Now, however, broadcasting has become electronic mass media, and instructional materials are needed to reflect the new directions of the industry while at the same time preserving the traditions that have served well. The purpose of this book is to provide the electronic mass media fundamentals needed by the student of the new millennium.

Like almost every other American institution, the businesses of broadcasting and cable—the electronic mass media—have undergone tremendous change in the past several years. Many retired pioneers of broadcasting and cable would be hard pressed to understand the industry that they knew so well just a few decades ago. The once stable structures that governed the electronic mass media have been shaken by such recent trends as developments in technology, major changes in regulatory philosophy, increased competition, unprecedented business mergers, international aspirations, and radical changes in our societal structure. The arenas of change are all interrelated and can be traced to the current upheavals in many other industries. The college student beginning study of the field of electronic mass media must be provided with information relevant to this new business as well as an understanding of what we once knew as broadcasting.

In an electronic world of burgeoning technology, inconsistent regulatory policy, intense competition, business mergers, international aspirations, and a deteriorating social order, the student about to embark on the study of the electronic mass media might feel a bit confused and uncertain about the future.

Some students may wish to pursue a career on the air, and others see themselves behind the scenes in a journalistic or production capacity. Other students have the entrepreneurial drive to obtain a sales or eventually a management position in the media. Some want to create advertising messages; still others wish to examine the media from a sociological perspective. No matter what the student's professional goal, or the twists and turns of our society and business, or what effect regulation will have on shaping tomorrow's media systems, it is important for each student to have as complete an understanding as possible of the antecedents of the business. Tomorrow's successful *intercaster* (a word that may replace *broadcaster* as we combine the functions of broadcast systems and computer networks) must know the history, technology, and journalistic and business practices of traditional broadcasting as well as the new interpersonal and mass communication technologies.

The challenge for the student is to understand the changes as radio, television, and cable have been reinvented to meet the demands of the increased competition that the new technologies have created. In this text the authors help the student meet that challenge by providing a broad overview of the electronic media, ranging from a theoretical to a professional perspective. The text is designed to provide the information needed by the student in an introductory broadcast course at the freshman/sophomore level in a community college, college, or university.

This text aims to accomplish the following goals:

- To serve as a foundation for the professionally oriented student by defining terms and organizational patterns of the business
- To provide fundamentals for upper-division courses that the student might take later in a broadcast or mass media curriculum
- To serve as a basic electronic mass media consumer education tool for students in the liberal arts curriculum

ORGANIZATION

This text is organized to best meet the needs of students taking courses in the fundamentals of electronic media. Our first task, therefore, is to establish the historical and analytical foundation to support the ever changing evolution of the industry. Because the student must be prepared to face constant change, the text emphasizes the following:

- Traditional, current, and evolving electronic mass communication organizational structures
- New technologies that are creating both a convergence of media and increased competition
- The future, which holds potential and challenge in the electronic media as purveyors of global information and entertainment within a competitive environment

To best accomplish the goals of the text, the book is arranged in four parts:

- Electronic media history, foundations, and the law
- The business of electronic media: systems, sales, and programming
- Electronic media technology
- Electronic media and society

Part I, Electronic Media History, Foundations, and the Law, provides the historical and legal background necessary to place into perspective the current business structure of radio, television, and cable. History of the electronic media in the United Stated is divided into three chapters. Legal considerations of broadcasting are treated in a similar manner—one chapter examines regulation from a theoretical and historical point of view, whereas the second looks at current regulation.

Part II, The Business of Electronic Media: Systems, Sales, and Programming, is devoted to business structures, sales and economics, radio and television programming, broadcast news, and audience research. In addition to these traditional topics, the text anticipates trends toward convergence of the electronic mass media and interpersonal media. It includes chapters devoted to nonbroadcast corporate and organizational video.

Part III, Electronic Media Technology, is divided into two chapters—one for audio and video and one that examines transmission technologies. Although it provides a foundation for broadcasting, cable, and telecommunications, this technical information is placed near the back of the book, because it is recognized that some instructors will choose not to incorporate it into the classroom curriculum.

Part IV, Electronic Media and Society, examines international media and media systems in other countries as well as research into effects of the media.

The text was written by three authors with extensive academic and professional experience in the mass media. Specific chapters were assigned to take advantage of each author's expertise. Professor Godfrey contributed chapters 1, 4, 5, and 10, detailing early history, regulation, and news and information programming. Professor Leigh authored chapters 2, 8, 11, and 16 and contributed to chapter 13, covering broadcasting and society, recent media history, radio programming, audience measurement, and media effects. Professor Craft wrote chapters 3, 6, 7, 9, 12, 14, and 15 and contributed to chapter 13.

The material covered in those chapters includes telephone communication systems, electronic media ownership, economics, television programming, corporate and instructional video, technology, and electronic media distribution systems.

Unique Attributes of the Text

Two unique chapters are included in this text:

- Chapter 3, Wired and Wireless Interpersonal Telecommunication Systems

- Chapter 12, Corporate and Instructional Electronic Media

Chapter 3 provides a historical and contemporary examination of wired telecommunication systems such as the telephone, as the trend in electronic communications seems to be toward convergence of mass media and interpersonal electronic media. Chapter 12 provides a look at nonbroadcast video communication systems and practices such as corporate video and distance learning. It is anticipated that students will be able to identify employment opportunities as these electronic communication fields develop.

The text is also unique in its descriptions of today's telecommunication issues. The themes of new technology, business, global information, and change are interwoven throughout the text. And though the authors deal with new technology in part III, digitization, for example, has permeated every aspect of the field and is covered is various chapters. Broadcast journalism is another example—while delivery systems are undergoing rapid evolution, the fundamentals of information gathering and communications remain at the core. Again, both subjects—delivery and reporting—must be addressed.

Structure of the Text

Because today's "television age" student obtains information in a highly fragmented style, the structure and layout of the book conform to the needs and lifestyle patterns of the student while moving quickly from topic to topic. The book is liberally illustrated with photographs, figures, tables, and sidebars. The graphic elements are designed to complement the text and contribute to fluid and comfortable reading. Running text is succinct, and factual information is often presented in sidebars. The layout and structure of the text are designed to be the print equivalent of the contemporary television newscast, with its collection of 20-second sound bites.

The authors want the book to be interesting for the student, yet they realize that the professor—who likely learned about the broadcast business at the feet of Sydney Head—is the person who will actually select the text. It must therefore be comprehensive and organized to fit within a traditional class structure, contain testable detail, and not be too radical in design so as to alienate current instructors of introductory courses. This textbook is original in design and organization, but it includes the traditional content areas for an introductory course. And though the text is ordered logically, the four parts are topically organized and stand independently, providing instructors flexibility in adapting reading assignments to their syllabi.

For the convenience of both instructors and students, end-of-chapter summaries are provided. We've also included at the end of each chapter helpful Internet addresses—Web Watch—and even online exercises based on *InfoTrac College Edition*. A glossary, providing definitions of the many key terms that have become the jargon of the electronic telecommunication industry, as well as extensive references and suggested readings are included at the back of the book.

Acknowledgments

Many people have contributed to the creation of this text. We would like to thank our spouses and families for their understanding of the time commitment required in the preparation of this work. In addition, we thank those many colleagues, both at Arizona State University and at other universities throughout the country, who patiently answered many questions. We would also like to thank those in the profession who provided much valuable information needed for the completion of this textbook.

We would especially like to thank Janet Soper of the College of Public Programs Publication Assistance Center, her colleagues Mary Fran Draisker and Roisan Rubio, and former colleagues Chrys Gakopoulos and Jan Nagle for their work in preparing the graphics and formatting the manuscript. Thanks also go to Wadsworth personnel for their able assistance and helpful comments: Karen Austin, radio-TV-film editor; Deirdre Cavanaugh, executive editor; Cathy Linberg, project editor; Stacey Purviance, executive marketing manager; and Dory Schaeffer, former editorial assistant.

All textbook manuscripts go through several critical evaluations by those teaching similar courses. The review process serves as an important check on not only the authors' ideas, but also on the validity of the information the work contains. We are grateful for the efforts of the many reviewers who have provided their time and expertise to make this a more accurate text. Those that have contributed to the review process include the following:

    Virginia Bacheler, SUNY-Brockport

    Warren Carter, Golden West College

    Joseph Chuk, Kutztown University

    Robert Finney, CSU–Long Beach

    Richard Goedkoop, LaSalle University

    Anne Hoag, Pennsylvania State University

    Jerry Howard, University of Oklahoma

    Val Limbugh, Washington State University

    Thomas Lindlof, University of Kentucky

    Barry Litman, Michigan State University

    Edward Morris, Columbia College, Chicago

    Michael Murray, University of Missouri–St. Louis

    Cathy Perron, Boston University

    William Ryan, Marist College

    R. Brooks Sanders, Tompkins-Cortland Community College

    Douglas Sudhoff, University of Kansas

    Patricia Turner, Angelo State University

About the Authors

John E. Craft, Ph.D., is director of graduate studies at the Walter Cronkite School of Journalism and Telecommunication at Arizona State University (ASU). He teaches courses in telecommunication management, media and society, and new communication technology, as well as consults for local media corporations. Prior to moving to Arizona, Craft was director of instructional broadcasting at WOUB-TV in Athens, Ohio, and director of educational television for Hancock County School District in West Virginia. He is currently producing documentaries that have been broadcast over numerous public television stations and have won national awards. Craft has served as president of the Arizona Chapters of International Television Association, the Arizona Cable Forum, and the National Academy of Television Arts and Sciences. He was trustee of the National Academy of Television Arts and Sciences. His doctorate degree in mass communication is from Ohio University.

Frederic A. Leigh, Ed.D., is associate director and clinical professor at the Walter Cronkite School of Journalism and Telecommunication at Arizona State University. In addition to school administrative service, he teaches courses in broadcasting and serves as adviser of the ASU campus radio station. Leigh's professional experience includes a decade in public radio programming and management. He has published numerous journal articles and several book chapters. He also co-edited a historical dictionary of American radio. He holds bachelor's and master's degrees in communication arts and a doctorate degree in higher-education administration. Leigh is a member of the Broadcast Education Association (BEA) and served two terms on its board of directors.

Don G. Godfrey, Ph.D., is a Canadian, born and raised in Alberta. A teacher, professional broadcaster, and published historian, he has worked for fifteen years in corporate communication and commercial radio and television, including KIRO-TV, KSVN-AM, and KEZI-TV. Godfrey retains his professional activity as a part of his teaching career through freelance consulting and corporate work. He has received numerous awards for his video productions and writing. Author of several books, Godfrey's current writing includes a biography of Philo T. Farnsworth (University of Utah Press, forthcoming 2001); he has also published in major scholarly journals. Godfrey is a member of the Broadcast Education Association, of which he was president from 1999 to 2000 and served on the board of directors from 1994 to 2001. He also served as president of the Council of Communications Associations (CCA) in 1999. He says the greatest reward of his career is teaching: "I simply enjoy watching my students grow."

Photo Credits

AP/Wide World Photos: p. 350

Archive Pictures: p. 30 right

Broadcasting & Cable magazine/Cahners Business Information, "Special Edition: The First Sixty Years," *Broadcasting* (December 1991): 56, 60: p. 40 top and bottom

Brown Brothers: pp. 6, 20 top and bottom, 30 left, 108

Patrick Chauvel/Sygma: p. 330

Corbis: pp. 12, 38, 98, 342, 344

Corbis/William James Warren: p. 118 and cover (radar dish)

John Craft: pp. 2, 8, 59, 61, 65, 74 left and right, 75, 126, 128, 129, 141, 149, 150, 154 left and right, 160, 167, 172, 183, 206, 213, 214 left and right, 218, 232, 255, 270, 271, 278, 284, 291, 294, 302, 316, 319 left and right, 348

CBS/Archive Pictures: p. 49

Culver Pictures: pp. 15, 23, 45, 47, 82

E. G. Farnsworth: p. 35

Index Stock: p. 118 and cover (hand with remote)

Keystone Paris/Sygma: p. 334

KPHO-TV, Phoenix, Arizona: p. 307

Pete LaSacco: p. 260

Denny Lehman/Corbis: p. 337

E. R. Murrow School of Communication, Washington State University: p. 32

Ohio Wesleyan University (reprinted with permission): p. 79

Photofest: pp. 194, 352

Reuters/Bruce Young/Archive: p. 78

Reuters/Jim Bourg/Archive: p. 355

Marc Van Horne: pp. 254, 257, 265, 267

Chris Ware/Keystone/Hulton Getty: p. 338

Electronic Media

INTRODUCTION

CONVERGENCE AND THE ELECTRONIC MEDIA

Long gone are the days of family television viewing and three network choices for programming.

Sunday evening, October 1968. The family is sitting down after dinner to watch television. There is a variety of programming from which to choose. At 7:30 they can turn to NBC for *Walt Disney's Wonderful World of Color,* or they can watch *Lassie* on CBS or *Land of the Giants* on ABC. At 8:00 the family favorite is *The Ed Sullivan Show* on CBS, and then the kids like to watch *Bonanza* on NBC at 9:00 (if their parents allow them to stay up that late). Sundays are great for family TV—so many good programs, and many of them in color!

Sunday evening, October 2000. Dad has just settled into his chair in the family room to watch a sporting event on pay-per-view. Mom is in the den, reading and listening to digital cable radio. Their teenage daughter is in her room, playing video games on an interactive cable channel, while her brother is doing homework on his laptop, with MTV on in the background. Earlier in the day, Mom did some banking transactions through their home computer, and Dad made some purchases through an interactive shopping channel.

ong gone are the days of family television viewing and three network choices for programming. Today we live in an electronic multimedia society with virtually unlimited choices of entertainment and information channels. We are immersed in media: radio, television, CD players, VCRs, computers, and cellular telephones. The cell phone is no longer simply a telephone; it has evolved into a *personal digital assistant* that functions as a wireless computer terminal as well. Where we once had three or four television channels from which to choose, we now have fifty to one hundred available on cable. Some of these channels are *interactive,* or two-way, allowing us to respond to or interact with various media. The remote control device has become our link to the world of information and entertainment. We learn early how to channel-surf through the seemingly unlimited choices. In fact, we even have a "smart" remote control that can be programmed to surf the channels for us and select programs that fit our interests.

The near future will bring additional changes in the home delivery of information and entertainment. The changes will be a result of the development of an electronic information network labeled the National Information Infrastructure, or the Internet, which consists of wired and wireless delivery systems. It is a two-way road. As part of the Internet, the world of multimedia will offer the consumer the ability to access a wide range of interactive services through digital technology, developed by the computer industry, which allows the relatively simple manipulation of many forms of data.

The concept of multiple television channels will give way to a single virtual channel in each cable home. Instead of surfing through hundreds of channels, the media consumer will select programs from a menu and schedule them on demand, thus moving scheduling from the broadcast/cable professional into the hands of the consumer. We will have completely left behind the traditional concept of broadcasting and evolved to a new concept: *pointcasting.* In other words, media will have evolved from "one to many" to "one to one." In fact, some have described the information superhighway as "mass to mass" communication, a convergence of broadcasting and pointcasting.

The Electronic Media Today

The business of broadcasting has changed radically in the past two decades, because evolving technology has introduced a new competitive environment. The traditional broadcast station now operates in a marketplace that enables an audience to receive hundreds of programming choices transmitted directly from a satellite stationed 22,300 miles above the equator, or to choose from thousands of movies from a video store just down the street. Viewers can choose from dozens of programming choices delivered by cable and DBS, as well as from many additional low-power and independent television stations. There are now six commercial broadcast television networks competing for audiences that only twenty-five years ago were not considered substantial enough to support three. If predictions bear out, the telephone companies will soon be major providers of television information and entertainment directly to the home through their fiber-optic lines. Audiences will pay for only the individual programs that they use and will need the services of their home computers to help them sort through all of the choices.

The technology that brought competition to the broadcaster can aid the industry. Using the digital technology that drives the computer, the broadcaster can deliver more programs through video compression; provide higher-quality, distortion-free sound and picture; and offer the wide-screen, home-theater advantages of high-definition television, as well as produce programs with visual interest provided through digital effects and computer-generated video. Satellite distribution technologies allow the broadcaster to provide live on-the-spot coverage of events from anywhere in the world as well as a few from outer space.

As the unprecedented changes in technology and competition developed, regulation of the electronic media seemed confusing and, in general, lagged developments in the industry. While the Federal Communications Commission (FCC) deregulated broadcasting, the Justice Department was in the process of breaking up the world's largest communications company—AT&T. In the past dozen years, Congress has written laws both deregulating cable television and reregulating it. In 1996 Congress rewrote the basic law under which all broadcasting operates—the Communication Act of 1934. The FCC, which was established to administer that act as well as to act as a traffic cop to regulate the use of the electronic magnetic spectrum, has given up its role of establishing broadcast standards and now auctions licenses to the highest bidder. Regulation is in transition as the new technologies and the Internet come online.

With an FCC philosophy allowing the marketplace to regulate broadcasting, many of the former restraints on concentration of ownership have been lifted along with the specific programming ownership regulations that discriminated against the large broadcast networks. In the kinder, gentler regulatory environment of the past few years, new networks have been established, and the older traditional ones have undergone changes of ownership. NBC is now owned by General Electric, CBS is owned by Viacom, and Walt Disney Co. has bought ABC. Many local broadcast stations have also seen ownership changes, as corporate groups have been allowed greater freedom in the number of stations they control. Changes of ownership have brought changes of network affiliation to what had been a very stable business relationship between the stations and the networks. The affiliation changes have brought intense competition, especially in local news, to a very confused audience. In this competitive atmosphere, some stations will be big winners in the ratings, and others will lose audiences at a time when direct broadcast satellite (DBS) might render the local station obsolete in the mind of the viewer.

While the fight for audiences intensifies in the United States, many of the same media conglomerates are just beginning to battle for viewers in the international marketplace. The Fox Television Network, owned by News Corp., has established international ties; owner Rupert Murdock, from Australia, has long owned the Sky Channel, a satellite service that broadcasts directly to homes in western Europe. In addition, he owns Star TV, which reaches DBS audiences from the Middle East to the Pacific Rim. NBC too is vying for the billions of potential viewers through its satellite-delivered CNBC in both Europe and Asia. Other cable networks, such as CNN, MTV, and ESPN, offer programming on a worldwide basis. The United States has long led the world in distribution of television programming to foreign markets through syndication, but now the media conglomerates are eyeing the potential of delivering programming directly to homes throughout the world. With international delivery, Marshall McLuhan's global village might become a reality—but complete with the 30-second commercial.

Historically, the business of broadcasting in the United States has relied on the sale of advertising time to turn a profit. To sell the standard 30-second commercial for the greatest profit, the broadcaster must attract the largest possible number of viewers while at the same time providing the lowest-cost programming. In recent years this practice has led to inexpensive reality-based programs and talk shows. Such shows, although cheap to produce, seem to attract large audiences only when they pander to the lowest common denominator by providing violence or outrageous topics. Programs that seem to rely on the deviance of guests have become the norm.

Society has changed radically from the Ozzie and Harriett days of the 1950s, and some sociologists question the role that television plays in shaping who we are as a people. Others are dismayed by the steady diet of video sex and violence delivered to America's living rooms. In today's world the latchkey child, with little parental guidance and in a home that watches an average seven hours per day of television, may be receiving more input from the mass media than ever before. The television junk-food diet of sex, violence, and the abnormal has a greater impact on today's youth than at any time in the past. Tomorrow's broadcaster may want to consider the effect of his or her programming on the audience as well as on the business's bottom line.

THE MASS COMMUNICATION MODEL

Communication is the critical process that holds our society together. We all engage in interpersonal communication every day between individuals and within groups. We learn early in our lives how to communicate using verbal and nonverbal languages. Interpersonal communication occurs on a one-to-one or one-to-several basis. Usually, we receive instantaneous feedback from those individuals with whom we are trying to communicate. But when we want to communicate with many individuals simultaneously, we become involved in the mass communication process, in which feedback is generally not instantaneous.

The *mass communication process* includes a source, a channel, and a receiver (see figure I.1). The source is the communicator, a reporter who is informing us about a news event, for example. The channel is the medium itself—a newspaper, a radio, or a television station. The receiver is the radio listener, television viewer, or newspaper reader. In preparing a message for mass communication, we need to encode it for a particular medium. If we choose an electronic medium, such as radio or television, we encode the message by writing it in a style designed to be heard. In other words, we write for the ear, not the eye. Yes, television is visual, but pictures are supplemented by words, or *copy*, read aloud by an announcer or reporter. This copy is quite different in style from copy in a newspaper or magazine. Electronic media messages are written in short sentences and in a conversational style. The words must be incorporated with video or around *sound bites*, or short voice statements by newsmakers.

Electronic media messages also must be encoded technically. In order to use radio (or television) waves to carry messages, sounds and pictures must be modulated electronically onto a carrier wave. This process is explained in more detail in chapter 14. Radio and television signals (waves) are subject to interference in the form of noise (static, electrical storms, or picture ghosting). But beyond physical noise, communicators also must deal with semantic noise, which is a barrier to understanding the message caused by poor writing, unclear symbols, inappropriate word choices, or other flaws in the message construction.

Feedback in the mass communication model portrayed in figure I.1 is usually not instantaneous. In the electronic media, feedback is obtained on a delayed basis through audience research. Surveys of audience

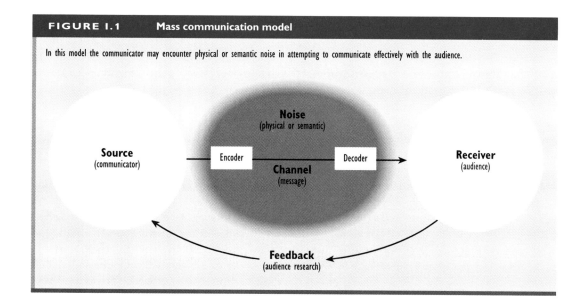

FIGURE I.1   Mass communication model

In this model the communicator may encounter physical or semantic noise in attempting to communicate effectively with the audience.

Advertising revenue has historically been the lifeblood of broadcast media in the United States. This set for a 1948 Jack Benny show prominently features the show's cigarette sponsor and their slogan, L.S.M.F.T.—Lucky Strike Means Fine Tobacco.

samples are the basis for program *ratings,* or estimates of audience size and makeup, for radio and television programs. As more homes become wired for cable services, two-way (coaxial) cable will allow instantaneous audience feedback. Audience research is examined in detail in chapter 12.

Media Functions

The functions of the mass media can be grouped into four general categories: information, entertainment, persuasion, and commerce. Of these, information is probably the most important. We rely on the gatekeepers of the news media—news directors, editors, reporters, and producers—to select and give us the most important information of the day. These are powerful individuals because they perform the agenda-setting function—determining what issues will receive the most attention by the media and, hence, by the public.

The entertainment function of the media is obvious: We are consistently entertained by books, magazines, films, and radio and television programs. Although this may not be the most important function of the electronic media in terms of society's needs, it certainly is the predominant one in the United States. Radio and television schedules are dominated by a wide variety of entertainment programs. Such programs, in fact, provide the major vehicle by which the third function of the media—persuasion—is accomplished.

Electronic media in the United States are, for the most part, supported financially by advertising revenue. Advertisers pay for access to radio and television audiences to persuade them to buy products and services. We, as media consumers, have seen and heard thousands of commercials. But the electronic media are used for persuasion in other ways as well: Politicians use the persuasive power of television and radio in their campaigns for election; religious and cultural leaders also use this power to persuade the public.

It is clear that the electronic media are critically important to U.S. commerce. Since the early days of radio at the turn of the twentieth century, electronic media have been developed and operated as private businesses. Communication legislation has defined radio and television as businesses operated in the public interest by private corporations for profit. Broadcast television and radio stations use public property—the electromagnetic spectrum—to provide programming and advertising. They are regulated, but not controlled, by the federal government. Within limits broadcasters are free to program their channels as they wish and sell the audiences attracted by their programming. Broadcasting was born primarily as an entertainment service, but it developed an information component (news and public affairs) as it evolved. Now broadcasting as we knew it throughout most of this century is changing. It is being challenged by a plethora of entertainment and information services delivered to the home primarily by wire.

As shown in figure I.2, dedicated professionals will use multimedia technology in advertising, graphic art design, and music composition applications. Business users will apply it in videoconferencing, corporate communications, and desktop marketing

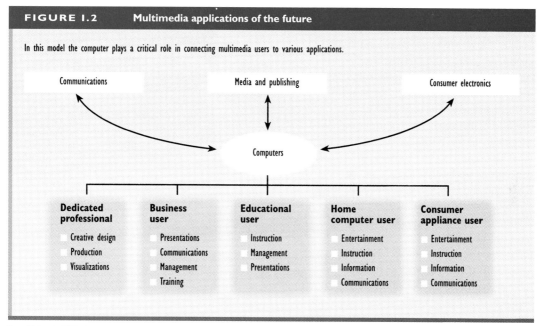

FIGURE I.2 Multimedia applications of the future

In this model the computer plays a critical role in connecting multimedia users to various applications.

From *Multimedia 2000: Market Developments, Media Business Impacts and Future Trends*, edited by Marcia De Sonne (Washington D.C.: National Association of Broadcasters, 1993), p. 11. Copyright © 1993 by Market Vision. Adapted with permission.

and sales presentations. Educational users will apply it in the classroom, libraries, museums, and distance learning (the delivery of education through media). But the largest growth area for multimedia technology is predicted to be among home computer and consumer appliance users, where consumers will use multimedia for video-on-demand, video games, interactive books, self-improvement programs, home shopping, and personal communications.

THE CONVERGENCE OF TELEPHONE AND TELEVISION

Among the oldest communications businesses in the United States are the telephone companies. For more than a century, AT&T has been providing telephone service as a common carrier, which means that AT&T provides the delivery system but not the information itself. Obviously, the public provides the content—phone conversations. During the 1980s AT&T divested itself of its regional operating companies that provided local telephone service. The divestiture left AT&T free to continue long-distance service while pursuing other aspects of information services. It also opened the way for competition in the long-distance market from such companies as MCI and Sprint. (See chapter 3 for a more detailed discussion of wired communications.)

The regional Bell operating companies (RBOCs) are pursuing business opportunities beyond local telephone service. They see future business growth in consumer information packages that will include telephone, computer, audio, and video services, provided via phone lines to existing telephone customers. They expect to compete with local cable companies, which are expected to get into the local telephone business as well. The consumer would therefore have at least two sources from which to choose for total information services. In another scenario, these services would be provided through a merger of the two companies.

Traditional common carriers like phone companies, which have been regulated by the federal government as monopolies, are evolving into huge information providers. This represents a drastic departure from the traditional separation of information and entertainment services. Cable and broadcast television have provided entertainment services for most of their history. Operationally, they have been separate from the telephone companies since the 1920s. Now they are faced with the prospect of competing or merging with regional telephone companies.

What does this mean for the future of broadcast stations and cable systems? These services must adapt to the evolving media marketplace and compete. For broadcast television it means the primary function must shift from over-the-air delivery systems to program producers. Broadcast networks will continue to provide entertainment and information programming, but will consider a variety of delivery systems to the home, primarily on cable and telephone systems. Where does this leave local television stations? They will concentrate on local services such as news and public affairs. Most local stations already have news and programming staffs; news programming has value no matter how it is delivered to the home. Many stations already have agreements to provide news services to cable companies and are experimenting with different program packages.

A convergence of technologies has resulted in a huge expansion in media types, production tools, and delivery options.

Historically, broadcast stations have made a profit from the sale of advertising time. They sold commercial time, or *spots,* to advertisers who wanted access to audiences. Profit margins were high for many years because the competition was limited. In the 1970s the three television networks claimed more than 90 percent of the viewers during prime time. In local markets three or four television stations typically divided the viewing audience. There was more competition in radio because major markets had many more radio stations than television stations. Still, broadcasting was a profitable business for much of the twentieth century.

Today the single revenue stream provided by the sale of commercials is no longer sufficient. Television networks and stations not only face rapidly increasing costs, but they also have much more competition for the advertising dollar. Cable systems are competing for advertisers and they compete with an advantage: They receive additional revenue from cable subscribers. This has forced broadcast television stations to explore other revenue sources. Initially, they turned to cable systems for cash in exchange for local broadcast signals or retransmission consent. Most cable systems refused to pay cash for signals, but did offer additional cable channels to be programmed by television stations. These cable channels bring in additional revenue from cable subscribers and advertising sales. Another revenue stream can be established through the sale to cable systems of local programming, such as news. Like television networks, broadcast television stations will survive as content providers. They will provide programming that will be delivered primarily by cable and telephone companies.

Multimedia

The future of information and entertainment services appears to be in the successful integration and delivery of content to the consumer. As noted, the integration, or convergence, will take place through digital manipulation of data—the computing function. Content providers will face the challenge of compatibility in product delivery systems. Digital technology will allow consumers to access a wide variety of information and entertainment content ranging from television programs to computer programs. Consumers are already familiar with digital technology through compact disc (CD) players, but technology is only part of the multimedia future.

Each service provider brings different strengths and weaknesses to the multimedia marketplace (see table I.1). For example, the computer industry has experience with operating systems, but is still developing its consumer marketing and distribution areas. The consumer electronics industry is strong in marketing experience, but still has to build consumer understanding of digital technology. Content or programming expertise is strong in the media and publishing industries, but they do not have the digital expertise of the telecommunications industry. The multimedia future will bring mergers of these industries to allow strengths to overcome shortcomings.

TABLE I.1	Multimedia industry strengths and weaknesses	
SERVICE PROVIDERS	**STRENGTHS**	**WEAKNESSES**
Computer industry	■ Accustomed to working in different operating systems and on several platforms at once ■ Digital production studios—experts in graphic images, character animation, digital music, and interactive entertainment ■ Experience with software authoring tools	■ No understanding of consumer marketing and distribution ■ Little innate creative talents for storytelling
Consumer electronics	■ Understands consumer needs and marketing requirements ■ Extensive distribution infrastructure	■ Consumers have little understanding of digital appliances
Publishing and entertainment	■ Understands mass distribution of low-cost products ■ Expert at telling stories and entertaining consumers ■ Analog video and programming expertise	
Telecommunications	■ Cash-rich ■ Experienced in personalized point-to-point communications ■ Expert with switched media—most media will become switched ■ Only two wires entering home—great for voice and perhaps two-way video channels	■ High copper-to-fiber replacement cost (estimated up to $500 billion in the United States) ■ FCC allows transport but no programming content ■ FCC restricts purchase of cable operators—joint ventures will proliferate
Cable companies	■ Bandwidth-rich ■ Coaxial infrastructure passes over 90 percent of U.S. homes with more than 60 percent of homes subscribing to the service ■ Only 25 percent replacement cost to go to digital	■ No switching capabilities at present ■ Minimal financial leverage—lacks cash and has heavy debt burden
Graphic arts	■ Highly talented in visual design ■ Gaining expertise in digital design	
Educational institutions	■ Establishment of long-term learning approaches	■ Slower development and deployment of educational tools

From *Multimedia 2000: Market Developments, Media Business Impacts and Future Trends*, edited by Marcia De Sonne (Washington, D.C.: National Association of Broadcasters, 1993), p. 11. Copyright © 1993 by Market Vision. Adapted with permission.

Future Media Businesses

Information and Entertainment

The multimedia model of the future links the major information and entertainment providers to consumers through computers. In figure I.2 we see that the communications, media, publishing, and consumer electronics providers will serve five user categories: dedicated professionals, business users, educational users, home computer users, and consumer appliance users. Through digital computer terminals, users will be able to interact with a broad range of information and entertainment services. Once in digital form, virtually any type of information ranging from simple numbers to graphics, audio, and full-motion video can be combined, processed, and stored.

The most likely storage format for information in digital form will be the *CD-ROM* (compact disc-read-only memory). The CD-ROM can store combinations of different types of data—text, audio, and video—which can be accessed through a home computer. The CD-ROM primarily has been used by the corporate world for a variety of specialized text databases. Corporations still represent the largest market for CD-ROM data, but utilization is increasing among libraries and publishers. In the early 1990s, the CD-ROM began to generate more interest in the consumer market. Computer software, video games, and general entertainment products such as movies are now available on CD-ROM and DVD for home use.

Preparing for the Future

Media professionals of the future will need to be prepared for the multimedia world. They will need to be familiar with the digital technologies available to the consumer, but content will continue to drive the demand for information and entertainment services. Creativity and good communication skills will be even more important in the media future. Hundreds of channels in the home will result in a huge demand for programming of all kinds. Consumers will have easy access to news, entertainment programs, movies-on-demand, and a wealth of information services. The challenge for media professionals of the twenty-first century will be to create, market, and deliver content to meet consumer demand.

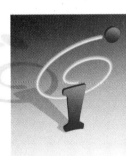

InfoTrac College Edition Exercises

Today we live in an electronic multimedia society with virtually unlimited choices of entertainment and information channels.

I.1 Many radio and televisions stations are putting their broadcast signals on the Internet—a process called "streaming." There are concerns, however, about making program content so widely available on the Internet. Using InfoTrac to access *Broadcasting & Cable,* read the article "Gauging TV's 'Net Effect" (February 21, 2000) and answer the following questions:

  a. The author of the article suggest several ways that content owners can protect their copyrights on the Internet. What are they?

  b. Why are content owners concerned that their copyrighted material would be available in digital form on the Internet?

I.2 The Internet brings the consumer a wealth of information and entertainment, but it also brings unsolicited material such as "junk e-mail." Using InfoTrac to access the *Journal of Broadcasting and Electronic Media,* locate and read an article about unsolicited commercial e-mail and the First Amendment.

  a. Write a paragraph outlining your opinion of the First Amendment issues related to potential bans of junk mail on the Internet.

WEB WATCH

Here is a list of a few URLs (Internet addresses) for some of the organizations or corporations discussed in this chapter. Please explore these Web sites and follow the links to learn more about the complex business of the electronic media. Add your descriptions and your own favorite sites at the end of the list. Please keep in mind that the dynamic nature of the Internet allows sites to come and go but also allows organizations to update information about themselves very quickly.

Address — **Description**

http://www.commercepark.com/AAAA/AAAA.html _____

http://www.att.com/whatsnew.html _____

http://www.primenet.com/~laig/cdrom _____

http://alnilam.usc.indiana.edu:1027/sources/programs.html _____

http://www.mci.come/home _____

http://execpc.com/~dboals/media.html _____

http://viswiz.gmd.de/MultimediaInfo/ _____

http://fourier.dur.ac.uk:8000/mm.html _____

mailto:majordomo@marketplace.com _____

http://www.bookwire.com/links/publishers/publishers.html _____

http://www.2sprint.net/sprintltd.html _____

Other favorite sites: _____

CHAPTER 1

FOUNDATIONS OF ELECTRONIC MEDIA

Technological history has taken communication from wire to wireless and back to wire, and early business patterns provided the foundation for today's industry.

From Wire to Wireless to Wire

The foundation of today's electronic media is an evolutionary history of a revolutionary technology. It has transformed our information gathering, our entertainment processes, and our social environment. Telegraphy eliminated the Pony Express, wireless communication replaced the wire, radio was transformed by an over-the-air telecast of the moving image, and today we have come almost full circle from wire to wireless to a new type of wire—the fiber-optic cable.

The science that made radio the "whispering gallery of the skies" began in the mid-1800s. Then as now, revolutionary and technical developments were part of the social, political, and industrial environments of the age. Traditional historians who study the spectrum of ancient history consider the electronic media hardly old enough to have a history. Perhaps they are correct—technology is still evolving, industry is still rapidly developing, and its cultural ramifications are still under investigation.

The history of broadcasting is a reflection of the social periods during which it grew. It is within this historical context that the developments of electronic media are discussed. The electronic media brought the industry and the audience to the brink of a communication revolution. The purpose of this chapter is to examine the foundations of our electronic media within the general historical context so that we may understand the precedents and organizational patterns that evolved.

The Industrial Age: 1850–1900

The earliest historical period of significance to the foundations of broadcasting technology ranges from the mid-1800s to the turn of the century. This prebroadcast period was a fascinating time in U.S. history. It followed the Civil War, it was a time of massive population growth and urbanization, and for the first time in its history the nation's manufactured goods were worth more than its agricultural products.

This era of U.S. history is known as the industrial age. The vast resources of the country seemed inexhaustible, mass production was a new idea developed for growing industry, and a steady supply of new labor arrived from Europe to fill the assembly lines. This was an era of ingenuity, filled with optimism, individualism, and few government industry regulations—three conditions that foster rapid growth.

The industrial age was a picturesque era of contrasts. For the first time, individual opportunity seemed equal to the desires of society. It was an age of conspicuous waste. The buffalo of the West were almost wiped out during a twenty-year period historian V. L. Parrington called "the great barbecue." Lavish parties celebrated the newly rich; platters of tobacco juice contrasted with the diamond-studded teeth and dog collars. Despite social disparity, the heroes of the age were the millionaires who had lifted themselves to the top of industry. John D. Rockefeller, reflecting on the era, noted, "The growth of a large business is truly a survival of the fittest . . . working out the laws of nature and of God."

Most industrialists declared themselves devoutly religious, attended church, and taught Sunday school, thus sanctifying their newfound riches. These people were considered first-class citizens, each with a rags-to-riches story. The industrial age had such an effect on our society that many of its dominant names are still familiar today: Andrew Carnegie, Russell Herman Conwell, Andrew Mellon, J. P. Morgan, John D. Rockefeller, Joseph P. Kennedy, William Randolph Hearst, and Joseph Pulitzer.

Adding the names of the electronic media pioneers who lived in the same era—James Clerk Maxwell, Heinrich Hertz, Guglielmo Marconi, Reginald A. Fessenden, and Lee de Forest—communicates some idea of the environment within which radio began. These radio pioneers, whose names are obviously not so well known, worked in the shadows of the industrial giants. The development of radio was new compared with other evolving industry (such as manufactured goods), but it began within the ideology of the same industrial age.

The Telegraph

The telegraph (discussed in more detail in chapter 3) was the most important development of the electronic media in the industrial age. It is too often thought of in terms of its western-movie image, as a small, isolated, and often deserted office used only to send for the cavalry. This perception overshadows the more important mass communication function telegraphy provided for the times. During the mid-1800s, *telegraphy*—the transmission of coded signals—provided the world's first instantaneous news service. Information had moved slowly to much of America before the telegraph. Although the country was becoming urbanized, there remained a significant rural population. The telegraph was the first practical medium that kept the agrarian and the growing urban communities throughout the nation in touch with the rest of the world. The telegraph enhanced the currency of the frontier press by overcoming the obstacles of time and distance: Information gathering and transmission were no longer hindered by the time required to physically carry the news from place to place. The telegraph rapidly replaced the Pony Express and reported the events of the Civil War and of the developing frontier. The public's interest in rapidly delivered information inspired the growth of commercial enterprises to relay it.

Samuel Finley Breese Morse is credited with the development of the telegraph system. He acquired his first patent in June 1840 and received financial assistance from Congress (see chapter 3).

The frontier success of the telegraph naturally led the way for voice communication—*telephony*. Alexander Graham Bell is credited with developing the analog transmission of the human voice over wire. Bell's early experiments led to several important contributions: the carbon microphone, the magnetic receiver (the basis for loudspeakers), and the electronic tube amplifier. Bell announced his successful voice transmission experiments in 1874 and patented his work shortly thereafter. Bell not only produced important technological developments for electronic media, but he also founded the American Telephone and Telegraph Company (AT&T), which would later make significant contributions to the foundations of broadcasting.

BROADCASTING'S PREHISTORY

Early telecommunication experimentation was not limited to telegraph land lines. Telegraph grew to include transmission of the human voice—telephony—and wireless telegraphy evolved into radio telegraphy. In broadcasting's prehistory, there were several inventors whose individual contributions would be combined to produce an over-the-air radio signal.

MAXWELL AND HERTZ

James Clerk Maxwell, a Scottish physicist, was first to publish a theory of radiant energy, which remains the basis of the modern concept of electronic media. Maxwell's ideas attracted the attention of German physicist Heinrich Hertz, who first demonstrated Maxwell's theories by projecting the signal into the air—paving the way for radio. The physicist's achievement is recognized by the use of his name as the unit of measurement for radio frequency—Hertz (Hz).

MARCONI: INVENTOR AND ENTREPRENEUR

Guglielmo Marconi was the most prominent and well-known experimenter in the industrial history of radio. Marconi, however, was more than an inventor; he was also an entrepreneur. He realized the commercial and public value of his inventions, took out patents, and organized corporations to profit from each of them. By the turn of the century, Marconi was establishing international corporations. His work progressed rapidly. After his own government in Italy seemed to have little interest in his work, he approached the British and Canadian governments, which gave him encouragement and financial assistance. He established the British Marconi Corporation, the Canadian Marconi Corporation, and the American Marconi Corporation. In 1901, in his most famous experiment, he succeeded in sending a signal through the windy skies from Cornwall, England, to Newfoundland, Canada. Marconi was actually the first person to use radio as a device to both send and receive information.

Shortly after these successful experiments, Marconi began constructing wireless telegraphy stations along the Atlantic coast to replace the aging telegraph cable systems and to transmit storm warnings, the dangers of ice flows, and so on to ships at sea. Their purpose was primarily ship-to-shore communication. These stations were also organized to earn a profit. The key to Marconi's success was surrounding himself with good managers who advanced his corporate interests while he continued his experiments with the projection of electromagnetic energy. The Marconi companies controlling these stations were subsidiaries of the Marconi Wireless Telegraph and Signaling Co. Ltd. Together they were a dominant force in maritime communication until after World War I.

Precedents: Electronic Pioneers

Marconi was the most prominent experimenter during the industrial history of radio. He was an inventor and an entrepreneur. Known for his 1901 transatlantic experiment, he also established several corporations and took out many patents on his work.

Fessenden, previously an experimenter, was first to succeed at superimposing voice on a high-frequency carrier.

De Forest is known as the father of radio because of his development of the Audion tube, which facilitated voice transmission.

Fessenden: Voice Experiments

Reginald A. Fessenden, a Canadian, took his experiments to the United States. He was like most of the earlier radio pioneers who preceded him—primarily an experimenter, not an entrepreneur. While Marconi was sending wireless Morse code signals, Fessenden was first successful at voice and music transmission. His first broadcast was from Brant Rock, Massachusetts, on December 24, 1906. The programming took advantage of Fessenden's multiple talents as well as his telecommunications knowledge: He narrated the program opening, played a phonograph of Handel's *Largo,* performed a violin solo, sang, and ended by wishing the men at sea (to whom the broadcast was aimed) a merry Christmas. He conducted a similar broadcast on New Year's Eve. Although Fessenden's first broadcasts were advertised for a general audience, the only audience at that time consisted of the radio operators aboard the ships that were in the harbor and a few newspapers that by this time had invested in radio receivers as a means of informational relay.

De Forest: The Father of Radio

Lee de Forest is often referred to as the "father of radio"—a title he gave himself. He developed the Audion tube, a three-electrode vacuum tube that (with additional circuitry developments) facilitated voice transmission. In his most famous experiments, he projected speech via radio. These were laboratory tests, and de Forest conducted a number of them in New York and in Europe—his most famous from the Eiffel Tower. This transmission, produced in 1908, was reported to have been received as far as five hundred

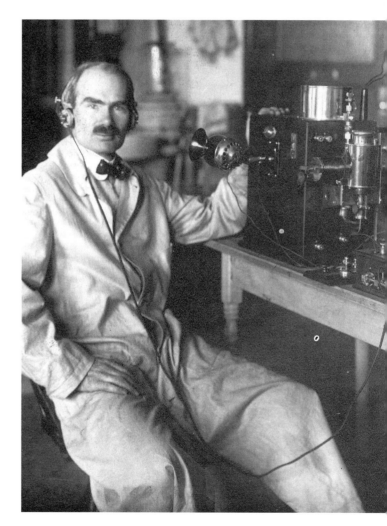

Lee de Forest held some three hundred patents in his name, among them the Audion tube, which facilitated voice transmission.

miles away. De Forest was like many of his forerunners: an inventor and not a businessperson. Although he formed his own company to develop his interests, lawsuits with Fessenden, financial problems, and an audience that was just not ready for general broadcasting eventually forced him to sell his investments to the American Telephone and Telegraph Company.

Corporate Players Emerge

At the turn of the century, several major corporate players were beginning to emerge, including the General Electric Company, whose engineer, Charles P. Steinmetz, developed the alternator to assist Fessenden in his first voice experiments; AT&T, which eventually acquired the Audion tube from de Forest; and the Marconi companies, whose aims some claim were nothing short of a world monopoly on radio. These corporations were primarily interested in the commercial value of the patents. They had the financial resources to see the patents developed into systems—a goal beyond the reach of most of the individual experimenters, who had the vision but lacked adequate financial backing.

The prehistory of broadcasting was a complex period of lawsuits, countersuits, litigation, financial development, competition, and experimentation. Everyone—inventors and corporate interests—seemed to hold patents to one or another critical element of radio technology, and few were willing to share. Radio at this stage was still a laboratory toy, but its importance as a means of point-to-point informational communications—particularly in marine and ship-to-shore communication—was becoming increasingly apparent.

Maritime Radio: The Era of American Expansion

At the turn of the century, the nation underwent a transition from an industrial age to an age of imperialism. The new mission for the United States was to expand. As the growth of industry continued, the country found itself with an empire and a growing need to communicate.

The roots of the new imperialistic empire were in the Spanish-American conflict, the closing of the western frontier, the rapidly developing industries that needed new markets for manufactured goods, and the politicians who were anxious to deflect the issues of the poor and agrarian revolt. All of these elements fostered a very emotional climate, a society looking to expand and take advantage of new opportunities.

Radio Technology and Sea Power

The era of U.S. imperialistic expansion was comparatively short but intense, and the development of maritime radio was a part of its ideology. The imperialists, reflected in Josiah Strong's (1885) rationale in *Our Country: Its Possible Future and Present Crisis,* argued the need for Christian socialism: It was the duty of this country to spread Christianity to those who had "the lesser law," namely the Philippines, Hawaii, and Cuba. "Manifest destiny" was the popular slogan used by the imperialists to suggest that the United States had a global responsibility to spread the ideals of its superior civilization to lesser countries. The problems developing between Asia and the United States, and consequently in the Philippines and Hawaii, were of strategic importance to the country. The role of radio in the imperialistic era was not broadcasting an ideological debate; it was maritime communication—ship-to-shore connection related to the sea power of the nation. Imperialists contended that a nation's greatness rested upon its prowess at sea. The formal debate on imperialism ended in a treaty signed with Spain in December 1898, but its ideology did not die. Its influence, which declared the need for dominant sea power to preserve the security of a supreme United States merchant marine fleet, can clearly be seen in the development of electronic technology from the turn of the century into World War I.

International conferences in Berlin in 1903 and 1906 also illustrate the growing importance of radio. The first conference was attended by eight nations; the second by twenty-seven, including the United States. These Berlin conferences produced protocol for ship-to-ship and ship-to-shore communication and established *SOS* as the international distress call.

The *Titanic:* A Catalyst

The first dramatic illustration of wireless radio as a maritime technology was produced by the sinking of the Royal Mail Ship *Titanic* on April 14, 1912. There was a ship near the *Titanic,* but its radio operator was not on duty when the *Titanic* struck an iceberg. By the

Precedents: The Titanic

The sinking of the *Titanic* on April 14, 1912, provided an opportunity for the first dramatic illustration of wireless radio as a maritime technology. It led to the passage of the Radio Act of 1912, which established SOS standards, call letters, and comprehensive legislation placing the secretary of commerce in charge of regulation.

time contact was established, the airwaves were jammed with irrelevant signals. All stations except those involved in the rescue were ordered closed April 16 so that there would be no interference with the disaster messages, but it was too late. Many of the passengers drowned in the Atlantic, and the *Titanic*'s radio operator died at the transmitter. There were 1,490 men, women, and children lost, many of them from prominent families. This disaster riveted the nation's attention on the new technology, which was thus catapulted into prominence.

The Radio Act of 1912

The Radio Act of 1912 was a direct result of the *Titanic* disaster. Its predecessor, the Wireless Ship Act of 1910, was simple legislation acknowledging the potential of the new medium as a ship-to-ship and ship-to-shore communication tool. It was less than one page long. The 1910 act was an outcome of the second international conference on radio in 1906 and required large passenger ships to carry radio equipment capable of sending and receiving within a distance of one hundred miles. The conference, attended by twenty-seven nations, was an attempt by the participating countries to achieve cooperation regarding the transmission of maritime information.

The Radio Act of 1912 was created following another international conference that had been called, just months before the *Titanic* disaster, to strengthen the previous declarations. The new law was still very simple, but it recognized the importance of maritime wireless and put the secretary of commerce in charge of licensing and specifying *wavelengths*, a measure of

Evolution of a Myth

An interesting but undocumented story resulted from the *Titanic* disaster. Gleason L. Archer, in his 1938 edition of the *History of Radio*, related that David Sarnoff—who was then the young manager of the American Marconi Corporation's station at Sea Gate, New York—failed to show up one day for his course at the Pratt Institute. As Archer tells it, Sarnoff was at the Marconi station relaying messages from the *Titanic*:

> For seventy-two hours of unceasing vigil the young operator sat at his instrument board . . . and picked up the heart-rending details of the *Titanic* disaster. "The S. S. *Titanic* ran into iceberg. Sinking fast" was the first message that was plucked from the ether. For three agonizing days, young Sarnoff stuck to his task while rescue ships combed the seas for survivors (Archer 1938, 110–112).

This story, which has made its way into most of the media textbooks of our day, is unsubstantiated. Even Sarnoff's biographer Kenneth Bilby has called this account into question (Bilby 1986). The coverage of the disaster by the *New York Times* and the *Wall Street Journal*, both of which carried volumes of material on the story, do not mention Sarnoff. It is true that Sarnoff was working as manager of the American Marconi Corporation station at the time; it is also true that the *Titanic* disaster did more to increase public awareness of wireless maritime communication than any event to date. The extent of Sarnoff's participation in the drama of this event, however, is not yet documented.

More important than the question of Sarnoff's involvement at the time, the disaster did focus attention on the practical utility of wireless. The story surrounding Sarnoff, true or false, has historically drawn a great deal of attention to David Sarnoff as well as to maritime radio.

the difference from peak to peak of the cycles of an electromagnetic wave. It prioritized the transmission of distress signals and legislated the SOS standards. It mandated the use of specific call letters to identify senders and gave priority to shipping communication versus commercial experimentation. The 1912 act was still primarily a maritime law dealing with shipboard and harbor receiving and transmitting stations, but it marked technological progress, recognized a practical utility for the new technology, and established several precedents important to legislation of the next decade. The Radio Act of 1912 was the first comprehensive U.S. governmental action dealing with radio. It did not, however, envision the discretionary standards that would become necessary for continued development as broadcasting grew into its commercial character during the Roaring Twenties. It lasted fifteen years.

WORLD WAR I

At the turn of the nineteenth century, radio was primarily a maritime tool. The utility of the invention was clear and, after the *Titanic* disaster, it was focused. Experimentation continued, as did the issues of patent interference and the struggles of development, into the early years of the twentieth century.

As World War I approached, the applications for electronic technology shifted. Business and industry were nationalized and focused on war production. On April 6, 1917, when the United States entered World War I, all wireless stations were closed. On April 7 they were reopened under the control of the U.S. Navy. President Woodrow Wilson had authorized the navy to take over all wireless stations that were not already operated by the military or the government.

The End of Patent Rivalry

The most important result of the navy's involvement in the electronic media was the development of the emergency patent pools. As technology grew through the industrial and imperialist progressive eras, the experimenters had continued to take out patents and licenses on their inventions. Each individual and corporation did so independent of the others in an attempt to gain the competitive edge. Marconi, for example, had the habit of boycotting any non-Marconi enterprise, a practice that aided his growth but created ill will. World War I, however, forced rivals into cooperation, as the navy took control of both commercial wireless and experimental operations and pooled the heretofore rival patents. The demands of World War I pushed government into an active role in the development of wireless and, spurred by its military importance and with rivalries set aside, the technology advanced rapidly.

THE ROARING TWENTIES: ANYTHING GOES

World War I brought electronic media experimentation into a new era. The patents had initially been pooled to facilitate the war effort, but the practical result brought together previously competitive ideas and set the stage for commercial development following the war. The move in the 1920s transformed the nature of the radio from experimentation and maritime communication into commercial broadcasting. Electronic media for public communication would grow rapidly during this decade, producing increased chaos on the air. Rival stations interfered with one another's signals by alternating wavelengths, increasing power, and changing hours of operation at will.

This decade was a time of genuine idealism, reflecting individual celebration, personal freedom and growth, an easing of traditional restraints, and a new era of governmental cooperation with business. It was an era of optimism, of rugged individualism, and of growing national industry. The twenties also reflected a revolution in the morals and manners of society: The popularization of Sigmund Freud led to a loosening of conventional standards and open discussion of sex. The burgeoning motion picture and radio industry in Hollywood began to standardize U.S. speech. The automobile and the railroads mobilized the populace, a social revolution that provided people with a new sense of personal authority and individualism. Accompanying these trends was a wave of religious skepticism, reflected in the broadcast of the Scopes "monkey" trial. The people were tired of the war, tired of its debate, tired of sacrificing—they wanted to go back to "normalcy." The 1920 election of Warren G. Harding was a popular landslide. "American business" in the twenties was "everybody's business," Harding declared. Bankers were regarded as the cream of society, and most national institutions were operated as businesses. The business ethic ran through all areas of life. Bruce Barton, an advertising executive, even tied religion directly into his business: "If advertising is worth doing at all it is worth doing all the time, for every day gentlemen, the 'king' dies and there arises a new 'king' who knew not Joseph." Business and its practitioners were the driving forces of the era in industry, government, and radio.

The business entrepreneurs of the 1920s, in contrast to those of the late 1800s, were shrewd, college educated, and operating at a time when government had never been more cooperative. Politics were dominated by the conservative influence, with business as its captain. Harding was elected for his willingness to support free enterprise with all the government assistance that could be made available. High tariffs were passed to curtail foreign competition, taxes were reduced to stimulate business, and statesmanship gave way to special interest groups. The era was described by Frederick Lewis Allen as a "self-conscious experiment," an era of the "prosperity bandwagon" with radio as the "youngest rider."

The twenties were not without their contrasts. Warren G. Harding, who was essentially a newspaper man from Marion, Ohio, was nominated and supported by big business as a president who would take orders. His administration came to a scandalous end with the investigations of the Teapot Dome and Elk Hills oil exploitations. He had abused the public trust and the public's natural resource—oil. The public trust and national resources were principles Congress would later remember when it discussed the question of who owned the airwaves—or "the ether." In reaction to the trends of skepticism and individualism, religious fundamentalists fought the changes of the new modernist thinkers and those who preached evolution.

Radio After World War I

The U.S. electronic media developed as the result of a series of social and developmental conflicts. All of these factors—a social revolution, a friendly government, governmental scandals, and technology—had a direct impact on radio in the 1920s, and the resolution of these conflicts set precedents for radio and television as we know them today. A friendly government provided big business (radio and others) a growth-friendly atmosphere. The issues of society—religion and censorship—were not only debated on the radio but can also be clearly seen in the governmental regulatory hypothesis of the 1927 Radio Act, which clearly forbids censorship in any form.

The era of the Roaring Twenties actually began for radio on July 11, 1919, when the navy returned the transmitters and stations to their private owners. These operators were anxious to renew their efforts. There was now a significant difference in their level of experimentation, however. The release of the patent pool meant that everyone was to have a piece of the action in the field of radio, and according to Sterling and Kittross (1990) that was "nearly two thousand patents which had been pooled." Although the specifics surrounding the patent pools, the cross-licensing, and the competitive rivalry are unclear, from the pool emerged cross-licensing agreements that, although they lasted only a short time, helped clear the tangled web of patent and licensing claims heretofore held by various individuals and corporations. The release of the pooled patents was a catalyst for the growth of the electronic media.

This marked an end to the individual patent rivalry in radio and the beginning of an age of corporate competition. The legal and financial patent struggles made way for the most important industry advance of the 1920s—the development of the communications business corporation, its key station, and the resulting networks.

The years following World War I mark an important transitional period for radio. While other national governments, such as Canada and Britain, were opting for governmental monopoly of radio, there was serious debate in the United States over governmental versus commercial control. In 1916 a government committee had proposed amending the Radio Act of 1912 to permit both government and commercial interests to compete for private stations. The American Marconi Corporation had fought diligently against this legislation. When the war erupted and the navy took over, the question of control was sidelined; but with the end of the war, the question of government versus commercial interests again arose. The difference this time was the size of the corporate operations and the political climate of the twenties. In an effort to protect itself after the war against a growing British monopoly in radio, which was controlled by Marconi, the U.S. government pushed for the sale of Marconi's American interests to General Electric Co. (GE). With that sale GE, on October 17, 1919, organized the Radio Corporation of America (RCA) to manage what were theretofore Marconi investments. In other words, a British monopoly was exchanged, with U.S. government approval, for an American monopoly. It was shortly thereafter, when RCA formed alliances with Western Electric and its parent corporation AT&T, that the patent pools were organized. These corporate relationships were blessed by the government and, though rocky at first, they continued until the government attitude toward monopoly began changing toward the mid-1920s. The worries about the growing media monopoly seemed to parallel the growth of RCA.

Leading the lobby for private corporate owners of electronic media was General Electric, RCA, AT&T, Westinghouse, and a large group of amateur radio enthusiasts.

Corporate Stations

Corporations were now a part of the electronic media landscape, and each operated a pioneering station. The most prominent stations were KDKA, owned by Westinghouse; CFCF, owned by the Canadian Marconi Corporation; WEAF, owned by AT&T; WJZ, WJY, and WRC owned by RCA; and WGY, owned by General Electric. There were other stations, to be sure, but these stations, owned and operated by corporations, played key roles in the development of network broadcasting.

KDKA: "The first commercially licensed station in the United States."

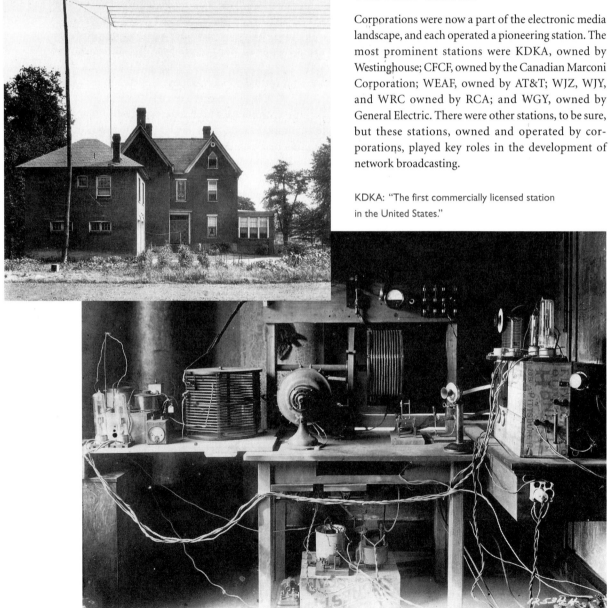

KDKA: "First in the United States"

KDKA Pittsburgh earned a place in the history of electronic media with its broadcast of the November 2, 1920, election returns of the Cox-Harding race, which illustrates the programming efforts of the era's struggling stations. It was election night, and the station was covering the results, which were being relayed via telephone in an agreement with the *Pittsburgh Post* newspaper. The announcer, Leo H. Rosenburgh, even asked anyone listening to communicate with the station. KDKA's historic broadcast riveted the attention of the nation on the possibilities of radio.

KDKA claimed that this broadcast was "the world's first scheduled broadcast," but other stations were experimenting at the same time. Charles D. Herrold pioneered station KQW in San Jose, California, with intermittent broadcasts beginning as early as July 1909. Professor E. M. Terry of the University of Wisconsin set up station WHA (with call letters 9XM, which designated experimental status) to broadcast weather and market reports. Station WWJ, owned by the *Detroit News,* went on the air August 22, 1920, with voice and music. CFCF, in Montreal, Canada, and PCGG in the Netherlands both began broadcasting in November 1919. Historian Asa Briggs noted that during 1920, "regular concerts began to be broadcast in Europe from the Hague" (1961). So although the focus is generally on KDKA, other stations were claiming "firsts"; wireless experimentation was evolving throughout the world.

CFCF Montreal is historically significant because it was owned by the Canadian Marconi Company. The CFCF broadcast that focused the attention of both the Canadian radio audience and the mariners in the St. Lawrence Seaway occurred on May 20, 1920. It was not unlike the KDKA broadcast of the Cox-Harding election. CFCF advertised the broadcast and even prearranged for an audience of Canadian dignitaries.

Popularizing the New Medium

An important historical function of experimental broadcasts such as those conducted by KDKA and CFCF was to popularize the new medium. This was aided by Chautauqua activities that sent radio enthusiasts to theaters and other receiving locations to hear the first public broadcasts. The *Chautauqua movement,* launched in 1874, aimed to bring culture to small-town America. It was originally a popular assembly, generally held outdoors under a tent, and was at one time a common means of adult education through lecturing and entertainment. Performers developed traveling circuits, moving from town to town, much like the circus, but with an educational or religious focus. According to a public relations release from CFCF, these radio theatrical showings often drew larger audiences than the motion picture. Chautauqua performers throughout both the United States and Canada got into the act and obtained receiving sets, which then became a part of their entertainment novelties. They would spend the day setting up the antenna, and the evening performance was merely the importation of the distant city signals. Ironically, it was radio that ultimately ended the need for Chautauquas.

The enthusiasm for radio was growing rapidly during the 1920s, as was the number of radio stations. By 1925 the Department of Commerce had issued more than 1,400 authorizations for new stations. In 1922 there had been a reported 60,000 radio households; by 1930 that number had grown to more than 13,750,000 (Sterling and Kittross 1990, 656). Sales of radio receivers boomed (see figure 1.1).

WEAF: AT&T Advances

Significant advances were made in technology and operational patterns by WEAF, the flagship station for AT&T. Its technical operations contributed to the development of a tool that today we take for granted—the *control board,* which routes, balances, mixes, and controls the audio. The New York station made two more advances that would have a national impact: It started to sell advertising and it was the first station to conduct network broadcasting.

Toll Broadcasting

WEAF was licensed to operate a toll station on June 1, 1922. The station, according to Lichty and Topping (1975, 166), was "available for hire by those wishing to reach the public by radiotelephony." AT&T's approach was described by Barnouw (1966, 105–115) as the "phone booth of the air." On August 28, 1922, WEAF conducted the first commercial program. It was a ten-minute speech for the Queensboro Corporation,

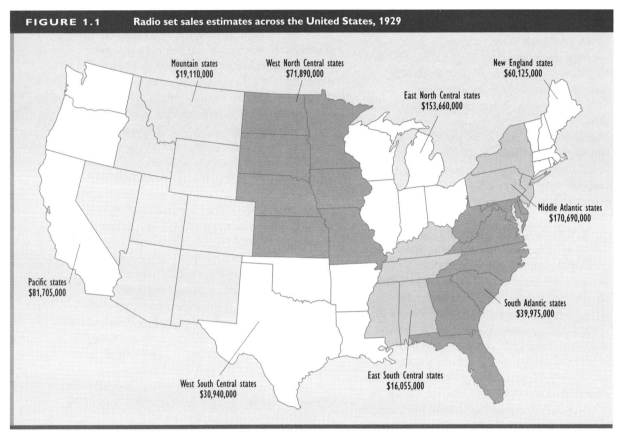

FIGURE 1.1   Radio set sales estimates across the United States, 1929

Mountain states $19,110,000
West North Central states $71,890,000
New England states $60,125,000
East North Central states $153,660,000
Middle Atlantic states $170,690,000
Pacific states $81,705,000
South Atlantic states $39,975,000
West South Central states $30,940,000
East South Central states $16,055,000

Source: Data from T. A. Phillips, "An Estimate of Set Sales," *Radio Broadcast* (September 1929): 270–272.

which wanted to sell apartments in Jackson Heights. The broadcast cost $50. During the weeks that followed, Queensboro continued to pay a toll to broadcast its messages, and it was said that sales did increase.

The broadcast was so controversial that the trade magazine *Radio Broadcast* editorialized against it. David Sarnoff at the time was advocating nonprofit public service programs. This advertising experiment also caused serious congressional debate. Some saw it as outrageous and threatened with bills that would prohibit radio advertising. Secretary of Commerce Herbert Hoover commented at the Third National Radio Conference in 1924 that "the quickest way to kill broadcasting would be to use it for direct advertising." His vision of advertising was more like what is now seen on Public Broadcasting Service (PBS).

Herbert Hoover on Radio Advertising

I believe that the quickest way to kill broadcasting would be to use it for direct advertising. The reader of the newspaper has an option whether he will read an ad or not, but if a speech by the President is to be used as the meat in a sandwich of two patent medicine advertisements there will be no radio left. To what extent it may be employed for what we now call indirect advertising I do not know, and only experience with the reactions of the listeners can tell. The listeners will finally decide in any event. Nor do I believe there is any practical method of payment from the listeners.

Source: From a quote by Herbert Hoover in Third National Radio Conference, *Proceedings and Recommendations for Radio Regulation*, Washington, D.C.: U.S. Government Printing Office (October 1924): 3.

Advertising Controversy, Yesterday and Today

Representative Celler, New York, proclaimed that radio advertising was an "annoyance to radio fans" and declared that "deceptive advertising is really worse than the above would indicate." Representative Celler's satire pokes fun at the advertising issue. His remarks reflect a program direction of his time and are not far from many experiments being conducted today.

> Permit me [Rep. Celler] to burlesque the situation and give you some of the radio pabulum and disguised advertising given out for radio consumption:
>
> This is BLAA broadcasting station of the Giant Peanut Co., Newark, N.J. You will now have the pleasure of listening to the "Walk-Up-One-Flight Clothing Co.'s Orchestra." Their first number will be, "You Don't Wear Them Out If You Don't Sit Down." Should any of our radio fans desire to communicate with the "Walk-Up-One-Flight Clothing Co.'s Orchestra" they can do so by communicating with BLAA Station.
>
> This is station MEOW. Mr. T. Cat speaking, and he is happy to announce that next Tuesday at 8 P.M. folks will hear Professor Bunion, well-known chiropodist. He has taken corns off all the crowned heads of Europe. We urge our invisible audience to write us what they think of Doctor Bunion.
>
> This is KOKO station. Doctor Bunkum's Sanitarium of Cripple Creek, Mich., Doctor Bunkum announcing. Folks we will receive with interest the news that I shall lecture on "My Pink Pills for Pale People." I shall be pleased to see any nervous, anemic person and show how to build him up with "Pink Pills for Pale People."

Source: *Congressional Record* (1926), vol. 67, pt. 5:5488.

At the 1924 conference, he also commented, "I had the foolish thought that if an advertiser at the opening of a broadcast announced that he was contributing this program to the public interest . . . such a practice would commend itself more to customers than annoying the public."

Representative Emanuel Celler of Brooklyn added his view of advertising to the debate in the House of Representatives when he spoke satirically about radio advertising (see above).

Despite the debate over the issue of advertising, little seemed to stem the tide. WEAF ignored the criticism and continued the practice. By January 1923, six months after the first toll broadcast, WEAF had sixteen commercial clients. The idea that an advertiser could operate a station and/or talk about his/her own products on another station was attractive for many radio stations that were underfinanced, and the trend was growing.

No station during the 1920s was well financed by advertising revenue, but WEAF's toll broadcast gave the fledgling broadcast industry an impetus—a dollars and cents reason to improve. By the end of the decade, a precedent had been established, one that continues today: advertising support for commercial media development.

The First Network Broadcast

Besides inaugurating the toll broadcast, WEAF was first to provide network broadcasting. AT&T, which owned WEAF at the time, already had telephone lines

Elaborate studios were common among the larger radio stations of the era. This is WEAF's studio in 1924.

Precedents: The Roaring Twenties

- **Legislation** The 1927 Radio Act reflected contemporary concerns and provided the "public interest, convenience, and necessity" standard for the regulation of radio, naming the Federal Radio Commission the regulatory body.
- **License** A patent license is granted by the inventor to a manufacturer, or another entity, in exchange for financial royalties on the use of the patent.
- **Networks** WEAF was first to provide network broadcasting. Linking chains of stations together seemed only logical for this key AT&T station.
- **Patent** A patent is the legal means whereby the inventor of the apparatus is guaranteed exclusive right of ownership and revenue generation for a period of seventeen years. An inventor's or corporation's patent portfolio is a potential source of income.
- **Patent interferences** Cases brought before the U.S. Patent Office and/or the courts involving one inventor or patent holder challenging another's patent.
- **Patent pools** These marked an end to individual inventor and corporate rivalry and the beginning of corporate business competition.
- **Royalties** Monies paid the patent holder in exchange for the rights to use an inventor's patents.
- **Toll broadcasting** Station WEAF was "available for hire by those wishing to reach the public." This first commercial broadcast gave the fledgling industry a dollars-and-cents reason to exist and invest.

spreading all over the country. Linking chains of stations together for purposes of programming seemed only logical. AT&T's first experiment was to link two stations—WEAF New York and WNAC Boston—on January 4, 1923. Other network experiments followed, but the one that focused attention was a twenty-two-station hookup that linked stations coast-to-coast. The broadcast occurred in October 1924 and featured a speech by President Calvin Coolidge. By the end of 1925, AT&T had twenty-six stations linked into the network.

WJZ: RCA's Key Station

At the same time AT&T was making its debut into network broadcasting, RCA was starting its own system. RCA had its beginnings in 1919 in negotiations over patents and patent pools. The first RCA network broadcast was in December 1923, between stations WJZ, the RCA-owned New York City station, and WGY of Schenectady, New York, owned by General Electric. As AT&T lines were competitively unavailable for such an experiment, RCA used the lines of Western Union. The RCA system did not grow as rapidly as the AT&T system because, under the cross-licensing agreements, AT&T had the exclusive right to sell broadcast time, and it had also prevented its competitors from using AT&T telephone lines for network purposes. In 1926, however, AT&T's involvement in broadcasting came to an abrupt halt.

The Birth of NBC

The rivalries among stations and their competitive parent corporations had historically been fierce. Marconi, AT&T, RCA, Westinghouse, and GE each envisioned having a monopoly—a dominant position in a rapidly growing industry. AT&T had a telephone monopoly, and its monopolistic philosophy is evident throughout its history. The patent pools were negotiated attempts to carve monopolistic corners out of a growing industry. Under these agreements, AT&T claimed the rights to manufacture and sell transmitters, conduct toll broadcasting—which it compared to a toll or long-distance phone call—and the rights to network the stations with existing telephone lines. All of these rights became increasingly important—and controversial—as the industry grew. Lines were drawn between the Radio Group, which consisted of General Electric, RCA, and Westinghouse, and the Telephone Group, which was AT&T and its subsidiaries. As the conflict raged, AT&T eventually grew tired of its

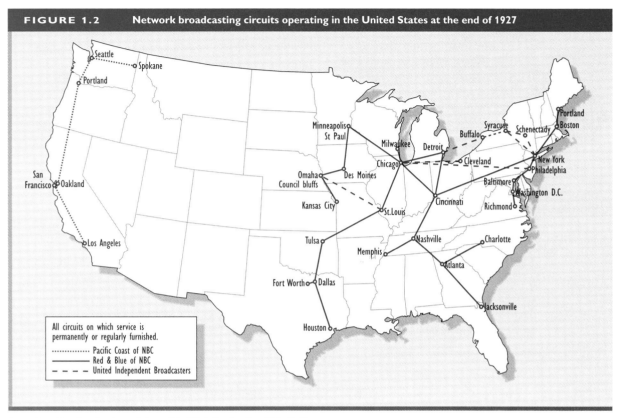

FIGURE 1.2 Network broadcasting circuits operating in the United States at the end of 1927

Source: From "How Chain Broadcasting Is Accomplished," *Radio Broadcast* (June 1929): 65–67.

competitive position—it was obviously not going to be able to create a monopoly in radio as it had done with the long-distance telephone. As a result, it decided to sell its radio interests and bow out of the business.

In September 1926 RCA formed a separate unit to conduct its broadcast and network operations: the National Broadcasting Company (NBC). Shortly thereafter AT&T sold its broadcast interests to RCA in an attempt to improve relations with RCA, Westinghouse, and General Electric. The sale immediately placed RCA in the dominant position. With the combination of its own operation based on its station WJZ and the newly acquired and financially successful WEAF, NBC now had two major chains. It had nineteen permanent stations on its network in 1926 and by 1928 had established the first permanent national network. The second, newly purchased WEAF-based RCA network became known as the NBC Red Network, and the older WJZ-based RCA network became known as NBC Blue. Although the two networks would become similar in size during the mid-thirties, the Red Network held the dominant position: It had the pick of programming and the most advertising revenue simply because of its AT&T roots. AT&T had been the first to experiment with the toll broadcast, and thus the Red Network had the lead advantage (Sterling and Kittross 1990, 634).

The entrepreneur behind the operation at NBC was David Sarnoff. According to his biographer Bilby (1986), Sarnoff "had more abilities than any other industry leader of the time." He not only had a vision, but also, according to Bilby, "the implacable determination to fulfill it." New technology had created new possibilities, and Sarnoff was there to influence the patterns. His vision of radio transformed the world. His fight with AT&T over network radio led to one of the strongest networks in broadcast history. In short, Sarnoff, according to Bilby, "was one of the giants of the industrial history, a leader of perception and courage."

SARNOFF

David Sarnoff (1891–1971) was a Russian-born immigrant to the United States. After his arrival in 1900, his first job was with the Commercial Cable Company, concerned with transoceanic undersea cable communications. His climb up the corporate ladder began with his employment at the American Marconi Corporation. He started as a junior wireless telegraph operator and shortly was the manager of the wireless station at Sea Gate, New York.

As mentioned earlier, at the sinking of the *Titanic,* Sarnoff was reportedly the only one transmitting on the air. The limelight from the story, its accuracy notwithstanding, provided Sarnoff with a focus of attention that would not dissipate for decades.

When American Marconi was sold to General Electric in 1919, forming RCA, Sarnoff moved to RCA with the title of commercial manager. In just ten short years, he was elected to the Board of Directors and then appointed RCA president in 1928. When General Electric and RCA were forced to split in 1932, Sarnoff was at the helm. He would remain there for the next three decades.

Sarnoff provided executive leadership through RCA's success and controversy. He initiated the NBC network and RCA's venture into television. He battled with CBS over color television standards and won. He also battled with FM and television standards—although he won and lost lawsuits with Armstrong and Farnsworth, he won the industry battle in establishing the FM radio and television industry.

Sarnoff died in 1971. His biographer would describe him as an "American success story of classic proportions." He was the man who most influenced industrial patterns of American electronic media.

THE BIRTH OF CBS

The creation of NBC's chief rival of the time, the Columbia Broadcasting System (CBS), began with George A. Coates, a promoter who had taken up radio's cause and became involved in the anti-ASCAP controversy. The American Society of Composers Authors and Publishers *(ASCAP)* is a nonprofit organization used to collect and distribute music royalties. During the mid-twenties it was demanding more money for music royalties than the newly established radio industry thought fair. Coates, along with the newly formed *National Association of Broadcasters (NAB)*, was seeking to free the struggling broadcasters from the financial demands placed upon them by ASCAP for music rights on material performed on the radio. Following a speech before the NAB, Coates began traveling and soon had contracted with a number of stations. He teamed with Arthur Judson, the business manager of the Philadelphia Orchestra who had been turned down trying to sell programming to RCA. The two of them formed a network—the United Independent Broadcasters, Inc. (UIB)—whose goal was to provide programming for the stations, purchase time from the stations, and sell advertising within the programming. UIB's programming debut was September 18, 1927, but its financing was weak, so it was soon looking for additional backing.

The joining of UIB and Columbia Phonograph Corporation was motivated by Columbia's desire to sell records. The phonograph manufacturer wanted to ally itself with radio because of the decline in the sales of phonographs and records. Because people were able

PRECEDENTS: THE NETWORK CONTRACT

William S. Paley, CBS pioneer and entrepreneur, developed the basic concept, still used today, for the network/affiliate contractual relationship. The network furnished a program service free to the affiliate, guaranteeing stations programming and contracting to pay the affiliate for the commercial time used by CBS. The affiliate guaranteed to make the commercial time available.

to listen to music on the radio, the move seemed logical. Columbia was also afraid that RCA would merge with the Victor Talking Machine Company and then dominate the record industry (RCA did merge with Victor in 1929). UIB and Columbia merged on April 5, 1927, creating a sixteen-station lineup. The agreement gave UIB a temporary financial boost and a name change—to the Columbia Phonograph Broadcasting System (CPBS-UIB). Advertisers still did not flock to CPBS-UIB, however, in contrast to RCA, which was growing rapidly. Because the RCA/Victor merger didn't seem to be immediately forthcoming and because competition was stiff, Columbia Phonograph reconsidered and eventually placed its interest up for sale.

WILLIAM S. PALEY CREATES NETWORKS

One party interested in the sale of UIB was a CPBS advertiser—the Congress Cigar Company of Philadelphia—whose vice president, the son of the owner, was William S. Paley. Paley was born September 28, 1901, educated in finance, and as vice president of his father's company was in charge of production and advertising. His introduction into radio advertising came at station WCAU-Philadelphia, where he signed on to do some advertising. The first program created was called *La Palina Boy*—named after a Congress Cigar brand. As it turned out, the program attracted quite a bit of publicity, including some criticism from his family, who felt that the program advertising was out of order. The program was canceled, but when listeners began complaining about the program's disappearance, Paley decided to continue with radio advertising. By this time the CPBS-UIB network had been organized, and Paley broadened his radio campaign. Sales rocketed from a mere four hundred thousand to a million cigars per day—the advertisement was a success. Paley and his family purchased the network in September 1928, and by 1929, thanks to some creative financing, the company was showing a profit. The Columbia name was retained. It purchased its first key station, WABC–New York, in 1928 and, ironically, a decade later CBS purchased the stock of the American Record Corporation—which included the Columbia Phonograph Company and several other record labels.

Paley had a good nature, a flair for finance, and a talent for negotiation. It was he who developed the concepts of network "sustaining" and "sponsored" programs, which created the foundation of the network/affiliate contract. Before Paley took over, the CPBS-UIB was obligated under contract to buy its programming from the affiliate stations. As advertising revenues were not always sufficient for the young network, this practice became a serious drain on resources, and contract reversals did not seem to be helping. Paley decided to revise the contract again, but he did so with the affiliates in mind as well as the reduction of the financial burden of the networks. The network's sustaining-program service was to be free to the affiliate. He guaranteed twenty hours of programming per week for the commercially sponsored programs and contracted to pay the stations $50 per hour for the commercial hours used. The local affiliates were thus tied tightly to the network. The new contract proved beneficial to both the affiliate and the network. The network provided programs a local station could not produce and, in exchange, the local gave back commercial time, which was financially beneficial to the network. Radio through the 1940s was dominated by the networks and their tight contractual affiliations.

THE MUTUAL BROADCASTING SYSTEM

There were many attempts by other network operations to compete. The Quality Network began in 1929 (later to become part of Mutual), the Amalgamated Broadcasting System debuted in 1935, and several regional networks tried to organize on the West Coast, but the next most significant network with any longevity was the Mutual Broadcasting System (MBS). In contrast to NBC and CBS, it was not developed from a key station and in fact had no studios or central ownership. Mutual was a cooperative venture, as the name suggested, built around program sharing. The program central to its popularity was the *Lone Ranger*. Mutual later promoted itself as "the largest network in the world." It is true that it did have many affiliate stations, but MBS was never able to attract the more powerful stations already affiliated with NBC and CBS. Its affiliates comprised the smaller urban and rural stations throughout the nation. NBC and CBS created monopolistic affiliate contracts, and their stations dominated the prime frequency assignments—they were firmly entrenched. Mutual was an exception because of the unique nature of its program-sharing concept among the smaller stations.

The historical importance of these developments is not so much in the names, places, and dates, although they are fascinating, but in the patterns that developed. Little has changed from these early network patterns as they evolved from WEAF/AT&T, WJZ/RCA, and WABC/CBS. They are the foundation of today's network operations.

Programming

The programming offered by these new networks and the individual stations was irregular at first, but it grew with the stations and the audience during the 1920s. Educational institutions had expressed an interest in the new technology as an educational outreach program. Such programming emphasis included the establishment of *Columbia's American School of the Air, An Invitation to Learning,* and *Little Red School House.* These programs and the discussion of public issues were the types of programming Secretary of Commerce Herbert Hoover had in mind when he declared that radio would die in twenty-four hours if it were limited to the playing of phonograph records.

Programming, despite what Secretary Hoover and the nation's universities had fostered, was primarily music—live music. Performers were willing to appear in hopes that the publicity would increase their own popularity. The networks even had their own live orchestras in the studios, and the programs were designed for studio performance. Large studios were draped with curtains, and although the performers would not be seen, they dressed in formal attire for the program events. All the programs were broadcast live. Programming schedules occupied primarily the evening hours and expanded with the increased audience and the capability of the technical operation. A breakdown of the programming of stations in New York, Chicago, and Kansas City in February 1925 indicated that 71.5 percent of the programming time was music (most of which was classical); 11.5 percent information oriented—education, market reports, politics, news, and sports; 6.8 percent children's and women's entertainment; 0.1 percent drama; and 10.1 percent made up the remaining portion of time, which was devoted to religion, health, and the miscellaneous (Spalding 1963–64; Sterling and Kittross 1990, 73, 120–276).

The Thirties: A Need for Escapism

The emotions of the Great Depression were in complete contrast to the optimism of the 1920s. Business was now running at a loss, fortunes were lost overnight, and the prestige of the U.S. business operator, especially bankers and brokers, was sharply lowered. The fortunes of individuals seemed to be interlocked with the national spirit. The 1932 election was not a contest between two parties or even two candidates; it was between two contrasting philosophies, both of which had the attention of a growing radio audience. President Hoover clung to the strength of the individual in a philosophy of rugged individualism, which his opponents dubbed "ragged individualism." Hoover believed that hard work would pull the individual out of the Depression and that such success would "trickle down" and buoy up the individual and economic spirits. Franklin D. Roosevelt, on the other hand, promised a "New Deal" for the public. The result of the election was a landslide for Roosevelt, and during the first few years following the 1932 election, he was able to assume a bipartisan

Herbert Hoover on Radio Programming

When broadcasting first started, the phonograph was a sufficient attraction to the radiotelephone listeners, who were swayed chiefly by curiosity and marvel at the new discovery. Public interest has long since passed this stage. Broadcasting would die in 24 hours if it were limited to transmission of phonograph records. We have made great improvements in material transmitted. Original music, speeches, instruction, religion, political exhortation, all travel regularly by radio today. Program directing has become one of the skilled professions. I have, indeed, a great feeling for the troubles of the director in his efforts to find talent and to give to his audience the best that lies at his command. He has done extraordinarily well.

Source: From a quote by Herbert Hoover in Third National Radio Conference *Proceedings and Recommendations for Radio Regulation,* Washington, D.C.: U.S. Government Printing Office (October 1924): 3.

leadership. He established new government programs putting people to work, thus providing employment and giving hope for the end of the Depression.

President Roosevelt's New Deal was not without its critics. Senator Huey P. Long (D-Louisiana), who had supported Roosevelt in the 1932 election with the slogan "Roosevelt or Ruin," later changed his mind and reversed the slogan in 1934 to "Roosevelt and Ruin." The policies of the Roosevelt administration came under growing fire in the mid-thirties, and a new issue evolved to replace the New Deal—isolationism versus one world. Although the Depression had created social upheaval in the United States, in other countries it had provided the opportunity for revolution, and people like Hitler and Mussolini promised reform in their respective countries. It took World War II to change the issues of the debate from the Depression—and the success, or lack of it, surrounding Roosevelt's New Deal. There was, however, one significant difference in these debates compared with others—the growing mass popularity of radio.

REPLACING THE HEARTH

Radio had a significant effect on those living during the Great Depression and into World War II. Broadcast historians most often call this period radio's "golden age." During the thirties and into the early forties, radio was beginning to attract more and more advertisers while other industries continued to struggle. According to Douglas (1987), "The depression was the making of radio.... Smaller incomes, recession, and deflation meant that more Americans had to watch their pocket books and spent more evenings at home before the fireplace. Now they would sit before the radio."

In the politically charged climate of the Depression and during the approach of the war, radio was a popular source of respite and entertainment and the major platform for the ideological discussion of issues of the Depression and the European conflict. It was a window on the world, a break from a provincial existence and the difficult challenges of the day. President Roosevelt used radio to inspire confidence in a crushed banking system and, just minutes after his March 5, 1933, radio address on the banking crisis, telegrams flooded the White House, expressing the public's support for his programs. Roosevelt used the magnetism of radio to project an intimate personal quality. People listened to his "fireside chats" and felt like he was talking to them personally. He had a passion for clarity and an ability to simplify the most complex situation for the millions in his radio audience.

In response to Roosevelt's fireside chats, critics Huey P. Long and Father Charles Coughlin of Michigan also used radio to broadcast their political ideologies of social justice. In his home state of Louisiana, Long owned a radio station and a newspaper and he used them to recruit more than 5 million people for his Share Our Wealth Society. Long's cure for the Depression was to take the capital from the rich and redistribute it to the poor. Father Coughlin was a well-known radio priest after only a few months of periodic broadcasting. As World War II approached, Coughlin recruited millions for his Radio League of the Little Flower. Coughlin hammered at Roosevelt for getting America involved in foreign entanglements, and Roosevelt responded, preaching for his "arsenal of democracy."

THE THIRTIES' FAVORITE RECREATION

The ideological issues of the 1930s were all hammered out on the powerful new informational medium of radio. Reporting on the growing impact of radio on public opinion during the period, *Fortune* magazine noted that "The nation's favorite recreation was listening to radio . . . [and] newscasts ranked third among favorite radio programs." A similar article in a 1938 *Scribner's* magazine underscored the impact of radio as it discussed the dramatic character of Edward R. Murrow and the importance of radio news to the American public. Written just a few months after the first CBS *World News Roundup* began, it noted three advantages that Murrow's radio programs had over "the greatest American newspaper": "First, he [Murrow] beats the newspapers by hours. Second, he reaches millions who otherwise have to depend on provincial newspapers for their foreign news. And third, he writes his own headlines."

War of the Worlds

The episode of Orson Welles's *Mercury Theater of the Air* broadcast on October 30, 1938, was an adaptation of H. G. Wells's *War of the Worlds*. Broadcast on the radio, it was a dramatic example of entertainment and news-styled programming. To fully understand the impact of Welles's program, the historical context surrounding it must be taken into account.

Between 1935 and 1937, the United States formulated the U.S. Neutrality Acts, which were

Programming of the late 1930s captures an eager audience. Entertainment—whether comedy like *Fibber McGee and Molly* (left) or drama like *Mercury Theater of the Air* (right)—provided respite from everyday concerns, and commentary followed the public debates about the Depression and the conflicts in Europe.

calculated to keep the nation free from European conflicts—the "European phony war," as it was called by some critics. In 1936 Hitler marched into the Rhineland. In 1937 Franco came to power in Spain, fighting began between the Japanese and Chinese forces, and Mussolini and Hitler signed their unification treaty. In the spring of 1938, Hitler annexed Austria. In short, prior to Welles's broadcast, a very rapid succession of momentus international events occurred. People were anxious to receive news bulletins and know what was happening. The radio audience had learned to use the medium during the Depression: Radio news bulletins and commentators were a part of their lives, right along with the fireside chats and the variety entertainment programming. The audience was emotionally charged by the events of the era and by the ideological exchanges that had been broadcast by radio commentators and politicians discussing everything from the Depression to predictions of World War II. Because of these factors, the American listening audience believed the *War of the Worlds* doomsday broadcast was real, even though it was presented as fiction.

The *War of the Worlds* broadcast realistic "news bulletins" announcing an invasion from Mars in the area of Grovers Mill, New Jersey. Listeners reacted to the program with panic—many called their local police, left their homes, and thought there was an actual attack. Even though it was an innocent Halloween prank and Welles had broadcast disclaimers before and during the program announcing that it was a production of the Mercury Theater, the Federal Communications Commission soon let it be known that such programming was out of order.

Programming Matures

During the 1930s, radio programming matured. Now 64.1 percent of the evening network's programming lineup was music (with 23.5 percent dance music and 4.5 percent phonograph records); 13.3 percent women's, children's, and feature entertainment; 12.1 percent informational, education, news, and politics; and 6.5 percent drama. The comedy and drama programs, such as *Suspense, Amos 'n' Andy, The Shadow, Little Orphan Annie, One Man's Family,* the

March of Time, the *Lone Ranger,* and a host of others, continued to propel the popularity of radio. In 1938, the number of the network's quarter-hours of programming were reported in the following program types: There were 36 vaudeville and comedy, 30 general and talk variety, 52 music variety, 48 concert music, 17 light dramas, 23 thrillers, and 23 news quarter-hours on the air during a typical week (Sterling and Kittross 1990, 642). The era of the thirties contributed to the impact of radio in the maturation of programming—radio's golden age.

Radio News Grows Up

The development of radio news was an important natural outcome of the demands of international events and media developments during the Depression. During the 1930s and particularly as World War II approached, radio news took its present form. William S. Paley indicated in his foreword to *History in Sound* (1963) that "Radio news grew up with World War II." News and information programming during the twenties and early thirties resembled programming that today would be considered special-event coverage: the speeches, the fireside chats, the political debates, the religious sermons (which were often debates of the political issues), and celebrations. News staffs consisted of only a handful of people at the networks' stations and virtually no one at most of the local stations. The network newspeople were not really a news staff by today's description; they were commentators and producers who arranged for the broadcast of public events. These staffs grew slowly. Some participated directly in spot news with commentary, others were busy arranging speeches, choirs, and sporting events. As Shirer (1984) described the days he and Murrow were in Europe, they were busy, "putting kids' choirs on the air for . . . Columbia's *American School of the Air*." As the events of the thirties evolved, however, pressure was on the networks to cover the Depression debates and the unfolding world conflict.

The Press-Radio War

The press-radio war was not so much a war as a business rivalry, created because of the increased competition between newspapers and radio during the Depression years (Pratte 1993–1994). In 1933 newspapers forced the wire services (Associated Press, United Press, and International News Service) to withdraw the subscriptions of the radio stations and networks. In Cincinnati and Washington, D.C., newspapers even refused to publish the radio program listings. At a meeting at the New York Hotel Biltmore on December 11, 1933, representatives of newspaper and radio forged the Biltmore Agreement: Newspapers would continue to publish a radio schedule, but CBS and NBC would eliminate their news services; stations would limit the amount of news they broadcast; and a Press-Radio Bureau would be established to provide stations with a news source. With the Biltmore Agreement, CBS and NBC gave up their original news services. The Press-Radio Bureau was to supply two 5-minute news summaries each day, but these were not to be broadcast until 9:30 A.M. or until after 9:00 P.M.—in other words, until the morning and evening newspapers had been distributed. Commentary, however, was exempt from the Biltmore Agreement, and both NBC and CBS continued with their commentaries, issuing guidelines for the growing staffs who, according to White (1947), were to "elucidate and illuminate the news of common knowledge and to point out the facts on both sides."

The commentators and their staffs provided the foundations that led to the development of radio news programming as we know it today. When the Biltmore Agreement fell apart in 1938, other agencies arose to sell news services to radio. During the Munich crisis, when England and Germany were

Precedents: The Depression and World War II

- **The Communications Act of 1934**   This legislation broadened regulatory powers of the Federal Communications Commission.
- **News patterns**   Programming conventions established during radio's coverage of World War II established the spot news and round-up style formats still used in both radio and television today.

negotiating for Europe, radio was there proving its value even without the Press-Radio Bureau. Commentators eventually began using spot news, or news from the field, in their programs, and the Press-Radio Bureau ceased to serve the networks. Because of this the networks were ready in December 1938, with commentators and support staffs, to transform commentary program operations into news organizations that would cover the events of World War II. As the war progressed, the purpose of these commentators and staffs, who were in the field arranging special events, evolved from commentary, to public affairs, to news. During the war the names of the individual commentators became household words, and the network news staffs grew into webs of news bureaus that covered the globe. As the commentators became news reporters, providing eyewitness reports and commentary about the war, the local radio news operation slowly followed the national networks. The Depression and World War II marked the beginning of the spot news program format still intact today.

Radio During World War II

Radio communications during World War II were different than during the First World War. During the Depression radio had become a consumer medium, growing in its technological sophistication and popularity. The Depression had established an emotional, politically charged climate in which relatively low cost receivers made radio a powerful source of information and communication. The Depression had spawned revolution in several countries and had also turned America's attention inward to its own economics.

In the meantime Japan seized the Chinese province of Manchuria in 1931; Hitler became chancellor of Germany in 1933; Italy's dictator Mussolini invaded Ethiopia in 1935; the United States signed the Neutrality Act in an attempt to keep out of the European "phony war" in 1935; Hitler marched into the Rhineland in 1936; Japan and China were at war by 1937; and Mussolini and Hitler signed a pact in 1937. President Roosevelt went on the air with his famous "Quarantine" speech, declaring that "when an epidemic of physical disease [meaning the war in Europe] starts to spread, the community [meaning the United States] joins in quarantine of patients in order to protect the community against the spread." The messages were noncommittal but direct. In 1938 Hitler annexed Austria; in 1939 Great Britain and France entered the war to defend Poland, and Roosevelt revised the Neutrality Act to allow the sale of arms; in 1940 France fell, and Roosevelt, utilizing radio again, called on the United States to help. In 1941 the philosophical debate of one world versus isolationism came to an abrupt halt when December 7 became "a day which will live in infamy," as the Japanese bombed Pearl Harbor and the United States entered the war.

The significance of these events was reflected directly in radio history, because radio had become the platform for these ideological debates. Through the medium of radio, Roosevelt inspired confidence, Hitler instilled fear, and a host of politicians and commentators debated the issues of the Depression and scared the audience with sounds of the approaching war.

During the war itself, radio grew slowly. FM was still in its pioneering stages, and both FM and television growth were frozen by the Federal Communications Commission. Television was stalled because materials utilized in the manufacture of goods went toward the war effort. The golden age of radio programming, however, continued into the war years, although the programs were noticeably more patriotic and actually took part both by urging listeners to recycle scrap materials and sell war bonds and, in general, by keeping the audience informed about the events of the war.

Radio news was the major program innovation of the era: Newspeople filled the airwaves with reports from the front. Edward R. Murrow took the sounds of the war into every American home, Elmer Davis carried messages of isolationism, and H. V. Kaltenborn was accused of being a fascist. A host of reporters—Robert Trout, Douglas Edwards, William L. Shirer, Chet Huntley—and a group of commentators turned to reporting the events, portraying the war as they saw it. It was the nation's first

Edward R. Murrow is considered the dean of electronic journalism. He reported on World War II from the field and continued with the CBS radio and television network for years, working on documentaries and news programs.

eyewitness radio news. As the war expanded, so did the news organizations at NBC and CBS. They established news bureaus around the globe, their staffs expanded, and the number of program hours dedicated to news grew dramatically. By the end of the war, CBS radio alone had grown from a mere handful of commentators to almost 170 reporters and stringers who filed almost 30,000 broadcast reports. World War II marked the beginning of a new era for electronic media journalism and information gathering, establishing the organizational principles that radio and television networks use to this day.

In addition to the development of the organizational patterns that are still in use today, the twenties and thirties saw two significant technical innovations in their nascent stages, both of which would again revolutionize electronic media: frequency modulated (FM) radio and "visual radio" (television).

Armstrong and FM

Edwin Howard Armstrong, considered the father of *FM (frequency modulation)* broadcasting, was among the last of the "attic inventors." He was born December 18, 1890, but his biographer, Lawrence Lessing (1969), called him "a young man of the twenties." He had the "sensitivity of his romantic generation . . . shared the great creativity of the era and [had] more than a touch of its madness."

Early in his career, Armstrong had been involved in the rivalry of the patent lawsuits. He and de Forest were working on similar circuitry during the early part of the century and wound up in a bitter patent conflict. Armstrong's primary concern, however—and his contribution to the science of radio—was his effort to eliminate the static that interfered with the transmission of *AM (amplitude modulation)* radio.

Armstrong's experiments with wireless date back to the turn of the century, but it was during the twenties that he conducted his primary FM experimentation. He even shared the excitement of his work with his good friend David Sarnoff, whose children called Armstrong "the coffee man," because he was always dropping by their home in the morning to discuss his technical experiments with Sarnoff over a morning cup of stimulation. RCA had purchased several of Armstrong's inventions with both cash and RCA stock options.

Armstrong, who had been working on FM throughout the twenties, applied for patents on FM in 1930; the patents were granted in December 1933. His FM radio demonstrations were impressive: There was no static in his signal.

The first public presentation of Armstrong's system came as a part of a paper he was presenting to the I.R.E. (Institute of Radio Engineers), an engineering conference in New York on November 5, 1935. The test was planned to be a surprise, and it did produce moments of suspense for Armstrong as well as surprise in his audience. Armstrong had arranged the demonstration for the conference with a friend, Randy Runyon. Runyon was still setting up the transmitting apparatus as Armstrong was presenting his paper. At the point in the paper where the experiment was to be presented, a colleague stepped to the podium and whispered, "Keep talking. Runyon has just burned out [the] generator" (Lessing 1969, 170). Armstrong continued reading until he received

Armstrong's FM Demonstration

The new station was tuned in with a dead, unearthly silence, as if the apparatus had been abruptly turned off. Suddenly out of the silence came Runyon's supernaturally clear voice. "This is amateur station W2AG at Yonkers, New York, operating on frequency modulation at two and a half meters."

A hush fell over the large audience. . . .

A glass of water was poured before the microphone . . . it sounded like a glass of water . . . not the sound effects on ordinary radio. A paper was crumpled and torn; it sounded like paper and not like a crackling forest fire. Sousa marches were played from records and a piano solo and guitar number were performed by local talent in the Runyon living room. The music projected with a "liveness" rarely if ever before heard from a radio "music box."

Source: Lawrence Lessing, *Man of High Fidelity: Edwin Howard Armstrong*, New York: Bantam Books (1969): 169–170.

the signal that everything was ready. It was an impressive demonstration.

Armstrong was dedicated to his system—he promoted it as a replacement for AM. The impact of FM was not immediately confrontational. Some scientists saw it as an improvement of the existing signals and gave little thought to its replacing AM. Sarnoff and RCA opposed FM for radio because they were beginning to experiment with television at this time and viewed FM as an opportunity within the development of television. As Armstrong pushed his position, however, those who had a financial investment in AM were not willing to consider anything new, and they soon began to fight back. The conflicts prompted legal delays in the allocation of frequency space and, as corporate engineers began developing other systems, more conflict resulted. Armstrong's personal relationship with Sarnoff was strained and broken. He spent most of his fortune defending his FM system as a revolutionary technology that would make AM obsolete. Sarnoff's biographer describes the embittered Armstrong as a man who "lacked the Sarnoff stomach for a prolonged battle" with the large growing corporate broadcast industry. FM's development was so slow that Armstrong became despondent and, in 1954, took his own life. FM would be delayed several decades before it achieved Armstrong's vision and replaced AM some years after his death.

Television: RCA Versus Farnsworth

The story of the technological development of television, "visual radio," is similar to that of Armstrong and FM. Television's technological history began in the late 1800s, when Maurice Leblanc developed scanning theory and Paul Nipkow developed the scanning disk. On January 26, 1926, Englishman John Baird conducted the first public demonstration of mechanical television in London. Its two most successful pioneers in electronic television were Philo Taylor Farnsworth and Vladimir K. Zworykin. Here, again, the dominant inventor was pitted against a major corporation. This time it was Philo T. Farnsworth who battled with RCA and David Sarnoff.

Zworykin was a Russian-born immigrant who came to the United States in 1919 and went to work for Westinghouse as a member of its research staff. Because Westinghouse's primary interest was still radio, Zworykin worked on his own with little support for his television experiments until they attracted the attention of Sarnoff at RCA. When Sarnoff asked Zworykin what it would take to develop television, according to Bilby (1986), Zworykin put the price tag at $100,000. Sarnoff later joked with friends that Zworykin was "not only a great inventor, but the greatest salesman in history. He put a price tag of $100,000 on television and I bought it." This investment would grow over the years as Zworykin, working for RCA, took the corporate lead in the development of television.

The youthful Philo T. Farnsworth was the independent inventor of television. Farnsworth was born August 19, 1906, in Utah. He milked the cows on the family's farm and took care of the family's source of electricity, a Delco generator. At age fifteen, while attending high school in Rigby, Idaho, Farnsworth sketched on the blackboard his dream for television. It was a discussion he and his high school teacher Justin Tolman would never forget, because Tolman had to recall the drawing years later as part of his testimony in the patent suits between Farnsworth and RCA.

When the Farnsworth family moved back to Utah for better educational opportunities for their children, Philo met George Everson, a community chest fundraiser. Seeing an opportunity, Everson asked Farnsworth how much and how long it would take to develop the dream. Farnsworth responded, "six months and five thousand dollars." This, however, was the Depression, and success did not come as soon—or as cheaply—as either the inventor or the financier expected.

Precedents: FM

Edwin Howard Armstrong was granted a patent for FM in 1933. Its first tests were impressive but led to confrontation with RCA. FM's development was delayed several decades before it achieved the inventor's original vision of FM dominance.

Philo Taylor Farnsworth was among the first to experiment with electronic scanning in the development of television.

The first Farnsworth laboratory was established in 1926 on Green Street in San Francisco. Among the first pictures transmitted were a dollar sign and a black triangle. These were laboratory experiments; the first public demonstration took place September 7, 1927. The invention might have been demonstrated months earlier, but investors wanted to wait until a photograph could be transmitted, thinking it more impressive than triangles and dollar signs.

Farnsworth's television system went unnoticed by RCA until the early 1930s, after which they became fierce competitors. Zworykin and Sarnoff visited the Farnsworth lab in San Francisco, and RCA tried to purchase some of Farnsworth's patents, but he refused. He had royalty payments in mind, which RCA did not want to pay. RCA was used to purchasing patents outright and collecting its own royalties. In the end Farnsworth won a court confrontation over the contested patents, and RCA was forced to come to agreement with him. But like Armstrong, Farnsworth did not have the stamina for a prolonged corporate war.

Hundreds of patents are filed under Farnsworth's name. In his conflict with RCA, he won the court battle but lost the public relations war. Philo Taylor Farnsworth was the last of the independent inventors: He was first to develop the electronic technology, but it was RCA that led in industrial and commercial manufacturing.

Wartime Transition

World War II was a transitional period for radio and television. Radio news and the golden age of programming peaked in popularity, but all the developments of technology were diverted to the war effort. Radio station licensing expanded little; television development and frequency allocations were frozen. Entertainment programming poked fun at Hitler, while the news of war scared the U.S. audience. Radio personalities encouraged their audiences to buy war bonds, recycle materials, and donate to the war effort. All levels of radio were affected, and the industry would again be revolutionized after World War II.

Summary

The importance of the developments in the history of electronic media will not be grasped by the memorization of names and dates, but by understanding the patterns and precedents established within specific historical periods. Radio and television did not evolve in isolation, but amid individual and corporate conflicts and questions of public policy. The precedents established during the Roaring Twenties include the foundations of the networks' operational patterns and organization, the development of public programming, and the beginnings of advertising as we know it today. The precedents of the Depression and World War II years include refinements in network operations and programming, the development of news programming, and the eyewitness radio news format. These eras reflect the growing strength of corporate control of electronic media and encompass the lives of

Precedents: Television

The youthful Philo Taylor Farnsworth first demonstrated electronic transmission publicly in Philadelphia in August 1934 at the Franklin Institute. RCA and Farnsworth became fierce competitors. In 1939–40 the first Farnsworth mobile unit went on countrywide tour. In 1939 RCA demonstrated television at the New York World's Fair.

the last independent individual inventors: Armstrong and Farnsworth.

▶ The industrial era fostered wireless technology and provided the environment for the lives of prominent pioneers: Maxwell, Hertz, Marconi, Fessenden, and de Forest.

▶ Marconi sent his first wireless transcontinental signal in 1901.

▶ Patent rivalries began among prominent radio pioneers. There were many experimenters, but few entrepreneurs.

▶ The importance of American sea power gave prominence to maritime radio and ship-to-shore communication. The Radio Act of 1912 reflected the significance of radio at sea as did the *Titanic* disaster.

▶ World War I pushed the government into active involvement in radio. Spurred by radio's military applications, the U.S. Navy assumed control. The war provided an important transition period for radio as it grew from a maritime tool into a public medium. Following the war, patents were returned to their private and corporate owners, and the commercial era of radio began.

▶ Radio was the youngest rider on the Coolidge prosperity bandwagon of the Roaring Twenties. Corporations began to take key roles: Westinghouse station KDKA broadcast the first election returns in 1920; Marconi station CFCF broadcast its first public transmission in 1920; and AT&T station WEAF experimented with advertising and network broadcasting.

▶ RCA formed NBC to conduct network operations and then purchased AT&T. CBS was formed as RCA's chief competitor. Mutual's *Lone Ranger* provided the foundation for a different kind of network.

▶ The Depression increased radio's role as an information medium. After the press-radio war and with the utilization of radio as a platform of ideological debate, news and information programming grew in importance.

▶ World War II established radio news as a staple consumer product and evolved the news patterns still in existence today.

▶ The last of the independent inventors worked in the eras of the Depression and World War II: Armstrong's experiments with FM eliminated the static from the radio signal, but he fought a losing battle to develop it. Farnsworth developed visual radio and likewise fought a losing battle with the corporate giants.

▶ World War II provided another transition period, as technology was diverted to the war effort, providing the foundation for the new era of television.

INFOTRAC COLLEGE EDITION EXERCISES

The history of radio and television is important not simply because of people and politics and their environment, but because of the precedents that were set forth and continue in our operations today.

1.1 The broadcast industry has its own periodicals. *Electronic Media* is relatively new and has a user-friendly programming orientation. *Broadcasting & Cable* (earlier called *Broadcasting*) is the industry periodical of record. The *Journal of Broadcasting and Electronic Media* is a publication of academic research. In a November 2, 1970, issue of *Broadcasting*, you'll find a chronology of events and articles on the importance of "who's on first." In the *Journal of Broadcasting* 21:1 (Winter 1977) is an examination of the stations claiming to be our nation's first. Find and read these articles.

    a. What was the first U.S. station on the air?

    b. What evidence do you have for that declaration?

    c. Identify "firsts" where you can see a precedent for today's operations.

    d. Can you identify "firsts" from any other countries?

WEB WATCH

Here is a list of a few URLs (Internet addresses) for some of the organizations or corporations discussed in this chapter. Please explore these Web sites and follow the links to learn more about the complex business of the electronic media. Add your descriptions and your own favorite sites at the end of the list. Please keep in mind that the dynamic nature of the Internet allows sites to come and go but also allows organizations to update information about themselves very quickly.

Address	Description
http://www.spot.colorado.edu/~rossk/history/phone.html	
http://www.songs.com/noma/philo/intro.html	
http://www.ge.com/annual rep	
http://www.promet12.cineca.it/htfgm	
http://www.wings.buffalo.edu.SBF/punctuation.html	
http://www.waroftheworlds.com	
http://www.westinghouse.com	
http://www.rca.com	
http://www.invent.org/book/book-text/71.html	
http://www.monviso2.alpcom.it/hamradio	
http://www.compulink.co.uk	
http://www.uspto.gov/~westernunion/index.htm	
http://www.ascap.com/ascap.html	
http://www.kdkaradio.com	
http://nab.org	

Other favorite sites:

CHAPTER 2

Electronic Media: World War II to the Telecommunication Revolution

By 1945 radio had become the most popular medium for both entertainment and news, but looming on the horizon was a new medium that threatened to render radio obsolete: television.

It is the summer of 1945. World War II is over, and people are celebrating in the streets. The radio industry has cause to celebrate. It has emerged from the war at the peak of its popularity.

During the 1930s the radio networks developed entertainment programs to an unprecedented level of popularity. Then during the war years, radio matured as a news medium. By 1945 radio had become the most popular medium for both entertainment and news. In the evening, families gathered around the radio receiver to hear their favorite programs. Many of the program genres on television today were developed by the radio networks. Network radio programs were broadcast live from studios in New York City. The immediacy of live radio was considered its major advantage over the print media. Recorded music was available, but recordings were not of high quality and they were considered beneath the standards of networks at the time.

Another characteristic of radio in 1945 was program sponsorship: Radio programs did not have spot commercials; programs were sponsored by a single advertiser. Major advertisers bought a block of time on the network and produced their own programs. But these characteristics would change drastically within the next decade. Looming on the horizon was a new electronic medium that threatened to render radio obsolete: television.

This chapter examines historical events important in the development of the electronic media after World War II, including the transition of radio from a national medium dominated by network programs to a local medium dominated by music, the development of network television, and the emergence of cable television.

Radio Growth and Transition

Radio emerged from World War II the most popular entertainment and news medium in the country: The number of radio stations on the air had increased dramatically from 765 in 1940 to 2,231 in 1950. As the number of stations grew, network advertising revenues also increased until 1950, when they began to decline. Part of the reason for the growth was that during the war, radio networks had received more than their usual share of advertising because print media experienced a materials shortage and could not produce as many pages. After the war these shortages disappeared, and more advertising went back into print.

Although competition from print media was considerable, television was the major factor in the changes evident in radio by 1950. Television had been introduced to the public at the 1939 New York World's Fair, but the U.S. entry into the war in 1941 brought television development to a virtual standstill. After the war the major networks once again turned their attention to television, and radio began to change.

By 1948 there were so many applications for new television stations that the Federal Communications Commission (FCC) had to put a freeze on license applications, but the freeze did not stop the transition of programs from network radio to network TV. One by one, beginning with the Lux Radio Theatre, network radio programs became television programs. As this emigration of programs progressed, local radio stations began to drop their affiliations with the networks. With fewer network programs available to local stations, many predicted the death of radio. After all, who would want to listen to a program on radio when one could hear and *see* it on television?

But radio was not about to roll over and die. By this time recorded music had become more prevalent on independent radio stations, which were not affiliated with networks and thus had to create their programming or procure it from some other source. Because a program of recorded music was relatively inexpensive to produce, this type of programming was already common on the independent stations. Typically, these stations would program a *sweep*, or uninterrupted segment, of music along with some news and weather (Routt, McGrath, and Weiss 1978). They depended on music programming for the bulk of their broadcast day just as do the radio stations of today.

The radio music program had originated years earlier on the *Grand Ole Opry* and *Your Hit Parade* (Bunzel 1991). On these programs announcers introduced musical performances by popular singers and musicians. As network radio affiliates began to lose programs to television, they too turned to recorded music to fill in the gaps. Radio stations used announcers to introduce selections of recorded music, and they soon became integral parts of the programs: The **disc jockey** was born. As the networks abandoned them, radio station owners used a tried-and-true programming source as the basis for a radio revival: music.

PIONEERS OF TOP 40 RADIO FORMAT

"It became necessary to find a formula that would succeed in spite of television and network radio. The formula found was one that succeeded because of network radio and TV."

TODD STORZ

Todd Storz's "formula" was the format now known as Top 40, and he "found" it in an Omaha restaurant. As legend has it, he was sitting in a restaurant in the late 1940s, listening to people play the same few songs over and over on a jukebox, when he got the idea to program a radio station by spinning only top-selling records. He bought independent KOWH-AM Omaha in 1949 (taking a chance by getting into a market where there was stiff competition from new TVs and established network radio affiliates) and programmed it with a select group of songs, upbeat jingles, and no dead air. The technique, which today has evolved into contemporary hit radio, quickly caught on; it soon featured fast-talking disc jockeys aimed at a young audience, and a playlist confined to top-selling records. His station was a success, and he did the same thing with two more. Combined billings of the three stations grew from $100,000 in 1949 to more than $2 million in 1955. Storz died in 1963 at age thirty-nine. His father, Robert H. Storz, ran Storz Broadcasting until the last of six stations, WQAM-AM, was sold in 1985.

"It all begins with creativity and programming. You can have the greatest sales staff and signal in the world, and it doesn't mean a thing if you don't have something great to put on the air."

GORDON McLENDON

Gordon McLendon was a radio format pioneer who by 1965 recognized the profitability and desirability of niche formats and superserving an audience. During the 1950s and 1960s, McLendon built up his family-owned group of five AM stations, seven FM stations, and one TV station. He used his stations to test new format possibilities, including Todd Storz's idea of collecting lists of the most popular records and adding colorful disc jockeys, flamboyant promotions, and contests. That technique evolved into modern Top 40 radio. In Chicago, McLendon was the first to try an all-news format on WNUS-AM.

That format would soon be adopted by Westinghouse Broadcasting and evolve, with great success, at stations that would become longtime market leaders: WINS-AM New York and KYW-AM Philadelphia. He is also credited with being among the first to editorialize on his radio stations. McLendon sold his last station in 1978 and invested in the precious metals business; he died in 1986.

Source: "Special Edition: The First Sixty Years," *Broadcasting* (December 1991): 56, 60.

By the mid-1950s network television had emerged as the dominant medium for entertainment programs. Most of the comedy and drama programs that started on radio had made the transition to television. Radio stations were also in the midst of a transition; they were changing from network programming aimed at mass audiences to local programming aimed at specific audiences. Some radio station owners clung to the notion that they could still attract mass audiences, but most recognized that they could better attract segments of the listening audience with a particular style of music. This was the philosophy of radio formats: People watch television programs, but they listen to radio stations. People would identify with a particular style of music played on a station and listen on a regular basis.

Top 40

Two individuals are generally credited with the creation of one of the first radio formats: *Top 40.* In the early 1950s, in Omaha, Nebraska, Todd Storz used the jukebox as the genesis of his version of Top 40. After noticing that people tended to play the same songs over and over on the jukebox, he decided to program his radio station the same way. He developed a rotation or roster of hit songs that would be repeated throughout the day.

At about the same time, Gordon McLendon was beginning to program his Dallas radio station with a Top 40 format. He too developed a rotation system for the music, but he added heavy on-air promotion and personalities to the mix. Disc jockeys with high-energy announcing styles achieved celebrity status because of the Top 40 format. The audience for this new format was teenagers, a portion of the population that was growing rapidly. And, fortunately for radio, a new music style was emerging at about the same time that would take teenagers by storm: *rock and roll,* a term thought to have been used first by disc jockey Alan Freed. The term itself was inspired by a blues lyric written in 1922: "My baby rocks me with a steady roll" (Routt, McGrath, and Weiss 1978).

No longer did young people want to listen to the old *Hit Parade* artists like Eddie Fisher, Perry Como, and Rosemary Clooney. They wanted to hear the new artists who were emerging with rock music—Chuck Berry, Bill Haley, and, of course, Elvis Presley—whose music was a blend of rhythm and blues, country, and gospel. Teens loved it and wanted to hear more. Top 40–rock radio stations rose to number one in markets across the country by playing only hits.

The idea that program content could be repeated throughout the day was almost revolutionary in radio. Even the early independent stations that played music much of the day did not repeat musical selections. Their program directors tried to provide something for everyone with middle-of-the-road (MOR, measuring no extremes in musical styles) music programming. This philosophy reflected the mass audience orientation that had been so common throughout the golden age of network radio. But the format philosophy conceded the mass audience to television and turned to specialized audiences, or *demographics,* defined by age and music preferences.

Payola

The success of the Top 40 format in the 1950s saw a revival of radio programming and increased advertising revenues, but success also brought some problems. Not the least of these was the emergence of *payola*—the practice of bribing a disc jockey to play a particular record. As the role of disc jockeys expanded in Top 40 radio, they became very powerful individuals, able to determine which records became hits because they selected the music to be played. As a result, the growing recording industry attempted to influence disc jockeys with expensive gifts, a percentage of record profits, and even cold, hard cash. In this manner they literally could buy a hit record, because DJs would "put their records in heavy rotation" (play them often during the broadcast day).

By 1959 payola had become so widespread that it drew the attention of Congress. That year Congress amended the Communications Act of 1934 to make the practice of payola illegal. Essentially, the amendment forced disc jockeys to disclose payments received from anyone other than their employers. The amendment did curb payola, but the practice continued on a smaller scale for years.

FM Emerges

By 1960 it was obvious that the future of radio was in formats geared to local audiences. Network radio programs that were the basis of the golden age of radio had all but disappeared. Network radio advertising revenues had leveled off, while national spot (ads placed directly with stations by national advertisers) and local revenues increased dramatically. Because of the development of transistors in the 1960s, radio receivers had become more portable and personalized. Based on the success of the Top 40 format, station owners began to experiment with other formats designed for specific audiences.

Much of the early success of format radio occurred on AM stations even though FM was superior in sound quality because of the absence of static. FM did not develop during the 1930s and 1940s primarily because of interest in television at that time. The radio network chiefs saw the future of broadcasting in television, not in FM radio.

During the 1960s the FCC made three rulings that were, in large part, responsible for the growth of formats on FM radio.

1. **Approval of stereo broadcasting on FM in 1961.** By this time the recording industry was producing records in stereophonic sound, and listeners were more interested in higher-quality recordings. Because FM already surpassed AM in sound quality, the availability of stereo sound made it even more attractive to listeners. Specialized formats, such as classical and easy listening, began to appear on FM stations during the 1960s.

2. **A partial freeze on new AM stations.** This second ruling also spurred the development of FM. By 1962 the FCC was aware that the AM frequency band was becoming saturated and that interference problems might result. The commission's response was to place a freeze on new AM applications which, in turn, encouraged new FM applications.

3. **Prohibition of duplicate programming on AM and FM stations owned by a single licensee.** The third ruling by the FCC probably had the most dramatic effect on the development of new FM formats. In 1965 the FCC ruled that AM and FM station combinations (operated by the same licensee) in cities with populations of a hundred thousand or more could not duplicate programming for more than one-half of the broadcast day. Until this time many of these joint licensees had been simulcasting (duplicating the entire broadcast day) on their AM and FM stations. The new FCC ruling, requiring different programming for one of the stations for at least part of the day, encouraged the development of new formats on FM stations.

FM Growth

By 1966 nearly 50 percent of the homes in the United States had FM radio receivers, and this would increase to 93 percent by 1975 (Routt, McGrath, and Weiss 1978). During this period the number of FM stations also increased dramatically, and experimentation with new formats continued. Because FM offered high-quality sound and a stereo signal as well, the emphasis was on music. Audience research indicated that some listeners wanted longer segments of music and less patter from disc jockeys. Some of the new music formats that emerged on FM during the late 1960s and early 1970s included progressive rock, AOR (album-oriented rock), disco, and "beautiful music."

Contemporary music was expanding during the 1960s as well. The rock-and-roll roots of the 1950s were being interpreted in the "British invasion" by the Beatles and the Rolling Stones. In San Francisco psychedelic, or acid rock, was emerging as a reflection of the drug culture. New FM formats reflected these changes in musical styles and presentation. In the AOR format, longer album cuts were played rather than the three-minute hits heard on Top 40 radio, and the DJs introduced the music in a style quite opposite that of the high-energy Top 40 "jocks." AOR DJs were laid-back with cool attitudes; they attempted to communicate more intimately with their listeners (Routt, McGrath, and Weiss 1978).

By the mid-1970s the FM share of the listening audience had reached 40 percent, and it increased to more than 50 percent by 1980. Even though AM radio still had a larger share of the advertising dollars spent in radio, the trend was clear: FM would surpass AM in music format listenership. AM station owners were already exploring different formats that might allow them to compete for listeners in the 1980s and beyond. Just as observers predicted the death of radio in the 1950s, there were prognosticators of the demise of AM in the 1980s.

During this period the FCC reacted to AM's struggle by authorizing AM stereo. It unfortunately did not designate a common system of AM stereo, leaving it to the marketplace to sift out the best system. This

Music Formats Defined

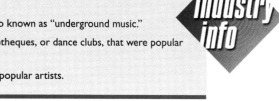

- *Acid rock* refers to rock music emanating from the drug culture of the 1960s. *Acid* refers to LSD.
- *Beautiful music* is primarily instrumental, orchestral versions of popular songs of the day. It is sometimes referred to as "elevator music."
- *Progressive rock* is a new form or interpretation of music; also known as "underground music."
- *Disco* is danceable music; the term originated from the discotheques, or dance clubs, that were popular in the 1970s.
- *AOR* (album-oriented rock) emphasizes album selections by popular artists.

---

led to compatibility problems with the few AM stereo receivers sold to consumers. Although stereo was not the savior of AM radio, specialized formats such as news/talk, all sports, big bands, and golden oldies have been successful for AM stations. The future of AM radio appeared to be in niche formats that would attract smaller but consistent audience shares.

As the 1980s began, the transition from AM domination of radio markets to FM was complete. FM stations now offered a mix of music formats geared to a variety of demographic groups. But radio now had a new competitor for contemporary music audiences: MTV. Music Television, a cable channel offering music videos and introducing the video DJ, or *VJ*, was introduced on Warner Cable TV in 1981. In some ways MTV became a partner to radio, because it introduced new artists that FM programmers were hesitant to play. It encouraged the revival of Top 40, which had declined in audience shares during the 1970s because of competition from AOR and other music formats. MTV introduced new artists and hits that were then played by Top 40 or CHR (contemporary hit radio) stations.

Television: Growth and Transition

As discussed in chapter 1, television technology was developed during the 1930s, and its growth was brought to a halt in 1941 when the United States entered World War II. The FCC had authorized commercial TV operation on VHF channels 1 through 13 in 1941. Channel 1 was soon dropped to make room for FM radio services (which were later reassigned to their present position, 88 to 108 MHz).

When the war ended in 1945, applications for new television stations again were accepted by the FCC. Renewed interest in television caused applications for new TV stations to increase. The number increased slowly at first because of a shortage of materials, but by 1948 applications were pouring in. Even though the FCC had authorized VHF channels 2 through 13 and a 525-line system, it was not prepared with a channel-allocation plan for cities in the United States.

By 1948 the emergence of television faced other problems. Both CBS and NBC were working on different color TV systems. CBS introduced its system first, but it was mechanical (not electronic) and not compatible with existing RCA black-and-white sets. This meant that a CBS television signal broadcast with the mechanical color system could not be viewed in black-and-white. The RCA labs had developed an electronic color TV system for NBC that would produce a compatible black-and-white picture. The FCC had to review both systems and choose one to become the industry standard.

Another issue that confronted the FCC as demand for new television stations increased was whether to make UHF channels available. It was evident by 1948 that the VHF channels would not provide enough allocations to meet the demand, so the FCC placed a freeze on new TV applications to allow time to work out these problems. The freeze was to have lasted up to nine months, but four years passed before the FCC was prepared to lift it. A final issue concerned educational television and lobbying efforts to reserve channel space for its use. During this period the FCC examined four major issues:

- VHF channel allocations
- Addition of UHF channel allocations
- Selection of a color TV system
- Channel reservations for educational TV

The Freeze Is Lifted

In April 1952 the FCC issued the *Sixth Report and Order* that lifted the freeze on new TV applications. The report included a TV channel allocation plan for more than twelve hundred U.S. cities. The allocations incorporated VHF and UHF channels (channels 14 through 83), but more than two-thirds of the channels were in the UHF band. This initially presented problems because most of the TV sets at the time could not receive UHF channels. Viewers could purchase UHF tuners, but they were difficult to use. It would be nearly a decade before the FCC required manufacturers to produce TV receivers with built-in capability to receive both VHF and UHF.

Another problem with the channel plan was that many cities were allocated only one or two channels, which meant that viewers in these cities were limited to one or two network stations—usually NBC or CBS affiliate stations, even though two other television networks, ABC and DuMont, also were providing programs. ABC had emerged after the FCC issued rules prohibiting the simultaneous operation of two networks by one company. NBC had been operating two radio networks, the Red and the Blue, until new FCC rules in 1943 forced NBC to sell one (see chapter 1). The new network became ABC. The DuMont Network was established in 1946 but could not compete and was out of business by 1956.

During the freeze the FCC also had to consider channel reservations for educational TV stations. Because of the efforts of an educational lobby group and FCC Commissioner Freida Hennock, 242 channels were reserved for educational television. Most of these were UHF channels, which later caused problems for educational stations because UHF requires more power. Early educational stations experienced many financial problems, and the expense of operating a UHF channel simply added to them.

The color TV issue was resolved in 1953, when the FCC adopted the RCA/NBC electronic system as the industry standard. The FCC had first adopted the CBS color system in 1950, but RCA filed a suit against the decision. After an independent engineering group studied both systems, it recommended that the RCA system be adopted. Despite the RCA system's adoption in 1953, color television grew slowly. The networks did not offer complete color schedules until the late 1960s. One reason for the slow growth was the expense of color TV sets: As the sets became less expensive, of course, more homes converted to color. By 1969 about one-third of the TV households in the country had color sets.

The Golden Age of Television: The Fifties

While the FCC temporarily froze the development of new TV stations, the networks continued to build their program schedules. NBC began feeding a schedule of programs to TV stations on a limited basis in 1946 from its New York TV station. Other network series that began in 1946 included a news magazine, a game show, a cooking show, and the first network soap opera (Brooks and Marsh 1979). The first regularly scheduled network drama series, *Kraft Television Theatre,* began the next year and continued on NBC for more than a decade.

Most of the programs that aired on the networks in the late 1940s and early 1950s came out of New York and were broadcast live. The influence of the live theater was evident in early TV dramas like *Kraft Television Theatre, Studio One,* and *Playhouse 90.* Indeed, the first network TV programs borrowed heavily from New York theater and vaudeville. One of the first stars of TV comedy, Milton Berle, came from the vaudeville stage. His program, *Texaco Star Theater,* developed one of the largest audiences for NBC and ran for nearly twenty years. Another television comedy legend was Sid Caesar, who, along with Imogene Coca, offered viewers ninety minutes of live comedy every week on *Your Show of Shows.*

CBS started feeding television programs to affiliates in 1948. It took a few years, but CBS surpassed NBC in numbers of viewers with such TV classics as *I Love Lucy, Arthur Godfrey, Jack Benny,* and *The Ed Sullivan Show.*

I Love Lucy is perhaps the best-known TV comedy ever produced. It started on CBS in 1951, but was not broadcast live and did not originate in New York. The program was a victory for Lucille Ball, who wanted to produce the show on film in Los Angeles. She also wanted her real-life husband, Desi Arnaz, to play the role of her TV husband. William Paley, head of CBS, was concerned that Arnaz's heavy Cuban accent would be a distraction. But Lucille Ball was persistent, and the program, preserved on film for all time, has become a television comedy classic.

Lucille Ball and Vivian Vance in *I Love Lucy.*

Although CBS and NBC produced some of the most popular TV programs, DuMont is credited with two programming firsts: the first situation comedy, *Mary Kay and Johnny,* in 1947 and the first soap opera, *Faraway Hill,* in 1946. The soap opera lasted only two months, but it established a genre that would be among the most popular in television. The DuMont Network struggled from its inception due, in part, to the fact that it did not have a radio network. Having established radio networks allowed NBC, CBS, and ABC ready

THE FIRST TV SITUATION COMEDY

The DuMont Network is credited with broadcasting the first situation comedy on November 18, 1947, with the premiere episode of *Mary Kay and Johnny.* The program featured young newlyweds Mary Kay and Johnny Stearns (their real names), who lived in New York. Mary Kay, the comedian, was rescued from dilemmas by her husband, Johnny, the straight man.

The series lasted for three seasons, from 1947 to 1950. It aired the first year on the DuMont Network and then switched to NBC in 1948. It stayed on the NBC schedule through March 1950, with the exception of three months in 1948, when it was broadcast on CBS.

The program was sponsored throughout its network life by Anacin, the pain reliever. Because there were no audience ratings available during that time, the sponsor offered a free mirror to the first 200 viewers who wrote in. Anacin had no idea how many letters to expect, so it ordered a total of four hundred mirrors to give away. The offer generated 8,960 letters! The power of the television situation comedy had been established.

Source: T. Brooks and E. Marsh, *The Complete Directory of Prime Time Network TV Shows, 1946–Present,* New York: Ballantine Books (1979): 384–385.

Gunsmoke: From Radio to Television

Gunsmoke debuted on the CBS radio network in 1952. The program featured the adventures of Matt Dillon, a marshal in Dodge City, Kansas, in the 1880s. The role of Matt Dillon was played on radio by William Conrad (who played detective Frank Cannon on the TV series *Cannon* in the 1970s). CBS added the television version of *Gunsmoke* in 1955, and the radio program with Conrad as the voice of Matt Dillon continued until 1961.

The original choice for the role of Matt Dillon on television was John Wayne, but Wayne was not interested in a weekly TV show and suggested his friend James Arness for the role. An unknown at the time, Arness (six feet seven inches tall) turned out to be the ideal choice to play the larger-than-life Matt Dillon. During its first TV season, *Gunsmoke* was not a top-rated (top ten) program, but it had jumped to number one by its third season. It stayed at number one for four years. Audiences for *Gunsmoke* had decreased somewhat by the mid-1960s, but the program stayed on CBS until 1975. When it finally was canceled after a twenty-year run, *Gunsmoke* was the only western left on network television.

Source: T. Brooks and E. Marsh, *The Complete Directory of Prime Time Network TV Shows, 1946–Present*, New York: Ballantine Books (1979): 243-244.

access to talent and affiliates. This, coupled with a shortage of revenue, brought about the demise of the DuMont Network in 1956.

The Move to Hollywood

By the late 1950s, another program genre had made its way to the top ten shows on TV. The "adult western" had surpassed comedy and variety in popularity by the 1958–59 season; seven of the top ten shows were westerns (Bunzel 1991). Many of the westerns were shot on film in Hollywood and distributed to TV stations or to the networks. As early as 1950, the *Cisco Kid* was available in color as a syndicated series.

Although the early western heroes such as Gene Autry and Hopalong Cassidy came from Hollywood movies, the TV western itself had its roots in both network radio and Hollywood movies. Two programs are credited with the introduction of the adult western: *Gunsmoke* and *The Life and Legend of Wyatt Earp* (Brooks and Marsh 1979). *Gunsmoke* started as a radio series on CBS in 1952 and became a TV series three years later. *The Life and Legend of Wyatt Earp* originated as a TV series, and both programs served as models for TV westerns for years to come. They also contributed to the shift of network television from New York to Hollywood by the early 1960s. The adult western reflected a growing demand among TV viewers for more action-and-adventure drama, which was a shift away from the theater-based dramas produced live for television in New York. This trend, coupled with an increasing demand for television programs produced on film, brought about the migration of network television from New York to Hollywood.

Initially, the movie industry saw TV as a threat. Many film producers wanted no part of television and even attempted to prevent movie stars from appearing on TV. Many studios prohibited the showing of their films on television. In 1953 ABC merged with United Paramount Theaters, a Hollywood film company. This brought new capital to a struggling television network and, more important, it signaled the beginning of a new relationship between movie companies and the TV networks.

With its new partner in Hollywood, ABC initiated a new series with a relatively small company, Walt Disney Studios. The series, titled *Disneyland,* premiered on ABC in 1954 and went on to become the network's first top ten program and it opened the door for television to enter Hollywood. Over the next decade, more network TV shows made the migration from New York to Hollywood. This marked the end of live programming and of the golden age of television. From that point network TV was dominated by filmed series produced in the movie studios of Hollywood.

The Game Show Scandal

Like radio, the development of early television was not without problems. Another type of program that achieved popularity very quickly on TV was the game show. The genre was born on network radio in the

1930s and grew in popularity throughout the 1940s. A number of game shows did make the transition from radio to TV in the early 1950s—perhaps the best known were the Art Linkletter programs *People Are Funny* and *House Party*. These were not strictly game shows, but they did have contestants and some element of chance in the program mix (Goedkoop, in Rose 1985).

By the mid-1950s game shows were near their peak in prime-time popularity. Two programs in particular, *The $64,000 Question* and *Twenty-One*, were bringing the TV networks unprecedented ratings for game shows. During the 1955–56 season, *The $64,000 Question* achieved a 47.5 rating (meaning that 47.5 percent of all TV households watched), the highest in the history of game shows (Goedkoop, in Rose 1985). But ratings were kept high by fraudulent activities such as the disclosure of questions and, occasionally, answers to popular contestants. The producers of these two programs recognized that the audiences would identify with contestants who kept winning for weeks. Indeed, some contestants were being given answers to quiz questions before the programs so that they were guaranteed to win.

The game show scandal broke when one of the contestants went to the authorities with evidence that producers were supplying quiz answers. In 1959 a congressional subcommittee was appointed to explore the fraudulent practices on game shows. As a result of the bad publicity, the networks backed away from game shows during prime time. In fact, the scandal led to another major change in network TV programming practices: The networks decided to take more control of programs rather than continue allowing major sponsors to produce them. This shift eventually led to the end of program sponsorship and brought about "participation advertising," wherein advertisers simply bought commercial time during programs rather than being the sole sponsor of a program.

Television Grows Up: The Sixties

As the decade of the 1960s unfolded, television began to mature as an entertainment and information medium. The innocence of the golden age was gone, and TV now reflected the turbulent changes of the decade. The tremendous power of the medium became evident during the 1960 presidential campaign, when

The first televised presidential debate, moderated by Howard K. Smith, was crucial for both John Kennedy and Richard Nixon.

a young, photogenic senator named John Kennedy took advantage of television to win the election. Televised debates worked well for Kennedy and proved detrimental to a nervous, perspiring Richard Nixon.

A few years later, television brought the nation together to mourn the death of an assassinated president. When President Kennedy was shot in 1963, TV's immediate and intimate coverage not only demonstrated its power to focus the attention of a nation, but also that it could reflect reality as well as fantasy—no matter how painful that reality was.

Television grew as an information medium during the sixties. The networks had established television newscasts in the late 1940s, but they were only fifteen minutes long. In the early 1960s, the network newscasts were expanded to a half-hour, bringing us coverage of the Vietnam War, the civil rights movement, and the assassinations of Robert Kennedy and Martin Luther King Jr. By the end of the decade, TV had come into its own as a news medium, and the most popular program in the history of television, *60 Minutes,* had debuted.

In the meantime entertainment TV had changed as well. The 1960s brought the development of police and private-detective shows. The success of *Dragnet* in the fifties was followed by *Naked City* and *The Untouchables.* Hollywood also produced glamorous detective shows like *77 Sunset Strip* and *Peter Gunn* early in the decade. NBC found success with a comedy program about police officers in *Car 54, Where Are You?* By the end of the decade, police programs such as *Ironside* and *Hawaii Five-O* dominated prime-time television and the adult western had all but disappeared.

Government Regulation and Public Pressure

By 1970 television had matured. Network program content had made a transition from the violence of *The Untouchables* to a stronger focus on character in *Ironside* and *The Mod Squad.* Although an element of violence remained in these programs, it was less gratuitous because it related the consequences of the violence to the characters. Part of the reason for this transition was pressure from the government and consumer groups during the 1960s.

The tremendous popularity of violent network programming drew criticism from a number of groups. In 1961 then FCC Chairman Newton Minow referred to television as a "vast wasteland." The term became the rallying cry for many others who were unhappy with network television. The government appointed several commissions to examine research on the effects of television violence. The general conclusion was that televised violence did contribute to aggressive behavior in viewers, particularly children.

At the same time, the FCC was concerned about network control of television programming, particularly the networks' domination of prime-time television at the expense of local stations. As a result, the FCC issued the prime-time access rule (PTAR) in 1971, which gave back an hour of prime time to the local network affiliates. It stated that the networks could program three of the four hours of prime-time television (7 to 11 P.M. or 6 to 10 P.M., depending on time zone).

Because many stations were already programming local news during the first half-hour of prime time, the focus was on the half-hour after the news. This was the prime-time access period that the FCC hoped would be filled with local, issue-oriented programs. For better or worse, this was not the case. Local stations turned to syndicated programs to fill this slot, because they were relatively inexpensive to acquire. Game shows became very popular because they were readily available and had not run on a network earlier in the day. Another stipulation of PTAR was that off-network programs or reruns (programs that have been broadcast on a network) could not be aired on network affiliate stations during the local hour. This part of PTAR applied only to network affiliates in the top fifty markets in the United States.

The Rise of Syndication

The business of selling or leasing television programs and series to individual television stations for broadcast in a local market is known as *syndication.* Syndicated programs are those purchased or leased directly from the producer by a local station. Local stations that are affiliated with a network and still have some hours during the day to fill use local or syndicated programs.

There are two types of syndicated programs available to local stations:

- *Off-network* (or *off-net*) *programs* originally aired on a network (also called "reruns") and

- *First-run programs* produced to sell directly to television stations (see chapter 9).

The production of first-run, or original, syndicated programs grew dramatically during the 1970s in part because PTAR prohibited network and off-network programs on local affiliates during the access period, and reruns could not be aired. This established a large market for first-run programs for the half-hour slot between local news and network programming. One of the first programs designed specifically for this slot was *PM Magazine,* which was followed in the late seventies by the highly successful program *Entertainment Tonight.*

The prime-time access rule not only stimulated the production of syndicated programming, it also helped independent television stations. Because they had no network programming, these stations had problems competing with network stations during the 1950s and 1960s. PTAR allowed independents major access to off-network programs during the first hour of prime time. Because network affiliates couldn't schedule off-network programs during this hour, independents began to dominate the access period with network reruns. Eventually, network rerun programs like *M*A*S*H* and *Happy Days* became consistent access-period ratings winners for independent stations. Independents expanded this success into other time periods as more syndicated programming became available during the 1980s. In 1995 the FCC determined that PTAR was no longer needed and proposed its repeal.

The Noncommercial Alternative

While commercial broadcasting was developing, its noncommercial counterpart, educational broadcasting, was struggling to survive. The history of educational broadcasting dates back literally to the beginning of broadcasting itself. WHA, a radio station at the University of Wisconsin at Madison, may very well have been the first educational station. Faculty in the physics department were experimenting with wireless telegraphy as early as 1902, and the institution actually licensed a wireless telegraphy station in 1915.

Educational radio stations like WHA began quite logically as instructional services—means by which to deliver courses to those who could not get to the classroom. Because of the growing interest in radio in the 1920s, more educational organizations, such as school systems and even churches, applied for station licenses. By the late twenties, nearly a hundred noncommercial radio stations were on the air, but many found it difficult to maintain a broadcast schedule as operational costs rose. As the decade of the 1930s and the Depression unfolded, the number of noncommercial stations had dropped by more than half.

Noncommercial educational broadcasting was not originally recognized as a service separate from that of commercial broadcasting. The Communications Act of 1934, the legislative foundation of broadcasting, includes the term *nonprofit* in section 307(c), which proposes that "Congress by statute allocate fixed percentages of radio broadcasting facilities to particular types or kinds of nonprofit radio programs" (Kahn 1968). After studying the proposal, the FCC recommended in 1935 that no fixed percentages of radio frequencies should be allocated to nonprofit radio. The rejection was based on the argument that educational radio programs would be adequately provided by commercial radio stations.

Within a decade the FCC did allocate specific channel reservations for noncommercial educational (NCE) radio stations. In 1945, when the FCC re-

M*A*S*H has been one of the most successful programs in syndication since its network run on CBS.

allocated the FM frequency band to 88 to 108 MHz, the first twenty channels were reserved for use by NCE stations. By this time it was apparent that commercial radio stations would not provide any educational alternatives to entertainment programs. Three years later the FCC added a new class of low-power FM stations, Class D, which allowed FM stations in the reserved portion of the band to operate with as little as 10 watts of power. Although coverage area (the area in which the station could be heard) was limited, the cost of operation also was minimal. This factor made the Class D stations attractive to colleges and universities interested in training broadcasting students.

It was not until 1952 that NCE television stations were recognized as separate from commercial stations. As noted, in that year the FCC lifted its freeze on new TV licenses and set aside channel reservations for NCE stations. In its *Sixth Report and Order,* the FCC ruled that in all communities with three or more TV channels assigned, one must be reserved for educational television. The first NCE TV station to be licensed on a reserved channel was KUHT, which signed on the air at the University of Houston in 1953. Over the next decade, another sixty signed on the air—many of them located on college and university campuses.

During its early history, educational television (ETV) was not well or even adequately supported by licensee institutions. In addition to a lack of funding, ETV stations had little access to high-quality programming. The ETV broadcast day was usually more limited than the commercial station schedule. There were some local productions, but they tended to be of poor quality. Independently produced programs were limited to documentaries and other educational programming on film.

The first nationally produced programming for ETV stations came from National Educational Television (NET), which began in Michigan in 1952 as a production and distribution center for ETV programming. These programs were distributed on tape or film via mail, not by live interconnection (as were their commercial network counterparts). By 1959 NET provided a few new programs for ETV stations, but still not enough to fill their broadcast schedules. It was obvious by this time that if ETV was to survive as a viable alternative to commercial television, additional funding was necessary.

In 1962 ETV broadcasters took their case for more funding to the federal government. They secured the first federal funding for educational broadcasting: the ETV Facilities Act. This piece of legislation authorized federal funds to contribute to the construction of ETV stations. It established a precedent for further federal funding of ETV and, unfortunately, ignored educational radio entirely. Sensing that they might be left out of federal funding permanently, educational radio broadcasters organized to lobby for their rightful piece of the funding pie.

Transition to Public Broadcasting

The National Association of Educational Broadcasters (NAEB) was formed during the 1920s when educational radio was beginning to develop. By the 1960s it had become an active Washington lobby focusing primarily on educational television. In 1963 the board of the NAEB proposed that radio membership should have its own division within the organization. The new radio division, to be named National Educational Radio (NER), began operations in 1964. As of that year, both educational radio and television were organized with two goals: long-term federal funding and the establishment of national networks. As a result of NAEB-organized efforts, two studies were done over the next few years that would be instrumental in the creation of new legislation authorizing funding for a new era in educational broadcasting.

The Carnegie Commission on Educational Television conducted the first study and released its report in 1967. In general, the report proposed a major reorganization of ETV, including the establishment of a new agency, the Corporation for Public Broadcasting (CPB), to disburse federal funding to ETV. At that time additional funding was to come from an excise tax on television sets. The Carnegie Commission also recommended strengthening local stations, more funding for national production centers, and live interconnection of ETV stations.

Throughout the report the Carnegie Commission used the term *public television* rather than *educational television.* It recommended the term *public* because *educational* television had a negative connotation among viewers. The commission also suggested that *public* would better portray the broader mix of instructional and cultural programming to be offered by stations in the future. In the report the commission was careful to point out the difference between instructional and public television: Public television programs are "in general not economic for commercial sponsorship, are not designed for the classroom, and are directed at audiences ranging from tens of thousands to the occasional tens of millions" (CCET 1967). (See chapter 12 for information on instructional television.)

In the meantime NER was conducting its own study of educational radio, released just months after the Carnegie report, titled "The Hidden Medium: Educational Radio." In general, the report recommended that educational radio's "ambitions toward national and international coverage should be encouraged and supported" (Land 1967). The members of NER desired that educational radio be set apart from commercial radio and justified as a unique broadcasting service. Certainly, educational radio broadcasters saw the potential for federal funding outlined in the Carnegie report and wanted to be included.

In 1967 Congress passed the Public Broadcasting Act. It is important to note that the title of the act used *public broadcasting* rather than *television*. Educational radio broadcasters were successful in their lobbying efforts to be included in the legislation virtually at the last minute. The Public Broadcasting Act was based on the recommendations of the Carnegie report, and therefore included the following:

- Creation of the Corporation for Public Broadcasting
- Federal funding for high-quality programming
- Network interconnection
- Development of new public broadcasting stations

The Corporation for Public Broadcasting was to act as the distribution agency for the federal funding, but it was not to be involved in station ownership or operation. Congress aimed to ensure that stations maintained control over their local operations and program schedules. The act did not include the excise tax proposed by the Carnegie Commission, however; it simply authorized funding for one year at a time, leaving public broadcasting without a permanent source for funding.

The transition to *public broadcasting* had begun. New federal funding for facilities, network interconnection, and even national programming had been authorized. A new agency—the CPB—had been created to disburse federal funds and insulate public broadcasting from political control. It was the beginning of a new era for noncommercial broadcasting, one born of the social climate of the 1960s. By the end of the decade, the CPB had established both public radio and television networks: National Public Radio (NPR) and the Public Broadcasting Service (PBS).

Beyond the fact that one was radio and the other television, the two networks were fundamentally different. PBS was established as an interconnection network only; it was not to be a program production agency. This reflected the congressional concern that public television remain a local service, with local stations controlling programming decisions. NPR, on the other hand, was established to provide both interconnection and national program production. Because of the lack of funding for radio programming, station operators supported a centralized program producer.

During the decade that followed the passage of the Public Broadcasting Act, the public broadcasting system grew dramatically. In 1967 there were 125 ETV stations and nearly 300 educational radio stations on the air. Nearly half of the radio stations were the low-powered (10-watt) stations noted earlier. By 1978, 276 public television stations and 200 public radio stations were on the air.

The radio stations were those that qualified for federal funding through the CPB. To distribute funds efficiently, CPB had established qualifying criteria. In essence, it required that the station be a full-time, professional station serving the community. This excluded student training stations, many of which were 10-watt facilities. During the 1970s, many institutions upgraded their student radio stations to qualify for CPB funding.

After its first few years of operation, PBS began to draw criticism for its dependence on British programming. Some critics quipped that *PBS* stood for "primarily British shows." In the mid-1970s PBS members formed the Station Program Cooperative (SPC), an organization devoted to the selection and funding of public television programs. Each year, programs are submitted to the SPC, and station representatives vote on what will be funded and carried on the network. In this manner the stations maintain local control of network programming and funding through their annual dues to PBS. The SPC arrangement resulted in more U.S.-made productions on public television.

PBS INTO THE MARKETPLACE

As public broadcasting moved into the 1980s, it faced a new era in its history. With the Reagan administration came cuts in federal funding for the public broadcasting system. At the same time, licensee institutions (i.e., universities and colleges to whom stations were licensed) were experiencing financial difficulties, which meant reductions in local funding.

Hence, fund-raising emphasis shifted to the third major funding source for public stations: donations from viewers, listeners, and private corporations.

From their beginning in 1967, public stations had sought underwriting grants from major corporations and had solicited viewers and listeners for donations. Under FCC rules, stations were required to identify corporate underwriters but could not mention any of their products or services. In reaction to pleas from public stations for funding help, in 1984 the FCC modified noncommercial broadcasting rules to permit "enhanced underwriting," which allowed underwriters to identify products and services in their announcements.

The rules stipulated that these allowances be "value-neutral." For example, a station could identify "General Motors, makers of Chevrolet, Pontiac, and Cadillac," in announcements, but it could not identify GM as a "maker of fine automobiles." The use of the term *fine* would violate the value-neutral stipulation.

In a further effort to bolster funding shortages, the FCC allowed an experiment with television commercials. In 1982 ten public TV stations were selected to test a series of commercials as a source of funding. The stations could actually sell and schedule commercial spots, but they could not interrupt programs with them. The experiment succeeded in raising additional funding for the stations that aired the commercials. The CPB and the National Association of Broadcasters (NAB), however, opposed using commercials, voicing concern that they would eventually erode the alternative-programming mission of public broadcasting. Of course, the NAB, as representative of commercial broadcasters, was also opposed to tax-supported competition for advertising dollars. After hearings on the experiment, the Senate Communications Subcommittee decided that commercials were not in the best interest of public broadcasting and set aside the proposal.

As public broadcasting entered the 1990s, station operators found themselves thrust into the marketplace. Because of reductions in federal and local licensee funding, stations must depend much more on private funding sources. This inevitably leads to marketing of station programs and to competition for audiences with other television stations in the market. Public stations now have creative marketing staffs that prepare and present sales proposals to potential underwriters. To listeners and viewers, they offer a wide range of products as incentives to donate to the station. Programming has been affected as public television stations offer classic film series and even situation comedies during on-air pledge drives.

These changes in public broadcasting have drawn criticism. Critics question whether the alternative mission of public stations is still valid. With the proliferation of specialized cable channels like Arts and Entertainment and the Discovery Channel, the need for a federally supported television system to deliver programs that are not available on commercial television networks is under further scrutiny. Another issue is that public broadcasting attracts very small audiences compared with the commercial networks. Should all taxpayers support a system that relatively few people actually use?

On the other hand, supporters of public broadcasting claim that only public stations can continue to provide local programming that serves underrepresented populations such as ethnic minorities and small children. They point to the rising commercialism of the cable channels and programming that caters to large advertisers. They suggest that, even in an era of deregulation with emphasis on the marketplace, a need still exists for alternative programming that can be provided only through a noncommercial system.

NEW COMPETITION: CABLE TV

The roots of cable TV go back virtually to the beginning of television in the late 1940s, when many areas of the country did not have access to the TV signals in large cities. This was often because of geographical barriers such as mountains between viewers and a broadcast station (see figure 2.1). Some rural areas used what was called a *community antenna television (CATV) system* to bring TV signals in from larger cities. As early as 1948, CATV systems were bringing TV signals to rural or mountainous areas that were unserved by or had poor reception of local stations.

At first these CATV systems were seen as extensions of existing TV signals: TV station operators supported the systems because they brought in more viewers for their programming. CATV systems grew slowly during the 1950s as more TV stations went on the air. Even by the midsixties, less than 5 percent of U.S. homes had cable TV. By this time, however, CATV systems were offering more than just nearby or local

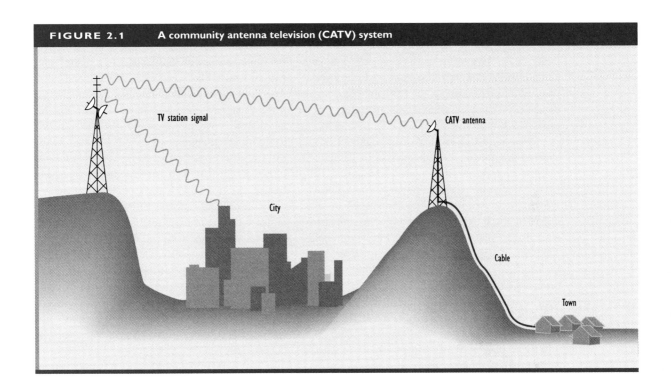

FIGURE 2.1 A community antenna television (CATV) system

stations. They also carried distant signals—programs from cities in other parts of the country. The addition of distant signals meant that cable TV was evolving into something more than an extension of local TV stations. Cable became competition for broadcast TV.

Until the early sixties, the FCC did not regulate CATV systems because they were simply extensions of broadcast TV. This changed in 1962, when the FCC moved to protect a local TV station from increased competition from a CATV system. In the Carter Mountain decision, a CATV system in Wyoming was denied a microwave application to bring distant TV signals to its subscribers. The local TV station owner petitioned the application on the grounds that the CATV system would duplicate its network programming. On the basis of potential economic injury to broadcast stations, the FCC established a pattern of protection for broadcast TV that would continue for more than a decade.

A few years later, the FCC passed its first regulations aimed at cable TV systems. By this time it was obvious that CATV systems had become more than TV signal extensions—they were cable TV systems offering their own programming. To protect TV stations and the public interest, the FCC approved a rule requiring cable systems to carry all local TV signals. The rationale for the *"must-carry rule"* was concern that cable systems might choose not to carry local signals, and subscribers would not have access to them.

At this time the FCC passed another regulation that prohibited cable systems from carrying programs that duplicated those carried on a local station. This regulation became the "syndicated exclusivity rule," referring to the right of local stations to protect the exclusive contracts they held for their syndicated programs. Again, the rationale for the rule was protection of the local TV station from competition from cable systems. The FCC further protected broadcast television stations in 1966 by placing a freeze on new cable systems in the top one hundred broadcast markets in the country. The commission was particularly concerned about the impact of cable competition on struggling UHF stations in these markets.

The FCC expanded its cable rules in 1972 to include channel requirements for cable systems and limits on the number of distant or imported signals. These rules also lifted the 1966 freeze and opened the way for cable companies to expand into cities that already had a number of local television signals. With these rules in place, TV station operators felt they could handle competition from cable systems in larger cities. As a result of the changing political climate in Washington and increased consumer demand for programming, the FCC embarked on a deregulation campaign that eliminated most of its cable rules by 1980.

The elimination of many of the federal rules spurred the growth of cable television. The large multisystem owners (MSOs) quickly recognized that money was to be made in the large cities. CATV systems that served the small towns with off-the-air television signals were less interesting when compared with the potential of broadband delivery of multiple channels and pay television. Cities, however, had learned that cable operators were willing to offer extensive services and perks in exchange for the franchise to wire the city. This time of cable operators' competing for the important city licenses became known as the "franchising wars."

Cities became demanding in their franchise requirements, and cable operators promised to provide anything that was requested. The early eighties saw the rapid growth of cable in urban areas, but the cable operators complained that they couldn't pay for the systems and services that the franchises required; the operators were especially concerned when the city franchise regulated rates charged to the customer. The customers, however, came in droves, because only cable could offer HBO, MTV, ESPN, CNN, and the superstation from Atlanta—WTBS.

While building cable plants and adding more satellite-delivered networks, the cable operators were lobbying through the National Cable Television Association (NCTA) for relief from the city-imposed franchise requirements. In response to the cable industry, Congress enacted the Cable Communications Policy Act of 1984, which, in effect, deregulated cable television; the cities were no longer able to dictate channel lineups or cable rates.

Cable channel offerings expanded during the late 1980s as did cable penetration, but rates charged to the customer grew as well, and by the early nineties many customers were complaining that they were paying too much for too little. Congress stepped in again, and the Cable Television Consumer Protection Act of 1992 became law. The act regulated cable rates through the FCC's newly organized Cable Bureau, but also redefined the relationship between the cable operator and the broadcaster. Under the 1992 rules, the cable operator once again had to provide "must carries," which had been struck down by the court. After 1992 a broadcaster could either demand that the station's signal be carried on the cable system or negotiate for payment for the over-the-air signal.

The Telecommunications Act of 1996 revised the sixty-year-old Communications Act of 1934. Although touted as the policy necessary to implement the information superhighway of the next millennium, its primary effect on the cable industry was to increase competition. While cable was allowed to offer telephone services, the telephone companies could also offer cable television to their customers. In addition, the deregulation of ownership rules encouraged mergers and buyouts of both cable companies and telephone providers. Soon the largest of the cable MSOs were owned by telephone providers while Internet service was being provided by the cable companies.

By the end of the millennium, nearly 75 percent of American homes were subscribing to a multiple-channel television service. Although most were wired services with fiber-optic cable providing the broadband backbone, the fastest-growing segment of the multi-channel market was provided by direct-broadcast satellite (DBS), with almost 10 million customers.

THE VIDEO MARKETPLACE: THE SEVENTIES

As competition for viewers grew in the 1970s, the television networks found themselves subject to further restrictions. Along with PTAR—the prime-time access rule—the FCC implemented the "financial interest and syndication (fin-syn) rules," which prohibited the networks from syndicating programs in the United States and from ownership of independently produced programs. The rationale for these rules, like that of PTAR, was concern that the networks were dominating television programming and that independent program producers did not have enough access to TV station schedules. The new rules were intended to stimulate the development and scheduling of independently produced programming.

Indeed the sources for and variety of television programs did increase as the seventies unfolded, and new trends in programming were reflected in network shows. One such trend was "issue" programming, an outgrowth of the social and political upheaval of the 1960s. Socially relevant issues and a focus on characters were common elements in entertainment programs about doctors, lawyers, and police officers. ABC had its first number one prime-time series during the 1970–71 season with *Marcus Welby, M.D.*

Perhaps the most innovative comedy series of the seventies was *All in the Family.* Based on a British situation comedy, the program departed from the conservative sitcoms of the 1960s. For the first time on network television, writers poked fun at racial and ethnic stereotypes through the ultraconservative character Archie Bunker. When the series premiered on CBS in 1971, network executives were anxious about public reaction, but viewers were ready for a change in situation comedy, and *All in the Family* became one of the most popular programs of the genre.

As television programming evolved in the 1970s, pressure on the networks continued from conservative consumer groups, the FCC, and Congress about sexual and violent content in programming. This caused the National Association of Broadcasters to add a "family viewing standard" to its Television Code in 1975. The standard stipulated that "entertainment programming inappropriate for viewing by a general family audience should not be broadcast during the first hour of network entertainment programming in prime time . . ." (*Writer's Guild v. FCC,* 423 F. Supp. 1064 [1976]). The networks adhered to the policy, but it lasted less than a year. When challenged in court by a group of television writers and producers, the standard was ruled unconstitutional.

The National Association of Broadcasters established the first radio code in 1929; in 1952 a TV code was added. Both codes covered the areas of programming and advertising. Adherence to the codes was voluntary for broadcast stations, but it was considered prestigious to be a code subscriber. In the early 1980s, advertisers challenged the code time limits on advertising for stations. The Justice Department threatened antitrust action against the NAB, claiming that the limits were keeping the price of broadcast ads artificially high. In 1983, upon advice of lawyers, the NAB eliminated its code entirely.

In the 1990s the NAB developed new voluntary programming principles restricted to four main areas:

children's TV, indecency and obscenity, violence, and drug abuse. Other codes used in the industry are those of the Radio-TV News Directors Association (RTNDA) and the Society of Professional Journalists (SPJ). These codes cover such topics as fair play, accuracy, objectivity, and press responsibility.

Cable Expands

In 1975 a company called Home Box Office (HBO) began distributing first-run movies by satellite. It was an important year for the cable industry. It brought pay or premium channels to cable outlets and gave cable subscribers another service that they could not get on broadcast television. HBO became particularly important for cable systems in large cities, because it was a selling point to gain new subscribers: Viewers did not need cable TV to receive a variety of TV channels or to improve reception. Programming services like HBO provided a reason for these viewers to subscribe to cable.

Other Technologies Emerge

During the 1980s a number of other technologies emerged in the video marketplace, including multiple, multipoint distribution systems (MMDS), direct-broadcast satellite (DBS), low-power television (LPTV), high-definition television (HDTV), and satellite-master antenna system (SMATV). All are video delivery systems; most are supplementary or alternative services to those provided by broadcast and cable television (see chapter 14).

Summary

▶ After World War II, radio was the most popular entertainment and news medium in the United States. But the golden age of radio soon gave way to television. By 1948 the FCC had received so many applications for new TV stations that it placed a freeze on them. The freeze was lifted in 1952 after the FCC had outlined a channel plan for VHF and UHF allocations and had evaluated new systems for color TV.

▶ During the 1950s many network radio programs made the transition to television. This left it up to radio stations to create new local programming to fill the gaps left by departing network programs. Music formats such as Top 40 emerged and brought radio back from the brink of extinction. In the meantime network television developed situation comedies, dramatic programs, and variety series in New York. During the 1960s television programming shifted to Hollywood coincident with the development of westerns and police programs.

▶ Noncommercial educational (later public) broadcasting began to emerge in the 1960s. The Public Broadcasting Act of 1967 established federal funding for noncommercial broadcasting and led to the development of public radio and television networks. Other competition for broadcast television emerged during the 1970s and 1980s, including cable TV, high-definition TV (HDTV), and direct-broadcast satellite (DBS).

InfoTrac College Edition Exercises

The history of broadcasting since World War II was dominated by the emergence of television and the evolution of radio.

2.1 You read in chapter 2 about the emergence of game shows on prime-time network television in the 1950s and the scandal surrounding them. Due to bad publicity resulting from fraudulent practices, game shows disappeared only to reemerge in prime-time programming schedules in 1999. Read the article on game shows in the November 1, 1999, issue of *Broadcasting & Cable* and answer the following questions.

   a. Which ABC game show became a summer hit and was then added to the network fall prime-time program lineup?

   b. Based on the success of the ABC game show hit, producer Dick Clark created a game show very much like it. What is the title of Dick Clark's game show that was broadcast in prime time on the Fox network?

   c. In 1999, NBC and CBS announced that they were getting back into the game show business by reviving two famous 1950s game shows. What were they?

2.2 The history of radio reflects its evolution from a network-dominated program service to a locally oriented format service. Radio continues to evolve as stations explore other forms of distribution such as satellite and the Internet. Read the article in the November 8, 1999, issue of *Broadcasting & Cable* on the love affair between radio and the Internet and then answer the following questions.

   a. Radio Advertising Bureau President Gary Fries says that there are two main reasons why radio loves the Internet. What are they?

   b. On the other hand, Fries says that there are several reasons why the Internet loves radio. What are they?

   c. What challenges do Internet-only radio stations present?

WEB WATCH

Here is a list of a few URLs (Internet addresses) for some of the organizations or corporations discussed in this chapter. Please explore these Web sites and follow the links to learn more about the complex business of the electronic media. Add your descriptions and your own favorite sites at the end of the list. Please keep in mind that the dynamic nature of the Internet allows sites to come and go but also allows organizations to update information about themselves very quickly.

Address **Description**

http://www.hofstra.edu/nacb/ _____

http://www.fcc.gov _____

http://www.cpb.org _____

http://www.ncta.com/ _____

http://www.expage.com/page/thellc _____

http://www.kuht.uh.edu/ _____

http://www.rockhall.com/ _____

http://www.mbcnet.org/mbenet/dumont.htm _____

Other favorite sites: _____

CHAPTER 3

Wired and Wireless Interpersonal Telecommunication Systems

Long before broadcasting we communicated by wire; now with cable, fiber optics, and the Internet, we are returning to the wire.

any people born in the twenties, thirties, and forties think of broadcast radio when describing telecommunications, and those born in the fifties, sixties, and seventies assume broadcast television. The generation of the eighties and nineties will be inclined to think in the broader terms of wired communication systems such as cable television, the Internet, and telephone delivery of mass communication messages. Traditional broadcast systems seem, in many cases, to be taking a backseat to wired systems as the delivery of more and more information is required by the ever increasing diversity of audiences. This chapter examines in detail the early wired telecommunication systems of the telegraph and the telephone that were first mentioned in chapter 1.

The technology has changed over the past 150 years, but the developmental patterns of the early telecommunication systems, such as telegraph and telephone companies, are similar to those we see today in cellular telephone and cable systems. It might even be said that the early codes that made the Morse telegraph system practical are not much different from the binary codes that drive computer systems. Certainly the computer networks of today could not exist without the transistor, which was developed by Bell Labs as a necessary ingredient to intercontinental voice transmission.

This chapter also looks at the integration of wired communication systems in the delivery of mass communication information. The role of the telephone companies, a history of AT&T, and the cable television systems in the projected 500-channel universe are also examined.

Visual Communication Systems

In the simplest terms, *telecommunication* means to send and to receive meaningful messages over distance through the use of some medium or device. The term *telegraph* means to write at a distance. One early device used to send information over a distance was the *semaphore,* or visual telegraph. In the waning years of the eighteenth century, three French brothers named Chappe developed a system of communicating with wooden paddles attached to the top of a tall pole. Semaphore systems were set up in France, England, Sweden, and Russia. Similar systems were quickly developed in Denmark and Prussia. One of the great weaknesses of the semaphore system was the weather and, of course, the inability to see the signals at night. Although the semaphore didn't receive the widespread acceptance in the United States that it had in Europe, one system was set up between Martha's Vineyard and the Boston harbor, which enabled ship owners to have some forewarning of the arrival of cargo in port.

Visual and semaphore signals lasted well into the middle of the nineteenth century. San Francisco's Telegraph Hill was so named for the semaphore that announced to the forty-niners the arrival of a ship. The quickest way that a message could be sent across the continent, however, was by physical means. The Pony Express was established in 1859 and could carry a letter across the western half of the continent in about ten days.

In the 1880s a communication system based on sunlight was used briefly by the U.S. Army in the vast spaces of the southwest United States. The *heliograph,* a complex set of mirrors mounted on a tripod, allowed a beam of sunlight to be reflected from mountaintop to mountaintop through the clear, sunny desert atmosphere. By interrupting the beam, flashes of sunlight could be converted into a code to form a message.

The Telegraph

The world, however, needed a communication system that would work day or night, in rainy weather or sunny, provide instant communication, and be reliable. Sir Charles Wheatstone was able to control the direction of a magnetized rotating needle to spell out words at a distance by sending electrical pulses through wires. Thus the first practical electrical telegraph system was born in England early in the nineteenth century.

The Invention

In America the development of the telegraph took a different turn. The well-known artist Samuel F. B. Morse visited a semaphore telegraph office in France and was impressed with its ability to send messages over long distances. Morse used the principles of electromagnetic energy in his telegraphic system: A pulse of electric current through the wire would

history

THE PONY EXPRESS

The years between the discovery of gold in California in 1848 and the coming of the transcontinental railroad in 1869 was a time of tremendous growth in the West—and of isolation from the rest of the nation. The gold fields on the western slopes of the Sierra Nevada were nearly two thousand miles from the Missouri River, which was then the western frontier of the United States. By 1860 war between the North and the South was inevitable. Both sides wanted the support of California, which had entered the Union as a free state in 1850. Rapid communication was considered the key to keeping California, as well as the territories of Washington, Oregon, Utah, and Nevada, loyal to the Northern cause. Mail delivery commonly went by ship from New York to Panama, by land across Panama, and then again by ship to San Francisco. It often took three to four weeks for a letter to reach its destination. The overland route, via the Butterfield stage, took even longer and had the additional disadvantage of traveling through the Southern states of Tennessee, Arkansas, and Texas.

In 1860 the western freighting firm of Russell, Majors, and Waddell formed a corporation—the Central Overland California and Pike's Peak Express Company—to operate what became known as the Pony Express. The combination of horses, riders, and stations could deliver a letter from St. Joseph to Sacramento in ten days or less, with riders traveling in relays around-the-clock. The route across the center of the continent was soon divided into two sections: The eastern section, between St. Joseph and Salt Lake City, was operated by the Overland Mail Company, with its government mail contracts. The Wells Fargo Express Company served as an agent for the operation of the western section of the route.

Just a few weeks after the first rider left St. Joseph, Congress authorized money to build a transcontinental telegraph line. The Overland Telegraph Company of California and the Pacific Telegraph Company of Nebraska were incorporated to construct the line. As the two lines were constructed toward a meeting in Salt Lake City, the distance traveled by the Pony Express become shorter each week as letters were ferried between newly established telegraph offices. The Pony Express ceased operations only eighteen months after it started, the victim of a faster and cheaper communication technology—the telegraph.

Although the Pony Express was never a financial success, it did provide news of a dividing nation and an impending civil war, which was reprinted by western newspapers. In later years the Pony Express fired the imaginations of numerous dime-novel authors, artists, and filmmakers. The image of the brave, lone rider dashing across the plains is a familiar one to most Americans.

Source: Waddell F. Smith, ed., *The Story of the Pony Express*, 1969, San Francisco: Pony Express History and Art Gallery.

---

activate an electromagnet at the other end. In 1837 Morse filed for a patent on an electrical telegraph system that would send a message over ten miles of wire. He convinced Congress to fund an experimental telegraph system that would connect Baltimore and Washington; by May 1844 the system was complete. With Morse's transmission of the now well-known quote from the Bible *What hath God wrought!* the world embarked on the era of electronic telecommunication.

The key to the success of the telegraph was the development of *Morse code*, formed by a series of dots and dashes sent via the telegraph key used to represent each letter of the alphabet, not unlike the binary digital code of 0's and 1's used in computer language today. Although Morse code might not have been the first code developed, it certainly became the most widely known and eventually evolved into International Morse Code, used to talk between nations and ships at sea.

THE EXPANDING BUSINESS

By the late 1850s most of the eastern states were crisscrossed by telegraph wires, many running along railroad tracks. In 1860 Congress passed the Telegraph Act, which advertised for bids to connect the eastern cities with San Francisco via a telegraph line running west from the state of Missouri. In exchange for some $40,000 and a good many sections of public land, the operator was enticed to build an intercontinental telegraph system that the government could use for

SAMUEL F. B. MORSE

Born in Charleston, Massachusetts, on April 27, 1791, Samuel Finley Breese Morse was educated at Yale University. His dream of becoming an artist led him on many voyages to Europe to study art. In fact, it was on a return voyage from art school in 1832 that Morse came up with the idea of an electromagnetic signaling system. But long before the invention of the telegraph vaulted him to lasting fame, he opened a portrait studio in New York City.

Dividing his time between his duties as the first president of the National Academy of Design (1826–1845) and his desire to create his signaling system, he worked on the development of the electrical telegraph. This telegraph was based on the electromagnet. With the help of his friend Alfred Vail (who some feel was the real inventor of the instrument as well as of "Morse code"), Morse improved the receiver he had created and perfected a printed dot-and-dash symbol alphabet for the telegraph system.

In 1844 Morse set up the first U.S. telegraphic link, which ran between Baltimore and Washington, D.C. In 1854 Morse received Supreme Court recognition for his patent. Samuel F. B. Morse died on April 2, 1872.

Source: Lewis Coe, *The Telegraph: A History of Morse's Invention and Its Predecessors in the United States*, Jefferson, N.C.: McFarland & Co. (1993).

Tapping the key sent a dot; holding the key sent a dash. Combinations of dots and dashes create each letter needed to send a message. Incoming codes were imprinted on the tape to the operator's left; the operator then translated the codes and transferred them to a telegram form.

free. Under the direction of Western Union, the line was completed and opened for business in 1861, eight years before a railroad was completed across the same area. As mentioned, the immediate effect of the line was the demise of the Pony Express.

By the end of the Civil War, the telegraph had become essential to U.S. business operations, and Western Union was considered one of the most successful companies in America. Although instant communication was possible in the United States and in many countries in Europe, it still took weeks or even months for information to travel across the oceans. Financier Cyrus W. Field became convinced that an underwater cable could be laid between Newfoundland and the British Isles. The venture was successfully completed in 1866, and for the first time in history the Old World could talk directly to the New, and news from Europe could be printed in the major U.S. newspapers of the day.

Although the telegraph wire provided instant communication, Europe and the United States were not destined to "talk" directly to one another over a wire for almost another hundred years.

EARLY COMPETITION

In business mergers not unlike those executed in the cable television industry of today, large telegraph companies took over smaller ones. By the late 1870s, Western Union was the largest telecommunication

Cyrus W. Field

New Englander Cyrus West Field (1819–1892) retired from the business world at the age of thirty-three to fulfill his dream of connecting the United States with Britain by telegraph. To this end he formed the New York, Newfoundland, and London Telegraph Company with British and American government charters. The company was financially backed by both British and American promoters.

After three failed attempts to lay telegraphic cable between the two continents, success came in 1858. This link between Newfoundland and Ireland made Field a hero until it malfunctioned and failed. In 1866 Field chartered the largest ocean liner of the day—*The Great Eastern*. Although a commercially unprofitable passenger liner, the huge ship was perfect for lowering the immense weight of two thousand miles of cable to the ocean floor, thus connecting continent to continent with reliable communications. This amazing feat brought Field the fame and fortune he sought.

Source: Arthur C. Clarke, *Voice Across the Sea*, New York: Harper & Brothers (1958).

company, with nearly two hundred thousand miles of wire in operation. This was out of less than three hundred thousand miles of communication wire in the entire United States at that time.

A great deal of intrigue accompanied the business machinations of the telegraph company owners. Stock issues of questionable value were printed again and again. Jay Gould, the railroad robber baron, took control of Western Union and, with it, much of the electrical communications in the country. Some independent systems survived the Gould monopoly, often companies of little consequence, such as the Marion and Scio (Ohio) Telegraph Company, with two offices, one employee, and seven miles of wire. Other independents included the Good Intent Tow Boat Company and the Pennsylvania Railroad Company. One organization that was destined to provide some competition was the Postal Telegraph Company. Counted among its backers were James Gordon Bennett, publisher of the *New York Herald,* and John W. Mackay, silver bonanza king of Virginia City. Both hated Gould and probably had as much interest in hurting him in the pocketbook as they did in making a profit. Rate wars ensued, and the cost to send a telegram dropped to all-time lows. Postal Telegraph remained a strong competitor until its 1943 merger with Western Union.

Telegraph technology had become so common by the 1880s that amateur telegraphers could order keys and sounders (needed to hear the dots and dashes) from mail-order catalogs, and many backyard systems linked house to house. In Wisconsin the Bear Valley line extended almost ten miles, linking farms along the way. Trade schools were established to teach the elements of telegraphy as well as Morse code.

The Industry Peaks

One of the earliest commercial users of the telegraph was the newspaper. Only four years after telegraph lines were first placed in service, James Gordon Bennett of the *New York Herald* met with other newspaper owners to form a system of sending newspaper stories over the wire to member organizations. His Associated Press changed the composition of the newspaper to allow for timely inclusion of national and international stories. By 1875 the Associated Press owned its own network of telegraph wires. Amid newspaper accusations of monopoly in the news wire business, competing wire services were formed by other companies in the late teens: United Press was started by Scripps, and the International News Service was started by Hearst.

In addition to the real news of the day now being available almost instantly to every subscribing newspaper, some wires even transmitted sporting events play-by-play so that the paper could put an extra on the streets as soon as the score became final. Until 1917 most newspaper copy and telegraph messages were sent by Morse code, but after that date the teleprinter came into use, signaling the end of the era of the Morse operator. Thereafter, messages were printed directly on paper tape and glued to the telegraph form.

Western Union and its chief rival, Postal Telegraph, established a variety of services in the first half of the twentieth century to maximize the use of the telegraph and thereby create greater dividends for their stockholders. The wire was quickly recognized as an ideal way to send stock market information to clients. Many wealthy homes were equipped with a ticker, or paper-tape printer, which would bring current stock quotes directly to the armchair of the master of the house. Banks, and sometimes even the less wealthy, turned to Western Union to transfer money from one place to another. In the early decades of the twentieth century, Western Union offered FTD, or Floral Telegraphic Delivery, which makes possible the quick delivery of flowers to a distant city. Another service provided by the telegraph company used a pulse to synchronize clocks across the country, thus giving us standard time. Telegraph lines even carried network radio programs for a short time in the 1920s. The lines were not balanced for voice, and the resulting quality was very poor. Agreements wrung from American Telephone and Telegraph (AT&T) during antitrust investigations allowed voice messages of competing radio networks to eventually return to telephone wires.

In 1920 a small Puerto Rican telephone system became the nucleus of the International Telephone and Telegraph Company (ITT). The company developed a telephone system for Spain. Romania followed, then the cities of Istanbul, Shanghai, and then many countries in Central and South America. The company quickly worked out a traffic deal with Postal Telegraph, which had completed the first Pacific Ocean cable in 1902, to gain access to its international telegraph cables.

Control of Postal Telegraph was taken over by ITT in 1928, which then gave that company control over submarine cables across both oceans. Fifteen years later, Postal, the company that was built on the profits of the Virginia City Comstock Lode, disappeared forever, finally swallowed up by Western Union.

The 1930s marked the beginning of the end for the telegraph. The economic climate of the Depression years was not kind to many businesses, including Western Union and Postal Telegraph. But the real problem was the spreading acceptance of a new technology—the telephone.

Alexander Graham Bell had first offered the rights to the telephone to Western Union for $100,000, but there was little or no interest in sending voice by wire in the 1870s. After second thoughts, Western Union enlisted the aid of Thomas Edison to develop a telephone free of the Bell patents.

After merging with Postal Telegraph and into the 1990s, Western Union developed many corporate strategies to fend off the inevitable. Obsolete wires were replaced, first by the new wireless technology, then by shortwave radio, and finally by microwave communication systems. In 1974 Western Union initiated the Westar domestic communication satellite business, which provided tremendous capacity for every form of electronic communication, including television delivery. Westar was sold to GM Hughes Corporation in 1989, when the Western Union Telegraph Company was reorganized as a company that handled only money transfers. Western Union was out of the telegraph business.

The dots and dashes of the Morse telegraph are now heard only on movie sound tracks, but the technology, the organization, and the utilization have provided the base upon which all modern telecommunications is built. The first hundred years of the communication revolution was driven almost solely by the telegraph.

THE TELEPHONE

In many ways the development of the telephone parallels that of the telegraph. Its invention was built on the prior discoveries of principles by a variety of scientists. Several inventors developed systems for sending sound through wires at about the same time. The establishment (government, in the case of the telegraph, and the telegraph company in the case of the telephone) was not interested in the development of the new invention. Many small organizations were formed to provide services, and, finally, in each industry one powerful organization emerged to develop monopoly power.

THE INVENTION

The word *telephone* comes from Greek, generally meaning "distant sound." The first reference to sending voice through an electrified wire seems to come from Germany in 1860, when Professor Philipp Reis developed an electromechanical ear that would transmit sounds over a wire. Even though he built ten or twelve improved models and lectured extensively on its principles, his invention was not practical and was never accepted as anything more than a curiosity.

Alexander Graham Bell

Born in Edinburgh, Scotland, in 1847 to a deaf mother, Alexander Graham Bell's childhood was filled with instruction by his father and grandfather in elocution and in teaching the deaf to speak. After tuberculosis caused the deaths of two brothers, the Bell family moved to the warmer climate of Ontario, Canada, to spare Alexander from a similar fate, because he, too, suffered from the disease.

In the early 1870s, Bell moved to Boston and became a professor of vocal physiology at Boston University. While there, he developed an interest in transmitting speech electronically through his invention—the musical telegraph. By 1875 Bell developed the concept of sending speech by wire. He applied for a patent for the telephone, and it was granted on the same day: March 7, 1876. Only three days later, on March 10, his historic first telephone message was transmitted: "Mr. Watson, come here, I want you."

In 1877 Bell married Mabel Hubbard, who, like his mother, was deaf. His father-in-law, Boston lawyer Gardiner G. Hubbard, was impressed by Bell's invention and provided much of the financial support needed to develop the telephone and to fight the lawsuits that developed during the next eleven years, as other inventors of the telephone came forward. These included claims by Elisha Gray, who had filed for a telephone patent on the same day as had Bell, and Daniel Drawbaugh, who claimed to have developed a "workable" phone in 1867.

Bell's associates formed the Bell Telephone Company and opened telephone exchanges in several U.S. cities within a year. In 1882 Bell, who had been transplanted from Scotland to Canada, became a U.S. citizen. In his later life, Bell served as president of the National Geographic Society and was made a regent of the Smithsonian Institution. His interests turned to aviation, and he helped invent the hydrofoil. Alexander Graham Bell died August 2, 1922.

Source: John Brooks, *Telephone: The First Hundred Years*, San Francisco: Harper & Row (1975).

At the time of Reis's death in 1874, two men in the United States were working on an electrical telephone: Elisha Gray, who went on to found the company that was the predecessor of the Western Electric Company, and Alexander Graham Bell.

Bell hired Thomas Watson, a machinist with an interest in electricity, to make the devices used in his search for an instrument that would send voice through an electrically charged wire. In the spring of 1876, human speech was transmitted through the wire by electrical means, and Bell filed for a patent on the telephone in that year. On the same day that Bell filed, Elisha Gray also filed for a patent for a similar device. Because the exact times of each filing were not recorded by the patent office, a legal door was left open, and immediate suits to determine the real inventor of the telephone were filed.

It is often a long road between the invention of a device and the profitable marketing of it. Gardiner G. Hubbard, Bell's father-in-law and an inveterate promoter, persuaded him to enter his new device in the 1876 Centennial Exposition in Philadelphia. The judges, including the emperor of Brazil (who was said to exclaim "My God, it talks!"), were much taken with the invention.

Bell's "talking telegraph" didn't lead to immediate fame and fortune, however. When Western Union turned down the rights to buy all of Bell's telephone patents for $100,000, Bell, Watson, and Hubbard went on a lecture tour to raise money. Bell was even invited to demonstrate his telephone to Queen Victoria of England. Back in the States, Hubbard was promoting the telephone, and under his direction the Bell Telephone Company was formed. The company advertised extensively, but would only lease the instruments, not sell them. In 1878 the New England Telephone Company was established as the first subsidiary of the Bell Telephone Company. Amateur telephone hobbyists, like their telegraph brethren before them, quickly improved telephone devices, and the patent office was deluged with applications. Western Union saw the error of its ways and established its own telephone company in the late 1870s. Its American Speaking Telephone Company used equip-

> ## Theodore N. Vail
>
> Cousin of Samuel F. B. Morse's partner, Alfred Vail, Theodore N. Vail was destined to work in the new field of electrical communication. Born in 1845 in New Jersey, Vail first worked for Western Union and later apprenticed as a telegrapher in New York City. After marrying in 1869, Vail moved to Washington, D.C., where he accepted the job of assistant general superintendent of the Railway Mail Service.
>
> Through an acquaintance with Gardiner G. Hubbard, Alexander Graham Bell's father-in-law and partner, Vail became increasingly interested in the telephone business. In 1878 Hubbard offered Vail the job of general manager of Bell Telephone Company. Displaying his excellent business skills, over the years Vail turned the Bell System into a safe, respected company. He left the company in 1887.
>
> After the merger of Bell Telephone with AT&T, Vail was called back to use his leadership skills to strengthen the company. During the years of his presidency at AT&T (1907–1919), Vail saw American telephone service more than double (rising to 12 million customers). Under his direction nationwide phone service began, long-distance service was created, and telephone company competition ended, turning the telephone business into a monopoly. Theodore N. Vail worked hard until his retirement in 1919.
>
> Source: John Brooks, *Telephone: The First Hundred Years*, San Francisco: Harper & Row (1975).

ment developed from Gray's patents. Thomas Edison was employed by Western Union interests to develop a better transmitter for their telephone.

Candlestick-style telephone of the 1920s.

Hubbard was destined to spend much of his time in Washington, fighting the mammoth Western Union telegraph company in court over patent infringements. While there, he persuaded a young government employee—Theodore N. Vail—to leave his job and become the general manager of the Bell Telephone Company. Vail was instrumental in leading the reorganized Bell Telephone Company to a final victory over Western Union. In 1888 the Supreme Court found for the Bell interests. Alexander Graham Bell was acknowledged as the true inventor of the telephone, and Elisha Gray faded into history. In less than five years after the invention of the telephone, the groundwork had been laid for it to catapult the Bell Telephone Company into what was destined to become the world's wealthiest corporation. Alexander Bell turned to other interests and was not actively involved in the further development of the huge communication corporation that bore his name.

The Expanding Business

In 1885 the American Telephone and Telegraph Company was formed to interconnect the many local Bell System franchises. Up to that time, telephone conversations were limited to only those homes and businesses served by the local companies. By 1892 customers in New York City could talk to those in Chicago. Growth of telephone franchises, in which local companies rented equipment from Bell, was rapid during that period. The telephone had caught the fancy of the public, but even its early use was mostly for business purposes. Physicians were among the first in a newly wired area to lease a telephone, followed quickly by druggists, hospitals, lawyers, livery stables, and liquor stores. The only residences equipped with telephones were those of doctors, business owners, or managers.

Early Competition

In 1894 Bell's patent advantages expired, and thousands of telephone companies sprang up overnight. Even small towns had competing telephone systems, which were often not interconnected. Businesses would thus be forced to subscribe to each of the competing companies to have telephone links with customers who were connected to different systems. Many of the independents were marginal, at best, in both service quality and financial stability. Some were scams that existed primarily to attract the funding of unwary stockholders; others were mutual systems created to serve the communication needs of a small local population. Some in rural areas were even constructed with barbed-wire fences as conductors. Marginal as they were, the independents were a thorn in the side of the former telephone monopoly. Before the turn of the century, the Bell System fought the independents in every possible way, and many were forced out of business because Bell had enough money to undercut others' rates for long periods of time. Some competitors were subject to sabotage and frequently found their wires cut. Bell mounted a publicity campaign against the independents and refused to sell them equipment that was manufactured by Western Electric, now a member of the Bell family. Bell also used its monopoly on long-distance lines in the war against the independents, denying access to all except the Bell subsidiary companies.

On the last day of 1899, the Bell Telephone Company's long-distance subsidiary, American Telephone and Telegraph, became the parent company of the Bell System and moved corporate offices to New York. Theodore Vail, who had resigned as president of the Bell Telephone Company in 1887, when it had been a very small company, was brought back as president and quickly became the architect of the communications monopoly. Financier J. P. Morgan, who had played a major role in the telegraph business, provided the funding needed to take the new corporation to even higher levels of power. AT&T's strategy to deal with competition was simply to purchase as many of the competitors as possible. In 1909 AT&T purchased Western Union stock, and by 1910 Vail was president of Western Union as well as of AT&T. This was the proverbial straw that broke the camel's back. Authorities in Washington, urged by the antimonopoly factions, focused their antitrust powers on AT&T. The telephone company was able to forestall action for about fifty years by issuing the Kingsbury Commitment, wherein AT&T promised to give up ownership of Western Union and permit independent telephone companies to interconnect with the AT&T long-distance lines.

As discussed in chapter 1, the development by Lee de Forest of the Audion vacuum tube made broadcasting possible. It also enabled AT&T to construct a coast-to-coast long line in 1915. The tubes served as amplifiers that allowed the speech signals to be boosted again and again after traveling through the wires for miles. Bell engineers also experimented with sending speech through the airwaves on the newly discovered Hertzonian (radio) waves.

The need for interconnection and long-distance service made the telephone a business necessity in the United States. By 1918 there were 10 million telephones in service in America. As telephone systems became popular around the world, governments assumed control of their operation. AT&T and U.S. independent telephone companies were free of government ownership and of substantial government control until 1918, when, under a war powers act, President Woodrow Wilson took control of all communication systems. The companies were returned to private ownership within a year, and the telephone business along with most other businesses boomed during the Roaring Twenties.

Between World Wars I and II, phone service improved considerably. Public telephones and cheaper party lines made the service so affordable that even middle-class city dwellers and farm families could enjoy the business and social advantages of instant voice communications.

AT&T Moves into Mass Media

With much of the development of radio fueled by the needs of the military during World War I, and the radio companies' coffers filled with profits from an inflationary economy, AT&T was ready to make money from their radio experiments. Western Electric, the Bell manufacturing subsidiary, had operated three experimental stations in New Jersey and New York City with an eye toward transatlantic radio-telephone service. At the urging of the government, AT&T entered cross-licensing agreements with General Electric and Westinghouse, which made it possible to use the patented inventions of the other companies to construct the most technologically advanced radio facility of the time. AT&T put station WEAF on the air in 1922 with the novel idea of making a profit by selling access to the radio waves in much the same way that an individual could buy time on a telephone line by putting coins in the proper slot in a telephone booth.

In 1923 AT&T opened its second radio station in Washington, D.C. Within three years of establishing WEAF, the telephone company was operating a national radio network of seventeen broadcast stations interconnected by telephone long-line service, which interconnected distant cities. Each used a transmitter manufactured by Western Electric; in fact, access to the network was denied to those stations that did not buy transmitters from the AT&T manufacturing subsidiary. Access was denied to AT&T's patent partners as well, who were also its competitors in broadcasting. The Radio Corporation of America (RCA), General Electric, and Westinghouse were all required to use lower-quality telegraph lines to interconnect broadcast stations. AT&T was accused of attempting to monopolize broadcasting.

Even as AT&T was embroiled in lawsuits with its radio competitors that would eventually force the company out of the broadcasting business, Bell Labs, its research subsidiary, was developing the technology that would make motion pictures talk. The Vitaphone Company, using equipment manufactured by Western Electric, produced the first full-length talking picture—*The Jazz Singer*—in 1927. AT&T entered the motion picture business with the same enthusiasm it had devoted to radio just five years earlier. It was estimated that by 1930, 90 percent of the talking pictures were made using Western Electric equipment and shown only in theaters that used the same style of equipment. The company even operated a small movie production studio and was involved with the casting, direction, and distribution of movies, as well as the financing of entertainment films created by other studios.

Bell Labs was hard at work developing another product in the late 1920s. Although television had already been demonstrated by the Scottish inventor John Baird and was emerging from the laboratories and workshops of corporations and inventors throughout the Western world, Bell Labs was developing a system for sending electrical pictures through a telephone wire. By combining a photoelectric cell and a vacuum-tube repeater, Bell engineers could, in the early 1920s, send photographs to newspapers via telephone wires. This opened the door to facsimile transmission and to network television. Near the end of the decade, a Bell Labs experiment allowed then Secretary of Commerce Herbert Hoover to talk on a telephone in Washington and be both heard and seen in the AT&T offices in New York. By 1929 the company was transmitting color television pictures over telephone wires, but the commercial introduction of television would have to wait for the Depression to run its course and for World War II to culminate in a victory for the Allied powers.

The World's Largest Corporation

Bell Labs, which would later develop both the coaxial cable and the transistor, making computer-switched modern broadband telecommunications networks possible, became a star in the AT&T corporate family. The research and development firm, coupled with Western Electric, which produced the only electrical equipment that could be connected to the telephone network, led to a vertically integrated corporation that was completely self-sufficient. Members of this single corporate family engaged in the invention of telecommunications devices, their manufacture, integration into a national long-distance network, and connection through the locally owned telephone company to the home. Even the home telephone instrument had to be a Western Electric model and was owned by the telephone company, not by the customer. A nervous government passed the Communications Act of 1934, which included provisions to regulate common-carrier wired communication systems as well as broadcasting.

The Federal Communication Commission (FCC) was established to administer the new law, taking over some of the regulatory powers of the Interstate Commerce Commission. The first communication commission was staffed by New Deal reformers of the Roosevelt administration. Virtually the first action of the new agency was to investigate the monopolistic structure of the telephone company. The growing threat of World War II interrupted the government's action against AT&T, and natural disasters, including floods in the Midwest and a massive hurricane along the East Coast, gave the telephone company the chance to act as the hero in times of national emergency. Even though the $5 billion corporation was the largest single company ever assembled in the history of private enterprise and controlled nearly 85 percent of all telephones and virtually 100 percent of all long-distance lines and transoceanic radio telephony in 1939, public opinion did not call for the divestiture of the valuable national resource offered by AT&T.

The fortunes of the telephone company took a dramatic leap forward after World War II. Postwar prosperity, fueled by a consumer demand for goods and services, led to a boom in the telephone business. With Western Electric producing telephone instruments at the rate of 4 million per year and with residential service installation occurring at an estimated twenty-five homes per minute, it seemed that everyone in the country wanted to talk to one another. Bell Labs innovations met the demand with coaxial cable that could carry many times the long-distance load as the older, conventional twisted-pair telephone wires, and with direct dial service, which replaced the thousands of operators that had been needed to connect both local and long-distance calls. By the 1950s microwave radio relay links carried much of the long-distance load and were to become the backbone for the interconnection of the national television networks to their affiliated stations. The two-hundred-foot-tall concrete structures that held the electronics and antennas for microwave relay began sprouting every thirty or forty miles across the northeastern quadrant of the country as New York was connected to Chicago. In 1953 the system was completed, and for the first time people in almost every part of the country could watch the live NBC, CBS, or DuMont television programs from New York City. Just three years later, the first undersea telephone cable to Europe was completed.

Satellite Communications

The idea of creating an orbiting space station that could retransmit radio signals back to the earth was first proposed by Arthur C. Clarke in a 1945 article in the British technical journal *Wireless World*. German rocket technology, the development of the transistor, and the launch of the Russian *Sputnik* in 1956 spurred the scientists at Bell Labs to develop the fundamentals of a space-based communication system. By the late 1950s, Bell Labs had completed prototype communication satellites. At the same time, the Bell System, through its new subsidiary Bellcom, was busy developing most of the communication and guidance equipment needed for the entire U.S. space program.

Early in 1961 AT&T was finally given permission to launch its own satellite. *Telstar I* was to be the first active communication satellite. It carried an on-board radio transponder system, which could receive radio signals from the ground and then retransmit them back to earth. In Washington, however, debate raged over who would own the domestic satellite communication system. AT&T argued that it was simply an extension of the existing telephone system, but others felt that space should belong to the public. When the satellite was launched in July 1961, the success of *Telstar I* hastened the passage of the Communication Satellite Act of 1962, which created a quasi-governmental corporation, Comsat, to oversee the development of a domestic satellite communication system. Stock in Comsat was offered to the public, and AT&T became part owner by buying nearly 28 percent of the shares. The Bell System quickly became not only a major owner of Comsat but also its prime customer.

Early Bird, the first geosynchronous satellite, was launched in 1965. *Geosynchronous* means that, when placed in an orbit 22,300 miles over the equator, the satellite travels at a speed that matches the rotation of the earth. The effect is that the satellite appears to remain stationary, thus eliminating the need for the expensive tracking stations that had been required with the lower-orbiting satellites, or "birds." Within a short period, nearly half of all European telephone calls were routed through the communication satellite.

In recent years other communication companies have entered the satellite business. MCI Communication Corporation purchased Satellite Business Systems from Comsat, IBM, and Aetna; and Southern Pacific Communication Company (owners of Sprint) and GTE, the largest independent telephone company, are partners in Spacenet. GTE Spacenet operates a series of satellites that carry voice, data, and television traffic.

THE BEGINNING OF THE END FOR MA BELL

Since the very beginnings of the telephone industry, the Bell Telephone Company, later to become AT&T, saw itself controlling the voice telecommunication business. After all, it simply made good economic sense to operate the only telephone company in a region. Early consumers quickly saw the folly of having two competing telephone companies in the same city: People on one block could not call those on the next, and merchants had to subscribe to both services. AT&T's answer to that early competition was first to use patent rights to limit competition, then, if necessary, to buy out the competition. Theodore Vail's vision was for one telephone system to provide a national universal service from end to end, which meant that the Bell System would lease the equipment to the customer as well as provide the local network with appropriate switching and, when necessary, also provide the long-distance lines. No inferior equipment manufactured by an outside company would be attached to the network for fear of degrading quality. With only one provider serving the nation, the cost of the universal service could be spread over the entire system, thus lowering the unit cost to each subscriber. The model worked with state regulation of rates and services by public utility commissions. Interstate regulation was provided first by the Interstate Commerce Commission and later by the FCC. The Bell System prospered and paid consistent dividends while building the most efficient and reliable telephone service in the world. The majority of its customers and certainly all of its stockholders looked upon "Ma Bell," as the corporation came to be known, with a certain degree of affection. Telephones almost always worked and they were cheap. Almost every home had one to be used for emergencies, for business, and for socialization. Businesses found them essential to survival.

Although there was justification for operating as a regulated monopoly in order to provide the most efficient *POTS* (plain old telephone system), the corporate actions of the Bell System attempted to extend the monopoly model to every new business that came along. One example was its development of sound for the motion picture industry; another, its attempted control of the communication satellite business.

AT&T's attempt to maintain monopolistic control of the manufacture of all telephone equipment and of all long-distance telephone service probably contributed greatly to its eventual breakup. Before 1968 no non–Western Electric equipment could be attached to the telephone network. The reasoning, of course, was that inferior equipment might create electrical problems in the system and an interruption of service.

In 1968 the battle between the corporate giant and the small-equipment manufacturers came to a head with the Carterfone decision. The Carter Electronics Corporation had invented a device that could connect private two-way radio-telephones to the telephone system. It was also acoustical but did contain an electrical connection. The FCC ruled that it was a legal device. In 1970 the FCC went one step further in challenging AT&T's domain: It authorized a common carrier—Microwave Communications Incorporated (MCI)—to construct a point-to-point radio microwave system between St. Louis and Chicago. The company, upon completion of the system, could sell private-line services to business users. It could, in effect, compete directly with AT&T's long-lines division by skimming the profitable, high-volume business users from the Bell System's long-distance lines. What the FCC didn't know was that William McGowan, president of the upstart communications company, planned to offer a long-distance service to any telephone user on a national basis. The FCC ruled that the planned MCI service was illegal, because it was in direct competition with AT&T's FCC-approved service. A federal appeals court overturned that decision, however, and by 1977 local telephone operating companies were required to provide access and connecting links for MCI.

MCI was in business, and the door to unrestricted competition for long-distance telephone service was

WILLIAM McGOWAN

biography

William "Bill" McGowan was born in Ashley, Pennsylvania, a coal mining town. He attended the University of Scranton and later Kings College, where he majored in chemistry. After graduating from Harvard Business School, McGowan went to New York, where he formed COMAC, an investment company, with Ed Cowett, a Harvard Law School graduate. After Cowett's death from a heart attack, McGowan spent four years as a business consultant in New York and gained a reputation for helping companies about to go under.

In the 1960s a small company called MCI came to McGowan, asking for help raising money to get out of debt. McGowan liked the company because he found it to be honest. He was also excited about its ideas of challenging the monopoly of AT&T. On August 14, 1969, the FCC voted to accept MCI as a carrier. McGowan's interests with MCI grew, until finally he took control of the company. McGowan promoted a new idea as head of MCI—discount long distance, and under McGowan's direction MCI became a huge company. McGowan believed in treating his employees in such a way as to make them appreciate their company, a lesson he learned back in college.

Source: Larry Kahaner, *On the Line: The Men of MCI—Who Took on AT&T, Risked Everything and Won!*, New York: Warner Books (1986).

open. Access was granted grudgingly and very slowly. McGowan began an attack on the Bell System with lobbying efforts in Congress and with the FCC, as well as by filing actions in the courts. Congress passed antitrust legislation that made it easier for the Justice Department to investigate large corporations. Tiny MCI had found a powerful ally in its fight against the giant telephone monopoly.

The Breakup of AT&T

In 1974 the Justice Department filed an antitrust suit against AT&T—probably the largest and most complex filing on record. It was estimated that more than 1 billion pages of information were entered into evidence during the six and one-half years that it took for the case to reach the courtroom. By default (the judge originally assigned to the case died of cancer) the case fell to Judge Harold H. Greene. AT&T lawyers were not happy with the selection of who they considered a liberal activist judge.

Judge Greene made it clear that he was going to prove that the court system could swiftly and fairly handle an antitrust case of the magnitude of *U.S. v. AT&T*. The suit was pushed toward trial. Both sides attempted to negotiate a settlement prior to the trial but were unsuccessful.

The case continued into the early years of the pro-business Reagan administration. The military,

including the joint chiefs of staff, felt that the dismantlement of AT&T would be a disaster for the country, which was still in the grip of the Cold War. President Ronald Reagan, however, declined to ask the Justice Department to drop the case. He generally wanted consensus among his advisers prior to announcing a decision, and there was none. If the suit were dropped at the president's insistence, it might look as though someone had been paid off.

Meanwhile, behind-the-scenes negotiations were continuing between AT&T and the Justice Department. AT&T's new president, Charles Brown, accepted the inevitable breakup of the Bell System and began looking to the future for whatever advantage might be wrung from a compromise. He wanted to protect, at all costs, Western Electric and Bell Labs, which were the stars of the system. The twenty-two operating companies were not as profitable and could be divested without fatally wounding the corporation.

A three-thousand-word agreement, which came to be known as the Modified Final Judgment, was worked out, much of it on the ski slopes of Park City, Utah. The agreement was the blueprint for the breakup of Ma Bell, the world's largest corporation. Its nine provisions called for the separation from AT&T of the Regional Bell Operating Companies (RBOCs) but allowed Western Electric and Bell Labs to remain with the corporation. In January 1982 the announcement was made that the hundred-year-old Bell System was history.

The Original Baby Bells

When AT&T ("Ma Bell") was required to divest its local telephone services, seven regional telephone companies were formed. These regional bell operating companies (RBOCs) are often called "Baby Bells." After the Telecommunications Act of 1996, some were bought out or merged with others.

- Ameritech, serving the upper midwestern states
- Bell Atlantic, serving the mid-Atlantic states
- Bell South, serving the southern states
- Nynex, serving New York and New England
- Pacific Bell, serving the Pacific coast states
- Southwestern Bell, serving the lower midwestern states
- U S West, serving the Rocky Mountain states

The Outcome: Bell Today

In his role of determining the public-interest suitability of the Modified Final Judgment, Judge Greene became the telecommunication czar. He monitored the divestiture from the corporation of the RBOCs, or Baby Bells, and interpreted the provisions of the judgment as the new systems were put into place.

The telecommunication structure that emerged in the decade since the breakup of the Bell System originally included seven Baby Bells, consisting of the original twenty-two regional telephone systems that the corporation owned. Two additional companies that were not solely owned by AT&T now are independent. There are about twelve hundred independent telephone companies in operation, and these serve about 20 percent of the U.S. population.

In April 1996 Nynex and Bell Atlantic, two of the seven Baby Bells, announced a merger. This followed a merger of two additional Baby Bell companies—Pacific Telesis and Southwestern Bell—earlier in the year. By 1999 the complexion of the telephone industry had changed. Mergers and buyouts had combined Bell Atlantic with GTE, one of the larger independents, and Ameritech with SBC. The Baby Bells were no longer babies.

AT&T was left with the long-distance service, the manufacturing company, and the research and development arm (Bellcore). Other common carriers, such as MCI and Sprint, gained equal access to the LATAs (local access and transport areas) and provide competition for long-distance service.

The Baby Bells complained bitterly when cut off from the corporate giant. With new fee structures, access payments, and new business ventures, however, they each have become immensely profitable. All have ventured out of the regulated services to take advantage of new telecommunication opportunities. Bell Atlantic, for example, operates regulated telephone service in a huge area on the East Coast. In addition, it operates Bell Atlantic Mobile Systems, supplying beeper and cellular telephone service in a variety of locations throughout the country; Bell Atlanticom, to install and maintain telecommunication equipment for businesses; MAI Canada, to market computers in Canada; Bell Atlantic Properties, to develop real estate; Bell Atlantic International, to provide telecommunication network services in other countries; and Technology Concepts, Inc., to provide computer software and design services.

Converging Systems: Telcos and Cable

In 1993 Bell Atlantic and TCI (Telecommunications Incorporated), the largest cable television operator in the United States, announced a merger. The planned merger was never completed and each went its own way, but it was a harbinger of things to come. Before the TCI/Bell Atlantic announcement, the cable television and the telephone industries were struggling to gain regulatory advantage in a race to offer enhanced communication services to U.S. homes.

The telephone companies wanted to offer video programming on dial-access telephone lines, and the cable companies wished to enter the data transfer and telephone bypass businesses.

Telephone bypass means that the home or business user could access long-distance companies without using the local circuits of the regional operating companies. Presently, about forty cents of each long-distance dollar goes back to the local telephone company to pay for access charges. Cable systems believe that they can offer a similar service for a lower cost and still make a considerable profit.

Telcos (telephone companies), originally barred from the cable television business in their own telephone service areas by both the Modified Final Judgment and the Cable Communication Act of 1984, successfully lobbied for the elimination of those rules. The Telecommunications Act of 1996 opened the door for the telcos to offer video programming to the home. Telephone companies are preparing for the future by projecting the five-hundred-channel television universe and interactive video that will be possible with the completion of a fiber-optic network into the home. In addition, telephone companies are buying and building cable television systems in foreign countries and conducting interactive television experiments in selected locations nationally. For example, U S West purchased the holdings of Post-Newsweek Cable in Great Britain and has entered into a partnership arrangement with Time Warner. And in 1999 the nation's largest cable MSO (multisystem owner), TCI, was purchased by AT&T. Cable ownership will now allow AT&T back into the local telephone business, thus coming full circle.

The economic engines that are envisioned to fuel the proposed electronic superhighway are **video on demand (VOD)** and the Internet. Video on demand, a **pay-per-view (PPV)** concept, would allow subscribers with the necessary technology to order movies or special-events programming on a program-by-program basis. Video on demand would use a computerized video server to allow each subscriber to order programming to start when desired. This technology will become practical when computerized video storage replaces the mechanical videotape players now in use. In addition, a navigation system will be required to allow users to select the desired programming from hundreds or even thousands of choices and to order it through a computer terminal with a user-friendly interface. Compressed video and a fiber-optic distribution backbone may also be necessary to serve the projected volume of a large-market VOD system.

DATA COMMUNICATION NETWORKS

As the functions of interpersonal communications systems and mass media systems merge, the electronic networks necessary to carry information and entertainment must also evolve. The old analog POTS that interconnected almost every home and business in the United States for voice communication became hopelessly outdated with the advent of the personal computer. Although the first transmission of data over a telephone network took place in 1940, the technologies of communications and computing began to merge in the late 1980s. Now with more than 70 percent of all personal computers connected to networks, the need for sophisticated data networks is apparent.

As might be expected, military and governmental uses of data communication came first in the late 1960s, when computers were developed that could share information via a telephone wire. The binary code of the computer was imposed onto an acoustical signal that consisted of a series of audible tones by a *modulator*. At the other end of the telephone line, a *demodulator* converted the audible tones back into the code of on-and-off signals that allowed the digital computer to operate. The *modem* (modulator/demodulator) allowed the computer to move beyond its original role as a mathematical supercalculator into the realm of a memory and communications tool.

The speed and convenience of electrical interconnection between the large mainframe computers of the day and remote terminals were soon recognized. Early users included banks and financial organizations that needed to send large volumes of information to and from multiple branch locations, as well as travel agencies, insurance companies, manufacturers, and retailers.

In the late 1970s and early 1980s, hobbyists discovered the personal computer (PC). Like the radio, television, telephone, and telegraph before it, the computer industry owes much of its early development to the experiments of the amateur. Many developments in what was destined to become the PC field were worked out in garage workshops around the country. The introduction of the Apple computer and the IBM PC in the early 1980s commercialized the PC field. The hobbyist benefited from the new computer technology, and so did education and small businesses. Small and relatively inexpensive computers became available to almost all. Even corporations with large data processing departments and massive mainframe units began placing PCs on the desks of employees. In 1985, 80 percent of these units were operating in a stand-alone mode as word processors or as elaborate calculators, but within five years the power of sharing computer capacity and of communication was recognized, and most were connected to *local area networks (LANs)*.

Office automation had become a corporate reality, and by 1992 there were estimated to be in excess of 2 million LANs in operation, supporting more than 20 million workstations. Facsimile, databases, and electronic mail (e-mail) have become a way of life for corporate information exchange, as did bar codes and laptop computers for remote data entry.

Telecommunication Networks

One of the telephone company's latest entries into the digital communication field is *Integrated Services Digital Network (ISDN)*—a set of standards that allows digital telecommunication services. The ISDN systems would be capable of delivering all forms of communication services to business or residential users. The wire that brings telephone signals into the home would also deliver television signals and carry information that controls the environmental and security systems of the building. This single-wire technology is predicted to be the electronic super-highway of the twenty-first century.

The Internet

The *Internet* is a worldwide computer network comprising millions of interconnected local computer networks. What was originally a U.S. government telecommunication network now provides interconnection for PC users around the world. It was started in the late 1960s as ARPAnet—a name derived from the Advanced Research Projects Agency of the Department of Defense. The goal of the network was to allow the computers of universities, defense contractors, and the military to share information. In addition, ARPAnet was designed to study how decentralized computer communications could operate in case of a nuclear attack. By developing protocol standards, it became possible to interconnect an almost unlimited number of computers.

As the PC began to appear on almost every faculty desk in the 1980s, the interconnected networks allowed university researchers worldwide to share information and to talk to one another through e-mail. The National Science Foundation interconnected university supercomputers in a new network called NSFNet, which became the foundation for the Internet in the United States. In 1993 the U.S. Defense Communications Agency mandated that the *TCP/IP* computer protocol, or language, be the universal standard for computer transmissions. From that point the Internet grew rapidly. Now many home computer users access the Internet through commercial network services, such as Prodigy and America Online. Many organizations and individuals are developing *home pages* and operating as hosts to provide information to others who might dial into that home address.

Although much of the activity on the Internet amounts to "surfing," or browsing, an increasing number of commercial corporations are going online to advertise and sell their products.

Network television news programs have published World Wide Web, or Internet, computer addresses for some time to allow viewers to provide feedback on news stories. In the late summer of 1995, NBC announced that it would produce a new computer-delivered news service called NBC Supernet. This first cyberspace journalism program offered online access to NBC news information through the Microsoft Network. Out of this joint venture came the cable network MSNBC.

Compressed video and fiber-optic transmission (discussed in chapter 14) are also being quickly developed as systems of data telecommunication. Some personal computers already contain circuit boards that allow full-motion color video to be displayed onscreen. Multimedia interactive entertainment and information are possible with a PC terminal coupled to a *CD-ROM* system and a telecommunication network.

Wireless Telecommunication Networks

Cellular Telephone

Although most of the telecommunication systems described in this chapter have been wired—either using twisted-pair wire, coaxial cable, or fiber-optic cable—it must be remembered that information and data are also sent through the airwaves using a portion of the electromagnetic spectrum. Voice transmission from one continent to another was accomplished via shortwave radio prior to the mid-1950s, and microwave transmission has provided the backbone for long-distance service since then.

The wireless communication revolution, which has changed the way we communicate with one another, is based on older radio technology. The first mobile radio-telephone systems were operating as early as 1921. It was in that year that the Detroit Police Department began experimenting with radio communication. During the 1930s mobile radio was adopted by almost every city police force in the country. After World War II, radio dispatch was used extensively for delivery services and taxicab companies. The first public mobile telephone service was developed by AT&T for the St. Louis market in 1946.

Mobile telephone service was slow to develop. The reasons for this most likely include the limited spectrum space assigned to the service, the high cost of the service and equipment, and regulation.

In 1983 the first commercial cellular telephone system was developed in Chicago, opening up the mobile telephone business to thousands of new subscribers. It is not unusual to see people walking down a city street while carrying on a phone conversation or to see car after car stalled in rush-hour traffic, with each driver holding a coffee mug in one hand and a cell phone in the other. In addition to the more common voice communication, the cellular system can turn the automobile into a fully equipped mobile office, complete with fax transmission and computer database access.

Cellular architecture uses many low-power transmitters rather than a few high-power ones. The same frequencies can be used over and over to provide additional channels to serve various parts of a city. In theory a map of a cell phone system looks like a honeycomb superimposed over a city. In practice, however, the coverage patterns of the cells are not nearly so uniform. In fact, cellular telephone service is subject to fading and multipath signal distortion as a

Telephone microwave antenna relay tower.

The cellular telephone has extended electronic interpersonal communication to the car as well as to other remote locations.

user travels across the cell. An additional problem occurs as the user travels from one cell to the next. A complex computer program is used to determine to which cell to transfer the connection as the signal in one cell fades.

Many telecommunication companies are banking on wireless services to deliver data to laptop or palmtop personal computers. Sprint and MCI (long-distance companies) are buying wireless cable systems in the nation's largest cities to develop wireless Internet systems.

VSAT

The benefits of satellite communication systems have long been recognized for the international distribution of voice and data information and for the delivery of television and radio programming to both broadcast stations and cable television systems. *Direct broadcast satellites (DBSs)* allow entertainment to be delivered directly to the home. Although the greatest advantage of satellite communication is distance insensitivity, economy and superior signal quality are also important.

With the advent of *VSATs (very small aperture terminals),* the advantages of satellite communication became available to businesses requiring private networks for data transmission and teleconferencing. A large corporation with geographically dispersed locations can use VSAT technology to interconnect branch offices or plants with the company's headquarters. The main office terminal is often a standard communication satellite uplink, which is often leased from a common carrier such as a telephone company. The company's host computer can then broadcast a stream of digital information via satellite to the small VSAT installations at each branch office. The VSAT terminal is connected to a computer that receives and displays the data. VSAT technology can also be two-way: The terminal can originate data transmissions that can be received by the host terminal or by other VSATs.

Other countries, especially those considered to be Third World, are considering the use of VSAT for television broadcast to the public and for telephone service in remote areas where telephone wires are not practical. Intelsat, a global telecommunication consortium, is actively exploring uses for VSAT as a major telecommunications distribution network for underdeveloped countries. As prices come down and equipment becomes more reliable, it is inevitable that VSAT will carry a greater share of the world's data stream.

VSAT dish to pick up business communications via satellite.

SUMMARY

▶ The telegraph, which was the first practical electrical telecommunication system, was constructed by Morse in 1844. By 1861 the telegraph lines crisscrossed the eastern seaboard and connected California with the rest of the United States. Five years later Europe and the United States were connected by a transoceanic telegraph cable.

▶ Voice carried over a wire became possible with Bell's development of the telephone in 1876. The "talking telegraph" eventually led to the demise of the Morse code telegraph, which for more than a hundred years was the primary business of Western Union.

▶ In 1885 the American Telephone and Telegraph Company (AT&T) was formed to interconnect the many local Bell System franchises. AT&T eventually developed a monopoly position in electrical communications, even purchasing Western Union at one point. Government antitrust intervention, however, led to the Kingsbury Commitment, in which AT&T promised that independent telephone companies would be allowed access to long-distance lines.

▶ By the 1920s AT&T had become involved in radio broadcasting with the establishment of station WEAF and the development of a chain of broadcasting stations interconnected by telephone lines for simultaneous program delivery. This was the forerunner of the broadcast network.

▶ AT&T became the world's largest company with a vertically integrated structure that included Bell Labs as a research and development arm, Western Electric as a manufacturing branch, twenty-two operating companies that provided telephone service, and a long-distance carrier. Ma Bell had been forced to sell Western Union and its broadcast stations because of restraint of trade problems.

▶ Bell Labs developed coaxial cable and the transistor, which made the modern computer and high-speed telecommunications possible. AT&T was also instrumental in the development of satellite communications.

▶ Carter Electronics and later MCI challenged AT&T's near monopoly position in telecommunications. Their court challenges led the way for Justice Department actions that broke up AT&T in 1982. AT&T spun off the Bell operating companies and concentrated on offering long-distance service as well as moving into new technologies. The Modified Final Judgment created the seven Baby Bells that are now providing telecommunication services to their respective regions of the United States.

▶ By the 1990s telecommunication had moved far beyond simple voice communications. The Baby Bells were forming partnerships with cable companies and merging with one another to offer video on demand (VOD), as well as provide digital data links for computer networks, ISDN for integrated services, and cellular service for mobile telephones.

InfoTrac College Edition Exercises

With the advent of the personal computer and the interconnection of computer networks, the plain old telephone system has taken on a new dimension. The telephone wires of the past are becoming the new media of the future.

3.1 Chapter 3 traces the history and development of wired telecommunication systems from the telegraph to the Internet. Some feel that the Internet may have greater impact on our lives than have had all of its telecommunication predecessors combined. Broadcasters, not wishing to be left behind, are jumping on the Internet bandwagon.

    a. Using InfoTrac to access *Broadcasting & Cable*, track the number and names of broadcasting corporations that have developed an Internet presence over the past year.

    b. How many similar articles on broadcast/Internet relationships can you find in computer periodicals such as *PC User, PC Week,* and *PC/Computing*?

    c. What evidence can you locate to indicate that broadcasting may well be confined to Internet distribution at some point in the future?

    d. Many network television programs now suggest that the viewer go to a Web site to find further information on the topic of the program. Are broadcasters encouraging their audiences to abandon the TV in favor of the computer?

3.2 Chapter 3 covers the establishment, the development, and finally the breakup of AT&T by the Justice Department. Since the Telecommunications Act of 1996, however, the "Baby Bells" have been allowed to combine and to enter businesses that they had been barred from in the recent past.

    a. Using InfoTrac to access *Broadcasting & Cable,* trace the buyout of TCI by AT&T.

    b. Has AT&T, the country's largest cable operator, continued to grow its cable business? What other cable MSOs has it bought?

3.3 Many of the old media companies have become active on the Internet. Disney/ABC, for example, owns a large portion of Infoseek.

  a. Using InfoTrac to access the November 15, 1999, issue of *Broadcasting & Cable,* locate media ownership of ten to fifteen additional dot-com companies.

  b. Who owns ESPN.com? mtv.com? vh1.com?

WEB WATCH

Here is a list of a few URLs (Internet addresses) for some of the organizations or corporations discussed in this chapter. Please explore these Web sites and follow the links to learn more about the complex business of the electronic media. Add your descriptions and your own favorite sites at the end of the list. Please keep in mind that the dynamic nature of the Internet allows sites to come and go but also allows organizations to update information about themselves very quickly.

Address	Description
http://www.ameritech.com/	
http://www.bell-atl.com	
http://www.telcordia.com	
http://www.bellsouth.com	
http://www.cinbelline.com	
http://www.pacbell.com	
http://www.swbell.com	
http://www.uswest.com	
http://www.chss.montclair.edu/~pererat/telegraph.html	
http://www.att.com	
http://www.mci.com	
Other favorite sites:	

CHAPTER 4

Foundations of Electronic Media Regulation

After the 1927 and 1934 radio legislation, the industry enjoyed the advantage of elasticity in the laws that governed it.

he events of the twenties and thirties provide the foundation of broadcast regulation—the Radio Act of 1927 and the Communications Act of 1934. F. L. Allen (1931, 136–137), a contemporary twenties historian, called radio the "youngest rider" on the Harding and Coolidge "prosperity bandwagon." During these presidential regimes, the business of the nation was indeed business; the electronic media developed rapidly, and the legal foundation established is still the basis of regulation today. This chapter explores the law and issues that resulted in legislative action. This historical perspective provides not only an understanding of the law and how it was formulated, but why it was created. Although technology and the corporate players have changed, these early struggles bear a striking resemblance to issues of today.

Key Issues

The key media development issues of this era were public trusteeship, censorship, and business monopoly. The debate over these issues ran parallel to the technological developments of the industry and the social revolution of the 1920s (see chapter 1). The resolution of these issues has left a lasting impression on electronic media legislation.

The Airwaves as a Natural Resource

The desire for conservation of the airwaves as a public resource was an outgrowth of the Teapot Dome and Elk Hills oil scandals, which occurred during the Coolidge presidency and left federal legislators outraged at a few "public servants" and their exploitation of public resources. During the Harding administration, members of the president's cabinet were accused of peddling their public offices for profit by selling off two public oil reserves in Teapot Dome, Wyoming, and Elk Hills, California. Senator Clarence C. Dill, a co-author of the Radio Act of 1927, was a member of the Public Land Committee investigating the scandals, which strengthened his determination to prevent a similar abuse of radio. In the debate over legislation of the airwaves, "the ether" was defined as a valuable

The Business of the Nation Was Business

During the Harding and Coolidge administrations, the business ethic permeated all areas of American life. To quote President Harding:

> American business is everybody's business. . . . We must . . . repeal and wipe out a mass of executive orders and laws which . . . serve only to leave American businessmen in anxiety, uncertainty, darkness.[1]

In a speech before the Chamber of Commerce of the State of New York, President Coolidge reflects the government and business partnership that existed during the 1920s:

> This time and place naturally suggest some consideration of commerce in its relation to government and society. . . . The foundation of this enormous development rests upon commerce. . . . The great cities of the ancient world were seats of both government and industrial powers. . . . It would be difficult, if not impossible, to estimate the contribution which government makes to business. It is notorious that where the government is bad, business is bad.[2]

Senator Clarence C. Dill, commenting before the Senate during its considerations for law governing radio, mirrors the business and political atmosphere as it related to media.

> Large corporations have invested large sums of money with little return on their investments. They hope for bigger returns in the future. I am not sure that it would be wise . . . to put too many legislative shackles around the industry at this state of its development. . . .[3]

1. Source: Warren G. Harding, "Business and Government," in *Our Common Country*, Indianapolis: Bobbs-Merrill (1921): 20–23.
2. Source: From a quote by Clarence Dill in the *Congressional Record* (1927), vol. 68, pt. 3:3027.
3. Source: From Calvin Coolidge "Government and Business," in *Foundations of the Republic: Speeches and Addresses*, Freeport, N.Y.: Books for Libraries Press (1968): 317–332.

Clarence C. Dill

Clarence C. Dill's role in the history of broadcasting was limited by his late entry into the legislative arena. Secretary of Commerce Herbert Hoover and Representative Wallace H. White had been working on radio law since the early 1920s and were largely responsible for the content and formulation of the law. Dill's contribution to passage was his active leadership. He stimulated the legislative debate and provided direction within a heretofore disinterested Senate. He basically redrafted White's version of the radio bill, adding his proposal for a Federal Radio Commission for the control of broadcasting. This commission concept passed, was reinstated in the Communications Act of 1934, and continues today as the governing power.

Dill was born of Scottish-Irish ancestry in Fredericktown, Ohio, in 1884. He was an Ohio Wesleyan University graduate before he moved to Spokane, Washington, where he first became involved in politics. He was elected to the House of Representatives in 1915, but his career was cut short due to his vote against U.S. entry into World War I. After the war he was reelected, this time to the Senate. Tucker and Barkeley, chroniclers of the era, described Dill's ability as an astute politician: "He can smell the change of public opinion . . . a month or six weeks before anyone else in the chamber." Dill was a progressive politician, proudly associating himself with William E. Borah (R-Idaho); Robert La Follette (R-Wisconsin), Thomas Walsh (D-Montana); Burton K. Wheeler (D-Montana) and James E. Watson (R-Indiana). Dill described his own role as a leader of the Senate debate over radio legislation as "a one-eyed man among the blind." He recognized his own limited knowledge, but foresaw the future needs of radio.

Source: D. Godfrey, "Senator Dill and the 1927 Radio Act," *Journal of Broadcasting* 23(4) (1979): 447–489.

Clarence C. Dill was the Senate's chief architect of the 1927 Radio Act and the Communications Act of 1934. Courtesy of Ohio Wesleyan University. Reprinted with permission.

public resource (it was limited—there were only specific frequencies available) and a trusteeship not to be squandered by the large corporate monopolies or some squatters who were staking claims to scarce spectrum space much like homesteaders of the Old West. The legislators were not willing to allow the broadcasters' uncontrolled growth to corrupt the airwaves. Thus they declared in the 1927 and 1934 acts that "the radio waves belonged to the people."

Censorship

The answer to the question of who owned the airwaves was mingled with the theme of censorship. It was not as simple as the conservation viewpoint would suggest. This theme exhibited itself in two forms: political censorship and the general censorship of ideas. According to Allen (1931, 83–84), in the 1920s milieu of popular protests over "sex magazines, confession magazines and lurid motion pictures," radio censorship was a topic of considerable discussion. The question of general program censorship was reflected in the broadcast of the 1925 John Thomas Scopes "monkey trial," wherein Scopes, a Tennessee high school biology teacher, was charged with violating a state law that forbade teaching evolution. The trial turned into an

Herbert C. Hoover

biography

Herbert C. Hoover was the Secretary of Commerce from 1921 to 1928 and U.S. president from 1929 to 1933. He came into the Commerce Department with a notable record of humanitarian service and was highly praised for his work until the 1930s, when the public sought a scapegoat for the ills of the Depression. His critics chided, "The great engineer has drained the ditch and damned the country."

Hoover was born into a middle-class family and worked his way to fortune in less than twenty years as an engineer. He turned to public service in the early 1920s and gained a reputation for being a great humanitarian. It was within the 1920s that he and his department exercised control over radio broadcasting, and it was his initiative that began the regulatory process. Less than one year after taking office as secretary of commerce, he called the first Radio Conference. Subsequent conferences in 1923, 1924, and 1925 included representatives of government and industry; and although the first conference had fewer than twenty-five participants, by 1925 more than four hundred participated in the Hoover Radio Conferences. Hoover guided the conferences under the theory of "rugged individualism." He sided with the industry in opposing a tax on receiving sets, governmental ownership of stations, and governmental censorship of program material and insisted on a minimum of governmental control, because "the more industry can solve for itself the less will be the burden of the government and the greater will be the freedom of the industry."

Much of the early law owes its existence to Secretary Hoover. He campaigned for its passage and, although he didn't like the commission proposed by Senator Dill, "most of the present law of broadcasting was called for by Hoover years before Congress finally put it into law."

The historical record offers mixed reviews of Hoover's political career. He is praised for his public service before the Depression and unfairly blamed for the Depression after the stock market crashed in 1929. Throughout it all Hoover retained his belief that the federal government was not the bountiful father. He remained to the end a foe of federal relief measures and an exponent of the individual's responsibility for his or her own welfare.

Source: D. E. Garvey, "Secretary Hoover and the Quest for Broadcast Regulation," *Journalism History* 3 (1976): 66; and C. M. Jansky, "The Contributions of Herbert Hoover to Broadcasting," *Journal of Broadcasting* 1(3) (1957): 241–249.

---

ideological debate between the religious fundamentalists (who believed in a literal interpretation of the Bible) and the Modernists (who mixed a little science with their interpretations). The courtroom drama, which featured William Jennings Bryan and Clarence Darrow in the starring roles, drew national attention and was broadcast by WGDN-Chicago. In the congressional session of 1926, a ban was proposed on the broadcast of all discourse dealing with the subject of evolution. Senator Cole Blease of South Carolina went on the *Congressional Record,* declaring "I am willing for the world to know that on this proposition I am on the side of Jesus Christ."

Broadcasters and governmental regulators were not about to legislate censorship of the airwaves. They were, however, greatly concerned about political censorship. The discussion of political censorship was much more heated and sometimes personal. The new medium of radio sparked colorful descriptions of its persuasive powers. The magazine *Popular Mechanics,* for example, described the "political spellbinding of radio" and the control the radio operator exercised over the coverage of conventions and political events merely by the "manipulating [the] silent switch." Powerful progressives like William E. Borah, Norman Thomas, and Robert La Follette spoke out strongly. They insisted that their radio speeches be free of censorship and were not about to submit their prepared texts for radio station management approval, as some managers had requested. Norman Thomas, executive director of the League of Industrial Democracy, asserted that both sides of any question must be heard on the air.

The discussion of censorship by the political appointee was another major point of debate directed at Secretary of Commerce Herbert Hoover. Although Secretary Hoover had the respect of most legislators for his outstanding record of national service, he also had legal control of the wavelengths, under the Radio

Act of 1912, and was a presidential candidate in 1928. These factors worried some lawmakers: "With control of the air goes the possible control of the vote, the control of the spread of ideas and education," said M. L. Ernst (1926). There were no charges ever directly aimed at Secretary Hoover, just a lot of public comment. Hoover himself declared, "The ether is a public's medium and its use must be for the public benefit." The legislative decision—stated in section 326 of the 1934 act—was that no powers of control or censorship would be entrusted to a political appointee or a cabinet member, and the censorship of ideas would not be permitted.

Corporate Monopoly

Of all the regulatory issues, the power of a corporate monopoly was most prominent. Large corporations had invested heavily in the development of radio.

Wallace H. White Jr. worked with radio legislation longer than any other representative or senator. He was the chief legislative draftperson for the 1927 Radio Act and the Communications Act of 1934.

When toll broadcasting opened the opportunity for profits beyond the sale of receiving apparatus, these corporations zealously protected their rights. Secretary Hoover, in the 1924 Radio Conferences, acknowledged the work of these corporations and expressed a debt of gratitude to AT&T and Westinghouse for their experimental contributions to the development of radio. These experimental stations, however, had grown into large interconnected radio trusts. The sale of AT&T stations to RCA occurred at the same time as the 1926 legislative debates on regulation and drew the attention of government. The corporate growth of AT&T, Westinghouse, and RCA also worried many lawmakers who were as concerned about corporate censorship as they were about the political and ideological varieties. Even Senator William E. Borah, who really had little interest in broadcasting, submitted a radio bill aimed at controlling corporate growth. The large corporations responded by seeking to ease the fears of the legislators and push through favorable legislative policy—the proposed 1927 Radio Law—feeling it would foster competition and consequently curtail monopoly.

Smaller stations took a much more defensive stance. These stations were looking for any competitive edge to gain their share of the industry and thus denied that the airwaves were publicly owned. They simply declared their ownership of wavelengths. True to the spirit of the postwar decade, they asserted that business interests were identical with the public interest and, while pushing for the clearance of confusion on the airwaves, they took up squatters' rights on their frequencies.

Representative White and Senator Dill, authors of radio legislation, sought to provide a degree of regulation that would preserve both industrial freedom and the public interest. This was not an easy task. They established a regulatory standard defined by the phrase *public interest, convenience, and necessity* and legislation that fostered the continued growth of the industry while clearing the air of confusion (and overlapping signals) and exerting a minimum of governmental control.

The solutions to the problems of conservation, monopoly, and censorship of radio were a direct reflection of these themes from the 1920s: The airwaves belonged to the people and would not be exploited, the monopoly growth would be controlled under new legislation, and censorship would not be permitted.

Wallace H. White Jr.

Wallace H. White Jr. was a veteran of the U.S. House of Representatives (1917–1931) and later the Senate (1931–1949). During the last years of his tenure, he was described by *Newsweek* as one of the top ten senators in the nation, having served on Capitol Hill for thirty years.

White was born in Lewiston, Maine, in 1877. He graduated from Bowdoin College and studied law at Columbia University. He was elected to the House of Representatives in 1916. Described as a traditional Republican with undeviating loyalty to Republican politics, he was a practical politician with a humanistic approach to the issues of public concern.

It was White's work in the Merchant Marines and Fisheries Committee and the Committee on Commerce that placed him in early association with Secretary of Commerce Herbert Hoover and the issue of radio legislation. He was a key player in the authoring of the legislative drafts that would eventually become the foundation of radio law. His first legislative effort, drafted in 1919, bears little resemblance to the final legislation, but it represents the beginning point in a learning process that made White one of the more knowledgeable within the legislature.

Most historians associate the White Radio Bill with control of broadcasting licensing in the hands of the secretary of commerce. This was not the only issue of concern, however. Senator Clarence Dill described White as a legislator whose primary concern was "public service."

As the 1927 Radio Act wound its way through Congress, it became known as the White Radio Bill. Others authors, including Dill, drafted their legislation primarily from White's proposals. It was Representative White who, without the flair of oratory and after passage of the 1927 Radio Act, acquired the support of Secretary Hoover and the signature of President Coolidge. By 1928 he was even an outspoken advocate of the Federal Radio Commission. In a speech before the National Association of Broadcasters on October 26, 1931, White extolled the public-interest virtues of the law. It was the public-interest provision that, according to White, "gave it the virtue of flexibility" for a growing and dynamic new industry. The Supreme Court later supported this concept.

Source: D. Godfrey, "Senator Dill and the 1927 Radio Act," *Journal of Broadcasting* 23(4) (1979): 447–489.

The legislative founders sought a degree of regulation that they thought would preserve the industrial freedom of a new medium and protect the public trust. Ironically, the gravity of these key issues has not diminished today; only the people and the corporate names have changed. The final form of the 1927 Radio Act reflects the public and legislative discussion of the issues of growing industrial chaos that created the demand for legislation (see figure 4.1).

The 1927 Radio Act

The legal crisis that resulted in the passage of the 1927 Radio Act was a breakdown of the Radio Act of 1912. This earlier law placed the secretary of commerce in charge of issuing wavelengths and had served its purpose, but by the 1920s the business environment became conducive to revolutionary technical and developmental growth within the industry. It was this growth and shift in radio service (from maritime to public audiences) that required the formulation of a new broadcasting law, known as the White Radio Bill for one of its authors.

Secretary of Commerce Herbert Hoover had worked to develop legislation over the years besides establishing rules and regulations in an effort to control the new uses of radio. From 1922 to 1925, he led a series of annual legislative development conferences, dubbed the Hoover Radio Conferences, aimed at obtaining cooperation and laying the groundwork for radio law.

Although Hoover knew he had little power to control radio—and some broadcasters disputed his authority on other grounds—the period of chaos (1926–1927) came to a head in a case that pitted the

FIGURE 4.1 Petition to Congress signed by Lee de Forest

Found among Wallace White's papers in the Library of Congress, this was apparently one of many cards sent to Representative White during consideration of legislation to control radio. Note the signature of Lee de Forest (it is unknown whether the signature is genuine).

Sign This Letter to Congress!

Dealers, jobbers, manufacturers, read this letter to Congress, sign it and return it to the Editors of "Radio Retailing," 36th Street and Tenth Avenue, New York City. They, in turn, will present it, in person, to Congress along with a mass of other evidence that this industry wants legislation covering the three points below. Do it now; help safeguard your own interests.

To the Sixty-Ninth Congress;
Honorable Gentlemen—

As an active part of the radio industry, as a voter and a citizen of the United States, I respectfully ask you, gentlemen of the Congress, to enact radio legislation which will accomplish the following things:

1. In issuing station licenses, give the licensing authority the right to recognize the priority of stations already established and efficiently serving the public as of July 1st, 1926.

2. Permit men within the radio industry itself to have places on any radio commission created.

3. Establish radio authority in the Department of Commerce, co-operating with a commission of radio men as provided by the White Bill.

We respectfully submit that the White Bill, thus amended, safeguards the best interests of the millions of listeners who are our customers and your constituents and that they, too, are overwhelmingly in favor of the White Radio Bill. May we not urge you to give this letter careful consideration?

Sincerely yours,

(Name) Lee de Forest
(Company) DeForest Radio Co
(Street) Jersey City
(City & State) N.J.

---

secretary against the Zenith Radio Corporation. The conflict revolved around the licensing of radio station WJAZ and its unauthorized operation. When WJAZ requested a change in frequency and an increase in its hours of operation, the Department of Commerce refused permission. WJAZ moved anyway. The case drew a lot of publicity and the verdict, in favor of Zenith, clearly noted that the secretary could only grant licenses. Stations were able to freely use other wavelengths. Hoover sought the opinion of U.S. Attorney General William J. Donovan, who, according to Bensman (1970, 423–440), suggested that "if the

Chaos on the Air

This telegram to Secretary Hoover humorously illustrates the problems in his attempts to hold stations to a particular frequency during the period of chaos leading to the 1927 Radio Act. This telegram was received from evangelist Aimee Semple McPherson.

> Please order your minions of Satan to leave my station alone. You cannot expect the Almighty to abide by your wavelength nonsense. When I offer my prayer to Him I must fit into His wave reception. Open this station at once.

Source: Herbert Hoover, *The Memoirs of Herbert Hoover: The Cabinet and the Presidency, 1920–1933*, New York: MacMillan (1952): 142.

present situation requires control, I can only suggest that it be sought in new legislation, carefully adapted to meet the needs of the present and the future." This case made it obvious that Hoover was powerless to control either the rapid expansion of the stations on the air or their assigned wavelengths. This chaotic situation called for immediate action by Congress, and the result was the 1927 Radio Act.

The 1927 Radio Act established the Federal Radio Commission (FRC) to regulate radio. Among the first things the new commission did was revoke all licenses and force the stations to reapply. It established the government's right to make an elaborate set of regulations as well as a code of "Standards of Good Engineering Practices." During the years 1927 to 1934, the new commission restored basic order to the spectrum and began to organize itself.

The technological issue of channel allocation was a major challenge for the first Federal Radio Commission. It was not enough merely to dictate public ownership of the airwaves; a system of allocation, eligibility, and enforcement had to be established. During the early years, the commission worked out a series of tables representing different classes of all stations and determined a system of allocation. Today the tables have changed somewhat—FM and television allocations have been added—but the theory of equal distribution of the frequencies remains. Then as now, the prospective station licensee had to locate an available frequency and/or create a noninterfering one and then justify its inclusion. In applying for that frequency, the station licensee will designate technological facilities for its use (power, frequency, coverage pattern, location, facilities of operation, and hours of operation). Today, because most of the favored channel positions have been taken, a potential new licensee will most often purchase an existing facility or, as noted, may show through a new transmission coverage pattern that the proposed allocation can fit into the established table. The rules and regulations section 73 now provides the classification, frequency allocations, and table of assignments for communities across the United States.

A construction permit (CP), required prior to operation of any station, was the next step in the application process. The prospective licensee, having located a possible frequency, completes the construction permit (today it is the FCC CP, or FCC Form 301).

If the CP is granted, the licensee has a specific period of time to construct the station and conduct designated tests of its power and operation. Following successful testing, the licensee is generally given permission to be on the air until the license is formally approved.

With the CP and the license application, the licensee is also required to designate qualifications related to character, finances, and technical ability. The licensee must be a U.S. citizen and be reputable, meaning the applicant has no record of criminal activity, antitrust violations, or misrepresentation to the commission. Historically, the Federal Communications Commission (FCC) required that a prospective operator have enough financial resources to operate for five years. Today the licensee must certify possession of sufficient funds for three months of operation without relying on advertising revenue. Today the financial requirements can be met by investment brokers, corporate underwriting, or just proof of financial ability. To meet the technical requirements, most applicants hire an engineer for operational purposes and perhaps a consulting engineer during stages of the frequency search, construction, and testing.

The enforcement of FCC/FRC rules (discussed in chapter 5) has historically relied on the threat of losing the license. Published statements and speeches, however, known as the "raised eyebrow," alone have often been enough to keep most broadcasters in line with FCC policies that were not specifically translated into rules and regulations.

The current procedures of channel allocation, license application, eligibility, and enforcement all had their roots in the 1927 Radio Act and the resulting actions of the FRC. Part of the reason for the similarity is that the 1927 Radio Act essentially became the Communications Act of 1934, which governed the industry until 1996. The major provisions of the act remain in force today.

PRECEDENTS: BALANCED PROGRAMMING

Technology and the spectrum were not the only concerns of the Federal Radio Commission. The key regulatory standard of "public interest, convenience, and necessity" still governs the electronic media. It has become the guide by which the courts and the FCC regulate a continuously growing and dynamic industry.

Radio Act of 1927 Declares "Public Interest" Primary

Section 302(a). The Commission may, consistent with the public interest, convenience, and necessity, make reasonable regulations . . .

Section 303. Except as otherwise provided in this Act, the Commission from time to time, as public convenience, interest, or necessity requires shall . . . classify stations, proscribe the nature of service rendered, assign frequencies, regulate the kind of apparatus, . . .

Section 309(a). Subject to the provision of this section, the Commission shall determine . . . whether the public interest, convenience, and necessity will be served by the granting of such [broadcast license] application. . . .

The phrase, as it pertains to electronic media regulation, first appeared in the 1927 Radio Act. It remains the center of regulatory interpretation even with increasing attempts to draft new legislation to meet today's technological demands.

Almost immediately following the passage of the 1927 Radio Act, it soon became evident that the public-interest standard would be applied to programming as well as to technological development. This became clear in the Federal Radio Commission's 1928 decision, known as the *Great Lakes* opinion, which provided the commission's views on programming service. The FRC opinion resulted from a comparative hearing procedure regarding the application of the Great Lakes Broadcasting Company for a new station license. The application was denied based on the limited program service the station proposed. The commission held that the station's programming would be of limited interest to the total audience—servicing only a small portion of the audience living within the station coverage area—and denied the application. This was not originally seen as censorship, because no specific programs were noted by the commission before they were broadcast, but the label would be applied afterward. Thus the commission established the first of a broad-based set of programming expectations to define what was suited to "best serve the public."

The Communications Act of 1934

On June 19, 1934, Congress enacted the Communications Act of 1934. This placed the regulation of the electronic media within the hands of a newly configured commission—the Federal Communications Commission. The 1934 act was based largely on the federal 1927 Radio Act and was the result of a general restructuring of the federal government. Under the Roosevelt administration, the commission's powers were broadened. Passed during the height of the Depression, it seemed to have been one of the least significant laws of Roosevelt's administration. It has

The Great Lakes Opinion

The tastes, needs and desires of all substantial groups among the listening public should be met, in some fair portion, by a well-rounded program in which entertainment consisting of music of both classical and lighter grades, religion, education and instruction, important public issues, discussion of public events, weather, market reports, news and matters of interest to all members of the family find a place. . . . The Commission does not propose to erect a rigid schedule specifying the hours or minutes that may be devoted to one kind of program or another. What it wishes to emphasize is the general character which it believes must be conformed to by the station in order to best serve the public.

Source: *Great Lakes Broadcasting Company v. Federal Radio Commission* (1928), docket 4900, 1928.

proven over time, however, to be the single most important piece of electronic media legislation considered by Congress. Although it brought forward the basics of the 1927 Radio Act and retained the "public interest, convenience, and necessity" standard, it was not a carbon copy. The passage of the Communications Act of 1934 provided a larger role for the commission:

- It extended the commission's control over wire and wireless communication, both interstate and foreign

- It added two members to the commission, enlarging it from five to seven

- It changed the name from the Federal Radio Commission to the Federal Communications Commission

The purpose of the new act was to centralize control over the various branches of the growing electronic media beyond radio to include control over wire and wireless communications. It provided the elasticity to control the continued evolution of the electronic media growth from wire to wireless and back to wire communication. The philosophy of the Communications Act of 1934 has been tested over the years in the court systems, and until 1996 it remained the basis of today's electronic media law.

From the 1934 law, the FCC developed a large body of rules and regulations ("R&R"), which define the specific directives of the law. How the FCC accomplishes enforcement, how it designates the license, how it interprets the law and a host of technical directives are detailed in the rules and regulations. In modern stations, copies of these regulations are found in every chief engineer's office; the programming rules and regulations are generally kept in the manager's or lawyer's office or on the library shelves.

LOOSENING THE CHAINS

After the passage of the Communications Act of 1934, the pressure continued for greater control over the monopolistic expansion of the larger national networks. By the mid-1930s almost all of the stations across the nation were affiliated either with NBC's Red and Blue Networks or with CBS. Because of the restrictive affiliate contracts, it was difficult for the competition to establish themselves. The network required affiliates to carry their programming exclusively. Although the first network/affiliate agreements were rather informal and nonbinding, by the mid-1930s they were the opposite: The contractual relationship was so tight it restricted the affiliates' freedom.

The catalyst that brought the issue to the investigation stage was the Mutual Broadcasting System (MBS), which began as a small midwestern network with the *Lone Ranger* as its most successful premier program package. MBS was frustrated in its attempts to expand from a regional to a national network. As a result, MBS complained to the FCC about what it saw as competitively unfair treatment. The FCC began an investigation in 1938. Three years later, in May 1941, the commission issued its Report on Chain Broadcasting (Commission order no. 37, docket 5060).

The results of the investigation report were dramatic. The commission forbade network duopoly—single ownership of two or more networks—within a single market; it forced the networks to give up their owned-and-operated talent agencies (which were a monopoly over program talent bookings); it eased the networks' control over a local station's time and diluted the exclusivity of the affiliation contract. It gave the affiliates the opportunity to control their own programming and provided the means for competitive growth—notably that of the Mutual Broadcasting System.

The networks challenged the FCC's Report on Chain Broadcasting in the courts, predicting the demise of network broadcasting. The courts upheld the report in a battle that lasted until 1943 and forced the sale of NBC Blue to Edward J. Nobel. (The network was later to become ABC.)

Although the Report on Chain Broadcasting acknowledged the contributions of the networks to early broadcasting development, it was a major turning point in the network/affiliate relationship by providing more autonomy in local programming control—if the affiliates wished to exercise it.

PROGRAMMING AND THE LAW

Nothing in the Communications Act of 1934 refers directly to broadcast programming. It says nothing concerning such issues as sustaining programs, false advertising, overcommercialization, controversial issues, religious programs, minority programs, cultural programs, public service programs, news, entertainment, or the vitality of the marketplace. Indirectly, however, the 1934 act has been interpreted as giving the Federal Communications Commission a great deal of latitude concerning program regulation. In an attempt to direct the stations to program "public service," interpretations of the 1934 act resulted in a considerable body of administrative and codified law.

The provisions most often discussed are federal laws that made it a crime to broadcast certain content, such as lottery information, fraud, and obscenity. Additional codes can be found in the Freedom of Information Act, the Privacy Act, the Copyright Act, and laws pertaining to advertising and cigarettes, which are discussed in chapter 5.

THE BLUE BOOK

In addition to the legislative acts, the FCC's administrative law, and U.S. federal law, the commissioners have historically exercised an informal enforcement threat by hinting at regulation they felt would be in the public interest. This was accomplished in the form of notices of proposed rule making, memoranda, and hearings. One such case was the *Great Lakes* opinion, which was just that—an opinion. It merely indicated the commission's desire to balance program structure with serving the public and directed the stations accordingly. It was, so to speak, a voice of warning, a voice of direction, a "raised eyebrow."

The most striking example of such directive action exists in a 1946 FCC memorandum entitled "The Public Service Responsibility of Broadcast Licensees." It was more commonly called "the Blue Book," because it had a blue cover. The book was largely the work of Briton Charles A. Siepnamm, a consultant to the commission that year. The document was the most descriptive programming directive the FCC had ever released, and it evoked cries of protest from broadcasters, who claimed that it was an infringement of their freedom. The FCC simply responded that it was only a memorandum.

The Blue Book was neither vigorously enforced nor officially set aside by the FCC but remained for years an interpretative instrument defining program regulation and the public interest. Perhaps the most thoroughly reasoned expression of the public interest ever issued by the FCC, it defined sustaining programs and balanced programming and it indicated what programs were appropriate and inappropriate for commercial sponsorship. It encouraged service to minorities and nonprofit organizations. Live local programs were also encouraged as was the discussion of public issues. It took a position against overcommercialization by issuing criteria regarding the number of commercially sponsored programs and spot announcements that were appropriate for a weekly schedule.

For years following the release of the Blue Book, the programming and commercial guidelines it described were considered general standards. Between 1946 and 1960, however, the increasing number of radio stations, combined with economic marketplace considerations created first by television, then FM, and recession, led to the easing of Blue Book standards. Attempts to codify public interest were consistently defeated in favor of marketplace-based regulation. Even with the relaxation of the proposed Blue Book standards, broadcasters were learning to understand the commission's broad definition of public-interest programming while it avoided having such a definition codified into law. The result has been a significant public and legal debate, with broadcasters maintaining flexibility in interpreting the standards—written or not.

THE 1960 PROGRAM POLICY STATEMENT

The 1960 Program Policy Statement, issued July 29 of that year, ended active debate over the Blue Book. Both milder and more effectively enforced, the statement reinforced the idea that the licensee was a trustee with a public service responsibility. Within the statement, fourteen major elements were emphasized as necessary by the FCC to meet public interests, needs, and desires within the community. It is doubtful that the policy's program list was in any preferential order, but it is interesting to note where the "entertainment" programs appeared in relation to other programming. It is important to realize that in today's marketplace, the value of entertainment has risen considerably within that list as other forms of programming have notably decreased.

The provisions of the 1960 programming policy were enforced through the license renewal process, specifically section IV-B of the renewal form itself. This

Program Policy Statement of 1960

Issued fourteen years after the Blue Book controversy, this statement is considerably toned down. Although the enforcement provisions of license renewal have been eliminated, the statement still reflects broad-based programming standards.

> The foundation of the American system of broadcasting was laid in the 1927 Radio Act when Congress placed the basic responsibility for all matter broadcast to the public at the grass roots level in the hands of the station licensee. That obligation was carried forward into the Communications Act of 1934 and remains unaltered and undivided. The licensee, is, in effect, a "trustee" in the sense that his license to operate his station imposes upon him a non-delegable duty to serve the public interest in the community he had chosen to represent as a broadcaster.
>
> The major elements usually necessary to meet the public interest, needs and desires of the community in which the station is located as developed by the industry, and recognized by the Commission, have included: (1) Opportunity for Local Self-Expression, (2) The Development and Use of Local Talent, (3) Programs for Children, (4) Religious Programs, (5) Educational Programs, (6) Public Affairs Programs, (7) Editorialization by Licensees, (8) Political Broadcasts, (9) Agricultural Programs, (10) News Programs, (11) Weather and Market Reports, (12) Sports Programs, (13) Service to Minority Groups, (14) Entertainment Programming.

measured the station application based on an analysis of the tastes, needs, and desires of the station's own community and the manner in which it proposed to meet those needs. The renewal form required the station to ascertain its community's needs and provide the commission with lengthy descriptions of past and proposed programming. The renewal provided the commission the opportunity to compare promises made by the station in relation to the station's actual performance over the years of the license.

A classic example of promise-versus-performance regulation occurred just after the policy was adopted. The license renewal for KORD in Pasco, Washington, was set for an FCC hearing in March 1961 for failure to program the station in accordance with promises made when the license was first granted in 1956. In the original application, the station promised that 6 percent of its time would be devoted to live local programs and 10 percent to talk and educational programs. It also promised a limit of 700 commercial announcements each week. In its 1960 renewal application, however, it showed no live, local, or educational programs. It did have a total of 1,631 commercial announcements during the typical week. Following the formal notice of a hearing by the FCC, KORD was allowed to modify its renewal application and promised again to bring programming in line with the original promises. The commission recognized the promised improvements, canceled the hearing, and granted the station renewal. The FCC felt that the threat of the KORD case was of such general importance to the industry that copies of the KORD order were sent to all station licensees so that they could have the opportunity to understand and comply with the policy.

Today the policy remains on the record as a part of the regulatory philosophy—a definition, of sorts, of the meaning of "public interest." But the licensing enforcement requirements have changed substantially. The 1960 Program Policy Statement has given way to a marketplace interpretation of its standards as opposed to a singular commission view.

The Fairness Doctrine

In its concern to maintain broadcasting as a medium of free speech, the FCC began debating more questions involving free expression and censorship. The idea of unlimited access—providing freedom of speech or a platform for anyone—did not always seem possible because of historically perceived limitations in technology and frequency. Because not everyone who wanted to speak could be accommodated, it seemed in the public interest to provide regulatory groundwork to secure the free expression of public issues.

The fairness doctrine—*In the Matter of Editorializing by Broadcast Licensees*—issued June 1, 1949, reversed the FCC's 1941 policy discouraging editorial

The Fairness Doctrine

Excerpts from *In the Matter of Editorializing by Broadcast Licensees* (13 FCC 1246), June 1, 1949:

It is apparent that our system of broadcasting, under which private persons and organizations are licensed to provide broadcasting service to the various communities and regions, imposes responsibility in the selection and presentation of radio program material upon such licensees. . . .

[T]he Commission believes that under the American system of broadcasting the individual licensees of radio stations have the responsibility for determining the specific program material to be broadcast . . . however, [this responsibility] must be exercised in a manner consistent with the basic policy of Congress that radio be maintained as a medium of free speech for the general public as a whole rather than as an outlet for the purely personal or private interests of the licensee.

---

statements and instead encouraged stations to present editorials and different points of view. Stations now were asked to make "reasonable judgments in good faith on each individual situation." The doctrine has left its mark on broadcast regulation and on the stations' process of exercising reasonable judgments.

The doctrine was divided into three categories: controversial issues, political editorializing, and personal attack. Programming related to a public issue and classified within one of these categories called upon the station to make reasonable judgments as to whether issues had been balanced in their presentation. Political editorials specifically referred to political endorsements that involved all but the appearance of the candidate (that was a matter left to section 315 of the 1934 act). They too required balance. In the case of personal attack, the licensee was obligated to notify the person or group attacked and provide a reasonable opportunity to respond. The airing of controversial issues also required balance and reasonable opportunity.

The fairness doctrine did not obligate the station to editorialize; it merely indicated that if a station did air such material, it was obligated to afford opportunity for opposing points of view. The key phrases defined "reasonable judgment" as a station's responsibility to assess the balance of the issue and its presentation over the air and "reasonable opportunity" as a station's responsibility to seek out and balance the presentation and personal attacks.

Although the industry was not happy with the fairness doctrine, it remained in force for more than three decades until it was overturned by the courts in 1985. Broadcasters criticized the doctrine for failing to meet the demands of the changing times, which ran the gamut from simple political debate, civic discussion, and personal attack to a discussion of commercial advertising, public service campaigns, and obscenity. In one case, for example, a complaint was filed asking for broadcast time to respond to new car commercials. The complaint reasoned that because automobiles pollute the air, and air pollution was a controversial subject, as defined by the fairness doctrine the complainant should be given free time to respond. The commission ruled in favor of the broadcaster (48 FCC 2d1; 30 RR 2d 1261).

The major criticism raised against the fairness doctrine by the broadcast community was that it inhibited free speech. The pressure it created was said to have a chilling effect that made broadcasters reluctant to approve any controversial programming, documentaries, or hard editorial viewpoints.

In spite of the criticism, the fairness doctrine survived many challenges in the courts. The 1969 *Red Lion* decision affirmed its stability. The case involved the license application of the Red Lion Broadcasting Company to operate radio station WGCB. The incident involved the airing of a program prepared as a part of a *Christian Crusade* series. A book by Fred J. Cook was discussed by the program moderator, the Reverend Billy J. Hargis. In the course of the program, Hargis cited several accusations against Cook, among them that Cook had fabricated charges against city officials and that Cook was affiliated with a Communist publication. When Cook heard the broadcast, he applied for free time to reply under the personal-attack provisions of the fairness doctrine. He was refused. Cook then took his plea to the FCC, which ruled that Red Lion had failed to meet its obligations under the fairness doctrine to notify Cook of the attack and to offer reply time. The decision was upheld by the Supreme Court.

The demise of the fairness doctrine began in earnest in 1984, when the FCC issued a Notice of Inquiry to reassess the doctrine because of recent changes in the communication marketplace. Was there any longer a scarcity of voices and views? Was the fairness doctrine needed to foster the widest possible dissemination of information? Did the doctrine, in fact, dampen the free expression of ideas as broadcasters had historically argued?

The FCC completed its inquiry in August 1985 and adopted a report concluding that the fairness doctrine no longer served the public interest and did indeed limit the broadcaster's programming and free expression. The commission found that, because of the development of new technology, the issue of scarcity of information sources was no longer an issue. Although no longer in force technically, the fairness doctrine still constitutes the basic policy by which responsible broadcasters operate: fairness and reasonable opportunity.

It is important to note that although rules regarding the programming of controversial issues no longer exist, personal-attack and political-editorializing rules still exist. Over the years of various challenges, court cases, and debates, broadcasters have evolved a set of written and unwritten operational standards set within the public interest. The Radio and Television News Directors Association has established standards of balance and fairness, and stations' newsrooms have devised their own guidelines—all within a historically debated standard of public interest. Little has really changed in the stations' methods of operation and deliberate considerations of fairness within their programming. The fairness doctrine's demise has done little to affect the promotion of public-issues programming or the way in which a broadcaster deals with balancing such program situations. Congress has expressed its dissatisfaction with the FCC's actions, and legislators have threatened to write the principles back into a new communications law. Although the fairness doctrine is gone, its tenets (personal attack, political editorializing, and balanced programming) are still heavily etched into the day-to-day broadcast operation. The provisions pertaining to fairness (although unenforced) are now within section 73 of the rules and regulations (see FCC 73:1910, 73:1920, 73:1930, and 73:1940).

POLITICAL CAMPAIGN LAW: SECTION 315

Running parallel to the historical fairness doctrine and still enforced are the political provisions of section 315 of the Communications Act of 1934, which differ from the fairness doctrine in that they deal specifically with candidates for public office and require equal opportunities. In other words, an appearance of a political candidate personally attacking an opponent was within the jurisdiction of section 315; an appearance by that candidate's spokesperson, saying perhaps the same words, was within the jurisdiction of fairness.

Section 315 obligates the station that sells or provides time to any candidate who is a "legally qualified candidate" to afford "equal opportunities." Nothing in the law obligates the station to sell time (except to federal candidates); the station may refuse to provide time to all but federal candidates.

If the station allows one candidate to use the facility, however, equal opportunities must be made available to all other candidates for the same office. The station cannot discriminate in charges, practices, regulations, facilities, time, or services rendered to candidates. Notable exceptions to the law include bona fide news, news interviews, documentary, or on-the-spot news coverage. According to section 315, these are not "deemed use" of a station's facilities.

One significant provision of the law relates to the power of censorship. Once a station provides a candidate a forum, the station has no power to censor the material broadcast by the candidate. The candidate is held responsible for libelous or slanderous comments in any legal action resulting from the remarks, however. In such situations the FCC has indicated that if the context is illegal, the station might be permitted not to carry the program. There is no case yet on this issue. This is the only situation, under law, in which a station is not held responsible for its programming despite being its point of origination.

Sarnoff on Equal Time Law

If broadcasters were relieved of the need to offer equal time to the minor candidates, NBC would offer appropriate opportunities for network appearances to presidential and vice-presidential candidates of both major parties . . . there is no question but that, freed from the irksome requirements of 315, we would be in a position to make more time available and would do so.

Source: Thomas H. Guback, "Political Broadcasting and the Public Policy," *Journal of Broadcasting* 12(8) (1968): 191–211.

The debate over the utility of this law led to the Great Debates Law issued on August 24, 1960. This was the first and only suspension of section 315. The suspension provided legal permission for the first broadcast of presidential debates—between candidates John Kennedy and Richard Nixon. Broadcasters hailed the suspension as a step toward the elimination of the burdensome law, contending that it would be of greater public interest to delete section 315. The criticism was not dissimilar to that of the fairness doctrine in that broadcasters felt that more time could be *afforded* to candidates without section 315. NBC President Robert Sarnoff noted, "There was no question but that, freed from the irksome requirements of 315, we would be in a position to make more time available and would do so" (Guback 1968, 191–211).

There has been no further suspension of the law. Rather, the debates since 1960 have been organized to sidestep the issue and take advantage of the provision of the law that exempts a "bona fide news event." Under the sponsorship of the League of Women Voters and other nonpartisan organizations (including the networks), debates have been organized and scheduled as news events and are thus exempt from section 315.

It is unlikely that controversy over section 315 will lead to its suspension, at least in the foreseeable future. To do so, without other changes in the election systems, would seem to be denying the minority candidates equal access to the media. And after all, who would pass the new law—the candidates now in office? Broadcasters criticize its chilling effect, and the politicians who must draft the legislation view with suspicion the record of broadcasters in relation to political fairness. The issue remains a hot topic. With the mounting costs of campaigns, legislation has been proposed that would provide blocks of free time. (See the discussion of a station's obligations under section 315 in chapter 5.)

New Technology, Deregulation, and the Marketplace

Cable television, satellite delivery, high-definition television (HDTV), spectrum regulation, the Internet, and a host of technological developments—have all been matters of growing concern to the FCC. As discussed in chapter 14, cable transmits by means of coaxial cable and fiber optics, and a microwave satellite brings the signals directly into our homes. New technology has cleared heretofore restricted spectrum space. Do these scientific developments in electronic media require new legislation? The FCC and Congress have answered yes, with the commission's emphasis on deregulation in recent years and congressional passage of the Telecommunications Act of 1996.

The roots of deregulation and of the new law can be found in the 1960s during debate surrounding the emerging "marketplace philosophy" and attempts to rewrite the Communications Act of 1934 under the auspices of Senator John Pastore (D–Rhode Island). The marketplace philosophy was simply a move away from what the industry considered burdensome government regulations toward an attitude of letting the marketplace determine the regulation. The movement was in response to industry unrest over whether the existing regulatory structure was sufficient to handle the rapidly growing technological change and whether it favored the public or the broadcaster. Many of the old laws now seemed burdensome, the technological developments spurred new issues, and the marketplace itself developed into a mass audience. Many of the deregulation rules removed substantial paperwork—eliminating community ascertainment and radio logs (these operational regulations are discussed in chapter 5). All of these moves to deregulate the electronic media now came within the realm of the competitive marketplace.

The Telecommunications Act of 1996

Broadcast technology has evolved over the past six decades to include more than the Communications Act of 1934. The law has been interpreted in the courts and by the Federal Communications Commission, thus creating a large body of administrative law. Even the act itself has been amended numerous times over the years as necessary.

In the late 1970s, the demand for more than amendments and rewrites began surfacing with regularity in Congress. Senator John Pastore, head of the Senate Communications Subcommittee, and Representatives Torbert Macdonald and Lionel Van Deerlin held hearings on a new legislative proposal. These resulted in Senator Van Deerlin's 1977 proposed rewrite of the Communications Act of 1934, which was met with both anxiety and interest from those within Congress and in the industry. In the 1980s the rewrite proposals, which were never passed into law, led to many of the deregulatory efforts under FCC Chairman Mark Fowler. So, even though there was no new legislation, there was change in the law. The size of the commission dropped from seven to five members, spectrum fees were imposed, licensing periods and ownership limits were extended, ascertainment rules were dropped, and a lot of paperwork regulations were eliminated for the stations.

New technology was bringing with it challenges to the law and growing concern regarding the ability to direct change under existing law. Is there enough spectrum space for new technology? How do you regulate the new technology? What about the growing concerns over sex, violence, and indecent materials—specifically on the Internet? What about the voice of the people versus corporate monopoly? Is the process of legislation being auctioned off in a game of politics?

Historians of the electronic media smile at these modern-day questions, because they bear a striking resemblance to those asked by the industry and legislators who drafted the 1927 and 1934 acts. There is no denying, however, that the sophistication of today's electronic media technology has created a new information society. The speed, reach, and frequency with which today's technology can transmit a message has challenged the Communications Act of 1934. Legislators and industry members alike have been calling for reform, believing the 1934 act insufficient for development, policy making, or operation control. The answer: a new law.

The Telecommunications Act of 1996 was signed into law by President Bill Clinton on February 8 of that year. The most sweeping legislative revision of electronic media law passed in sixty years, it covers broadcasting, cable, telephony, and online services, touching on almost every aspect of electronic communication technology. Though the popular press call the act "sweeping," it is good to remember that although the new law is extensive, it retains a good portion of the 1934 act. Indeed, reading early drafts of the 1996 act necessitates having a copy of the old law close at hand, because the 1996 law contains amendments, repeals, references, and cross-references to the 1934 legislation. For example, the 1996 law retains the "public interest, convenience, and necessity" provision of the 1927 and 1934 acts as well as references to the criminal codes and the mandates for political broadcasting. At the same time, it repeals a lot of FCC administrative law, such as ownership and licensing provisions, and creates new law for the regulation of cable, telephony, and online services. The challenge for the FCC today is the implementation of the new law.

The Telecommunications Act of 1996, like all of those before it, is now subject to interpretation. The consideration process for interpretation includes legislative intent, the FCC, the industry, and the courts. Almost immediately after the passage of the 1996 act, the FCC initiated rulemaking proceedings which will eventually turn the legislative law into an interpretative set of rules and regulations governing the industry. The FCC will hear from the public, the industry, and lawmakers as it interprets the new law and implements its directives through its rulemaking process (discussed in detail in chapter 5). The 1996 law is also subject to the interpretations of the courts; its indecency aspects have already been challenged by the American Civil Liberties Union. It will be some time before we know what effects will result. Broad interpretations and commentary are going to be in the popular press for some time.

Ironically, the new law written to promote new technology and television has had its greatest effect on radio. The new ownership limits created a massive exchange of radio properties across the nation. More stations have been bought and sold since the 1996 act

Summary of the Telecommunications Act of 1996

BROADCASTING AND ELECTRONIC MEDIA LEGISLATION STANDARDS

Public Interest Requirement (Sections 201(d) and 204). Nothing shall be construed as relieving a station from its obligation to serve the public interest, convenience, and necessity.

Obscenity and Violence (Title V, section 502). This section applied to broadcasting, cable, telephony, and online services. It increases the fine for broadcasting obscene material from $10,000 to $100,000, requires cable operators to scramble adult and sexually explicit services at the request of the subscriber (Sections 505 and 640), and provides regulation of obscene materials carried through the use of computers (Section 507).

V-Chip (Section 5519(c)). This section mandates that television sets manufactured in the United States have the ability to block programming based on electronic codes. The industry is to establish the television rating code, thus allowing parents to control "violence, sex, and other indecent materials" in their homes.

Telecommunications Development Fund (Section 714). This establishes a new telecommunications fund to promote small business, to enhance competition, to stimulate new technology and employment training, and to promote the delivery of telecommunications services to rural and urban areas. The fund is administered by a seven-member board of directors appointed by the chair of the FCC.

National Educational Technology Funding Corporation (Section 708). This section establishes a private nonprofit corporation with the purpose of upgrading schools and education. The mention of this corporation in the act is to recognize it and "provide authority for Federal departments and agencies to provide assistance [to it]."

Radio (Section 202(a)). The law establishes new limits on the multiple ownership of commercial stations. There are basically no caps regarding national ownership or audience reach. There are, however, local restrictions in markets where there are

- 45 or more stations; ownership is limited to 8 radio stations with no more than 5 in the same service (that is, AM or FM)
- 30 to 44 stations; ownership is limited to 7 radio stations with no more than 4 in the same service
- 15 to 29 stations; ownership is limited to 6 radio stations with no more than 4 in the same service
- 14 or fewer; ownership is limited to 5 radio stations with no more than 3 in the same service

In general, no single owner may have interest in more than 50 percent of the radio stations within a market.

Television (Section 202(c)). National ownership is capped at 35 percent. This means no single owner may own stations reaching more than 35 percent of the national audience. The provision increases the cap from 25 percent.

Cross-ownership (Section 202(f)). Restrictions on the cross-ownership of broadcast stations and cable systems are to be lifted. The FCC rule banning such combinations, however, is still in place.

Television Duopolies (Section 202(d)). Television duopolies (ownership of two stations by one person) were not addressed in the bill but are subject to FCC rulemaking.

Local Marketing Agreements (Section 202(g)). Television local marketing agreements were approved so long as they are in compliance with current regulations.

License Terms (Section 203). Current television and radio license terms were extended from five and seven years, respectively, to eight years.

License Renewal (Section 204). License renewals are streamlined, provided the station has

- served the public interest, convenience, and necessity
- no serious violations of law or rules and regulations
- no serious patterns of abuse

> **Direct Broadcast Satellite** (Section 205). The FCC is provided with exclusive jurisdiction over direct-to-home satellite services.
>
> **Cable** (Section 301). Rates are deregulated for the basic services. Upper-tier rate regulation will be gone in three years. Rate regulations for smaller systems disappear immediately.
>
> **Common Carriers** (Section 301). It permits common carriers to provide video services as common carriers, cable operators, or via open video systems.
>
> **Access** (Section 531). Cable operators may refuse public access programs containing "obscenity, indecency, or nudity."
>
> **Telephone Video Service** (Section 302, part 5, and section 651). It permits local telcos to provide video in their service areas as common carriers, cable operators, or via open video systems. In other words, it allows phone companies to provide video programming in their service areas. Cable and phone companies in both business enterprises are regulated by both phone and cable regulations.
>
> **Telephone Open Video** (Section 653). This new entity allows the phone company to establish open video systems, under which the operators qualify for reduced regulatory burdens and do not need a local franchise in order to operate in exchange for being limited to programming one-third of the active channels. (An open video system is yet to be clearly defined, however.)
>
> **Online Services**. Provisions of the act relating to the definitions of common carrier and telecommunications place the FCC in the position of regulating online services. The specific goal of these regulations is the prohibition of the use of interactive computer services to transmit indecent communication to children.

than in several decades prior. In television there is cautious but determined progress toward digital and HDTV. Likewise, in the cable and television industries there has been only a little encroachment into the new territory now allowed by law. This is part of the democratic rulemaking process—a process of philosophical as well as legal refinements.

Summary

▶ The importance of the history of electronic media law is not so much in the memorization of the individual issues, cases, and regulations, but in the understanding of the circumstances that brought the issues to the forefront of discussion and therefore resulted in legislation. In other words, the legislation has been in response to a need, a problem, a technological advance, or an issue of public concern. The 1927 Radio Act was formulated in the 1920s and passed as a result of a legal crisis—the need was immediate. The Communications Act of 1934, formulated under the Roosevelt administration, renamed and broadened the powers of the Federal Communication Commission.

The chain-broadcasting investigation resulted from the Mutual Broadcasting System's attempts to compete on a national level with CBS and NBC as well as the government's continued interest in the question of monopoly. The fairness doctrine was created to deal with the questions of balance and reasonable opportunity to air differing points of view.

The historical perspective provides an understanding of how and why the law was formulated and aids in understanding current debate as it relates to often very similar issues. For example, have the issues of monopoly and censorship subsided? They have not. It is therefore important that in seeking to understand these issues, we are familiar with the platforms, circumstances, and laws from which they grew. Many issues of today are strikingly similar to ones raised in the past. Similar issues foster similar debates—only the faces and a few corporate names have changed.

▶ The electronic media developed rapidly during the 1920s. The legislative foundations established during this period were in response to both the technological challenges and cultural revolution of the era. The basic regulatory philosophy remains in force today.

▶ Key personalities responsible for creating the 1927 Radio Act and the Communications Act of 1934 were Hoover, White, and Dill. The period from 1926 to 1927 created a legal crisis that left Secretary Hoover powerless to control radio. This was a major catalyst in the creation of the 1927 act.

▶ The "public interest, convenience, and necessity" was the key phrase adopted into the regulatory language and remains the standard.

▶ The passage of the Communications Act of 1934 extended the commissioners' control over wire and wireless, and interstate and foreign communication; it added two members to the commission and changed its name to the Federal Communications Commission. The basic hypotheses of the 1927 and 1934 legislation have been court tested and remain the regulatory force today.

▶ The "public interest" standard has been interpreted by the FCC and the courts to include programming issues. The *Great Lakes* opinion, the Blue Book, the fairness doctrine, and the 1960 Program Policy Statement reflect this interpretation and provide a broad direction for a station's understanding of programming in the public interest.

▶ The fairness doctrine was eliminated by the FCC, but its basic tenets remain. The commission felt that the fairness doctrine had a chilling effect, discouraging broadcasters and free expression.

▶ The debate calling for the elimination of section 315 is similar to that surrounding the fairness doctrine, but it is unlikely that it will be eliminated—the legislators who drafted the law are not likely to trust the broadcasters for politically fair treatment.

▶ New delivery systems have brought new challenges to the law. The regulatory years 1927–1977 provided control and direction for a rapidly developing industry. Deregulation provided a maturing electronic media with freedom from some burdensome regulation. Much of the basic philosophy remains even in today's marketplace era.

▶ The Telecommunications Act of 1996 provided the most sweeping legislative change in the past six decades. While it retained the public-interest requirements of the historical law, it strengthened obscenity penalties, required a new V-chip, extended ownership limits, allowed cable and telephone companies into both video and telephone businesses, and moved into online regulation of indecent communication.

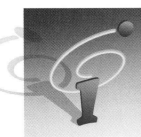

INFOTRAC COLLEGE EDITION EXERCISES

The legal issues surrounding the history of radio and television are ironically parallel to those of today. In fact, the basic fundamental hypotheses that set the foundation for the 1927 Radio Act have been folded over into the Communications Act of 1934 and the Telecommunications Act of 1996. You can identify these from the text.

4.1 In the 1920s and 1930s, the issues surrounding legislation and the industry were monopoly, ownership of the airwaves, a permanent communications commission, political access to the airwaves, the control of advertising, and censorship. In *Journal of Broadcasting* 23:4 (Fall 1979) and *Journalism Quarterly* 67:1 (Spring 1990), you'll find a historical perspective relative to these issues. Peruse any of the more recent *Broadcasting & Cable* publications for the current perspectives.

    a. Divide a notebook page in half vertically. In the left column, list the issues and write a sentence describing the concerns of those pioneers.

    b. In the right column, list those same issues and write a sentence describing the current concerns relative to them.

    c. What additional current issues can you identify? Be specific.

WEB WATCH

Here is a list of a few URLs (Internet addresses) for some of the organizations or corporations discussed in this chapter. Please explore these Web sites and follow the links to learn more about the complex business of the electronic media. Add your descriptions and your own favorite sites at the end of the list. Please keep in mind that the dynamic nature of the Internet allows sites to come and go but also allows organizations to update information about themselves very quickly.

Address **Description**

http://www.law.umkc.edu/faculty/projects/ftrials/scopes/scopes.htm _____

http://www.wwlia.org/us-home.htm _____

http://www.supct.law.cornell.edu/supct/justices/fullcourt.html _____

http://www.oyez.nwu.edu/ _____

http://www.rtnda.org/rtnda/ _____

http://www.fcc.gov/telecom.html _____

http://www.hoover.nara.gov/ _____

Other favorite sites: _____

CHAPTER 5

Operational Regulations for the Electronic Media

Law affects the daily management and operations of the electronic communication industry.

roadcasting's legal roots, like its historical foundations, lie within the 1920s and 1930s—in the 1927 Radio Act and the Communications Act of 1934. Both were passed during times when the industry was demanding regulation to bring order to the system. The Telecommunications Act of 1996 follows the same pattern.

The interpretation of the act has, over the years, provided the regulatory and deregulatory boundaries within which electronic media operate. It forms the body of law that affects all aspects of management and operational practices: what chatter the radio disc jockey can get away with on the air, how the reporter relates a news story, how the engineer designs the system, or how the manager directs the operation. This law, and now the Telecommunications Act of 1996, are integral factors in defining the limits and setting the direction for electronic media operation at all levels.

The purpose of this chapter is to discuss operational regulations that can affect daily practice and management decisions. We look at the source of the law and at the Federal Communications Commission (FCC) and how it operates, related statutory law, aspects of the criminal code as they pertain to the electronic media, and FCC rules and regulations. This discussion is illustrated with classic cases of FCC directorial efforts as well as the precedents and current status of the laws under discussion.

To understand electronic media law, one must first understand the sources of the law. There are three basic origins of media law in the United States: constitutional, statutory, and administrative laws. *Constitutional law* is the law that formulates the government. It provides the patterns and direction and is the final interpretation of all other law. *Statutory laws* are those passed by a legislative body. In the case of electronic media, the current law is the Telecommunications Act of 1996. *Administrative laws* are those created by the administrative order of a government agency—such as the FCC. These three sources provide the basic framework not only of media law, but virtually all law.

Discovery law, or *precedent law* as it is often called, is not a source of law, as such, but a critical component in the interpretation of law. A precedent case is one that provides an interpretation of law within a specific set of circumstances at a given time and within a specific environment. The value of this case law comes when a judge views a similar situation. He or she may judge in accordance with the precedent. The judge can accept the precedent and rule the same, or the precedent can be modified based on a different set of similar circumstances, or the decision can be overturned, establishing a new precedent.

Constitutional, statutory, administrative, and precedent case laws provide the foundation for the direction and interpretation of all law within the United States; as such, media operations are forced to act within that law. The alternatives are to lobby for constitutional amendments (difficult to achieve), lobby for change in statutory and administrative law (one of the primary purposes of the National Association of Broadcasters [NAB]), or seek change through the court challenge of precedent case law.

The Telecommunications Act of 1996 provides the statutory framework of electronic media law. The act continues the provisions of the FCC, providing it with the authority and responsibilities of spectrum management, station licensing, and a great deal of additional activity linked to this governance.

The Organization of the FCC

Working under the commission are numerous divisions that facilitate the functions of the FCC (see figure 5.1).

- The **Office of the Managing Director** is the administrative office handling the budget and personnel.

- The **Office of Plans and Policy** recommends budgets, coordinates research, manages research accounting, and is generally responsible for the long-range decisions.

- The **Office of Public Affairs** handles informational requests about the FCC. This office includes the News Media Division and a Consumer Assistance and Small Business Division.

- The **Office of Engineering and Technology** is the technical office within the FCC. This office houses a staff of engineers to deal with matters of spectrum management and the demands created by constantly changing technology.

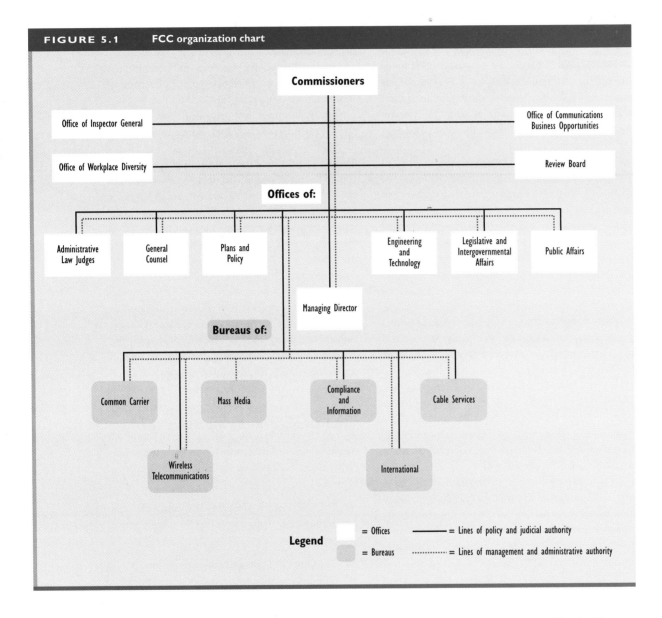

FIGURE 5.1 FCC organization chart

- The **Office of the General Counsel** serves as the FCC's legal representative in court when necessary and works on the development of legislation. The counsel's legal staff work in the preparation of administrative law, legislation, and litigation for the FCC.

- The **Office of Legislative and Intergovernmental Affairs** works closely with the legislature in the development, interpretation, and direction of new law. The office provides both directional and resource material to legislators and others who may be involved.

- The **Office of Administrative Law Judges** and the **Review Board** provide the appeals process within the FCC organization. A station that has been cited and desires a hearing before the FCC would first appeal through the Office of the Administrative Law Judges. The Review Board is the second step within the FCC hearing process—the Review Board is positioned between the administrative law judge and the commission itself.

A station contending for its license—assuming an administrative law judge's decision has ruled against the station—can appeal to the Review Board and then to the five commissioners, if necessary. The commissioners are the last appellant body within the FCC. Should the complainant desire to appeal an FCC decision, the challenge would go from the FCC appellate structure to the federal Court of Appeals. It is necessary to exhaust all administrative remedies, however, before appealing to the federal court. Most stations work hard to keep their legal matters in order, because this is an expensive process. Even a simple phone call to a station's lawyers and the staff hours required to investigate a specious complaint can be costly.

The bureaus of the FCC are the organizational arms performing the day-to-day operations. The decisions made by the commission and the respective offices are carried out by the bureaus.

- The **Mass Media Bureau**, formerly the Broadcast Bureau, handles the broad spectrum of electronic media management. This includes traditional broadcast, cable, and new technology. Matters concerning station licensing, policy, and enforcement all begin here.

- The **Compliance and Information Bureau** is the primary enforcement area of the FCC. This bureau maintains offices across the country that support large monitoring vans, which move from community to community, monitoring stations within the market. This bureau has the authority to monitor compliance with technical standards, enforce those standards, license engineers, and provide engineering support when requested for equipment specializations.

- The **Common Carrier Bureau** handles matters pertaining to telephone, telegraphy, safety, and mobile services, such as aviation, marine, and the growing industrial and business communications technology.

- The **Cable Services Bureau** deals with cable television as it relates to consumer protection, technology, and related services.

- The **Wireless Telecommunications Bureau** deals with commercial business wireless.

- The **International Bureau** focuses on international planning and negotiation.

Rule Making and Enforcement Process

The FCC's task is still to ensure that the electronic media operate within the "public interest, convenience, and necessity," so it is important to understand the process of rule making and enforcement, especially because the interpretation of the administrative law generated by the commission constitutes the definition of the public interest.

The generation of the new rules and regulations may come from several sources: the FCC, Congress, the president, the industry, or even the general public (see figure 5.2). An issue that suggests new rule making is always preceded by a good deal of public discussion—both formal and informal. This is followed when the FCC begins an action it calls a "petition for rule making." Such a petition, if it receives enough support, is followed by a published notice of proposed rule making. In effect, this announcement is a request for public comment on the subject under consideration. A wide response is sought from both the media and the consumer. Assuming the matter is considered worthy of rule making, the FCC will issue a report and order, thus formally taking action on the matter and creating the rule for operation.

The process, simplified here, is complex and includes at every point feedback from the industry and the consumer. Not all issues, discussion, and cries for new law result in a report and order. Some are handled by self-regulation; others are simply set aside, perhaps to be reconsidered. In addition to the formal process of considering a proposed new law, the FCC may show its concern merely by issuing a memorandum, a tactic aimed at focusing the attention of the industry on a particular issue and seeking resolution without creating a new law. The best historical example is the Blue Book (see chapter 4). In 1946 the issues before the FCC were promise versus performance in license renewal, tests of program service, and overcommercialization. Because the FCC did not want to create a new program law,

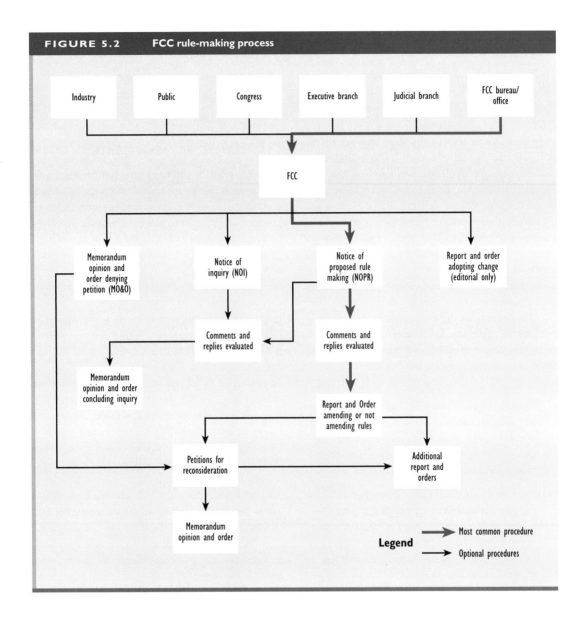

FIGURE 5.2 FCC rule-making process

it merely issued a memorandum—the Blue Book, which guided broadcast programming as the FCC desired. More recently, the FCC issued memoranda providing guidelines for children's programming and family-viewing times. These were issues of concern that never made it into the formality of law, yet the industry was directed by FCC memoranda.

The FCC is also well known for its "jawboning"—influencing the media simply through its public comments and personal contact from its attorneys or members of its field bureaus. Commissioner Newton Minow's public statement declaring that television was "a vast wasteland," a criticism made many years ago, still angers industry leaders. Similarly, a visit to a station or a corporate office by a legal representative of the commission is an effective means of directing a station. Such visits likely include thorough inspections of the station's technical and legal documents, prior to the station owner's realization that the visit resulted from a programming complaint received from a consumer. This implied threat of rule making is perhaps the most powerful tool among the FCC's enforcement powers—the raised-eyebrow technique.

The ultimate enforcement power of the FCC is the denial of a station's license to operate. The FCC can refuse to grant a license and may refuse to renew one previously granted. The fear of having a station's license revoked is generally sufficient to force compliance, but in reality this seldom occurs. The other tools of enforcement include the following:

- **Letters** from the commission are usually applied to the less significant violations, such as when a member of the audience complains to the FCC about a specific station practice. The viewer's letter will be forwarded to the station by the FCC and will require a response from the station to the FCC. This initiates a costly investigation in which the station must internally assess the complaint, then forward its response to its own lawyers, who will draft the formal response to the FCC. Such complaints are generally disposed of rather easily, but there are significant legal and personnel costs involved.

- **Cease-and-desist orders** prohibit a station from an activity under investigation. Such orders are rare, however, as other methods of compliance are generally more effective.

- **Forfeitures** are fines levied by the commission for technical violations. The amount of the fine is determined by the nature of the violation and the market and ranges from a few hundred to a thousand dollars. The more serious violations include falsification of reports, fraudulent billing, and running or advertising a non-state-sanctioned lottery. Fines for these infractions can reach $20,000 per occurrence.

- **Short-term and conditional renewals** are considered for the more serious infractions and may place a station on notice of probation. Short-term renewals provide a station that has a historically favorable record with time to correct the infraction. These sanctions are reserved for equal employment, fraudulent billing, and contest rule infractions.

From an operational standpoint, it is the body of administrative law that provides the most direct governance, but there is significant media-related and applied law within the standards of the Constitution, statutes in addition to the Telecommunications Act of 1996 and precedent law. These areas include: the First and Sixth Amendments, libel and slander laws, the Privacy Act of 1974, the Freedom of Information Act of 1966, the Copyright Act of 1976, and several criminal codes.

STATUTORY AND RELATED APPLIED LAW

The First Amendment guarantees our freedom of speech. The Sixth Amendment ensures a speedy trial and a free press. *Libel* (written defamation) and *slander* (oral defamation) laws protect the reputation of an

THE FIRST AMENDMENT

Congress shall make no law respecting an establishment of religion, or prohibiting the free exercise thereof; or abridging the freedom of speech, or of the press; or the right of the people peaceably to assemble, and to petition the Government for a redress of grievances.

THE SIXTH AMENDMENT

In all criminal prosecutions, the accused shall enjoy the right to a speedy and public trial, by an impartial jury of the State and district wherein the crime shall have been committed . . .

individual and require accountability of the press. The *privacy laws* are the other side of libel ensuring the rights of the individual versus the right of the press. The Freedom of Information Act was created to provide the press open access to public meetings and government records. The Copyright Act was created to protect the rights of the creator and it was updated in 1976 to include electronic media creations. Three sections of the U.S. Criminal Code reflect directly on electronic media. These codes prohibit broadcast lottery, fraudulent use of the airwaves, and obscenity.

The dynamics of the ever changing electronic media are dwarfed only by the dynamic of an ever changing law. No other industry is regulated in quite the same way as the electronic media. Indeed the media themselves are regulated differently. The print media are directed by the First and Sixth Amendments, libel and slander, privacy, and information and copyright laws, but these do not parallel the Telecommunications Act. These laws were founded in print media yet are applicable to the electronic media as well; they are identified here as their subject relates to operations. They are detailed in the case histories, legal writings, and law courses. The Telecommunications Act, however, is the foundation for all electronic media communications.

FCC Rules and Regulations

The administrative law created by the FCC is the largest body of law affecting the electronic media. Sections 326 and 315, the 1960 programming policy statement, the Blue Book, the fairness doctrine, and others already discussed deal with restricting programming and defining what is permissible. Frank Kahn (1984) described the station's situation as "that of an acrobat trying to balance himself on a slack rope suspended between public interest at one end and administrative law at the other." The FCC does make rulings that are often taken to court, where each decision is tested and the limits are drawn by precedent.

The industry has lobbied for some time to overcome what it considered burdensome regulations, and in recent years the FCC has attempted to deregulate the industry. In spite of these efforts, however, many of the regulations, policies, and laws dealing primarily with program content are still intact. The FCC has actually done little to promote substantial change in program content regulation, and it is these laws that direct some of our basic operations. They include laws based on the issues of politics (section 315 and fairness), fraud (news distortion, contests, and advertising), obscenity, access to the air (ownership and equal employment opportunity), children's programming, and new technology.

Sections 315 and 312-7a: Questions and Obligations

This simple-sounding law, the historical foundations of which were discussed earlier, provides equal opportunity for all political candidates. Probably no section of the Communications Act of 1934 receives more attention, because of the availability of the media within the political election process, and it remains unchanged in the new law. Interpreting the law is complex. What is permissible for the broadcaster is key. Refer to the box on the facing page for the text of the law that is discussed in the following paragraphs:

"If any licensee shall permit . . ."

What are the broadcaster's obligations to the candidates? First, a licensee is *not* obligated to provide time to a candidate and every political issue. A station can choose not to become involved. The only exception to this interpretation is section 312-7a, which states that a license may be revoked if a station repeatedly refuses to provide reasonable access to candidates for federal office. In other words, time may be denied a mayoral candidate, but time may not be denied a candidate running for the U.S. Senate. The station is under "no obligation" to provide time to any, except the federal, candidates.

"legally qualified candidate . . ."

Who is a legally qualified candidate? The answer to this question invokes local and state laws. The key question is: Can you (the audience) vote for the candidate? It is the local and state requirements that generally define who is a legal candidate. Basically, if you can vote for the individual on election day, he or she is generally considered a legally qualified candidate. It is important to know the specific state and local laws that pertain to this definition, however.

"use . . ."

What constitutes use? If you allow the legal candidate to "use" the station, you are obligated. *Use* is one of the more interesting terms. If the station's weather-

Political Broadcast Law

SECTION 315

(a) **If any licensee shall permit** any person who is a **legally qualified candidate** for any public office to **use** a broadcasting station, he shall **afford equal opportunities** to all other such candidates for that office in the use of such broadcasting station: *Provided,* that such **licensee shall have no power of censorship** over the material broadcast under the provisions of this section. No obligation is imposed under this subsection upon any licensee to allow the use of its station by any such candidate. Appearance by a legally qualified candidate on any

(1) bona fide newscast,

(2) bona fide news interview,

(3) bona fide news documentary (if the appearance of the candidate is incidental to the presentation of the subject or subjects covered by the news documentary), or

(4) on-the-spot coverage of bona fide news events (including but not limited to political conventions and activities incidental thereto), shall not be deemed to be a use of broadcasting station within the meaning of this subsection. Nothing in the foregoing sentence shall be construed as relieving broadcasters, in connection with the presentation of newscasts, news interviews, news documentaries, and on-the-spot coverage of news events, from the obligation imposed upon them under this chapter to operate in the public interest and to afford reasonable opportunity for the discussion of conflicting views on issues of public importance.

(b) The **charges** made for the use of any broadcasting station by any person who is a legally qualified candidate for any public office in connection with his campaign for nomination for election, or election to such office shall not exceed

(1) during the forty-five days preceding the date of a primary or primary runoff election and during the sixty days preceding the date of a general or special election in which such person is a candidate, the lowest unit charge of the station for the same class and amount of time for the same period; and

(2) at any other time, the charges made for comparable use of such stations by other users thereof,

(c) For purpose of this section—

(1) the term "broadcasting station" includes a community antenna television system; and

(2) the terms "licensee" and "station license" when used with respect to a community antenna television system mean the operator of such system.

SECTION 312-7A

The commission may revoke any station license . . . for willful or repeated failure to allow reasonable access to . . . a broadcasting station by a legally qualified candidate for federal elective office on behalf of his candidacy.

Source: Sections 315 and 312-7a are both part of the Telecommunications Act of 1996. Their text is also part of *FCC Rules and Regulations,* Section 73:1940.

person decides to run for public office, does that person's weather forecast constitute use? The answer is yes. If the person is identifiable, even if he or she is talking about the weather, the appearance constitutes use. There is no obligation upon the station to seek out candidates, but once use is permitted, section 315 applies.

Daily situations that are exempt from the use provisions are easily identifiable in the language of section 315:

1. Bona fide newscast
2. Bona fide news interview
3. Bona fide news documentary
4. On-the-spot coverage of bona fide news events

Examples of these situations include the coverage of presidential debates, because they are considered bona fide news events. The incidental appearance of a candidate in a documentary also is considered exempt. News conferences and news events have become one of the main stages of political campaigning, because they provide excellent platforms for the candidates and free the station from the obligation of section 315. Too much coverage of one candidate, however, is going to bring the station under investigation for slanting the news.

"afford equal opportunities . . ."

What constitutes equal opportunity? Notice that although the law says *equal opportunity,* and not *equal time,* it has come to be understood and interpreted as equal time. It also means equal facilities and comparable costs. If a station charges Candidate Melvin $1,000, all other legally qualified candidates should get equal time and comparable facilities for the same dollar amount. If Melvin buys morning time, his or her opponents are going to get morning time and facilities at that price.

"licensee shall have no power of censorship . . ."

Who is responsible for content? The candidates can talk about the weather, they can slander one another, or they can talk about the issues. This is the only place in the law where the broadcaster is not responsible for program content. The candidates may use the time as they wish. This has brought up some interesting challenges by "candidates" testing the law, using it to push a personal cause: Can a pro-life or pro-choice candidate run such spots within the campaign even if there is no hope of getting elected? Can a pornographic advertisement be aired under section 315 to push the cause of free speech? To date, the FCC has issued a policy statement saying no, and there have been only minor abuses of the law, but the questions pose interesting dichotomies. Perhaps this is where good judgment and reason should replace the letter of the law—or should it?

"charges . . ."

The fees charged cannot exceed those charged for comparable use of the station. Election year should not provide a financial windfall for stations. The cost charged to political candidates is the lowest unit rate charged for the comparable time period. The law defines the time period as forty-five days prior to a primary election and sixty days prior to a general election. If a station has conducted an advertising campaign lowering rates to attract business, that low dollar figure would also be the one charged to the candidate. Section 315, First Amendment issues, and politics are often at odds with one another. Although the NAB continues to question the constitutionality of section 315, the law is not likely to change, because political candidates who become legislators do not appear inclined to deregulate. The issue will have to be fought through the courts. Broadcasters feel that section 315 inhibits free-flowing discussion. Legislators, on the other hand, have historically been protective about their access to the media and, lately, their interests have broadened. Section 315 and the election issues are complex, and it is just as likely that before section 315 is deregulated, more regulation will be added to respond to current issues. Meanwhile, the broadcaster must continually study the law to understand what is permissible.

THE FAIRNESS DOCTRINE

As discussed earlier, the fairness doctrine was declared unconstitutional in 1987, so why is it a continuing object of discussion? Because it is still on the books (see FCC 73:1910, 73:1920, and 73:1930) and will remain so until all challenges have made their way through the court system. There are also individuals in the legislature who are refusing to let the doctrine die. In 1989 Congressman John Dingell and Senator Earnest F. Hollings both introduced bills designed to codify the doctrine.

The fairness doctrine is still an important consideration when dealing with political broadcasts. Although the general doctrine dealing with controversial issues was eliminated, this did not free stations from programming dealing with elections or personal attack. The fairness doctrine dictates that a licensee provide balance in the discussion of public issues to ensure that all sides of a question are given presentation. This is often referred to as the "Zapple doctrine," after Nicholas Zapple, a former staff member of the Senate Subcommittee on Communications. Established by the FCC in 1970, it states that if a station "sells time to supporters . . . of a candidate . . . the licensee must afford comparable time to the spokes-[person] for an opponent." Similarly, if a station airs a program on which a person's "honesty, character, integrity or like personal qualities," are attacked, it has the obligation of determining whether contrasting viewpoints were represented. If they were not, the station is obligated to seek out opposing views and provide "a reasonable opportunity... [and] reasonable time... no later than one week after the attack" for a differing point of view. Under the political editorials provision, if the station endorses or opposes a candidate, it has twenty-four hours to enable the opposition to access the station's facilities and provide "a reasonable opportunity" to respond.

On the surface it seems clear-cut—who could oppose a philosophy of balance and reasonable opportunity? Yet the stations and their lobbies—the National Association of Broadcasters and the Radio and Television News Directors Association (RTNDA), have fought diligently over the years to repeal the doctrine. The rationale is that the doctrine "diminishes the discussion of controversial issues and reduces diversity"—the very things it was originally meant to inspire. The legal obligations, according to the stations, have quelled public discussion; many stations have simply decided not to program such materials rather than risk a legal infringement.

The case that brought about the fall of the fairness doctrine involved the Meredith Corporation's station WTVH Syracuse in New York. In 1984 the FCC ruled that WTVH had violated the doctrine in a 1982 series of spots that had supported the construction of a nuclear power plant. The commission found that WTVH had "failed to afford a reasonable opportunity to the presentation of viewpoints contrasting to those presented." Meredith asked the commission for reconsideration based on the grounds that the fairness doctrine was unconstitutional. As the case wound its way through the courts, the NAB and the RTNDA joined in the arguments. Finally, the doctrine was declared unconstitutional by the FCC. The appeals court upheld the finding that the doctrine was contrary to public interest. An appeal to the Supreme Court is pending, as are congressional efforts to redraft and codify the law.

INDECENT, OBSCENE, AND PROFANE MATERIAL

Obscenity is one of the most complex legal issues in electronic media. Although prohibited by federal law, it has been a difficult term to define. In prosecuting such cases, the FCC has preferred to focus on the term *indecent* and has focused its decisions based on spectrum management and children in the audience.

The issue of broadcast obscenity was a concern as early as the 1930s, when Senator Clarence C. Dill threatened to have a station's license revoked for having used the profane language—"damn"—on the air. In 1937, NBC Radio aired a *Charlie McCarthy* program in which he and Mae West were doing an interpretative sketch of Samuel Clemens's *Adam and Eve*. The actual lines of the script included nothing objectionable, but Mae West's voice inflections made the lines sound suggestive. As a result, both the network and the commission received hundreds of letters from listeners expressing the opinion that the sketch was vulgar and indecent. This prompted the FCC to write a reprimand to NBC and each of the stations carrying the program. The commission took no further action, but it didn't have to at this time in history.

The broadcast parameters of indecency regulation have evolved over the years. The reprimands of Mae West and NBC would certainly not have been fact under today's standards. Even the "bleeping" of curse words on Johnny Carson's old *Tonight Show* seemed almost silly.

One of the FCC's first obscenity cases involved Mae West's voice.

The laws regarding indecency and obscenity are found in several places. The U.S. Criminal Code forbids obscenity, the FCC rules and regulations provide a fine, and the Telecommunications Act of 1996 provides an increased fine in addition to attempts and direct control. The interpretations of these laws, however, are found in case law.

Perhaps the most important of the case law is found in the *Miller* test. Marvin Miller was prosecuted for advertising and mailing erotic books and films. The 1973 *Miller v. California* case, in which Miller was found guilty, provides the definitional criteria for obscenity. The court declared the following parameters, known as the *Miller* standards.

1. An average person, applying contemporary local community standards, finds that the work, taken as a whole, appeals to prurient interest.

2. The work depicts in a patently offensive way sexual conduct specifically defined by applicable state law.

3. The work in question lacks serious literary, artistic, political, or scientific value.

In electronic media, the FCC broadened the *Miller* standards to include situations unique to the mediums where:

1. Unsupervised children have access to the medium.

2. The receivers are at home and people's privacy is involved.

3. The material can come into the home without warning.

4. The government has licensing power.

These two sets of conditions are obviously broad. Anyone who plans to work in electronic media, however, must know what is permissible and what may carry legal consequences. You find what legal philosophers call the "safe harbor" of operations within case history. The two cases that currently provide those boundaries are George Carlin/Howard Stern and Pacifica.

In Carlin's case, radio station WBAI was airing a Carlin monologue—it became known as the "Seven Dirty Words." The program aired in the afternoon. After complaints reached the FCC and the situation wound its way through the litigation process, the FCC ruled that WBAI was in violation of its standards, because children were in the audience, receivers were at home, and the material could be received with no clear warning. The ruling stands. More recently, "shock jock" Howard Stern was fined $600,000 by the FCC. Infinity Broadcasting appealed the fine, but finally agreed to settle the case—and paid the largest settlement ever—$1.715 million. Operational practices on this side of the harbor are unsafe.

The 1964 Pacifica case illustrates the other side of the safe harbor. Pacifica was challenged in its renewal process because of a complaint that one of its stations had broadcast obscene material. The programming under question was Edward Albee's play *The Zoo Story*. The FCC ruled in favor of Pacifica and renewed its licenses, indicating that the public had been served as the play was a serious dramatic work. Operational practices on this side of the harbor are safe.

News Distortion

The Federal Communications Commission takes a dim view of the falsification of just about everything, be it misrepresentations to the commission or the deliberate distortion of news programming. The biased reporting of news is clearly against the public interest. Walking that tightrope between the First Amendment, section 326, and programming regulatory statements, the commission's desire is to indicate a direction for the station's news operations rather than regulate them.

As tabloid journalism expands within the electronic medium, there will likely be more-active discussion related to the staging or restaging of news events. This practice is readily identifiable within the traditional newscasts where highlighted events, such as a "silent-witness" program, are preceded by a dramatization. The "silent-witness" program inserts are usually video news release segments, produced by local police departments, in which the crime is reenacted in hopes of gaining airtime on local stations and viewer participation in solving the crime. The identifications of many such dramatizations, however, are being hidden within the small on-screen print at the beginning or end of the tabloid broadcasts. Although the FCC would not fine the station for such practices, a charge of slanting the news would be taken under consideration at license renewal time.

Hypoing

The definition of *hypoing*, generally found in the back of ratings books, is "any unusual promotional or programming practices designed to distort survey results." The FCC, generally speaking, has a hands-off policy when dealing with hypoing and refers such matters to the Federal Trade Commission (FTC). The FCC speaks of the practice only in terms of the questions raised regarding a station's qualifications to retain a license if the practice of hypoing is discovered.

In actuality, the practice of hypoing is quite well regulated by the industry itself. Infractions or unusual program practices are also published in the front of each ratings book under the heading *Special Notices*. The most common notes are simply explanations of situations beyond a station's control, for example, being off the air for a portion of the rating period. Any unusual distortions in the rating will appear.

False or Deceptive Advertising

The practices of false or deceptive advertising are also under the jurisdiction of the Federal Trade Commission. The FCC considers the station's qualifications to retain its licenses when such practices are clearly identified. The FTC considers all specifics pertaining to the claims of "false, misleading, or deceptive advertising." The station, after deregulation, no longer has to determine the truth in advertising, but it is still responsible for programming that includes advertising content. If there are claims of deception, they are brought to the FTC through the advertising agency, the product representatives, or federal and local trade agencies.

In 1995 the FTC levied a $200,000 civil penalty against a Phoenix health products firm for falsely stating the benefits of its bee-pollen products. According to the FTC, the product advertisements airing in television "infomercials" claimed that the bee pollen sold would not result in allergies, but would, in fact, alleviate pollen allergy symptoms, slow the aging process, cure or prevent impotence, and cause weight loss. The FTC found the claims untrue.

The FTC and the FCC both take a dim view of false and misleading advertising.

Multiple Ownership

In the mid-1950s the FCC became concerned with the issue of corporate monopoly and what Commissioner Nicholas Johnson later called the "media barons." The general purpose of the ownership limitations at the time was to promote diversity in ownership and different points of view. The rule specified the limits of seven AM radio, seven FM radio, and seven television stations for each owner. The industry criticized the ruling as arbitrary, but the rule stuck for thirty years. By the mid-1980s, the commission decided that there had been dramatic change in the marketplace, and the limits were raised to twelve stations in each category, then to eighteen AM, eighteen FM, and twelve TV stations. Today, under the Telecommunications Act of 1996, the limits have been loosened even further, creating a new surge in buying and selling station properties. In radio there are now no national caps; there are, however, local restrictions. For television national ownership is capped at 35 percent.

Equal Employment Opportunity Regulations

The FCC became interested in equal employment practices in the late 1960s and began requiring that stations file employment records and develop equal employment opportunity (EEO) programs. The FCC reviewed every station's EEO program at the time of license renewal.

Stations have in the past been placed on short-term renewal for failing to comply with EEO guidelines. In 1984 station KDEN-AM in Tulsa was placed

FCC Rules on Contests

A licensee that broadcasts . . . information about a contest it conducts shall fully and accurately disclose the material terms of the contest, and shall conduct the contest substantially as announced or advertised. No contest description shall be false, misleading or deceptive.

Source: From FCC 73:1216.

on short-term renewal for such a failure. EEO reporting conditions were also imposed on the University of Tulsa's FM station that year, when it was found that no one hired within that year was a minority.

The law was dropped in 1998 after the U.S. Court of Appeals ruled that it "did not have a compelling interest in the stated EEO goal of diversity in programming." The ruling is being applauded by some, who see it as a relief from regulatory paperwork. The laws are being reworked, however, and broadcasters are well advised to continue to abide by the commission's EEO standards.

CHILDREN'S TELEVISION ADVERTISING

Advertising targeted at children has become an increasing concern. A group called Action for Children's Television (ACT) has been at the forefront of this issue. Although the FCC has illustrated its concern for children within the audience in its rulings regarding obscenity, the issue of children's television goes beyond obscene or indecent programming. ACT has pressed the FCC to place limits on children's advertising and adopt minimum standards for children's programs. In 1977 a suit was brought against General Foods for commercials thought to be "fraudulent, misleading and deceptive." Its television spots implied that kids who ate the candied cereals were "stronger, more energetic, happier, more invulnerable and braver" (*Committee on Children's TV v. General Foods*, 1977 Cal. Rptr. 783). In 1983 ACT filed suit against five television stations for broadcasting what it considered "program-length commercials." Programs, ACT claimed, that were targeted at selling products were being broadcast under the "guise of entertainment." It is worthy of note that it took six years for this case to work its way through the system. Although the FCC eventually ruled against ACT and the National Association for Better Broadcasting, the issue was before the public and did bring about operational change in the manner in which toys and programs were presented (*National Association for Better Broadcasting v. FCC*, 830 F. 2d 270 1989).

The FCC has resisted codification of children's television guidelines, and the NAB feels that there is no need for such regulations, but citizens' groups continue to pressure the FCC through the legal channels for improved practices. The FCC's concern remains in three areas of children's programming: the separation of program content from commercial content, hosts selling products, and the excessive promotion of products via program content.

In response to the FCC concerns, stations use "bumpers"—momentary interruptions—to separate commercials from program content in children's programming. The FCC policy prohibits a host from selling products when he or she could take unfair advantage of the "trust [that] children place" in a television personality. And program content appears to be improving in its efforts to separate product promotion from program content.

PROFESSIONAL ADVERTISING

Lawyers have become an important source of revenue for local stations. Between 1988 and 1989, television advertising increased 22 percent—a rate of almost three times other commodities. During the first six months of 1989, $39.8 million was spent by the legal profession on television advertising. It has been on the increase since 1977, as has been the controversy surrounding such advertising.

For years there was a ban on advertising by lawyers, doctors, and members of other professions. In 1977, however, in *Bates v. State Bar of Arizona* (433 U.S. 350 1977), that ban was essentially lifted when the Supreme Court decided that the prohibition was unconstitutional. Following *Bates,* the bar association and some state regulatory agencies have imposed their own regulations, but continual challenges have left only one prohibition in place: that against false or misleading advertising.

INTERNATIONAL ADVERTISING

This issue is less an FCC concern, rather it is a Canadian law that affects stations along the Canada-U.S. border. For years small U.S. stations in Washington and New York states have targeted the larger Canadian markets of Vancouver, Toronto, and Montreal—a $20 million annual business, according to the NAB. The Canadian attempt to protect its own media created legislation known as Bill C-58, which denied Canadian businesses a tax deduction for expenses incurred when placing advertising on a foreign station. In response, the United States adopted a mirror action by amending section 162 of the Internal Revenue Code to deny a tax deduction for U.S. businesses' expenses incurred by advertising on a foreign station.

It is unlikely that the Canadian government will change Bill C-58, but the law does remain a bone of contention, particularly to the NAB. The lobbies on both sides of the border are currently awaiting the results of the Canadian-U.S. trade agreements (NAFTA—the North American Free Trade Agreement) to see if these can effect any changes. Meanwhile, border stations broadcasting into Canadian markets have used the pragmatic approach—they have lowered their rates to offset tax deductions previously allowed. Broadcasting, after all, is a business.

Public Files

The public files of broadcast stations are widely misunderstood. Every broadcaster is required to maintain records for public inspection. These records contain a wealth of information and are open to anyone during normal business hours, but rarely does anyone ever look at them. The only exception is perhaps students who visit as part of a class assignment.

The industry had lobbied to rid itself of this burden, but the public files are still with us. Every station is currently required to maintain the following in its public file:

- A copy of the station's license application, including letters, documents, and exhibits as well as the application

- A copy of the ownership report

- Records that pertain to political candidates regarding purchases and requests for time

- A copy of the annual employment report

- A procedures manual (see FCC 74-942, 39 32288)

- Letters received from members of the public

- A list of programs related to the treatment of community issues during the preceding three-month period

- A copy of the notice used to inform the public that the station is filing for its license

Station Identification

All broadcast stations are required to identify themselves hourly, but it is a common practice to present the identification with the station promotional materials. The content of such identification (made either visually or aurally) includes a station's call letters and the community within which it broadcasts.

Sponsored Programs

A broadcast station is required by law (FCC 73:1212) to identify sponsored programs for which consideration is received. For practical purposes this means "the following program is brought to you by . . ." In the case of political announcements, the fine print on the end of the television commercial designates the sponsorship identification, for instance, "sponsored by the Committee to support Candidate Smith." Product advertisements require the corporate or trade name or the name of the sponsor's product. This sponsorship identification requirement basically prevents fraud by requiring those paying for the commercial time to identify themselves. You see the results of this regulation primarily in relation to political broadcasts, infomercials, and other program-length commercials.

Recordings

Telephone recordings, such as a reporter might make while gathering news for a story, are subject to law. The reporter or whoever is making the recording must let the person being recorded know that the recording is taking place and is intended for broadcast (FCC 73:1206).

Any material recorded wherein time is of significance must also be identified as a recording (FCC 73:1208). *Time* is the critical term. The FCC's concern is that the material not be misunderstood as part of a simultaneously broadcast event. In practical terms, this law is observed at the end of programs with the statement "portions of the preceding program have been recorded."

Payola and Plugola

Payola and *plugola* refer to the forbidden practice of accepting money, or anything of value under the table or as a bribe, in return for the promotion of a product or recording. Section 507 of the Communications Act of 1996 requires that such practices be reported to management prior to broadcast. Section 317 requires management to disclose payment for services rendered.

Payola was a practice during the 1950s, when on-air personalities accepted *considerations* for the promotion of specific music. Today it refers not so much to the number of plays a hit record or CD will receive, or the mentions on the air by the disc jockey, but any practice or activity seeking to influence presentation not disclosed as required under section 317. Here, again, the FCC can issue sanctions.

Fraudulent Billing

Fraudulent billing is not an over-the-air program content violation, but primarily one of sales/management accounting. Although the FCC eliminated its billing rules under deregulation, the matter is still of serious concern. Such an infraction of state or federal law can result in court arbitration and reflect negatively on the station's character—all of which affect the FCC's deliberations at license renewal time.

Cooperative advertising programs consist of an agreement between product manufacturers and retailers to share the costs of local ads, which then promote the manufacturer's product and the local retailer's store. Double-billing occurs when false ad schedules or ad invoices are used to demand payment for ads that never ran. Anyone who works in broadcasting should be wary of advertisers who ask for duplicate copies of a schedule invoice or who want falsified commercial-run schedules. These are signs of double billing.

Advertising Tobacco Products

Cigarette advertising was banned from the air in 1971. The Federal Cigarette Labeling and Advertising Act was specifically designed to prohibit broadcast advertising of tobacco and required product manufactures to carry the label *Caution: cigarette smoking may be hazardous to your health*. The law has been extended to include the advertising of small cigars, chewing tobacco, and snuff. Although it is still legal to advertise regular cigars and pipe tobacco, the practice is rarely seen on the air because the difference between products is confusing, and ads may open the station to litigation.

Cable and New Technology Regulation

The cable and new-technology regulations are perhaps the most constantly changing laws of the 1990s and are dramatically changed in the Telecommunications Act of 1996 (see chapter 4). Technological innovations demand deregulation, reregulation, and regulation. The Telecommunications Act expands cable's opportunities and allows it to enter into new business ventures, such as online and telephone services. The cable systems remain local franchises, however, and are still governed by the Cable Communications Policy Act of 1984 and the Cable Television Consumer Protection Act of 1992 as well as by FCC regulations.

Copyright Issues

A cable system uses (retransmits) three types of copyrighted signals: a local station's; distant signals, such as superstations WTBS, WGN, or WPIX; and nonbroadcast signals, such as cable networks—HBO, ESPN, and CNN. The Copyright Act of 1976 directs cable's royalty payments to the stations (both local and distant) who own these materials. The payments are to be made based on the system's gross receipts received from basic service subscribers. The *Copyright Royalty Tribunal (CRT)* is the body created for the purpose of collecting and distributing these royalties.

There have been a host of claimants for these fees, including the Motion Picture Association of America; professional sports leagues; the American Society of Composers, Authors and Publishers (ASCAP); Broadcast Music, Inc. (BMI); The Public Broadcasting Service (PBS); syndicated producers; and even Canadian broadcasters. Determining who gets what portion of the fees collected is a continuing source of debate. To date, the traditional TV stations have actually received only a fraction of the royalties collected, and the distribution of these fees lags considerably behind the broadcast dates as the CRT continues to struggle with decisions. The "superstations"—WGN, WTBS, and the like—currently receive a significant portion of the money distributed.

Broadcasters, through the NAB, remain an active lobby and continuing participants in the CRT proceedings.

"Must Carry" and Retransmission Consent

Early in cable's development, the FCC adopted regulations requiring the system to carry local stations. The rules were designed to maintain local station viability within a market: In some markets a station would lose audience members if it were not carried on the cable. These "must carry" rules were officially set aside for a second time in 1985, but the Cable Television Consumer Protection Act of 1992 has again opened the legal debate, and the Telecommunications Act of 1996 retains the "must carry" rule. The new provisions provide a "must carry" or a "retransmission consent" option to licensed full-power TV stations. The "must carry" option bears no obligation for an exchange of compensation—the station permits the cable operator to carry the signal in exchange for a perceived audience gain. The "retransmission consent" choice places the station and cable system operator in a position of negotiation; this could include financial compensation to the station and/or spot time availabilities in exchange for giving the cable system permission to carry the broadcast signal. If the station and cable operator cannot come to a negotiated agreement, the cable operator is obligated to supply consumers with an A/B switch, allowing them to choose reception of the cable or an antenna-captured over-the-air signal.

The Telecommunications Act of 1996 requires the FCC to act expediently on any complaints. The laws are new, and court challenges are under way.

New-Technology Issues

A host of new-technology issues faces the FCC: AM expansion, high-definition television, and the phone companies.

AM Band Expansion and Stereo

In the late 1990s, the FCC issued "notices of inquiry" requesting public comment on the expansion of the AM band. Such an expansion would allow daytime stations to expand times and move to the expanded band position. It would also accommodate new-station demand, including frequencies now reserved for noncommercial and minority operations. The outcome of these expansions is obviously an increase in commercial service.

The process of deregulation has hurt AM stereo. Historically, the FCC had mandated the newest technical standards, but with deregulation it backed away from the practice. Thus, as FM grew in popularity and signal quality and AM sought the alternative of adopting its own stereo equivalents, no one regulated the standards for AM stereo. Several alternatives were proposed, but the FCC refused to designate standards as it had done historically; the commission left it instead to the marketplace—its new regulatory philosophy. As a consequence, competing technologies were adopted by desperate commercial AMs who advertised themselves as AM stereo, but with no standards dictated, the marketplace has wavered and AM has continued to decline.

High-Definition Television

High-definition television (HDTV) appears to be one of the waves of the future. It provides wide-screen display and pictorial definition heretofore seen only on film. Although the FCC has declined to specify AM stereo technology, the opposite appears to be expected for HDTV. At first, U.S. interests watched the HDTV marketplace as international interests forged ahead, particularly in Japan and Canada. Within a few years, U.S. electronic interests in HDTV found themselves developmentally behind other nations. This posed the possibility that a Japanese system would be adopted in the United States, because the United States had no coordinated system. The thought panicked the industry. Today the FCC, the NAB, and major manufacturers find themselves struggling to retain a leadership position in the development of HDTV.

As exciting as it is to watch this development, there are challenges to consider. Marketable reality also mandates significant change in technical and regulatory challenges; existing service to viewers must be maintained during a transition period; the spectrum allocations are going to have to be changed dramatically to allow enough space for the HDTV signal; and the entire U.S. industry is based on the

MEDIA-RELATED FEDERAL LAWS

R & R 73:1211, BROADCAST OF LOTTERY INFORMATION

No licensee of an AM, FM or television broadcast station . . . shall broadcast any advertisement of or information concerning any lottery, gift enterprise, or similar scheme, offering prizes dependent in whole or in part upon lot or chance, or any list of prizes drawn or awarded by means of any such lottery, gift enterprise or scheme, whether said list contains any part of all of such prizes.

The determination whether a particular program comes within the provision of a lottery [is] . . . if as a condition of winning or competing for such prize, such winner or winners are required to furnish any money or thing of value or are required to have in their possession any product sold, manufactured, furnished or distributed by a sponsor of a program broadcast on the station in question.

Not applying [are] lottery conducted by a state[,] . . . fishing contest accepted under 18 USC 1305[, and] . . . gaming conducted by an Indian Tribe pursuant to the Indian Gaming Regulatory Act.

1304. BROADCASTING LOTTERY INFORMATION

Whoever broadcasts by means of any radio station for which a license is required by any law of the U.S. or whoever, operating any such station, knowingly permits the broadcasting of any advertisement or information concerning any lottery, gift enterprise, or similar scheme, offering prizes dependent in whole or in part upon lot or chance, or any list of the prizes drawn or awarded by means of any such lottery . . . shall be fined not more than $1,000 or imprisoned not more than one year or both. Each day's broadcasting shall constitute a separate offense.

1343. FRAUD BY WIRE, RADIO, OR TELEVISION

Whoever, having devised or intending to devise any scheme or artifice to defraud, or for obtaining money or property by means of false or fraudulent pretenses, representations, or promises, transmits or causes to be transmitted by means of wire, radio, or television communication in interstate or foreign commerce, any writing, signs, signals, pictures, or sounds for the purpose of executing such scheme or artifice, shall be fined. . . .

1464. BROADCASTING OBSCENE LANGUAGE

Whoever utters any obscene, indecent or profane language by means of radio communications shall be fined. . . .

---

traditional 525-line screen system. With HDTV everything will change, from the camera to the home screen and all the technology in between. All this has to be accomplished at a cost affordable to the industry and the consumer.

The Telephone Companies

No change on the horizon is quite as frightening to station and cable operation management as the telephone companies' entry into the business of video and audio delivery. The Telecommunications Act of 1996 permits this direct competition for the first time in history. When AT&T broke up into Bell Operating Companies, they were legally barred from originating "content-related services"—they were common carriers, not broadcasters or cable operators. With the Telecommunications Act, that restriction has been lifted, and even today's smallest Bell company is larger than the largest cable operator. Although Bell executives had steadfastly denied any interest in entering the field of electronic media, they were wiring the nation with fiber optics. The new law could allow not only an increase in the common-carrier versatility of your individual telephone, but the phone company's fiber optics could also easily carry a wide range of computerized information and an almost unlimited number of content-related audio and video services into the entertainment center of every home in the country. So what happens to the traditional networks, the stations, the production houses, and the U.S. system of broadcasting as we know it? It could be revolutionized. Today the question is not what can be done legally or technologically that will determine our direction, but what the consumer will pay for. The Bell companies have historically said that they were not moving into traditional broadcast television–type services, yet it is now legally permissible and technically feasible. This is a most significant challenge of the future, the marketplace, the industry, and the regulatory agents.

Summary

▶ Laws governing the management and operation of the electronic media originate from the Constitution, the statutory laws created by Congress and other legislative bodies, the Criminal Code, and administrative law. Within the electronic media, the Telecommunications Act of 1996 is the statute. This act includes the philosophical principles of the 1927 Radio Act and the Communications Act of 1934, which created the administrative body—the Federal Communications Commission (FCC). Today it is the FCC that creates the administrative laws directing electronic media. The constitutional law pertaining to the electronic media are the First and the Sixth Amendments. Statutory law of importance includes the Freedom of Information Act, the Privacy Act of 1974, and the Copyright Act of 1976. Coded law primarily relates to the criminal codes, which deal with lottery information, fraud, and obscenity.

▶ The administrative law is the largest body of law pertaining to the management and operation of electronic media. Here the FCC often finds itself in conflict among the laws themselves. For example, the First Amendment and section 326 mandate free speech and prohibit censorship, yet section 315 clearly dictates program regulation.

▶ The administrative laws governing media deal with a broad range of issues: The FCC's interpretation of section 315 requires equal time; multiple-station ownership rules limit corporate growth; equal opportunity regulations ensure minority representation; children's program directives protect minors in the audience; access to public files ensures the opportunity of community interaction; station identification and sponsored program announcements clearly identify the sender to an audience; and payola and disclosure laws control influence peddling. Additional administrative laws cover issues such as news distortion, hypoing, logging, and paperwork regulations that have been deregulated. This means that the FCC, while cutting down on paperwork for itself and the stations, can and does consider serious infractions of such law at times of license renewal.

▶ Laws governing the media include constitutional, statutory, coded, and administrative law. The Constitution provides the foundation for the legal system of the United States. Statutory law emanates from the Communications Act, the Cable Communications Policy Act of 1984, the Copyright Act, and the Freedom of Information Act.

▶ Precedent case law provides an interpretation of the law within a specific set of circumstances at a given time.

▶ The FCC's primary function is the management of the spectrum. The FCC is made up of thirteen offices and bureaus that facilitate the regulatory activities.

▶ The commission directs the industry by initiating formal law, jawboning, memoranda, sanctions, and short-term renewal. Its ultimate power is the ability to revoke a station's license.

▶ The First Amendment declares our rights to free speech, and section 326 forbids censorship, but these conflict with laws that dictate and control content such as obscenity, political broadcasts, and balanced content.

▶ Free press and fair trial are issues that contrast the rights of the press with the rights of the accused to a fair trial.

▶ Privacy laws exist to protect the individual.

▶ The copyright law gives legal rights and ownership of a work to its creator. These include the rights to copy, distribute, adopt, and perform the work.

▶ Lotteries consist of contests that include prizes, chance, and consideration. Broadcasters are forbidden by law from conducting an illegal lottery.

▶ The material terms of a contest define how the station conducts the contest.

▶ The FCC has declared its right to regulate indecent programming based on the following facts: There are children in the audience, receivers are at home, the material comes without warning, and the FCC has the ultimate licensing powers.

▶ Section 315 of the Communications Act pertains to broadcasts by candidates for a public office. It has been interpreted as the "equal time" law.

▶ The fairness doctrine was declared unconstitutional by the FCC. A Supreme Court appeal is pending as are congressional efforts to redraft and codify the law.

▶ The FCC's concern regarding children's programming is in the separation of program from commercial content.

▶ What the FCC did not deregulate is as important as what it did. The FCC still considers most practices at the time of license renewal.

▶ The new technology provides a host of issues: AM band expansion, AM stereo, and high-definition (HDTV) are now primary areas of technical regulation.

▶ The Baby Bell systems pose the greatest fear to the traditional broadcast operator, because they will soon have the technology to deliver unlimited signals into every home in the country.

InfoTrac College Edition Exercises

Laws governing the management and operation of electronic media affect the daily operations of a station. The wrong word choice in a news story can end up in the libel courts. The misuses of copyrighted information too can lead to court challenges. There are any number of examples where simple error can spawn legal entanglements.

5.1 The operators, practitioners, creators, and mangers of electronic media must be fully cognizant of the law and its constant evolution. List the major legal issues discussed in the chapter—the First Amendment, equal time, obscenity, and so on. In *Broadcasting & Cable, Journal of Broadcasting and Electronic Media, American Journalism Review, Electronic Media, Variety,* and many of the other InfoTrac-listed journals, you'll find discussion relating to these issues. Find and read a relevant article.

a. What is the broadcast industry doing about EEO?

b. What is the latest case surrounding section 315? Equal opportunity?

c. There are children in the audience: What difference does that make to the FCC?

d. Copyrights: How can we stem the tide of bootlegging?

e. Must carry: How is the cable industry reacting to the new satellite distribution of local signals? Are these "must carry" for satellite distribution?

CHAPTER 5 OPERATIONAL REGULATIONS FOR THE ELECTRONIC MEDIA   **117**

WEB WATCH

Here is a list of a few URLs (Internet addresses) for some of the organizations or corporations discussed in this chapter. Please explore these Web sites and follow the links to learn more about the complex business of the electronic media. Add your descriptions and your own favorite sites at the end of the list. Please keep in mind that the dynamic nature of the Internet allows sites to come and go but also allows organizations to update information about themselves very quickly.

Address **Description**

http://www.fcc.gov/aboutus.html _____
  _____

http://www.courttv.com/casefiles/ _____
  _____

http://www.whatis.com/hdtv.htm _____
  _____

http://www.fcc.gov/indstats.html _____
  _____

http://www.cme.org/intro.html _____
  _____

Other favorite sites: _____
  _____
  _____
  _____

CHAPTER 6

Patterns of Media Organization and Ownership

Although broadcast stations and cable systems serve local markets, they are likely to be owned by giant media conglomerates.

As the world becomes more crowded and complex, we rely increasingly on the electronic media to report and to analyze the events that shape our environment and dictate our future. Our view of the world has become increasingly defined by the messages that we absorb daily from radio and television broadcasts. This reliance on secondary images to create our perception of life, politics, history, and economics beyond the immediate bounds of daily activity has led some to question the agenda of the electronic media. Charges of media bias are leveled as one controversial issue after another captures the attention of the U.S. public. There is a suspicion that those who own and control the media are deliberately distorting coverage of events and even the content of entertainment programming in order to manipulate the values of the audience. Consequently, patterns of ownership and control of the U.S. electronic media are of interest to the public as well as to those who are interested in a media career.

This chapter examines the management structures and ownership patterns of the U.S. electronic media. Organizations such as networks, radio stations, television stations, cable systems, program producers, common carriers, and program distributors are described in terms of ownership, purpose, structure, and interrelationships. Although the emphasis is on the traditional broadcast and cable businesses, evolving media organizations are also examined. Even though many media organizations have become multinational in recent years, this chapter is primarily concerned with the structure and control of the electronic media in the United States.

The electronic mass media developed in other countries for the purposes of promoting culture, education, public service, and propaganda. Generally, the early development of the media in such countries as Canada, Great Britain, France, Italy, and the former Soviet Union was sponsored and directed by the government. In contrast, much of the electronic media in the United States was initiated by business and established as profit-making ventures. The development of the business of radio, and later television, in this country followed a pattern typical of the growth of any business in our capitalistic society. It generally involved the inventions of an individual that were perfected and marketed by a large corporation in search of a profit. This chapter also explores the profit motive, business diversification, vertically and horizontally integrated media businesses, and the ethics of the media organizations.

OWNERSHIP OF RADIO AND TELEVISION STATIONS

THE BROADCAST STATION: THE RETAILER

The retail outlet for the dissemination of broadcast entertainment and information is the local radio station, television station, direct-broadcast satellite system, or cable television company. In the process of developing programming and delivering it to an audience, the broadcast station is often the final step as it sends the information directly into the home. The station is licensed by the Federal Communications Commission (FCC) to serve the public in a limited and specific geographic area. Generally, the station competes with several other radio or television stations also licensed to serve the same area. The public should, therefore, have a choice of signals and information, providing several different points of view. Larger communities usually have a greater choice of programming, because economic structure dictates that larger markets can support more local stations. With cable and direct broadcast satellite (DBS), however, homes in both large and small markets can have a diversity of national programs.

As a business the traditional radio or television station is unique in that the consumer of the entertainment or information, which is the product of the station, does not pay the station for the use of its programming. Instead, the content of the station is bought and paid for by a third party—the advertiser. In exchange the advertiser has the eyes and ears of the audience for a few moments. In that commercial time, the advertiser attempts to present the advantages of its products and to encourage the audience to make the appropriate purchases. In effect, the traditional commercial broadcast station has two customers: The first customer is the audience, who does not pay the station but uses the station's product or programming; the second customer is the advertiser, who pays the station for advertising time but rarely has any interest in the station's programming beyond its ratings and audience *demographics*. To make a profit, the traditional local broadcast station must produce, obtain, and schedule programs that will attract the largest audience of a type, or *demographic composition,* that the advertiser wants to reach. Once the audience has been generated, or at least anticipated, the station can sell time to an advertiser that wants to reach that particular group.

OBTAINING A BROADCAST LICENSE

Although the process of making a profit from operating a broadcast radio or television station would on the surface seem simple, many factors complicate broadcasting and limit the number of those who actually profit from an investment in a station. Most broadcasters operate as simple retailers of programs that others have produced, but in order to do so a significant investment in complex electronic equipment is required. More important, the broadcaster must first obtain a license from the FCC to operate a broadcast station.

Because the *electromagnetic spectrum,* through which broadcasting takes place, is a limited natural resource, only a certain number of stations can broadcast at the same time in a given geographic area. To date, more than twelve thousand radio stations and fifteen hundred full-power television stations are licensed to operate in the United States. Most of the available *broadcast frequencies,* or channels, are already in use. Those that are still available are usually located in sparsely populated communities with limited economic potential.

The process of obtaining a license for a new broadcast station is laborious and might take several years and hundreds of hours of expensive consulting engineers' and communications lawyers' time. In addition, the applicant must be a citizen of the United States, with a clean character record and the financial resources necessary to operate the station at a loss for at least a year. If additional individuals or groups also file with the FCC for the same frequency, the already lengthy application process can extend even longer, until the other applicants withdraw from the process or the FCC makes a decision.

Once the applicant has identified an available frequency for a particular market through an engineering search, he or she can apply to the FCC for a construction permit. The permit, which allows the applicant to build the station and test the transmission equipment, requires that the public and other stations in the market be notified of the impending new license. Public hearings are scheduled prior to the issuance of the permit. The license is granted only after the applicant has proven that the station can be operated on the assigned frequency without interfering with other broadcast stations in the market. The Telecommunications Act of 1996 states that radio and television station licenses be granted for a term of eight years. Barring a serious violation of FCC rules, the station owners may assume that the license will be renewed. The station must keep documentation showing that it is serving the public interest of its broadcast community and provide that material to the FCC at renewal time.

Because of the limited number of available frequencies for broadcast stations and the complex and time-consuming application process, most new ownership of stations now comes from the purchase of existing facilities and the subsequent transfer of the FCC licenses to the new owners. Buying and selling of broadcast properties is a big business, with nearly 150 brokers serving the industry. Radio stations in small markets can be obtained in many cases for less than $100,000, and small-market television stations might cost less than $1 million. Large-market FM stations may cost several million dollars, and some large-market television stations have sold for several hundred million. A Los Angeles independent television station, KTLA-TV, sold in 1985 for more than $500 million. In 1999 KRON-TV in San Francisco sold for $823 million. In 1996, 671 radio stations were sold for a total of more than $2 billion, and the sales of 99 television stations totaled more than $10 billion. The high number of properties sold and the record number of dollars exchanged was a direct result of new ownership rules established by the Telecommunications Act of 1996. In 1999 more than $30 billion changed hands for 600 radio and TV stations. Normally, the larger the market, the more money the sale of the station should command. Table 6.1 shows the top twenty-five designated market areas (DMAs) based on the number of households that contain a television, or *television households.*

The cost of the real estate, television equipment, and programming rights associated with these stations does not justify the huge prices paid for broadcast facilities in recent years. The extra cost is

TABLE 6.1 — Top 25 DMAs ranked by television households

RANK	MARKET	ESTIMATED TV HOUSEHOLDS (IN MILLIONS)
1	New York	6.8
2	Los Angeles	5.0
3	Chicago	3.1
4	Philadelphia	2.6
5	San Francisco–Oakland–San Jose	2.3
6	Boston	2.2
7	Washington, D.C.	1.9
8	Dallas–Fort Worth	1.9
9	Detroit	1.8
10	Atlanta	1.7
11	Houston	1.6
12	Seattle-Tacoma	1.5
13	Cleveland	1.5
14	Minneapolis–St. Paul	1.4
15	Tampa–St. Petersburg–Sarasota	1.4
16	Miami–Fort Lauderdale	1.4
17	Phoenix	1.3
18	Denver	1.2
19	Pittsburgh	1.1
20	Sacramento-Stockton-Modesto	1.1
21	St. Louis	1.1
22	Orlando–Daytona Beach–Melbourne	1.0
23	Baltimore	< 1.0
24	Portland, Oregon	< 1.0
25	Indianapolis	< 1.0

Source: From *Broadcasting and Cable Yearbook 1998*, B-234.

for the transfer of the license that allows the station to remain in the business of broadcasting in a particular market. Without a license a broadcast station is out of business. The value of its used equipment and even its real estate is small compared with the value of stewardship of the license.

OWNERSHIP LIMITS

In the past the Federal Communications Commission has not condoned the indiscriminate buying and selling of broadcast properties to make profits from the licenses. The new owners of a station must receive approval from the FCC for the transfer of the license prior to completion of the sale. The FCC has imposed rules on the issuance of licenses to limit the *concentration of ownership*—the ability of one owner to control the access to information through a monopoly of all the broadcast channels in a single market. As stated earlier, prior to the fall of 1992 the FCC limited the number of stations that one individual or organization could own to a maximum of twelve AM stations, twelve FM stations, and twelve television stations (minorities could own fourteen). That number was increased in 1992 and again in 1994. In addition, past FCC rulings and Justice Department actions have limited the number of stations that one entity could own in a single market and have dictated that no one entity could own enough television stations to reach more than 25 percent of the U.S. public. Additional cross-ownership rules have forbidden telephone companies from owning cable television systems in their service areas and did not allow television stations to own cable television systems in the coverage area of the station.

The rules on cross-ownership and multiple-station ownership have changed several times since the 1980s. The current philosophy is that because of the increase in competition from the expanding number of traditional media outlets, as well as competition from the new media outlets such as cable television and video stores, the concentration of media ownership need not be so tightly controlled. The Telecommunications Act of 1996 again changed ownership limits. One entity could own an unlimited number of AM and FM stations, but no more than eight in a large market with forty-five total radio stations. One entity is limited to owning TV stations that reach no more than 35 percent of the U.S. national audience. In addition, one entity can now own two or more television networks.

One area of the ownership philosophy that did not change significantly in the 1990s was minority ownership of broadcast facilities. In the late 1970s, the FCC developed policies giving minority applicants some advantage in the acquisition of licenses for new stations. The dividends of this policy include a significant increase in station ownership by black and Hispanic groups, but minority ownership still falls far short of the proportion of minorities in the general population.

Women have been listed as owners in the broadcast industry for many years, though there is some question as to whether that ownership equals control. Many female relatives are listed as owning significant amounts of stock in male-dominated broadcast facilities. All other factors being equal, the FCC favors applications that indicate large percentages of female and minority ownership.

Network Owners

Broadcast stations recognized early that the programming that people wanted to hear, and later see, cost a lot of money to provide. They also quickly recognized that the high cost of programming could be spread among several stations if the stations shared the programming. *Chain broadcasting,* later to become the networks, developed when the costly programming that was produced in New York was simultaneously broadcast by other East Coast stations that were interconnected by telephone cables.

To take advantage of the economics of programming distribution, the early networks during the 1920s and 1930s—NBC's Red and Blue Networks and CBS—began buying stations in major cities. Soon each owned the then maximum limit of seven AM stations. Those stations, known as *O&O,* for "owned and operated" by the network, became extremely profitable. Later, in the 1940s and 1950s when television and FM radio developed, the three major networks—NBC, CBS, and ABC—built or purchased facilities in those services also.

Because of the economics of having one group manage several stations from a central location and the savings associated with spreading the programming costs over several broadcast outlets, the O&O stations soon became the profitable stars in network corporate holdings.

In recent years, with increased competition, regulatory changes, and new ownership, the networks have seen many changes in previously O&O stations (see table 6.2). NBC's radio services have been sold, and thirteen television stations are owned by NBC, which in turn is owned by General Electric. In addition, the networks have engaged in buying and selling stations in a constant search for more profitable markets, while not exceeding the limitation of owning stations that reach more than 35 percent of the population.

Currently, ABC, owned by the Walt Disney Company, has television stations in ten locations and also owns radio stations and cable networks. CBS, recently merged with Viacom, owns thirty-five television stations in twenty-nine television markets as well as more than 163 radio stations through Infinity Broadcasting Corporation (see table 6.3).

The Fox Network was created in 1986 when Rupert Murdoch, the Australian-born international media magnate, purchased seven television stations of the Metromedia group to add to his Twentieth Century Fox movie studios. Fox now owns twenty-three stations in twenty-two cities.

Interestingly, because the O&O stations are licensed to serve local communities, they are managed by a separate division of the owner corporation and can reject (or preempt) a network program to offer their own local programming. This does not happen

TABLE 6.2 Commercial television networks

- ABC Television Network Group, Capital Cities/ABC Inc.
  77 W. 66th St., New York, NY 10023-6298; http://www.abc.go.com

- CBS Television Network, CBS Inc.
  51 W. 52nd St., New York, NY 10019; http://www.cbs.com

- Fox Broadcasting Company
  10201 W. Pico Blvd., Los Angeles, CA 90035; http://www.fox.com/frameset.html

- NBC Television Network, National Broadcasting Co.
  30 Rockefeller Plaza, New York, NY 10112; http://www.nbc.com

- UPN, Paramount Television Group
  5555 Melrose Ave., Hollywood, CA 90038; http://www.upn.com/hmupn.htm

- WB, Warner Brothers Television
  4000 Warner Ave., Burbank, CA 91522; http://www.thewb.com/

TABLE 6.3 Owned and operated television stations, 1999

ABC	CBS/VIACOM	NBC	FOX
Chicago	Atlanta	Birmingham/Tuscaloosa, AL	Atlanta
Flint/Saginaw, MI	Austin, TX	Chicago	Austin
Fresno	Baltimore	Columbus, OH	Birmingham/Tuscaloosa, AL
Houston	Boston	Dallas/Ft. Worth	Boston
Los Angeles	Chicago	Hartford/New Haven	Chicago
New York	Columbus, OH	Los Angeles	Cleveland
Philadelphia	Dallas/Ft. Worth	Miami/Ft. Lauderdale	Dallas/Ft. Worth
Raleigh/Durham, NC	Denver	New York	Denver
San Francisco	Detroit	Philadelphia	Detroit
Toledo, OH	Green Bay	Providence, RI	Greensboro/High Point, NC
	Houston	Raleigh/Durham, NC	Houston
	Indianapolis	San Diego	Kansas City
	Los Angeles	Washington	Los Angeles
	Miami/Ft. Lauderdale		Memphis
	Minneapolis/St. Paul		Milwaukee
	New Orleans		New York
	New York		Philadelphia
	Norfolk, VA		Phoenix
	Oklahoma City		Salt Lake City
	Philadelphia		St. Louis
	Pittsburgh		Tampa/St. Petersburg
	Providence, RI		Washington
	Sacramento		
	Salt Lake City		
	San Francisco		
	Seattle		
	Tampa/St. Petersburg		
	Washington		
	West Palm Beach, FL		

often, however, due to corporate economic and political pressures.

GROUP OWNERS

Although the network O&O stations (especially television) are among the most prosperous and influential in the country, they are definitely in the minority in terms of numbers. Most broadcast stations are owned by corporations that also own other broadcast properties. These are called *group-owned stations*. The corporations that own broadcast groups range in size from a single individual who owns several small-market radio stations in rural areas to some of the largest multinational corporations in the world. NBC, for example, is owned by the huge General Electric Corporation.

There are more than five hundred broadcast groups in the country. These include many small ones such as the Rex Broadcasting Corporation, which owns three radio stations in Tucson, Arizona; and the McGraw Group, which operates radio stations in Cadiz, Ohio, and in Weirton and Elkins, West Virginia. Although these obviously are not major media markets, the stations in those communities undoubtedly serve important communication functions. The owners of the stations in these small groups can benefit from increased borrowing power from station equity and also by combining legal and engineering consulting services and centralizing accounting and management functions. By obtaining more-economical operating procedures as a benefit of group ownership, a radio station might be able to exist in and serve a community otherwise thought too small to support a station.

At the other end of the scale, the large group owners can take advantage of the same economics. In addition, some large group owners operate their own sales organizations that sell the time on all their stations to national advertisers. They also buy the rights to broadcast programs for all of the group's stations at reduced rates.

Some large group owners include: Paxon Communications, Tribune Broadcasting, Gannett Broadcasting, Clear Channel Communications, and E. W. Scripps Broadcasting. Many of the large groups specialize in a particular type of station: Trinity Broadcasting Network owns eleven stations and broadcasts religious programming; Telemundo and Univision are groups that broadcast Spanish-language programming; and the Home Shopping Network, through its Silver King Broadcasting, owns a dozen full-power television

TABLE 6.4 TV's top ten

COMPANY	NO. OF STATIONS	PERCENTAGE OF U.S. TV HOUSEHOLDS
CBS/Viacom	35	40.0
Fox	23	35.2
Paxson Communications	70	35.0
Tribune	23	28.9
NBC	13	26.6
Walt Disney/ABC	10	23.9
Chris Craft/United Television	10	18.8
Gannett	22	17.2
Hearst-Argyle	32	15.9
USA Broadcasting	13	15.5

Source: From "Top 25 Television Groups," *Broadcasting & Cable* (January 24, 2000): 72.

stations and many low-power stations involved in direct marketing to the public.

CROSS-OWNERSHIP

Many of the large group owners are also *media conglomerates,* meaning that they own other types of media outlets such as magazines, newspapers, or cable television systems. Ownership of two or more types of media outlets is called *cross-ownership.*

Many early radio stations were established by existing newspaper owners. The call letters of several stations indicate newspaper ownership: Station WGN in Chicago was named for the "World's Greatest Newspaper"—the *Chicago Tribune.*

Some of the largest group owners in the country have cross-ownership ties with major newspapers. Station KRON-TV in San Francisco is owned by the Chronicle Publishing Company, which is also the publisher of the *San Francisco Chronicle.* The Bonneville International Corporation publishes the *Deseret News* in Salt Lake City as well as owns KSL-AM-TV in the same city and stations in Los Angeles, San Francisco, Chicago, and Washington, D.C. Scripps Howard owns more than a dozen major newspapers as well as nine television stations.

Although corporations are constantly changing their holdings, the largest media conglomerate in the United States is the Gannett Company, which owns the national *USA Today,* some 80 daily newspapers, 52 nondaily publications, and 22 television properties, including major television stations in Phoenix; Denver; Washington, D.C.; Jacksonville; Atlanta; Minneapolis; Greensboro, N.C.; and St. Louis.

Many of the cross-ownership rules and limitations that have come and gone over the past thirty years were directed toward limiting the access of any one owner to the U.S. public through the electronic media. Not only has there been concern that one corporation might gain too much access to the national public, but there was fear that if one entity owned several media outlets in a single city or market area, only that owner's ideas or concerns would be presented to the public. The FCC's role was to encourage diversity of ownership and, therefore, diversity of thought and opinion in the communications process.

Although cross-ownership and absentee management through group ownership offer the potential for control of information, there has been little evidence of it happening. Broadcast stations are licensed to serve the needs of the local community, and audiences seem to respond best to those stations that do. Recent thought in public policy, therefore, has been to deregulate broadcasting and let the marketplace determine the content of a station. In keeping with this philosophy, the FCC has relaxed most of its former ownership limitations. Current rules allow one entity to own more than one radio station in a market, depending on the number of stations in the market. Because of this some formerly competing stations are now under the same ownership.

LOCAL OWNERS

Radio and television stations can be owned by a network, a group ownership, or a media conglomerate. Most television station owners fall into one of those three categories, but many owners of radio stations fall into a fourth category: the single-media entrepreneur. A few television stations in smaller markets are owned and operated as ma-and-pa stations, wherein the owner, and often his or her spouse, is directly involved in the day-to-day management of the broadcast property.

There is some advantage to an owner's direct involvement in the day-to-day management of a local radio or television station. First, in the past, one criterion of particular interest to the FCC when granting licenses was local ownership and local management. Second, the owner who is a resident of the community the station serves is likely to know the community well and take a genuine interest in its welfare. Programming and station operation will probably reflect a concern for the community. Third, because the economic health of the community will directly influence the profitability of the station, the local owner will undoubtedly have a finger on the pulse of the local market and operate his or her station to maximize profits while contributing to the growth of the community.

There are also disadvantages to local ownership. First, the economics of broadcasting dictate that programming resources be amortized, or spread over the largest possible audience. With a limited coverage area, a local station may have high programming costs per home. By sharing programming with other stations owned by the same group, the costs can often be divided among several markets, thus lowering the cost per home. In addition, the single station will not have the buying power that the group ownership enjoys when negotiating rights for syndicated programs. Second, the financial resources of the single station are probably more limited than those of the station belonging to a group. Given the equity available to the group owner, loans necessary to upgrade one of the group's stations are probably more readily available at a better rate than to the single-station owner. Third, the risk factor, when spread over several broadcast properties, is likely lower for the group owner than for the single-station entrepreneur. With the group, losses at one facility may be offset by profits from another, but the single-station owner has only one chance, with one property, to make a big profit.

With the passage of the Telecommunications Act of 1996 and its liberal ownership rules, a frenzy of broadcast property sales has led to even greater ownership concentration by a few media conglomerates. Local ownership, with its small ma-and-pa stations, is fast becoming a relic of the past. Radio group owners now have outlets numbering in the hundreds.

Whether owned by an individual, a limited partnership, a corporation, a broadcast group, a media conglomerate, or a network, the local radio or television station is the basic unit of the broadcasting industry. The station is the retailer that delivers entertainment directly to the audience and thus provides the market that allows the advertiser to reach potential customers. Most important, the local station serves the people of the community with information relevant to them.

STATION ORGANIZATION

To generate audiences and then sell them to an advertiser, the radio or television station must, at a minimum, perform four functions:

1. The technical process of creating a carrier wave on the assigned frequency and modulating the video and/or audio
2. The production and/or selection of the content or programming that is delivered to the audiences
3. The sale of commercial time to the advertiser to reach the audiences
4. The marshaling and control of resources and the accounting of revenues

The organizational structure of broadcast stations is designed to implement these four functions at a departmental level (see figure 6.1). Although each station is different in structure to accommodate the size of the staff, the market, and the philosophy of management, most stations have departments of engineering, programming, sales, and business. Several additional departments are becoming increasingly common in television stations. For example, now that greater importance has been placed on the profitability of the local news product, most have set up news departments as well. The promotion department is often included to advertise the programming of the station as well as to provide presentation materials for sales staff to promote the station to potential advertisers. Many television stations have also developed a separate production department that sells its services to the local advertisers for producing commercials.

Each manager of the four major departments of the local radio or television station must develop objectives and policies that contribute to the long-term goals of protecting the station's license and making a profit for the stockholders.

News is important programming for most network-affiliated local television stations. This "live" truck allows a microwave signal to be sent to the station from a reporter in the field to be incorporated into a live newscast.

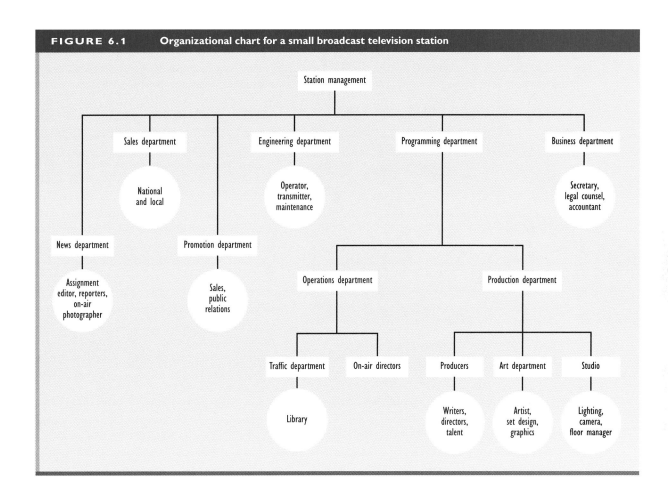

FIGURE 6.1 Organizational chart for a small broadcast television station

Ownership Patterns in Cable Television

The cable television system that provides television signals to individual homes is analogous to the local broadcast station because it too serves the consumer. But, although the broadcast station has both the audience and the advertiser as customers, only a few cable systems actively pursue advertiser dollars as a major revenue stream. The economic basis of the cable television system is the monthly subscription fee that the cable customer pays for the delivery of forty or more channels.

Cable television systems can be either locally owned single systems or operated by a *multisystem owner (MSO)*, which is similar to the group ownership in broadcasting.

There are currently almost 12,000 cable systems providing service to almost 35,000 communities. These systems range in size from 100 subscribers (the legal minimum limit to qualify as a cable system) to more than 1 million subscribers connected to a single cable plant. The biggest cable system, in terms of number of subscribers, is in New York and is owned by the large MSO Time Warner. Texas has the most cable television systems, but California, with almost 6 million cable homes, has the largest number of subscribers in the country.

Early Locally Owned Systems

Early cable systems, known as *community antenna television (CATV) systems,* were usually built by a technically oriented individual to serve a small community. Like many small-market radio stations, these CATV systems were most often locally owned and operated as small ma-and-pa businesses, with "Pa" serving as the engineer, installer, and construction crew and "Ma" running the office and billing the subscribers.

TVROs, or satellite dishes, needed to capture cable network signals.

This pattern continued as CATV systems proliferated in the country's rural valleys.

Generally, the small CATV systems began in the late 1940s and early 1950s in response to demand for television service in areas not served due to a four-year FCC freeze on the issuance of new broadcast TV licenses. The development of coaxial cable by Bell Labs in the 1930s gave the technically minded rural entrepreneur the means to capture the faint television signal from a nearby hilltop and transport it into the valley to the receivers of his friends and neighbors. Many early systems were funded by neighborhood associations; others were developed by businesspersons looking for a profit.

One interesting example is Tower Antennas in Sugar Creek, Ohio. This small town is in the hilly Amish country of north central Ohio. In the early 1950s, the owner of the town's general store decided to add a line of television receivers to his inventory of groceries, hardware, wood cookstoves, kerosene lamps, and buggy harnesses. Sales were slow. The Amish half of the clientele followed a religion that forbade not only the use of such devices but also the electricity to operate them. Most of the other citizens lived in the valley town, where the faint television signal from the Cleveland station was blocked by hilly terrain.

Not wishing to be stuck with a substantial investment in television receivers, the merchant asked the village council for permission to place an antenna on top of the village water tower. Cable was then strung on telephone poles from the tower to the store, and the flickering sets came alive with pictures of Uncle Miltie on the *Milton Berle Show; Kukla, Fran, and Ollie;* and test patterns.

Those that purchased a receiver could also, for a fee, tap into the ever expanding coaxial system. Tower Antenna's CATV system was born and soon expanded to serve other small towns in the area. By the mid-1960s, Tower was a multisystem owner with properties in Ohio, West Virginia, Pennsylvania, and Kentucky. Cable television was now the primary business of the general-store owner, and he became a member of the board of the newly formed National Cable Television Association (NCTA). The original Tower systems lost their identity when the company merged with Communication Properties of Texas, which was later purchased by the Times Mirror Corporation, which then sold its cable properties to Cox Communications, Inc., in 1994.

This scenario is typical of the development of the cable television industry. Many locally owned systems were absorbed by larger cable companies as the economics of computerized billing and inventory control were discovered. In addition, many of the small operators were unable to obtain the financing needed to rebuild the local systems as more and more television channels were demanded by subscribers. Although there are still many small MSOs, the locally owned cable television system that serves a single community is almost a relic.

The Multisystem Cable Owner

Almost all cable television systems today are operated by one of the more than 350 multisystem owners. These, like broadcast group owners, run the gamut from very small to very large. The largest MSO is AT&T, with more than 16 million subscribers. In its early years, the Denver-based TCI Communications developed by buying small cable systems serving local communities. In recent years the organization has grown by acquiring or merging with other large MSOs. In July 1995 TCI bought Viacom's cable systems with 1.2 million subscribers for $2.25 billion. TCI disappeared when bought by AT&T in 1999.

Antennas and microwave dishes used to receive over-the-air broadcast signals at a cable headend.

Ownership of many of the large cable MSOs has become interwoven as they buy cable systems and pieces of one another in the quest for expansion and increased profitability.

Many of the biggest MSOs are themselves owned by media conglomerates (see table 6.5). Some names familiar to those in the media include Cox Communications, Cablevision Systems Corporation, Comcast Cable Communications, and Time-Warner Cable. Many of the major providers of cable television services also have significant holdings in other electronic media properties. In addition, many of the MSOs have ownership ties with major newspaper interests.

Some media critics have suggested that the concentration of ownership now found among the MSOs is not conducive to the diversity of opinion demanded by a free marketplace of ideas that was envisioned by our system of control of the media.

VERTICAL INTEGRATION

By the early 1990s, cable television had become big business. One of the largest multisystem owners at that time, ATC, with 466 operating cable television systems, listed total revenues of $973 million in 1990. It later

TABLE 6.5   Top ten multisystem owners

RANK	MSO	TOTAL BASIC SUBSCRIBERS (IN MILLIONS)
1	AT&T	16.2
2	Time-Warner Cable	12.9
3	Comcast Cable Communications	5.3
4	Cox Communications Inc.	5.1
5	Adelphia	4.9
6	Charter Communications	3.9
7	Cablevision Systems Corporation	3.3
8	Falcon Cable	1.1
9	Insight Communications	1.04
10	Jones Intercable	1.007

Source: From John M. Higgins, "Top 25 MSOs," *Broadcasting & Cable* (May 24, 1999): 34.

merged with an even larger MSO. Even tiny MSO Dickerson California-Arizona Associates Ltd., with three operating cable systems, listed total revenues of nearly $3 million that year. The economics of the MSOs have become important to the development of the cable television industry. It is more cost-effective to centralize business, legal, engineering, construction, programming, and personnel functions for many cable systems than to allow each separate system to handle them locally.

As MSOs increased in size and economic power, they turned their attention to the development of programming resources. Cable subscribers indicated early on that they were interested in programming beyond that offered by the three major networks. With the development of satellite distribution, it became economical for programming services to be delivered nationally to cable systems. Home Box Office (HBO) and what is now superstation WTBS began delivering programming by satellite in 1976. By the early 1980s, several more cable networks were in existence, including CNN, ESPN, Nickelodeon, and MTV. Many of these and a multitude of later cable networks were owned, in part, by the major MSOs. For example, Time-Warner/AOL owns HBO and parts of CNN, WTBS, TNT, Cinemax, and Comedy Central. Viacom, another large MSO which merged with CBS, owned MTV, Nickelodeon, Showtime, the Movie Channel, and VH-1. In fact, most of the current cable networks are owned wholly or in part by one or more of the major media organizations.

Naturally, the cable systems owned by those MSOs are likely to carry the programming services owned by their parent firms. Generally, the price per customer for the carriage of these services is lower for the systems owned by the MSO. The process of one division of the corporation supplying services to another division is called *vertical integration*. This concept is not new. Cable operators associated with equipment manufacturers, such as Jerrold Communications, have been vertically integrated for years. In the late 1920s and 1930s, the major motion picture producers owned chains of theaters that displayed their products. The Justice Department later declared that practice to be in restraint of trade and required the studio-owned theaters to be sold.

Media Megamergers

In the fall of 1999, Viacom and CBS announced the largest media transaction up to that time. In a deal valued at $37 billion, Viacom (which owns Paramount Pictures, Blockbuster Video, MTV Networks, television stations, and other media properties) assumed ownership of CBS Corp. Formally owned by Westinghouse, CBS is best remembered for its traditional television network that brought the American public Edward R. Murrow and Walter Cronkite as well as *The Beverly Hillbillies*.

Interestingly, CBS owned Viacom about thirty years ago, but was forced to sell because the FCC formulated a ruling that forbade a television network (CBS) from owning a cable operator (Viacom). The recombining of the two media giants creates the second-largest media corporation in the country after Time-Warner.

Media giant Time-Warner became even bigger in early 2000. A $181 billion deal was announced that would combine that company with America On Line (AOL), to form an organization that would cover the media spectrum from news magazines to the movies and from CNN to Netscape. The huge media conglomerate would control many phases of media production as well as distribution via cable and the Internet. With AOL as the managing company, the new entity will combine the country's largest Internet provider with the largest producer of TV shows and movies as well as cable programs. Time-Warner's cable systems currently pass more than 20 percent of American homes.

Time, Inc., was founded in 1923 to produce America's first news magazine. The company combined with Warner Brothers Studios in 1990 to create the world's biggest media company. In 1996 Time-Warner merged with Turner Broadcasting to combine Warner Brothers, the WB network, and HBO with CNN, TNT, and TBS.

AOL was formed in 1985 to serve the few computer geeks who existed at the time. AOL's Steve Chase will be the chairman of the company, Time-Warner's Gerald Levin will be chief executive, and Ted Turner of Turner Broadcasting will serve as vice chairman.

Although there are obvious economic advantages to vertical integration, questions have been raised concerning the appropriateness of the large MSOs also owning the networks that serve them and, in the case of United Artists Entertainment, the production facilities that provide the entertainment and information product. The fear is not only concentration of ownership, but also that some members of the cable public may be denied access to programming that has not been produced and distributed by the system's own divisions.

Multinational Ambitions

Most of the economically viable cable markets in the United States and Canada have already been franchised and wired. Although only about 65 percent of U.S. homes currently subscribe to cable, the wire now passes nearly 90 percent of all the television homes in the country. To continue expanding, the cable television industry has had to look to other countries for new markets and opportunities.

In the mid-1980s Great Britain approved communication regulations that made cable television possible in that country. The U.S. MSOs were quick to investigate involvement in the cable systems that, for the first time, would bring a multitude of television signals to the British public, but cable did not take off in the United Kingdom. Cable opportunities also opened up in other parts of the world, including western Europe, Australia, New Zealand, and Hong Kong.

The cable MSOs of today are seemingly interested in becoming the multinational communication giants of the next century, offering telephone, high-speed Internet, and data communications as well as televised information and entertainment.

Cable System Organizational Structure

Like the radio or television station, the cable television system can operate only with the blessing of government. The broadcast station must obtain a license from

Media Merger Chronology

Spring 1986	Capital Cities Communication buys ABC, Inc.
Summer 1986	General Electric buys NBC
Fall 1989	Sony Corp. buys Columbia Pictures
Spring 1990	Warner Communication merges with Time, Inc.
Summer 1994	Viacom, Inc., buys Paramount Communications
Summer 1994	Viacom, Inc., buys Blockbuster Entertainment
Summer 1995	Seagram Co. buys MCA (Universal Studios)
Winter 1995	Westinghouse Electric buys CBS, Inc.
Winter 1996	Walt Disney Company buys Capital Cities/ABC
Winter 1996	Westinghouse/CBS buys Infinity Broadcasting
Fall 1996	Time-Warner merges with Turner Broadcasting
Winter 1997	Westinghouse Electric becomes CBS Corp.
Spring 1999	AT&T buys TCI (largest cable MSO)
Spring 1999	CBS Corp. agrees to buy King World Productions
Fall 1999	Viacom Inc. agrees to buy CBS Corp.
Spring 2000	Time-Warner and AOL agree to merge

the federal government, and the cable television system must apply for a franchise from a local government before it can begin to operate. Therefore, the cable system operator must place high priority on the protection of the franchise and compliance with the franchise's contractual obligations. As a business, the cable television system must also show a profit. Whereas the broadcast system must sell audience access to the advertiser, the cable system must appeal to the subscriber. The more subscribers who can be enticed to buy cable services, the greater potential for increased revenues and subsequent profits.

As in broadcast properties, the organizational structure of the cable television system depends on the size of the market served, the structure of the parent corporation (if applicable), and the management philosophy of the ownership. As in broadcast stations, department heads report to a system manager, who in turn usually reports to an MSO corporate office. The cable system usually has many of the same departments plus some additional ones to respond to the needs of the many individual subscribers (see figure 6.2).

Probably one of the most important departments in the typical cable television system has no counterpart in broadcasting: the customer service department. This department provides an interface between the cable television system and the customer. Customer service representatives (CSRs) answer calls from cable customers on a variety of topics, from billing questions to programming information. A larger cable system has several CSRs on duty at any given time. An effective customer service department can be vital in maintaining a satisfied subscriber base. When customer questions are answered quickly, pleasantly, and accurately, subscribers with complaints can often be pacified and retained as customers. The cable television system is above all a service organization and must be responsive to the customer in order to provide that service.

ANCILLARY BUSINESSES IN BROADCASTING

The broadcast station and the individual cable television system are basic units of the electronic media business. They serve as retailers for the electronic delivery of entertainment and information to the audience. Many ancillary businesses, however, serve

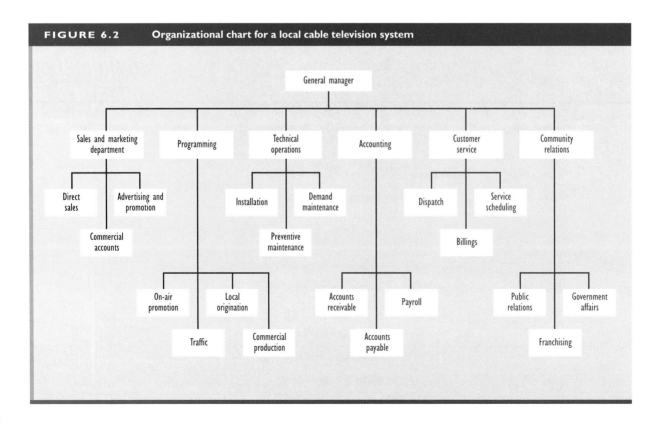

FIGURE 6.2   Organizational chart for a local cable television system

these retailers by providing programming and services. These businesses include the networks, the programming producers, the programming syndicators, consultants, unions and guilds, licensing organizations, professional associations, and lobbying organizations. Each of these businesses fits into a well-defined niche in the electronic communications industry and is necessary for it to function efficiently. All make it possible for the broadcast stations and the cable systems to operate, by supplying the programming and services necessary for the stations and systems to serve their customers. In the process of providing service, though, these ancillary businesses, primarily through economics, exercise some control over the information—product—that is delivered to the consumer. Thus when we criticize the station or the cable system for programming content, we must realize that although legal responsibility rests with those organizations, it is the supplier—generally the network or the production company—that has more actual control over the information that is broadcast or distributed. In addition, the content is shaped by legal requirements and popularity as measured by ratings.

The Networks: Then

Early radio stations recognized quickly that the process of broadcasting had an insatiable appetite for programming. Early stations, even with their abbreviated broadcast schedules, were constantly scrambling to provide programming by inviting amateur performers to come in off the streets to display their talents. Many early stations relied on poetry readings, bad sopranos, and scratchy, almost inaudible, phonograph records to fill the programming day. Good programming that the audiences wanted to hear was expensive but was quickly provided by the development of the networks.

The functions of the networks was twofold: providing a national programming service to affiliated broadcast stations and selling commercial time to sponsors that wished to reach national audiences.

For more than five decades, NBC, CBS, and ABC were the traditional electronic communications media networks in the United States. Two of the three, CBS and NBC, were controlled most of that time by those who founded them—William S. Paley and David Sarnoff. During the 1990s, however, many changes have reshaped the traditional broadcast networks. As they have changed, so has the entire electronic communications industry. In addition, three new television networks—Fox, United Paramount, and Warner Brothers—have emerged in recent years; thus the ownership structures have become very complex.

The Networks: Now

In 1985 the FCC approved the sale of the ABC-owned and -operated stations to Capital Cities Broadcasting. This cleared the way for the completion of the $3.4 billion sale of ABC, Incorporated, which was formed through a 1953 merger of the American Broadcasting Company and United Paramount Theaters, Incorporated. The corporation that emerged, New York–based Capital Cities/ABC Inc., is now owned by the Walt Disney Company and is composed of many divisions: ABC Television Network Group, Capital Cities/ABC Broadcast Group, and Capital Cities/ABC Publishing Group.

In early August 1995, Michael Eisner, the chief executive of the Walt Disney Company, announced the completion of a $19 billion deal to buy Capital Cities/ABC, Inc. This, at the time, was the second-largest takeover in U.S. corporate history, and would provide a distribution outlet for the programs of the Disney corporation.

The ABC Television Network Group includes the following operations: Research, to collect and analyze audience data; Affiliate Relations, to ensure that the more than two hundred affiliated stations will carry as many of the network programs as possible; Network Sales, to sell network time to national sponsors; ABC Television Network, to provide the operations of network program delivery; ABC Entertainment, to develop, obtain, and schedule programming for daytime, late-night, and prime-time delivery to affiliates; ABC News, to provide news, news magazine, and documentary programming; ABC Broadcast and Standards, to act as a programming censor; ABC Sports; ABC Productions; and Business Affairs for both the East and West Coasts.

ABC Incorporated also includes ESPN Cable Network, Disney/ABC International Television, and ABC Publishing Group.

The ABC Publishing Group operates Video Enterprises, which distributes product in international markets, and ABC Distribution Company, which sells programming to the home video market and to cable. The ABC Publishing Group also publishes specialized magazines for the fashion, financial, and medical industries. Daily newspapers are owned in several cities, including Ft. Worth, Texas, and Kansas City, Missouri. The Owned Television Station area controls the ten television stations that ABC owns, as well as National Television Sales, and Broadcast Operations and Engineering.

CBS, which was established, owned, and developed into a preeminent position by William S. Paley, was taken over in 1984 by President and Chief Operating Officer Laurence A. Tisch. Over the years, CBS has owned a variety of often money-losing enterprises, including a record company, a cable television network, a car rental company, a publishing house, and a major league baseball team—the New York Yankees. Tisch cut operating costs and changed the emphasis to network broadcasting, with more than two hundred television affiliates, three hundred radio outlets, and O&O broadcast properties. CBS, Inc., includes Operations and Administration, as well as the Broadcast Group, which comprises Affiliate Relations, Entertainment, Marketing, News, Sports, and Enterprise Divisions. The latter is involved in CBS Video, among other endeavors. The CBS Television Stations Division then operated the fourteen O&O television stations as well as an in-house sales, or national spot sales, operation. The CBS Radio Division at the time operated the seventy individual radio stations as well as the radio network.

A day after the announcement that Disney had obtained the ABC network, Westinghouse announced that it was buying CBS for $5.4 billion. With anticipated changes in FCC ownership rules, Westinghouse's Group W Broadcasting would, with the CBS purchase, operate thirty-five television stations and more than 160 radio stations. Mel Karmazin, formerly of the stations group, has become the outspoken CEO of CBS, Inc. Karmazin merged CBS with the major program syndicator Viacom.

The National Broadcasting Company (NBC), established in 1926 by RCA, was sold in 1986 to General Electric for $6.28 billion. The sale included RCA and all of its properties. The divisions of the company include: NBC Television Network with more than two hundred affiliated stations; NBC Sports Division; NBC Entertainment, which develops network programming for the different dayparts; and NBC Television Stations, which manages the O&O stations. In addition, NBC has departments of Cable and Business Development, Corporate Communications, Employee Relations, Finance, and Law. Its cable network, C-NBC, is operated by the Consumer News and Business Channel Department. MSNBC is a cable network operated as a joint partnership with Microsoft.

Fox Broadcasting Company is a recent addition to the list of traditional television networks. It was formed in 1986 by Rupert Murdoch, who combined the production and programming support of the movie studio Twentieth Century Fox with the Metromedia television stations. The network, owned by Murdoch's news corporation, has divisions of Entertainment, Advertising/Publicity/Promotion, Sales, and Affiliate Relations. The Fox Network owns twenty-three television stations and has nearly 175 primarily television affiliates. By offering innovative entertainment and competitive sports programming designed to reach younger audiences, Fox became a prime-time ratings leader by the fall of 1998.

United Paramount and Warner Brothers began programming feeds to affiliated stations in early 1995. Both networks have extensive programming libraries and production experience, because each is affiliated with a Hollywood film studio. Both Paramount and Warner Brothers have been among the primary suppliers of programming to the television industry for years.

By mid-1995 television networks, which had been declared a casualty of the cable industry boom in the late 1980s, had become hot properties. Much of the change was due to revisions in FCC rules on station and programming ownership, although audiences for network prime-time programs had dropped 30 to 50 percent since the mid-1970s.

Network/Affiliate Relationships

The network supplies programming to the local broadcast station in exchange for commercial time that the network then sells to a national advertiser. In the licensing process, the FCC requires that the local station be responsible for the content of all of its programming. The station can, therefore, preempt network programming, presumably to provide other programming thought to be more in line with serving the public interest, convenience, and necessity in its particular community. In the past the FCC had also enacted regulations that required the contract between the network and the affiliated station to last no longer than two years.

The FCC had enacted various other rules over the years to limit the power of the networks. For example, the prime-time access rule (PTAR) limited the amount of time that the network could have access to the local station in the four hours each night that are considered prime time. The *Financial Interest Syndication (FinSyn) rules* limit the amount of programming that the network could own. In 1995 the FCC eliminated the PTAR and FinSyn rules to the advantage of the networks (see chapter 5).

Within the limitations imposed by various FCC regulations, networks and stations have developed contractual agreements that are mutually beneficial. The affiliate relations department at each of the networks has the task of gaining clearances (or obtaining broadcast time) for as many network programs as possible on all of the stations associated with each of the networks. In the past when the networks paid large compensation fees for station time each month, the task was easier; but in recent years, there has been a trend away from substantial monthly payments. The networks wish to pay less each year to the stations for their broadcast time.

To help convince stations to carry network programming, affiliate relations departments often spend large sums to wine and dine the affiliate-station owners and managers and their spouses. Each year, the managers are brought to Hollywood for a week of luxury among the current stars and celebrities for a preview of the network's next prime-time season programming schedule.

Affiliate Contracts

Currently, most television stations, especially those in larger markets, have a primary affiliation contract with one of the networks. The contract usually states that the local station will have first call or first right of refusal on any network program that is on the schedule.

In addition, the network delivers the program, via satellite, to the station at no charge and provides the station with a variety of promotional materials that directly link the station with the network in the mind of the viewer. Most stations also receive payment for running the network programs, which is usually based on a formula that takes into consideration the market size, the popularity of the local station in that market, the station's rate structure, and the number of network hours that the station carries in the different dayparts.

In turn, the station provides assurance, several weeks in advance, that it will carry a particular network program uninterrupted. This means that the commercials sold by the network and contained in the programming will also be aired by the local station. The network can then sell the commercial time in the program to a national advertiser based on the projected size of the national audience. The more stations that carry the commercial, the more money can be obtained from the advertiser.

In the past, the affiliation contract with the network meant that the station was assured a large audience and therefore likely to be highly profitable. Most viewers were interested in the programs that the network provided, and, until the mid-1970s, 90 percent of television viewers in prime time were shared by the three original television networks—ABC, NBC, and CBS. The commercials that the local stations sold during the station breaks had the very large audiences of the network programs. The 1970s saw the pinnacle of the networks' success, and the affiliated stations rode a wave of profitability generated by network programming.

The Future of the Networks

During the past decade, increased competition from a multitude of cable channels and independent television stations, the emergence of Fox as a programming force, and the arrival of the home video market have eroded the ratings advantage that the networks once enjoyed. By the late 1990s, the combined prime-time audiences of the three traditional networks were down to about 45 percent, and the pundits were quick to forecast the demise of the broadcast television networks. The networks themselves were seemingly anticipating the end.

General Electric, not realizing the hoped-for profits from NBC, indicated at one time that the network was for sale if the right buyer could be found. The entertainment business was just too fickle when compared with the manufacture of light bulbs, locomotives, and military hardware. CBS contributed to the deathwatch by informing its affiliates that the golden days of station compensation were over. All three networks downsized, and layoffs became commonplace in the late 1980s. The prestigious award-winning news departments closed foreign bureaus, and belt-tightening was mandatory at all levels. The budget cuts became especially noticeable in the programming, as relatively inexpensive reality shows, such as *America's Most Wanted,* supplanted more-costly action adventures and situation comedies.

Nevertheless, the traditional television network is far from dead, especially when the bulk of prime-time viewing is still of ABC, CBS, NBC, or Fox. The role of the broadcast television network, however, will undoubtedly change in the twenty-first century, perhaps much as the motion picture industry, the radio networks, and phonograph record businesses have changed since television viewing became a force in the 1950s.

Satellite-Delivered Cable Networks

The distribution technologies available to the traditional broadcast television networks prior to the mid-1970s were the common-carrier microwave and coaxial cable links supplied by the telephone company. The cost of using this technology was high and effectively kept other, less affluent organizations from entering the networking business. Limits also existed on the number of distribution outlets or stations available as well as on the number of advertisers that could pay the high rates required to have their commercial messages distributed over the national broadcast television interconnections.

By the early 1970s, changes in regulation allowed cable television systems to build in the cities, while requiring that they carry additional channels yet limiting the number of network affiliates that could be carried. Often the cable system could carry only one ABC, one CBS, and one NBC affiliate. In addition, the public began to demand more variety in television fare.

It was inevitable that to meet the demands of the new cable television systems, the cable networks would be born. Early ones, such as Home Box Office, established in 1972, used microwave links to interconnect cable systems. Other large cable operators attempted to offer local-origination programming by establishing local cable studios or renting movies for playback on the newly developed inexpensive videocassette players. Both methods of providing subscribers with additional programming were costly and unsatisfactory to meet the demand.

Program distribution by geosynchronous communication satellite, available in the mid-1970s, proved to be the answer. Initially slow to develop because of the high cost of the receiving dish and electronics, HBO, the first cable network service, provided funds to local cable systems to help them purchase the receiving equipment. In 1976 HBO began distributing a pay movie channel nationwide over the satellite system. Ted Turner, owner of a UHF independent television station in Atlanta, was quick to realize the advantages of satellite delivery. HBO was not using all of the available time on the satellite transponder, because its movies were programmed only in the evening hours. Turner arranged to use the remaining time on the satellite for distribution of his channel 17 signal.

HBO set up a system of charging the subscriber for the use of the service. The cable operator collected a fee above the regular monthly subscription for each home that used the service. All but a small portion of that fee was returned to HBO to pay for the rights to the movies, the distribution costs, promotion, and to make a profit. The cable operator also paid a few cents per month out of the monthly subscription fee to Southern Satellite Service to cover the cost of satellite distribution of channel 17. The station anticipated making its own profit by charging the advertiser more for the increased audiences delivered by cable.

As cable television systems made the investment in the satellite receiving equipment, there was more incentive to establish cable networks. The great advantage to the satellite-distributed network was that the cost did not increase as new *TVRO (television receive-only)* dishes were added. The distribution costs to the network are the same with TVRO technology, whether six receivers are picking up the signal or

6 million. Once the cable operator makes the investment in the TVRO, the additional electronics needed to receive additional signals (assuming that they are on the same satellite) are small.

Other cable networks were quickly developed. A second superstation, WGN from Chicago, was launched in 1978. The Christian Broadcast Network was delivered to cable systems the year before that. C-SPAN and Entertainment and Sports Programming Network (ESPN) were established in 1979, and Showtime, Cinemax, Cable News Network (CNN), the Movie Channel, and the USA Network were established in 1980. MTV was launched the following year. Since that time dozens of superstations, pay channels, pay-per-view channels, and cable networks have sprung up. Some were undercapitalized and soon disappeared, some merged with other networks, some offered programming that no one wanted to see, and others competed head-on with programming similar to that offered by one of the earlier networks. Some of these have survived and now attract significant audiences and advertising dollars; others have fallen by the wayside. Table 6.6 shows cable networks and their subscribers.

TABLE 6.6   Top 25 cable networks

RANK	CABLE NETWORK	SUBSCRIBERS (IN MILLIONS)
1	TBS	76.6
2	Discovery	76.0
3	CNN	75.9
4	ESPN	75.8
5	TNT	75.5
6	USA	75.4
7	Nickelodeon	74.6
8	Fox Family	74.1
9	Nashville	73.9
10	A&E	73.8
11	Lifetime	73.4
12	Weather	72.0
13	MTV	71.3
14	CNN Headline News	70.8
15	AMC	69.9
16	TLC	68.9
17	CNBC	68.3
18	VH1	65.6
19	ESPN2	63.8
20	Comedy Central	57.8
21	History	56.3
22	Cartoon	56.2
23	BET	56.0
24	E!	55.1
25	Sci-Fi Channel	54.6

Source: From "Cable Network Subscriber Counts," *Cable World* 11(24) (June 14, 1999): 98.

Cable Network and Cable System Relationships

It took several years for the typical relationship between the local cable system and the cable network to evolve. Several different models were initially tried. For example, in its formative years when still owned primarily by the Getty Oil Company, ESPN proposed supporting its network through the sale of advertising with part of the proceeds then returning to the local cable system in the form of system compensation. Because early audiences were tiny, advertiser support was limited, and ESPN abandoned the model.

Cable television networks now are typically funded in one of several ways. Most common is the advertiser-supported network. The network sells commercial time to an advertiser who wishes to reach a national audience of a particular demographic that the specialized programming can generate. In addition, the cable network bills each local system operator several cents per month per subscriber to help pay satellite-distribution fees as well as some of the programming costs. The fee often varies in terms of the size of the system or the MSO that is contracting for the programming. The larger operators often pay a smaller monthly, per-subscriber amount than do the smaller systems and the independents. In return for carrying the network program on the limited "shelf space" of the system, the cable operator can sell several spots per hour in the network programming to local advertisers. The satellite-delivered signal comes with an inaudible tone that triggers automatic-insertion equipment to play local commercials loaded on videocassette recorders at the cable system head-end, or the technical heart of the cable system. Although local advertising is not yet a major revenue stream for the local operator, there is every reason to believe that it will become so in the future. Certain local advertisers find it advantageous to have their messages seen on CNN or ESPN in their communities.

A second method of support is more common with the pay channels. The cable operator acts as a sales agent for those channels by simply collecting the fees, returning most of them to the pay channel, and keeping a percentage for the use of the cable shelf space and the administrative effort of offering the channel and collecting the fees.

There are additional methods of support. Some channels are offered free to the local cable operator and carry no advertising. For example, religious channels are often supported by the fund-raising efforts of the programming and the subsequent donations by viewers. Tremendous sums have been raised in this way by some of the channel operators, as well as by some public broadcasters. Other cable channels pay the local cable system to carry the channel. The home-shopping channels, for example, pay the cable operator a percentage of each sale made in a ZIP code that the operator serves.

Each cable network must maintain affiliate relations departments to ensure that the network is carried on as many systems as possible. With more networks available than can be carried on the limited channel, or shelf space, of the cable system, the local cable system programmer or manager becomes the constant target of incentives designed by the cable network to keep the programming on the system. Representatives from the cable network call on the system managers frequently to offer marketing services, sales statistics, and premiums, such as HBO coffee cups or Showtime baseball caps, that can be passed on to new cable subscribers. Often the promotional packages that the networks offer can aid the cable system operator in obtaining more subscribers.

Satellite-delivered cable networks are a relatively new venture, with new entries each year as others go out of business. Currently, the local system manager can choose from more than a hundred satellite-delivered, basic cable networks, premium cable channels, and pay-per-view services to fill his or her complement of cable channels. The manager must find the mix of off-the-air signals, cable networks, and pay channels that will attract and hold the largest number of subscribers. In so doing, he or she must work within the limitations of the available shelf space and the local and national requirements for specific channel types—such as public and educational access. In addition, the system must receive value for the money that must be spent for programming copyright payments and channel charges.

Content Providers

Although the broadcast television station and the local cable television system can be thought of as the retailers of electronic entertainment and information, few of those outlets are in the business of producing or distributing programming (with the exception of local news, local public-affairs programs, and a limited amount of local sports programming). Even with the recent emphasis on news, locally produced fare constitutes a small percentage of the television broadcast day. Local-origination and local-access programming add up to even smaller percentages of the programs available on the cable television channels. Radio programming is different. Even though many stations use programming services from networks, and most of the music played is from mass-produced compact discs and minidiscs, radio programming is still considered local if a local announcer or disc jockey introduces the music.

The agency that produces the programming—be it distributed by radio, television, or cable—generally has control over both the creativity of the production and the message that is delivered to the audience.

Television Production Companies
If the programming of the television station is not local, where does it come from? Most of the programming, called *product,* found on a local affiliated television station is created and delivered by the network. The majority of that product is produced in either Hollywood or New York. The networks themselves produce much of the daytime as well as late-night programs, but were forbidden by FCC regulations from owning very much of what we saw in prime time. In the past they bought the rights for prime-time programming from a major production company, which then produced the program or series to fit the needs of the network. With the repeal of the FinSyn rules, television networks can now produce and own their own prime-time programs, although much programming is still produced by production companies.

Many television production companies have names that are familiar to moviegoers, including Sony Picture Studios, the Walt Disney Company, Columbia-Tristar, Universal, Paramount-Viacom, Twentieth Century Fox TV International, MGM, and Warner Brothers. Many of these are the big Hollywood studios that produced the motion pictures of the 1930s, 1940s, and 1950s.

Other television production companies have developed around the talent of a producer, writer, actor, or director. Some of the more recognizable names in past years include Cannell Studios; Dick Clark Productions; Bright, Kaufman, Crane; Merv Griffin Enterprises; Carsey-Werner; MTM Enterprises; Orion Television Entertainment; the Fred Silverman Company; Aaron Spelling Productions, Lorimar Telepictures; and Viacom International. Most contemporary prime-time television programming is produced by one of these studios or production companies.

In addition to these major players, thousands of smaller organizations around the country produce entertainment programming, educational programming, commercials, and corporate videotapes. Many local television stations own production companies that produce local commercials for advertisers. Much of the output of these organizations, however, is limited to commercials or relegated to small-market stations and cable television systems.

Radio Production and Distribution
Many radio programming services produce audio commercials, entertainment programs, or radio news or offer full-service radio networks. ABC owns and operates several radio networks. Other radio services include Associated Press Broadcast Services, American Urban Networks, USA Radio Networks, and Westwood One, which distributes the CBS Radio Network, NBC Radio Network, and CNN Headline News. There are about thirty additional satellite-delivered national radio networks, some of which offer as many as ten different program formats, each targeted to a different demographic. There are also more than a hundred regional radio networks, some specializing in foreign-language programming, others in news or regional sports. In addition, there are nearly seventy-five format providers that deliver weekly programming on reels of audiotape to automated stations.

Other producing agencies also make radio programming, including: the U.S. Army Reserve, American Chiropractic Association, American Stock Exchange, Beethoven Satellite Network, CNN Radio Network, the Good News Broadcasting Association, and the National Council of Churches Communications Unit.

Syndication
With about 60 percent of its programming provided by the network and an additional 10 percent produced by the network news department, the

network-affiliated television station must look to outside sources for the remaining 30 percent of its programming. The independent station, on the other hand, with no network affiliation, no network programming, and often without a local news department, must buy the rights to all of the programming it broadcasts. Most of this is obtained from a programming syndicator.

The programming *syndication* business is based on the syndicator's purchase, from the producer, of the rights to broadcast a program or series in television markets. The syndicator can sell the same program again and again to stations in each of the two-hundred-plus television markets throughout the United States. In addition, the same programs can be sold in the international markets to stations around the world. Although movie packages are sold to television stations in the same manner, syndicated television programs are of two types: *off-network programs* and *first-run programs*. Off-network syndication includes television series that were originally in prime time on one of the three broadcast networks. These become the reruns of *Friends, Seinfeld,* or *Mad About You.* If the program was popular on the network and stayed in production long enough to accumulate about a hundred or more episodes, local stations are often eager to show the reruns five days a week in early *fringe time*—usually 4 to 8 P.M. eastern standard time. Audiences are usually larger in early fringe than at any other time period except prime time.

First-run syndication is produced for sale directly to the local stations and has not been broadcast over a network. This programming includes such favorites as *Oprah, Judge Judy, Jerry Springer, Wheel of Fortune,* and *Jeopardy.* These new programs are sold market by market.

Syndicators range in size from the large distributors operated by each of the major studios, such as Paramount, Universal, and Buena Vista Television (the Walt Disney Company), to the small syndicators with one series in their catalogue, such as Phoenix Communications Group from South Hackensack, New Jersey; Northern Lights Communication of Minnetonka, Minnesota; or Fishing the West (both the syndicator and the program name). Although some companies produce their own product, others specialize in obtaining the rights of programs produced by others and distributing them both in the United States and abroad. Some of the international syndicators include Granada Media International (British programs), Octapixx Worldwide (Canada), Salsa Distribution (France), TV Azteca (Mexico), Carlton America (England), and Deutsche Welle/Transtel (Germany).

Given the number of producers, programming suppliers, and syndicators, it would appear that there is no dearth of product to fill the screens and speakers of the electronic media. But the programming diversity provided by the hundreds of producers, distributors, and broadcasters seems not to reach the customer. Because the advertiser wants to reach the largest possible audience, only the most popular programs ever reach the public—and thereby continue to be the most popular programs. Those that are successful in one market are the only ones considered for broadcast in another market. The producers that have had programs on the network—even programs that have failed—are those that are trusted to produce other programs. An unknown with fresh but untried ideas has a difficult time breaking into the business of broadcasting. Thus, although the industry boasts a multitude of media outlets, producers, and distributors, the public is often treated to the same old retreaded programming or warmed-over copies of whatever tops the ratings at that time. (See chapter 9 for a further examination of syndicated product in television programming.)

THE CONSULTANTS

Consulting firms that advise the broadcast and cable industry come in all sizes and specialties. Stations and cable systems hire consultants, who are outside experts, to examine and offer advice on a particular problem, from a technical question concerning the electromagnetic spectrum to a programming decision about attracting the largest possible audience. Some consultants serve as employment and talent agencies that locate on-air talent and executives for the local station. Others specialize in public relations and promotion for the broadcaster. As already discussed there is also the need for technical and legal consultants to help the station obtain a broadcast frequency and license. A Washington, D.C.–based communications

Local television anchor team on set. Local talent is constantly evaluated, both by program ratings and by research, on their believability and likability. Often focus groups determine the fate of on-air talent.

lawyer who is a member of the Federal Communications Bar Association is necessary when the station must formally argue a case before the Federal Communications Commission. Although not all are members of the Federal Communications Bar, nearly six hundred law firms represent stations and cable television systems in the nation's capital. In addition, almost every station retains an attorney in its city to deal with local, usually business, issues. Broadcasting has become a complex business fraught with many legal traps and loopholes.

About 250 technical consultants advertise their services to broadcast stations. Many offer full service, from engineering studies and studio design to equipment sales and installation. Some technical firms specialize in broadcast transmission maintenance and calibration; others have crews that will change the marker lights on the broadcast tower. Some cable television technical firms specialize in the construction of the cable plant. Their turnkey construction operations (that is, from groundbreaking to "turning the key" in the completed building's door) allow the system operator to simply walk into the new building and "flip the switch" to start cable service when the system is completed.

Management consultants for the electronic communications industry will analyze the administrative structure of the station or cable system to detect weaknesses and suggest changes to increase profitability. Financial consultants will arrange funding to establish a new station or system or will find money to enhance operations.

Although all of these consulting services affect the businesses of broadcasting and cable television, their impact on the content delivered to the audience is minimal when compared with that of the program consultants and research services. These groups are hired by broadcasters and cable operators to analyze the wants and needs of the community and to develop program services that will attract the largest audiences.

The best-known audience research firm in the industry is Nielsen Media Research, but many other firms also provide research services with information about program usage, program likes and dislikes, music preferences, and on-air talent likability and effectiveness. About 150 firms advertise broadcast research services for local and national markets; many more undoubtedly exist that provide strictly local audience measurement services. Generally, all that is required for a local researcher to collect and analyze data is a telephone, a personal computer, and knowledge of sampling and statistics. Specific methods of research are discussed in chapter 11.

Some local radio stations conduct music and audience research; others belong to group ownerships that support research divisions or subscribe to research organizations. No matter where the data is generated, it is used in an increasing number of stations to make every programming decision.

Many television stations use the services of programming and news consultants to help determine which syndicated series to purchase and whom to hire as the local news anchor.

One of the most controversial uses of consultants has been in the area of news. By the mid-1970s local television stations began pouring money into local news after discovering that audiences and advertising dollars could be generated with properly packaged information. (The evolution of news programming is discussed in detail in chapter 10.) Marketing experts, in the guise of news consultants, helped them develop the package. Two of the earliest and most influential firms were Frank Magid and McHugh-Hoffman. With a new emphasis on film or videotaped action, glitzy graphics, music, and warm and friendly anchors, audiences flocked to the television screen. The consultants multiplied as each local network affiliate contracted to have its newscast developed into the number one newscast in the community. The consultants gained power, and the local news directors rarely made a move without their advice.

News formats developed a sameness as the consultants sold their services from market to market. News content degenerated into human-interest features, happy talk, and tabloid journalism—but the result was a profitable bottom line for the stations. As the emphasis in television news moved from journalistic content to anchor personality, there is no question that the news consultants had significantly influenced what audiences watched on local stations.

Unions and Guilds

In many local markets and in network environments, labor unions and professional guilds exert some control over the operation and programming of broadcast stations and cable television systems. Personnel involved in the business of program production can be divided into *above-the-line personnel* and *below-the-line personnel*. These designations come from movie studio accounting procedures that separate budget items between those that appear on the screen (above the line) and those that are behind the camera (below the line).

The guilds, which represent professional, above-the-line personnel, have developed an array of rules that governs working conditions, credits, pay, benefits, and residuals. Some of the major guilds with which the networks and major production companies must negotiate include the Directors Guild of America (DGA), the Screen Actors Guild (SAG), the Screen Extras Guild (SEG), the Writers Guild of America (WGA), and the American Federation of Musicians (AFM). These organizations represent members primarily in New York and Hollywood production. The American Federation of Television and Radio Artists (AFTRA) represents members in the major production centers and at the local station level. In a union market, the news anchor and the booth announcer are probably both members of AFTRA.

Below-the-line personnel are covered by a variety of technical unions. Many of the production crafts are represented by the International Alliance of Theatrical and Stage Employees (IATSE). In Los Angeles twenty-four different locals of this union represent the range of crafts required for motion picture and television production: Local 659, for example, represents photographers and camera operators; Local 847 protects set designers; and Local 892 represents the costume designers. As with the guilds, the union contracts specify wage scales, hours, benefits, travel times, and working conditions. In addition to the IATSE members, the studios must also negotiate with several other unions, including the Studio Transportation Drivers, the Production Office Coordinators and Accountants Guild, the Studio Utility Employees, and the Police Officers and Firefighters associations.

Unionized engineers at local stations and at network locations in New York and Chicago usually belong to either the National Association of Broadcast Employees and Technicians (NABET) or the International Brotherhood of Electrical Workers (IBEW). Unionized cable employees are often members of a local of IBEW or belong to the Communication Workers of America (CWA). There are dozens of other unions and guilds that represent those working in the electronic mass media.

The guilds and unions have shaped the content of programming over the history of broadcast television. Certainly, the cost of production determines the level of sophistication that the producer is willing to provide in any program. Much of the cost is derived from the salaries and wages paid to both above- and below-the-line personnel. In addition, union and guild members bring their skills to the production of a television program. When television was produced in New York in its formative years, the look and style of the programs betrayed the stage backgrounds of those working on its production. When production moved to Hollywood, the film backgrounds of local workers found its way into the television production, and the programs produced there took on the look and style of Hollywood.

Music Licensing

Much of the programming on radio and television is composed of or includes music. To use a copyrighted script or musical composition, the radio or television programmer must obtain permission from the author or composer and pay whatever fees are required. For the radio station that broadcasts multiple musical selections during each day, the process of identifying and contacting each copyright owner for permission to use the work would be daunting. Luckily, stations are spared the task of clearing each piece of music separately by the dozen or so organizations that secure copyright clearances for broadcast stations. The major organizations used by U.S. stations are the American Society of Composers, Authors, and Publishers (ASCAP), Broadcast Music Inc. (BMI), and the Society of European Stage Authors and Composers (now SESAC, Inc.).

Most stations pay a small percentage of the gross revenues to these music licensing agencies for a blanket license. Radio stations that broadcast music must pay copyright for the records or CDs that they play. Television stations broadcasting syndicated programs that include music for bridges and background must also carry blanket licenses to cover the rights for the rebroadcast of that music.

Professional Associations

More than two hundred professional electronic media societies are national in scope. In addition, nearly every state has both a broadcasting association and a cable television association.

At the national level, the National Association of Broadcasters (NAB) and the National Cable Television Association (NCTA) have developed large staffs as well as many committees to work on developing and supporting policy positions concerning the complex issues facing the business and science of electronic communications.

Other associations are concerned with media content. The National Association of Television Programming Executives (NATPE) operates a convention each year that has become the marketplace for syndicated television programming. Programming syndicators and producers buy floor space at the convention to display their programming packages to station managers and programming directors from around the country. The Radio and Television News Directors Association (RTNDA) provides a forum for the discussion of issues of concern about the programming of news and public affairs.

Other organizations help stations and cable television systems sell commercial time by providing statistical information, sales data, and tips for use by research departments and account representatives. An organization that provides this service for the cable industry is the Cabletelevision Advertising Bureau (CAB). The Television Bureau of Advertising (TvB) and the Radio Advertising Bureau (RAB) serve the same function for television and radio stations. Much of the data used in the commercial sales process is based on audience ratings. The Electronic Media Rating Council is a watchdog organization that oversees the methods and statistical analysis used by the rating companies to generate information about the audiences of radio and television stations, as well as of the cable networks.

Several engineering societies provide educational services for the technical personnel employed in the electronic mass media. The Society of Motion Picture and Television Engineers (SMPTE), the National Television System Committee (NTSC), and the Advanced Television Standards Committee have played major roles in developing the technical standards the FCC has adopted for the transmission of television signals.

Other professional organizations encourage and reward excellence in programming. A variety of awards, from the cable ACE to the advertising Clios to the motion picture Oscars, recognize the best of the production for that year. Perhaps the most prestigious awards in television are the Emmys given by the Academy of Television Arts and Sciences (ATAS) and the National Academy of Television Arts and Sciences (NATAS). These awards for outstanding programming and production expertise are determined by blue-ribbon panels of producers, writers, directors, and others involved in the production process. Emmys not only are given at the national level, but are also used to encourage the production of outstanding programs at the regional level.

Professional societies, often made up of volunteers, have contributed in many ways to the electronic media professions and to shaping the content of the media. Each of the many different rules, sets of standards, and recommendations influences the programming the audience eventually receives.

Business Infrastructure

Although the many different societies, unions and guilds, networks, programming distributors, licensing agencies, and media conglomerates have input and some level of informal control over the programming content of the electronic mass media, the local station or cable system remains ultimately responsible for what the public sees and hears. The FCC charges the local station with the responsibility of serving its audiences. As the retailer of information and programming, however, the local station or system manager must serve those audiences to make a profit. He or she can steer the station toward profitability only by providing audiences with what they want and with the information that informs them about their community.

The organization of the local electronic media outlet is structured for the purpose of generating an audience for the programming of the station or cable system and then selling to the advertiser access to that audience. The advertiser is charged for the time of the station in direct relationship to the quantity and quality of the audience that views the commercial. The advertiser will pay more for access to larger audiences and more for the desirable demographics or for those viewers who, because of age, education, or income, are likely to be consumers of the services or products offered in the commercial.

One goal for the management of the station or system, therefore, is to provide the most popular programming to the greatest number of people. The cable operator can increase revenues by offering channels of programming that will attract the interest of the potential subscriber and therefore increase *penetration,* or the percentage of subscribers. The broadcast station manager can increase revenues by selling the limited commercial time for more money. The advertiser will pay more money for larger audiences or better demographic quality. The manager must therefore strive to increase the size of the audience by providing the programming to attract them.

Summary

▶ There are many influences on the content selection of the electronic media, including the ownership patterns of the 12,000 radio stations, 1,500 television stations, and almost 12,000 cable television systems that operate in the United States.

▶ Most radio and television stations are owned by broadcast groups. Those stations owned and operated by the traditional television networks in the top markets are still among the most profitable, even though network prime-time ratings have slipped considerably in the past decade.

▶ The local television or radio station is the retailer in the electronic media business and is responsible for its content. Local cable television systems are also retailers of television programming in that they deliver many television signals to the subscriber for a fee.

▶ Both the local cable system and the local broadcast station are organized from an administrative perspective to protect the license or franchise and to make a profit. The broadcaster does this (1) by creating the signal through the efforts of the engineering department, (2) by producing or providing the content within the programming department, (3) by selling the audiences generated by the programming through the sales department, and (4) by collecting revenues and controlling expenses through the business department. The news and promotion departments have gained prominence in the contemporary television station.

▶ The cable system often has departments similar to those of the broadcast station; but because their customers are different, cable systems often require additional staff, the most important of these being the customer service department, because it is often the only direct link with the cable subscriber.

▶ Many related businesses that are ancillary to broadcasting and cable also determine media content, especially those organizations involved in the production and distribution of programming for radio, television, and cable. These organizations include the traditional broadcast networks, cable networks, programming services, production companies, programming syndicators, and consultants.

▶ Other organizations that influence programming content include copyright licensing companies, trade unions and guilds, lobbying groups, and professional organizations and societies.

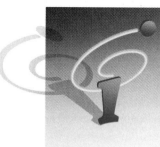

InfoTrac College Edition Exercises

In recent years, especially after the ownership limits were changed by the passage of the Telecommunications Act of 1996, the number of broadcast properties bought and sold has exploded exponentially. Convergence of new technologies, as well as concentration of ownership, has created a new universe of media conglomerates. The daily shifting of the broadcast ownership sands has one bottom line: get big or get out.

6.1 Chapter 6 provides some indication of ownership patterns in broadcasting and other media. The real power behind the new media companies, though, often comes down to a dynamic leader.

  a. Using InfoTrac to access *Broadcasting & Cable*, find articles that profile six network or broadcast group owner CEOs or other executives. Write a short report on each and their plans for their respective companies.

  b. Many of the names that you will find as you search *Broadcasting & Cable* articles have been involved in leadership positions in other large media companies. Using InfoTrac, develop a short biography for one of the persons you identified in (a) above.

6.2 Recent mergers, buyouts, and joint ventures have, in many cases, included one of the traditional television networks (ABC, CBS, or NBC). Using InfoTrac to access *Broadcasting & Cable*, find articles about the networks' expansion into other media businesses such as cable networks or dot-com companies.

6.3 Radio ownership has undergone dramatic changes in recent years. Use InfoTrac to access *Broadcasting & Cable* and complete the following.

  a. Locate articles that refer to the largest of the radio group owners.

  b. Can you find articles that question the concentration of ownership that has taken place in the radio industry since the Telecommunications Act of 1996?

CHAPTER 6　　PATTERNS OF MEDIA ORGANIZATION AND OWNERSHIP　　　　　**147**

WEB WATCH

Here is a list of a few URLs (Internet addresses) for some of the organizations or corporations discussed in this chapter. Please explore these Web sites and follow the links to learn more about the complex business of the electronic media. Add your descriptions and your own favorite sites at the end of the list. Please keep in mind that the dynamic nature of the Internet allows sites to come and go but also allows organizations to update information about themselves very quickly.

Address　　　　　　　　　　　**Description**

http://www.scripp.com/

http://www.cox.com/

http://www.sag.com/

http://www.tribuneinteractive.com/about/index.htm

http://www.dga.org/

http://www.emmyonline.org/

http://www.nab.org/

http://www.bmi.com/

http://www.tbn.org/

http://www.gannett.com/map/television.htm

http://www.chriscraft.net/

http://www.medialinkworldwide.com/

http://www.ktla.com/

http://www.ascap.com/ascap.html

http://www.sesac.com/

http://www.phoenix360.com/

Other favorite sites:

CHAPTER 7

COMMERCIAL AND ECONOMIC STRUCTURES

The electronic media operate to make a profit: Cable systems sell service to subscribers, and broadcast stations sell the audience to advertisers.

Broadcasting and other electronic mass media, such as cable television, direct broadcast satellite (DBS), wireless cable, and home video, developed in a capitalistic society in the United States. The economic structure that supports our electronic media systems is different than that of some other countries, where the broadcast media were financed, owned, and operated by government for the purpose of providing a nation's people with information and education that the government felt were necessary. In Great Britain, for example, early broadcast efforts were organized under the British Broadcasting Corporation (BBC), which is owned and operated by the British government. The public paid for the programming through a tax that the government levied on radio and later television receivers.

The U.S. system, however, was established by entrepreneurial individuals or corporations that wished to make a profit by providing communication services to the public. They were willing to invest in the development of the technology needed for an electronic mass communication system as well as provide the resources necessary to create information and entertainment that would attract large audiences. This chapter examines the economic structures that have evolved to support the U.S. electronic mass media.

When broadcasting started in 1920, no one really knew how to make it into a profitable business. Some eighty years later, radio and television stations were estimated to generate more than $40 billion in revenue from advertisers, and cable television systems were estimated to receive nearly $30 billion, most of which came from subscribers. In addition, a variety of businesses have grown to support the electronic media outlets, including advertising agencies, station representatives, media consultants, programming producers, audience rating firms, programming syndicators, equipment manufacturers, talent agencies, law firms, media buyers, financial services, and station brokers, to name just a few. Collectively, the electronic mass media and associated support services generate a substantial amount of revenue.

Traditional Economics

Commercial broadcasting is unique as a business. A business exists to make a profit on an investment, generally from the sale of a product to a customer. The

Station personnel at work in the newsroom of a television station. The local newscast is a very important element in attracting the audience to the station's programming. That audience is then sold to an advertiser, who hopes that the viewers will also watch the commercials and buy the advertised products.

customer perceives a need for the product and is therefore willing to pay more for it than the business spent for its manufacture and distribution. In broadcasting, the *product* is the information or entertainment delivered to the audience in the form of programming.

There are two customers associated with a commercial broadcast program. The first and most important is the audience; the second is the advertiser. Broadcasting is unique because the product, in the form of the program, is delivered to the first customer for free. Normally, it would be difficult to make a profit by giving away the product for free, but the broadcaster turns to the second customer for the money to pay the bills and to make money for the stockholders. The advertiser is willing to pay the broadcaster for access to the audiences who are listening to or watching the programming. The advertiser prepares commercial messages designed to create a need for the advertiser's product or service. Presumably, the commercial messages aired next to a program will be heard by the audience for that program. If the audience is large, more people will be exposed to the commercial message. If many people are exposed to the commercial message, the advertiser is willing to pay more money to the radio or television business to have the message broadcast. If a large audience also comprises persons who are more likely to buy the advertised product because of their demographic composition, the advertiser will be willing to spend even more.

The sale of broadcast station time to the advertiser is the way that traditional radio or television stations make the money needed to operate the station. It is also the basis of the profit that stockholders of the broadcast property wish to make from their investment. Although some stations have developed additional revenue streams from the sale of video production services, and others have made money from selling ancillary services such as broadcasting closed-circuit music and information to businesses, the real business of most broadcast stations revolves around selling commercial time, usually in the form of 30-second *spots,* or commercials. Prices can vary from a dollar for 30 seconds on a small-market radio station to a million dollars or more for 30 seconds on a television network during the Super Bowl.

The value of the time that the station sells to the advertiser is based on many factors, including market size, coverage area, competition, and audience demographics.

Market Size

The more people watching, the more the time is worth to the advertiser. The role of the broadcaster is to entice as many viewers or listeners as possible, to make the time of the station more valuable. Some factors that create large audiences are within the control of the broadcaster; others are not. For example, if many people live in the coverage area of the station, there will be more people to turn to the station. More people are crowded into the big cities of New York, Los Angeles, and Chicago than live in the rural areas of the country. A station serving a large city will almost always be guaranteed a larger audience, and therefore more money for a 30-second commercial spot, than one located in Flagstaff, Arizona, or in Lexington, Nebraska.

Coverage Area

Another factor that determines the number of people the station can reach is the technical facilities that the Federal Communications Commission (FCC) has licensed the station to use. For example, a 5,000-watt AM radio station will not reach as large a geographic area as a 50,000-watt station. Conversely, a television station licensed to operate at full power on VHF channel 2 will be able to cover a larger geographic area, and therefore reach more viewers, than a UHF station licensed to operate on channel 64 at maximum power (see figure 7.1). Generally, the lower the assigned channel, the greater the coverage area (given the same power and antenna height).

Competition

Also influencing the potential audience of a commercial is competition from other media outlets. If there are many radio and television signals to select from, the potential audience will be fragmented, with different viewers watching different stations. If there are only a few choices in the market, the percentage of available audience will be larger for each station. If a market has only three television stations, theoretically each would receive 33 percent of the available audience. If there were only two stations serving that market, each station could be expected to receive 50 percent of the audience. As late as 1980, each of the three network affiliates in a smaller market could expect to receive about one-third of the available audience. Since cable entered most markets, audiences now are divided among 20 to 100 possible channels.

Market size, coverage area, and media competition are beyond the control of the broadcaster, but the size of the audience can be influenced by the type of programming that is broadcast and by the amount of promotion provided.

Audience Demographics

To attract large audiences, the broadcaster carefully researches the market to determine who the available audiences are and what types of programming they are most likely to watch. Different types of programs attract different types of people. Advertisers not only want to reach large audiences, but they also want to make sure that the people who receive their messages are able and willing to buy the advertised products. The broadcaster designs the program schedule to attract the audiences that the advertiser is likely to want to reach. In addition, the programming schedule is designed to attract different types of audiences in different time periods: Audiences are composed of preschool children in the midmorning and early-afternoon hours, schoolchildren in the late afternoon, families in early prime time, and adults later at night.

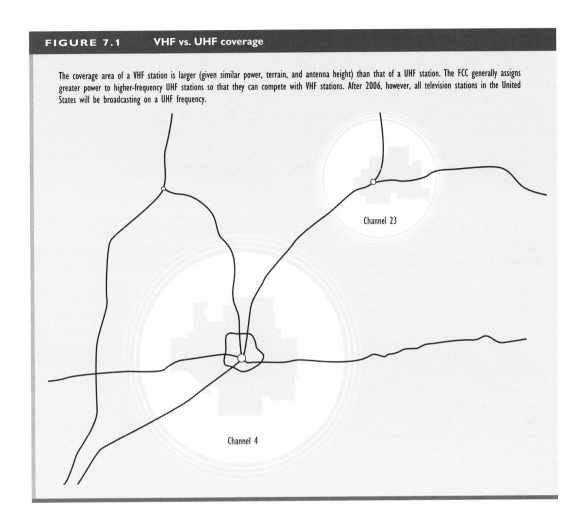

FIGURE 7.1 **VHF vs. UHF coverage**

The coverage area of a VHF station is larger (given similar power, terrain, and antenna height) than that of a UHF station. The FCC generally assigns greater power to higher-frequency UHF stations so that they can compete with VHF stations. After 2006, however, all television stations in the United States will be broadcasting on a UHF frequency.

Many broadcasters attempt to maximize the audience by carefully examining the programming schedules of competitors and then providing a different type of programming than that offered by other stations in the market. Such *counterprogramming* strategy often results in the maximum number of viewers watching different types of programming on each station.

The broadcaster must also attempt to create the largest possible audience by promoting the programs on the station. The promotion, or advertising, is needed to tell viewers or listeners to stay tuned to the station or to tune in at a specific time to find more of the entertainment and information they like.

The large audience is what adds value to the commercial time a station has for sale. Without that audience the advertiser would not be willing to buy time on the broadcast station. The advertiser is, in effect, buying the audience that the station has generated through its programming and promotion efforts. The size and quality of the audience is measured by a ratings company, such as the A. C. Nielsen Media Research Company, which estimates the number of homes watching a particular television program. In addition, the ratings company provides estimates about the composition of the audience. The data of audience age and gender give the advertiser an indication of the quality of the audience in terms of its ability and willingness to buy the advertised product.

The Advertiser as Customer

In today's competitive world, many different companies make similar products or offer similar services. For an organization to survive, its product or service must be *marketed*—actively sold to a consumer. The consumer must be told of the product and convinced of its advantages over the competition. In today's era of mass marketing, advertising through the mass media is a necessity for many companies.

The marketing goals of the advertiser vary, but most advertisers, in an attempt to sell more of their products, will tell consumers how the product will make life easier or happier. The advertiser also will often show the consumer how to distinguish its product from the inferior one offered by the competition. Another marketing goal is to tell the potential buyer where the product can be obtained and at what price.

Advertisers generally attempt to create a market position with advertising through the mass media. *Positioning* is the process of creating a uniqueness in the mind of the consumer through advertising by creating an image for a particular brand. The organization that creates that image can often become very successful in marketing its product. For example, one car manufacturer may position its product as the choice of upscale, single professionals (as in the Infiniti campaign); another may appeal to families by emphasizing safety features (as in Volvo ads). Successful positioning shows members of the target audience that the product fits their perception of themselves: People who like good times at the beach drink product A; people who appreciate the finer points of craft brewing drink product B. Positioning now must be used by other manufacturers to differentiate their similar products from these.

The brand image established for a product is usually the result of advertising and promotion over a long period of time, and the image is a combination of rational and emotional appeals. The consumer buys toothpaste because it is necessary to brush teeth to keep the mouth clean, prevent cavities, and have clean-smelling breath. Probably most toothpastes will do the job, and the rational appeal of the advertising message will tell the consumer the active ingredients that make the toothpaste effective in preventing cavities. The emotional appeal comes into play when the advertising implies that the consumer will be more popular if he or she uses the advertised brand of toothpaste.

Because the electronic mass media can reach almost every person in the United States on a daily basis, because they can effectively deliver both emotional and rational appeals, and because they can target specific demographic audiences, radio and television have become extremely popular advertising outlets. They are so popular that nearly 20 percent of everything that we see or hear through the electronic mass media is a commercial advertisement or a promotional announcement.

The Sales Department

The sales department is a major force in the radio or television station, because it is responsible for bringing money into the station to pay for operations and to make a profit by selling the station's commercial time to advertisers. In many stations the sales department is also an important member of the advertiser's marketing team to help make sure that the advertising time the client buys is effective in meeting marketing goals. A happy advertiser will continue to buy commercial time.

Sales departments vary in size, depending on the size of the station and the size of the market that the station serves. Generally, the sales department is headed by a general sales manager, who sets sales policy, develops sales strategies and client lists, directs the activities of the account executives, and administers the activities of the sales department.

Larger stations may employ a local sales manager and a national sales manager, who report directly to the general sales manager. The local sales manager directs the efforts of the sales reps who call on local advertising agencies and businesses. These local businesses, who might buy local spot advertising, could include an automobile dealership, a grocery store, or the neighborhood dry cleaner. On average, local sales revenues account for 80 percent of the income for a local radio station, 60 percent for an independent television station, and 40 percent for a network-affiliated television station.

The national sales manager usually works with a station representative company to obtain national spot advertising. The national advertising is usually bought by the advertiser through an advertising agency which works with the station representative company to place the commercials on stations around the country. Examples of national advertising include spots for large corporations that manufacture and sell brand-name merchandise. Ford, Nissan, Toyota, Coca-Cola, Pepsi-Cola, Maytag, Dial soap, and MCI telephone service are all national advertisers that buy commercial time on radio and television stations at the national spot level.

Many local radio and television stations rely on national spot advertising for a large portion of their revenue. Generally, national spots account for about 20 percent of the revenue for the local radio station, 40 percent for the independent television station, and 50 percent for the network-affiliated television station. Some network-affiliated television stations have a third source of income—station compensation, which is money paid to the affiliated station by the network for the local station's time that the network sells on the national advertising market. The network often bills the affiliated station, however, for production, promotion, and program-delivery costs. That cost often is charged against the expense of the station's airtime. Because of these offsetting bills from the network, very little actual income is derived by the affiliated station.

In addition to selling the station's commercial airtime, the sales department has several other related functions generally grouped under the term *sales service*. One of those major functions is keeping track of each second of the station's commercial time inventory. Sometimes called *traffic*, this department's function is to log each commercial spot that is sold, including the name of the advertiser, the source of that spot, and the identification number of the tape that will be used to play the commercial over the air. The traffic department provides a log of all the programming and commercial spots to the operation's department to air each day. The log not only serves as the schedule for the broadcast day, but also becomes a signed affidavit that provides proof to the client that the spot was broadcast at the time and day requested.

Keeping track of each advertiser's schedule of spots—called a *flight*—is a complex process. To reach a specific demographic audience, most advertisers have specific requirements as to the time of day that they want their spots to run. Most advertisers schedule a flight of spots up to six months in advance, and many have a variety of different commercials running on a station at different times of day. It takes a sophisticated computer program to track all of the information necessary to schedule the thousand or so 30-second commercials that the typical radio or television station handles each day. In addition to scheduling spots for multiple advertisers, the traffic department must keep track of unsold time so that the account executives can provide other clients with available times for running their advertising. The current unsold spot time periods are called *avails*, or spot availabilities.

At the close of each business day, the traffic department prepares a programming log for the next day. That log must account for each second of time in the broadcast day, including all programs, all sold commercial time spots, *station identifications (IDs)*, *station promotional announcements (promos)*, and *public service announcements (PSAs)*. Rules established by the Federal Communications Commission determine when station IDs must be run, and most stations have policies that determine how many PSAs and station promos will be aired in each hour of scheduled time. Many stations also sell lower-priced commercial spots called *run of schedule (ROS)*, which means they can be placed in the schedule at the last minute if a time slot has not sold before the log for the next day is prepared.

The sales service area is often asked to provide a library service for an advertiser who might have several versions of a commercial that are scheduled to air at different times. The various versions of the advertising spot are kept at the station for use when needed. Sometimes a station will provide copies of the commercial to other stations to air over their facilities. Many stations provide storage, duplication, and delivery services to the advertiser client.

In some radio and television stations, sales service can provide complete production services. The advertiser, especially in a smaller market, can rely on the creative talent and technical facilities at the local station to conceive, write, and produce the radio or television commercial. Some stations provide such services free to the advertiser, but most operate production services as a profit center. The advertising client pays for the production of the commercial in addition to buying the airtime to broadcast the message.

Many stations provide promotional services to their advertisers, including promotional tie-ins, on-air contests, and remote broadcasts from the sponsor's place of business. Such promotional activities are very common in the highly competitive radio markets.

Audience research service is almost always provided to the advertiser as well. It is often used by the account executive as a way of convincing the potential client that he or she should advertise on the station. Generally, the research data is simply a quick

Computer display of demographic data. Using a laptop, the account executive can display research information, including demographics, to the client during a sales presentation.

A sales rep pitches the advantages of placing commercials on cable networks to a media buyer.

interpretation of the ratings information that the station buys from an outside ratings company. Often it will tell the advertiser only about how many homes were listening or watching during the time period that he or she is considering. Other stations employ research directors to carefully analyze the ratings information so that demographic data can be obtained. Chapter 11 discusses ratings and research methodology in detail.

SELLING TOOLS OF THE BROADCAST STATION

Research results are among the tools that the account executive brings to the potential client when attempting to encourage him or her to advertise on the broadcast station. The ratings numbers are often used to show the advertiser how many homes or how many individuals of a particular age or gender are in the audience and are therefore exposed to his or her commercial message. The advertiser often measures the value of the advertising in terms of *cost per thousand (CPM),* a standard method of determining the effectiveness of a commercial on one station versus on another station or on radio versus television. To determine the CPM, the price of the spot is divided by the number of listeners or viewers. Thus a $500 commercial on a television station that is seen by 10,000 viewers equals a CPM of $50. If the audience for the spot were 100,000, however, the CPM drops to $5—obviously a better deal.

Another way to determine the cost of a flight of advertising spots is through a *cost per point (CPP).* Using the ratings information, the advertiser determines how many ratings points the commercials accrued during the week and then, by dividing the total cost, can determine a cost per ratings point. Gross ratings points are the total number of ratings points that are achieved by a flight of commercials. The method for determining CPM is as follows (for more information, see chapter 11):

$$\text{Cost of spots/gross impressions} \times 1{,}000 = \text{cost per thousand (CPM)}$$

Many advertisers are interested in the quality of the audience as well as its size. Demographic breakdowns give the advertiser an indication of the age and gender of the audience for a particular program. The advertiser can therefore buy time next to a program that will draw the type of audience most likely to purchase the product. It doesn't make sense for the advertiser to place a commercial for an automobile next to a Saturday-morning children's program or to buy time to tout a new action toy next to a Saturday-evening news broadcast. In addition, an inexpensive commercial may not be a good buy if it doesn't reach the correct audience.

Psychographics is a type of research often used to define the audience even further. This lifestyle information can give the advertiser a complete picture of the buying habits of the viewer or listener. Some audience "values" research can even tell the advertiser the income and buying habits of the audience by ZIP code.

Another tool that the sales account executive uses is the *rate card.* Because the audience size varies with the dayparts, the time that the station has for sale varies in value and therefore price. Prime time in television has the greatest number of viewers; therefore spots sold in that time period cost the advertiser more than those available during other dayparts. In radio, morning-drive time is considered prime time, and spots sold from 6 to 9 A.M. will probably cost more than at any other time of day. The rate card divides the broadcast day into dayparts and indicates the different rates for each. In addition, the card will usually offer a *frequency discount*—a lesser cost per spot if many spots are purchased at one time. Generally, the advertiser will do this because broadcast advertising relies on repetition to get the message across to the viewer or listener. The commercial must be repeated frequently to have the desired effect on the audience. Frequency of the spot also extends the reach of the advertising campaign: Because different people tune in at different times, it is likely that different viewers will see the commercial each time it is aired. Many advertisers are interested in the **cume** (short for *cumulative audience*)—the number of different viewers that are exposed to the commercial in the course of a week.

Generally, a broadcast station uses several different rate cards during the course of a year so that it can take advantage of the cyclic nature of advertising, which is a supply-and-demand business. Because time is the unit sold in broadcasting, there is a natural limit to the number of avails. No broadcaster can sell more than 60 minutes in an hour or 24 hours per day—and no broadcaster would want to: Viewers or listeners tune in to hear programs, not commercials. If too many commercials are sold, viewers quickly tune out. If the broadcaster places too many spots in a commercial break, the audience perceives *clutter* and remembers very little of any given commercial announcement. The advertiser will quickly take his or her business somewhere else if the commercial is buried in clutter. Thus the amount of time that the broadcaster has for sale is limited, and once the unsold time has passed, it can never be recaptured. The broadcaster must therefore make every available second of commercial time count and must sell that time for the greatest amount of money possible.

The broadcaster's business year is divided into four quarters. Most of the demand for station time is in the fourth quarter, when holiday sales and political campaigns are under way—often a greater demand for broadcast time than there is supply. The rate card for the fourth quarter will often reflect higher prices than those of the first-quarter rate card, when the demand for advertising time is lower.

In recent years a *grid rate card* has come into favor for many stations. It lists different prices for the different dayparts, but it also provides different prices for spots sold in the same daypart. If the demand for commercial time is high, the advertiser pays the higher rate. The advantage of the grid card is that it takes into account the day-to-day demand for the station's airtime.

Whereas most rate cards give prices for the standard 30-second spot in different dayparts, some offer other ways to buy advertising time. Some stations offer 60-second spots, as well as 20-second, 15-second, 10-second, and station ID spots. Other rate cards list *adjacencies*—spots next to popular network programs—at higher rates. Some cards offer sponsorship opportunities for local news or sports programs, and some even sell time by the hour for "infomercials." Many advertisers buy spots at lower rates by agreeing to ROS (run of schedule), thus allowing the station to schedule the spot at the convenience of the station or in an otherwise unsold commercial availability. A nonpreemptable spot will cost more, but the advertiser will be guaranteed that the spot will be broadcast at the specified time.

The role of the account executive is to sell the *inventory* (time) of the station as soon as possible, but it is also to offer value to the client. A client that reaches the target audience and sells the product will come back to the station to advertise again and again. Account executives develop a list of repeat advertisers that become their own clients. Often a personal relationship develops between the salesperson and the advertiser. Many account executives learn the client's business and work to provide service and support to help the advertiser market his or her product. The extra time and effort that the account executive spends with each advertiser ensures the station continued business.

The account executive, or sales representative, is often among the highest-paid employees of the station. Each station hires very few salespeople. A small-market station might need only one or two people to cover the market, and a large-market station might have a sales staff of only ten. The established account executive might expect to receive 15 percent or less commission on each sale. The new account executive, without an established list of clients, might work for the first year on a draw against commission in order to have a steady income. Often the new salesperson is pounding the streets and making *cold* (unannounced) *calls* to find a small business who is willing to advertise on radio or television for the first time. The often unsuccessful new account executive is paid for his or her efforts by drawing pay against commissions that are not yet earned.

Other Sales Strategies

Not all advertisers want to reach their potential customers in time segments of 30 seconds or less. Some feel that their message is best presented in a longer format. Others want their products associated with the sponsorship of a particular radio or television program rather than scheduled in a particular time slot. Some sponsors will buy all of the commercial time slots in the broadcast program, so only the products of that sponsor are advertised during the program.

Throughout the early history of both network radio and television programs, advertisers generally sponsored entire programs. *The Milton Berle Show* (at first called *The Texaco Star Theatre*), for example, was sponsored by the Texaco company. *The $64,000 Question* was sponsored by the Revlon cosmetic company. There is good reason for a sponsor to identify closely with a top program and its talent, because every time the audience member thinks of the program or star, he or she will also think of the sponsor. During the 1960s, however, the networks wrested program control away from the advertising agencies, resulting in the sponsorship of entire programs being replaced by the 30-second spot.

Although few television programs today are sponsored entirely by one advertiser, there are still some examples. Occasionally, the Hallmark greeting card company sponsors a high-quality drama on television, and some local sports broadcasts are sold, in part, to sponsors that become well known to the audience. Sponsorship of programming, even at the local level, can become very expensive.

More common now than sponsorship is the concept of **barter programming**. Many television stations buy programming in the syndication market (see chapter 9). Because stars can demand vast sums of money, popular television programs that attract large audiences are very expensive to produce. The broadcast rights, therefore, are also expensive, and many television stations cannot afford the better programs that attract the larger audiences. During the 1990s the barter concept became significant in the production of programs for the syndication market. The program producer presells a portion of the commercial spot time to finance the cost of production. The presold commercials are built into the completed program so that they will be shown by any station broadcasting the program. The presold commercial time allows the syndicator to offer the program to local stations at a reduced price, but the broadcast station must run the presold spots along with the program, often in dayparts specified by the contract. In addition, the station has lost part of its own inventory by allowing the syndicator to fill it with presold spots. Still, barter is often the only way that a small-market television station can afford to offer a top-quality syndicated program schedule.

Infomercials are program-length commercials. First appearing on cable television networks, they have migrated to commercial broadcast television. The cable network or the broadcast station sells the advertiser a 30- or 60-minute time period, usually late at night, when the small audience size will not justify a high spot rate. The advertiser then prepares a complete television program that extols the virtue of the product. The program can be a demonstration of a kitchen appliance, an exercise machine, or a miracle auto finish restorer. One infomercial, designed to appear to be a medical information program, actually is marketing a brand of sunglasses.

Per-inquiry spots are sold by stations at a greatly discounted rate or, in some cases, provided to the advertiser at no out-of-pocket cost. In exchange for the commercial time of the station, the broadcaster receives a payment for each product sold in the station's coverage area. Usually, the products offered in per-inquiry advertising require the customer to call an 800-number and order the product directly. The customer authorizes payment with a credit card number and usually waits six to eight weeks for delivery by mail or parcel service. As with infomercials, advertisements for kitchen items and exercise equipment constitute much of the per-inquiry market. Some cable networks, such as QVC and the Home Shopping Network, have combined infomercial and per-inquiry advertising techniques to develop a new and very profitable type of television programming.

Corporate underwriting is a type of institutional advertising that is limited to public broadcasting. To help pay the costs of program production and transmission, many public stations have turned to the corporate world for support. Because the public station, with a few exceptions, is not allowed to broadcast commercials, they provide underwriting announcements indicating that the program was funded in part with a grant from whichever large corporation provided the money. Enhanced underwriting allows the corporation to say something about its mission as well as provide a visual of the corporate logo or product. Corporate underwriting has become a good way of enhancing the image of a large organization while providing additional funding for public broadcasting.

Trade-outs are very common in smaller broadcast markets and with small-town radio stations. In a trade-out deal, the station provides commercial time to a local business in exchange for the product of that business. For example, the local service station would allow the radio station sales staff to fill their cars in exchange for five 30-second commercial spots each week in morning drive time. Or a client's product might be offered as a prize in an on-air station promotion in exchange for advertising time on the station. Of course, the time of the station should not be viewed as free, just as the product of the advertiser should not be considered without value simply because it is exchanged in a trade-out. Although no cash changes hands in the trade-out, careful records must be maintained to get full value for the commercial airtime, as well as to avoid problems with the FCC and the Internal Revenue Service.

Advertising Agencies

Advertising agencies were born in the late 1800s to help the advertiser design and place display advertising in the mass-distribution newspapers that came into existence at that time. As advertising grew as a business, so did the number of ad agencies. Now every city of any size has at least one advertising agency or marketing firm listed in the yellow pages of the local telephone book. *Broadcasting & Cable Yearbook 1998* lists more than 175 major agencies that work with radio and television stations. Many of the largest, such as Young & Rubicam; J. Walter Thompson; Foote, Cone & Belding Communications; BBDO; and Bozell, have branch offices in many of the larger cities in the country. Many agencies, such as Bernard Hodes Advertising, McCann-Erickson/A&L, and Saatchi & Saatchi have an international scope, with offices in the world's leading commercial centers.

Full-service advertising agencies perform three major functions for their clients. The first function is account management, which serves as the agency's sales force. Account executives work directly with the client to develop the advertising strategies. The second function is creative services, which develop and produce the scripts, artwork, and completed commercials to be broadcast over radio and television stations. The third area is media buying, which researches coverage areas and audiences of radio and television stations as well as newspapers and magazines to determine the most appropriate way to reach the demographic that the advertiser wants. The media buyers purchase commercial time from television networks, cable networks, or broadcast station groups or from station representative companies.

When buying time in the broadcast media, the advertising agency is often allowed a 15 percent commission by the media. For example, a $100,000 spot on a popular network program will cost the advertiser $100,000, but the agency collecting from the advertiser will keep $15,000 and send $85,000 to the network to pay for the time. In addition, the agency charges clients for creative services necessary to create and produce the commercial that will be aired in the time slot that the advertiser has purchased.

A variety of different types of advertising services exists. Boutique advertising agencies have become popular in recent years. They often do not offer full service to the advertiser, but will specialize in creative work or perhaps serve primarily as a media-buying agency. House agencies are set up by the advertiser to serve its own needs and to save paying the 15 percent fee to an outside ad agency.

Most broadcasters in larger markets work closely with local advertising agencies as both a buyer and a seller of services. Naturally, the station's account representatives call on media buyers at the agency to obtain placement of an advertiser's commercials. In addition, the station often sells the use of technical facilities to the agency to produce local television commercials. In many cases, the station promotion department is also a client, using the creative services and media buyers at the agency to prepare promotional material and to place print and billboard advertisements for the station.

Station Representative Companies

As mentioned earlier in this chapter, many local radio and television stations contract with station representative companies to sell a portion of the station's commercial time at the national level. Most local broadcast stations cannot afford to maintain a sales staff in New York City or in the large cities that are national sales markets, but because as much as 50 percent of the local station's revenue can come from national spot sales, it is important that the local station be represented in those markets. The station rep company calls on the national advertising agencies to sell the local station's time to major advertisers. The rep company receives a commission for each sale made for the local station. Commissions vary somewhat but are usually in the range of 6 to 12 percent of the cost of the commercial time. When combined with the commissions of the advertising agency, the 30-second commercial time slot that the station prices at $1,000 might actually bring in closer to $700 when sold through the station rep. But without the station rep, it is unlikely that the local station would receive much national spot advertising. This is especially true in a competitive market where several stations are vying for the advertiser's business.

Most of the station rep companies have offices in New York City, but many have additional offices in the major advertising markets of Detroit, Chicago, San Francisco, Los Angeles, and Atlanta. There are more than 100 rep firms operating in the United States. Some specialize in television, others in radio, and still others in cable television. Each of the traditional television networks has in-house rep firms to sell the time of their owned-and-operated stations. The largest of the rep firms are Blair Television, Katz Communications, and, the oldest of the rep firms—Edward Petry. Station representative companies sell the time of only one station in each market.

It is to the economic advantage of the rep firm to sell the time of the local station at a high rate, because a percentage of the sale goes to the firm as a commission. In recent years many of the larger rep firms have developed research departments as well as programming consulting services in an attempt to provide information to the local stations that would be useful in developing larger audiences.

The Audience as Customer

The advertiser pays the bills for the station by purchasing commercial time, and if the revenue for the time bought exceeds the cost of operating the station, the station stockholders make a profit. Because of this, many station managers consider the advertiser the most important customer of the station. The advertiser is essential, but without an audience watching the programs and the commercials, no advertiser would spend a penny to buy time on the station. The most important customer of the station is, therefore, the audience.

The broadcast station must continually court the viewer or listener to entice him or her to tune in to the station and, hopefully, be counted in the ratings. The broadcaster serves the audience in a variety of ways: (1) providing the entertainment and information programs that the majority of viewers will select, (2) providing information to the viewer about the programs that are offered, (3) providing public service information to the community, (4) providing marketing services to the community, and (5) serving as a good citizen in the community. Much of this service is provided by the broadcaster in the form of programming, advertising, and on-air promotion.

Programming for the Customer

Programming is the most important element in attracting the audience to the radio or television receiver. Chapters 8 and 9 provide detailed explanations of the process and theories involved in determining the best program schedule to attract and hold audiences; but when looking at the audience as a customer of the station, several important factors must be taken into consideration.

First, the potential audience must be examined through research. The unique characteristics found in every community must be identified to determine the types of programs most likely to appeal to the most people. Some communities are sports-minded; others are more concerned with politics. Weather information is vital to the economy of some communities, and in others it is only of passing interest. Many stations use ratings information only for research purposes, but ratings indicate only what the viewers selected from available choices; they don't tell the broadcaster what the audience *would like* to see or hear given an unlimited number of choices. The station must be in tune with the community in order to know or predict audience preferences. A variety of research techniques, such as *focus groups* and *call-out surveys,* can aid the station in knowing the community, but information can also come from the involvement of the station personnel in community activities.

Second, stations must not only select programs of interest to the community, but also reject programs that could offend a significant portion of it. Although there will always be an audience for gratuitous sex and violence on television and hate-mongering on radio, many listeners and viewers are offended by what they see and hear in the media. Not only will they change the channel, but they will often convince friends and neighbors to do the same. If sufficiently provoked, irate viewers have been known to write letters of complaint to Congress and the FCC. In extreme cases they have organized boycotts against sponsors. The commercial role of the station is to aid the advertiser in selling the product, not to embroil the advertiser in controversy.

Third, the many diverse segments of the community must be taken into consideration when selecting programming. The audience is not a homogeneous mass of humanity with common interests, heritage, desires, and needs. It is composed of unique individuals who have specific wants and needs when selecting radio or television programming. Demographic studies tell us that people of a certain gender, age, educational level, or income group are likely to select specific types of programs when turning to radio or television for entertainment or information. Those studies are useful in selecting programs that will appeal to the most people. What we must remember, however, is that there are many different publics that make up a community. All should be served by the media in some fashion. Many long-ignored ethnic or racial minorities are being discovered by advertisers. Although conventional wisdom has dictated that the advertiser target 18- to 49-year-old viewers, some have realized that teens, Hispanics, mature adults, African Americans, and Native Americans have money too and are willing to spend it to satisfy their wants and needs. If the advertiser is to reach diverse audiences, the broadcaster must provide the programs that they are willing to watch.

Fourth, audiences must be given not only what they want in programming, but also what they need to be informed citizens. All broadcast stations provide information as well as entertainment. It is obvious that television stations are providing information with local and national news programs, but even all-music radio formats provide information when the commercials are broadcast. It is imperative that the information, be it commercial, educational, sports, weather, or news, be truthful and honestly presented. Audiences quickly lose faith in a station that presents information later proven to be misleading or false. If any part of a broadcast is proven false, other commercial announcements on the station also can lose credibility.

Broadcast stations should be supporters of their communities in at least some of their programming. They should present news of importance and public service as well as serve as a bulletin board for the community. The broadcast station should be a cheerleader for its community, whose economic health has a direct bearing on the ability of the public to afford goods and services that the advertiser offers.

Promotion of the Program

Most broadcasters are sold on the value of advertising. After all, they must convince their advertisers that commercial time on the broadcast station is the best way to reach an audience. The broadcaster must also realize the value of advertising to reach the other customer of the station—the audience. The role of the promotion department of the station, which serves as the in-house advertiser for the station itself, has three goals: (1) promote the value of the station's commercial time to the advertiser, (2) promote the station's programming to the potential audience, and (3) promote the station's good deeds to the greater community.

Promotion that is directed toward the advertiser is accomplished by sales brochures that the account executive can leave behind after a sales call. Long after the salesperson has left, this information can continue to sell the advantages of advertising on the station, by graphically illustrating audience research. In addition, sales promotion can extol the virtues of the station through display advertising in the trade press. *Broadcasting & Cable* magazine often carries four-color advertising for individual television and radio stations.

Audience promotion takes many forms. Certainly, the promotional spots that ask the audience to stay tuned for the next program are important parts of generating a large audience. The promotion department also develops and places advertising for the station in other media. Newspapers often carry display advertising for upcoming radio programs, as well as a schedule of the evening's television fare. The promotion department makes sure that the program schedule is sent to the local papers and to *TV Guide*.

Often the department contracts with a local advertising agency to develop ad campaigns to reach the potential listener or viewer. The results of these campaigns can be seen in the form of billboards and advertising on the sides of buses. Some broadcast stations even accept advertising for other stations in the market. Cross-promotion between radio and television stations that are owned by the same company is common. In contrast, advertising for one television station is almost never seen on the airwaves of the competition, although that practice is quite common on cable TV.

It is important for the broadcast station to convey to the potential audience the advantages of watching the programs that the station offers and to inform them of the times that those programs will be broadcast.

The third goal of the promotion department is public relations. The station can improve its position in the marketplace by being perceived as a good citizen of the community. It can do this in several ways. Most important is the use of airtime for the benefit of the community. Most stations provide a small percentage of commercial time for the presentation of public service announcements, which are often informational spots for a charity or community organization. Many PSAs are sent to the station each year from national organizations such as the American Red Cross, the National Heart Association, the American Cancer Society, or the local United Way. These nationally produced spots provide the public with information concerning the particular organization, while encouraging the public to donate money to it. In addition to donating airtime to the national organizations, many of which have local offices in the community, the station may also promote local groups and

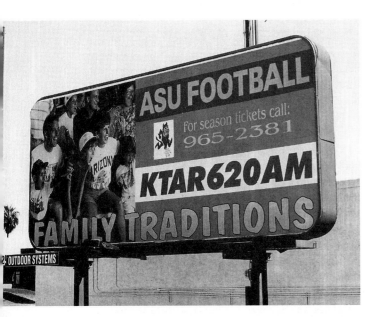

Billboards are often used by radio stations to promote their programs to motorists.

organizations. Other PSA opportunities exist to encourage viewers and listeners to stop smoking, to stay off illegal drugs, and to not drink and drive.

Many stations develop community service campaigns to address particular problems. Stations in the Phoenix, Arizona, market, where swimming pool ownership exceeds 50 percent, provide spot campaigns each summer that encourage parents to watch children around the pool. Other stations in the market warn the audience of the dangers of desert travel in the summer. Many stations encourage the audience to contribute to food drives or to provide toys for children during the holiday season. Stations also become the information source for the community in times of natural disasters. This, too, is public service.

The station's public relations efforts are intended to contribute to the public's positive perception of the station, which might translate into increased loyalty to the station's local news and public-affairs programs and increased loyalty by the advertiser when buying commercial time.

Cable Economics

Historically, cable television has operated very differently than the broadcast media. From the industry's earliest days in the 1950s, the cable television operator sold service directly to the subscriber, who was the only customer for the early cable operator. The product purchased was improved television reception and more television channels than could be received directly off the air.

As technology developed and regulation changed, the cable operator found that subscribers demanded more channels as well as programming that could not be obtained off air. In the late 1970s, the first pay channels appeared; Home Box Office, and later Showtime, the Movie Channel, and others were provided to the subscriber as pay services. In addition to buying basic cable television service, the viewer could order a channel devoted to first-run movies and special-events programming for an additional fee of $5 to $10 per month per channel. Although a large percentage of the fees collected for the pay services had to be sent to the pay service owner, the cable operator suddenly had a second revenue stream to bolster the profits of the local system.

As channel capacity increased, cable operators experimented with *tiers* to maximize subscriber fees (see table 7.1). Basic cable rates, which have been regulated both by cities and the FCC at different times, have often been set low. The cable operator quickly determined that the subscriber would pay additional fees for more programming. Cable networks, such as ESPN, CNN, and the Discovery Channel, were sold as tier 2 services at an additional cost. Pay services, such as HBO, the Disney Channel, and the Playboy Channel, were sold on tier 3. With tiering, a cable system could offer packages of channels to fit every subscriber's needs. Off-the-air broadcast service, whose rate was regulated, cost one price per month; additional cable networks and superstations (which had to be purchased or paid for through copyright fees) cost the subscriber an additional fee; and the expensive pay channels were available only to those willing to pay yet another hefty fee.

Cable operators quickly found other revenue streams. Some charge for installation of the service, others for cable radio service. Almost all cable operators charge for each additional television set hookup and for remote-controlled decoder boxes.

By the late 1980s, after several years of unregulated cable rate growth, consumer outcry reached Congress. In 1992 Congress passed the Cable Television Consumer Protection Act to again regulate the rate that cable operators could charge subscribers for basic service. The complex formula that the FCC devised for rate regulation confused operators and subscribers alike. Some customers, looking forward to slashed monthly bills, actually saw rates increase. Under the Telecommunications Act of 1996, cable rate regulation has been lifted.

It soon became apparent that the cable subscriber was willing to pay only a limited amount for an increased number of viewing choices. Pay services actually lost subscribers in the late 1980s, although now 75 percent of subscribers buy additional pay services. Many subscribers wanted additional programming but were unwilling to pay additional charges to get it. The answer was in advertiser-supported cable networks. The services delivered to the cable operator by satellite in the late 1970s and early 1980s were MTV, CNN, and ESPN. These networks, as well as the hundred or so that now exist, were probably successful because they developed two revenue streams needed to ensure

TABLE 7.1	Three-tier system of cable programming, typically used to allow subscribers to design service levels to fit their pocketbooks		
CHANNEL	PROGRAMMING	CHANNEL	PROGRAMMING
2	Showtime	23	Discovery
3	KTVK—Ind.	24	AMC
4	HBO	25	TNN/CMT
5	KPHO—CBS	26	TNT
6	Bulletin Board	27	VH–1
7	TBS	28	C-Span
8	KAET—PBS	29	MTV
9	WGN	30	The Family Channel
10	KSAZ—Fox	31	Lifetime
11	Government Access	32	USA
12	KPNX—NBC	33	ESPN
13	KASW—Ind.	34	ASPN/Prime Sports
14	The Movie Channel	35	The Disney Channel
15	KNXV—ABC	36	CNN
16	KUTP—Ind.	37	Nickelodeon
17	Cinemax	38	Headline News
18	QVC	39	Telemundo
19	KTVW—Spanish	40	A&E
20	Educational Access	41	C–Span II
21	KPAZ—Religious	42	The Weather Channel
22	Local Access		
		43*	Viewer's Choice PPV
		99*	Pay-per-view Guide

Tier 1, Limited Basic service level, includes channels 3 through 22 for $18.95 per month, but not 2, 4, and 17.
Tier 2, Standard service level, includes Limited Basic channels plus 23 through 42 for $32.95 per month, but not 2, 4, and 17.
Tier 3, Classic service level, includes all channels including premium, but not PPV.
*PPV programming available at an extra charge, per viewing or per channel.

support. First, the local cable operator pays a small amount per subscriber, usually between 10 cents and a dollar, to offer the network on a local system. Second, the cable network sells commercial time to national advertisers. The advertiser, although reaching only a small audience, can be assured that an audience of highly specific demographics will tune in to the specialized programming that the cable network offers. These combined sources of revenue provide the funding needed for the cable network to produce new programs, distribute the signal to cable systems nationally, and make a profit.

Competition, created by the increased number of cable networks and the limited channels on the cable television system, exists for shelf space, or carriage on the cable system. The advertiser on the cable network wants to reach a large but demographically specific national audience. That can happen only if the cable operator chooses to put the network on the system. Because the operator must pay to carry the network, some inducement must be offered to select one network over another. Some networks, such as CNN, MTV, ESPN, and the Weather Channel, are carried because of public demand. Others are carried because they are owned, in part, by the major cable operators. These large *MSOs (multisystem owners)* provide programming to their systems in a management structure called *vertical integration* (see chapter 6). Other cable networks, such as the home-shopping channels, have developed methods of payment to the cable operator for the carriage of the network on a cable channel.

One method of ensuring carriage on a local cable system is for the network to provide commercial availabilities for the cable operator to sell locally. During the 1990s, most cable networks made local advertising spots available adjacent to regular programming. Many cable systems invested in insertion equipment that automatically plays a local videotaped commercial when triggered by an inaudible tone in the satellite-delivered network signal.

In markets served by several cable systems, operators work together to sell local advertising in cable network programming. Cooperative selling ensures that the local advertiser reaches audiences in different geographic parts of the market. The two or more cable operators in a market are usually not in direct competition. Each is franchised, or licensed, to serve a specific geographic location, or portion of a city. The local advertiser is usually more concerned with reaching as many subscribers as possible and is not at all interested in the geographic boundaries of a particular operator.

With rates as low as $1 for a 30-second commercial, local businesses that could never afford broadcast television advertising can advertise on cable. Many cable television systems in the United States now accept local advertising to be inserted in cable network programming. Cable advertising revenue is growing each year, and many operators see it as an important revenue stream for the future.

Addressable technology developed for the cable television industry allows different television programs to be sent to each subscriber. A natural outgrowth of addressable technology was *pay-per-view (PPV)* systems, which allow the cable subscriber to order and pay for an individual television program rather than an entire channel of programs. Some pay-per-view dramatic and cultural programs have been produced, but the biggest market for pay-per-view to date has been sports and entertainment specials. Prizefights, wrestling, and professional basketball games have been the most successfully marketed programming content for pay-per-view.

REVENUE STRUCTURE OF THE NEW MEDIA

Traditional radio and television stations have been with us for more than eighty years. During much of that time, most have made a profit for their owners by selling commercial time to advertisers. Radio and television networks have been profitable by selling the time of their local affiliated stations to national advertisers who want to reach as many viewers or listeners as possible at the same time. During the past fifty years, local cable system operators have created a business by selling service directly to the consumer. Cable networks originated advertising on cable channels and later made local advertising time available to the cable operator. Times are changing, however, and the traditional revenue sources for the electronic media may evolve in the future.

Competition for audience attention is becoming fierce. As more electronic media outlets begin to broadcast or are delivered by coaxial or fiber-optic cable, the mass audiences are fragmented. With more choices for viewing, fewer potential audience members are available to watch each channel. Network programs that used to expect an audience share of at least 30 are now quite happy to receive national shares of 10. Cable network programs rarely receive shares of more than 1 or 2.

As the mass audience declines for any given station, the sales department begins to talk to the potential advertiser about the importance of demographics, arguing that the mass audience is inconsequential if most of them are not interested in buying the advertised product. A more important audience is a smaller one qualified by age, gender, or income to be interested in the product that the advertiser has for sale. As the mass audience is further splintered by competition from direct-broadcast satellite and cable delivery systems, even selling audience demographics may become difficult. It is possible that the electronic media of the future will be forced to rely on several revenue streams to deliver entertainment and information to the consumer for a profit.

The U.S. media consumer is becoming accustomed to paying for entertainment and information that is delivered to the home electronically. Twenty years ago about two dozen subscription television stations operated in the United States. These stations broadcast a scrambled signal to their customers during a portion of the day. The customer paid a monthly fee to lease a box that would unscramble the signal so that first-run movies and, later, soft R-rated movies could be seen in the home. HBO, Showtime/the Movie Channel, the Disney Channel, and pay-per-view have all taught the cable television viewer that premium programming can be seen in the home for a monthly fee. The home video market allows the viewer to select entertainment from thousands of titles for $2 to $3 per movie.

Although the concept of the viewer paying for electronic information and entertainment became well established in the 1990s, advertising revenues are still strong. Advertisers want to reach audiences that the electronic media can generate. Even with the knowledge that the viewer is likely to channel-surf through the commercials, the advertiser still perceives value in the electronic mass media. This is especially true if programming can attract and hold the specific audience demographic that will buy the advertiser's products.

The trend in generating revenue in the electronic media is toward combining advertising sales with viewer subscription fees. An economic model based on cable television seems to be emerging in other electronic media. Subscription fees are used to offset distribution costs, and advertising revenues pay for program production. This trend can be seen in the DBS systems that now broadcast programming from satellites to the home. If the telephone companies enter the mass home entertainment and information business via delivery of television programs through fiber-optic cable, the subscriber will be required to pay a monthly service fee. In addition, the subscriber will be billed on a pay-per-view basis for programs that are used in the home. Other video services will be supported by advertising and offered free, or at a reduced price, to the viewer. Over-the-air broadcasting will continue to be advertiser supported in the near future, but to hold the attention of the remote control–armed viewer, the entertainment value of the commercial will be greatly enhanced. Commercials will probably increase in length, feature regular characters, and follow a dramatic story line. Commercials of tomorrow may become the miniseries of their day and the most talked about programming on television.

Technology may offer the viewer a choice of using a program in one of two ways: If the viewer selected to actively view commercials, the program would be offered to the home free. But if the viewer decided not to view commercial material, a commercial-free version of the program would be delivered to the home on a pay-per-view basis, with the bill automatically tabulated and deducted from the customer's bank account by electronic fund transfer.

Summary

▶ Commercial radio and television stations in the United States developed their first revenue base by selling airtime to an advertiser. Because the airtime is valuable to the advertiser only if an audience is watching, the station, in effect, sells the audience to the advertiser.

▶ Intense competition in the electronic media has fractionalized audiences. Advertisers are now less interested in reaching the mass audience and more interested in having their commercials shown to a specific demographic group that will be interested and able to buy the advertised product.

▶ The sales department of the radio or television station tracks commercial time inventory and sells it to the local advertiser. National spot sales are usually arranged for the station by a station representative company. The advertising agency helps the advertiser develop the message and buy time in the media.

▶ Cable television sells a service, which includes more television channels and improved signal quality, directly to the subscriber. The sale of advertising time on cable as an additional revenue stream became popular after the cable networks were developed in the 1980s. Although local advertising on cable networks is now very profitable, most systems rely on the sale of advertising for less than 7 percent of their gross revenue. It is the third-highest source of revenue for the local cable operator, behind basic cable subscriptions and the sale of premium channels.

▶ As the new electronic information and entertainment media develop, the subscriber will probably pay to receive the service in the home, but will be able to choose between ad-supported and pay-per-view programs.

InfoTrac College Edition Exercises

In the United States, and increasingly in other countries around the world, the sale of advertising time drives the content of the electronic mass media.

7.1 Chapter 7 discusses the ways in which broadcasting and cable generate revenue and make profits.

a. Using InfoTrac to access *Broadcasting & Cable,* find articles relating to the most expensive commercials on network television. During which programs are these commercials shown? What are the record prices paid for airtime?

b. Which advertisers spend the most on network television advertising time?

7.2 The electronic media have sources of revenue beyond the sale of commercial time. Cable television and direct broadcast satellite, for example, rely on subscriber fees to generate income.

a. Using InfoTrac to access *Broadcasting & Cable,* find articles relating to cable subscriber rates. Have monthly rates charged subscribers increased or decreased since cable rate deregulation?

Web Watch

Here is a list of a few URLs (Internet addresses) for some of the organizations or corporations discussed in this chapter. Please explore these Web sites and follow the links to learn more about the complex business of the electronic media. Add your descriptions and your own favorite sites at the end of the list. Please keep in mind that the dynamic nature of the Internet allows sites to come and go but also allows organizations to update information about themselves very quickly.

Address	Description
http://www.rab.com/	
http://www.tvb.org/	
http://www.cabletvadbureau.com/	
http://www.petrymedia.com/petry/	
http://www.bozell.com/	
http://www.fcb.com/	
http://www.jwtworld.com/	
Other favorite sites:	

CHAPTER 8

Radio Programming

The role of the radio personality, or disc jockey, varies by format. Personalities usually play larger roles in the morning drive time.

"Radio is dead!" was probably uttered more than once in the late 1940s and early 1950s as television, the medium of the future, began to emerge. Surely, a medium that could offer both sound and pictures would replace one that offered only sound. But radio did not die; it went through a metamorphosis. After the advent of television, radio stations struggled with the loss of network programs and of the advertising revenue that went with them. But a few visionary radio station owners saw that radio still had a future as a mobile, personal medium. The future of radio lay in programming styles or formulas geared to specific groups of listeners, not in programs for the masses. The future of radio lay in *formats*.

This chapter examines contemporary radio programming, including the development of formats, radio dayparts, the radio clock, programming the local station, and radio networks. Current issues such as content regulation, the consolidation of stations under recent ownership rules established by the Federal Communcations Commission (FCC), and the effects of new technology are discussed.

Contemporary Radio Emerges

Broadcasting history documents the effects of television on radio as radio network programming gradually made the transition from sound only to sound and pictures as TV programs. Radio could not compete with television as a mass medium, and it became obvious by the mid-1950s that radio had to change or cease to exist. Fortunately, the radio medium still had some unique characteristics that would become major factors in its continued survival. Because radio was limited to sound only, it had the potential to become a mobile medium. By this time radio receivers had become smaller and more mobile. More important, they were becoming common in automobiles—one could listen to the radio while driving. Radio could cater to a huge audience in cars—a potential audience not available to television.

Second, radio had the potential to become a very personal medium. Programming for mass audiences had now shifted to television, leaving radio to develop programming for specific audience *demographics*. This led to formula programming, the development of a style of music or information that would appeal to only a particular demographic group. Radio stations could cater to the musical tastes of teenagers and young adults, for example, by playing a continuous rotation of popular music; thus the *Top 40 format* was developed. Other music formats emerged as radio audiences become more segmented. Audience research indicated that musical tastes among listeners tended to vary primarily by age group, so stations developed a sound or style that appealed to teenagers and adults 18 to 34 or to adults 35 to 54, for example. People began to identify with radio stations; in this sense, the stations became personalized.

In addition, radio in the 1950s became a more *localized* medium. Rather than offer network programs designed for national mass audiences, radio stations began programming more to the local community. Several elements contributed to this localization of radio: disc jockey personalities, local news, community events, and more local advertisers.

As discussed in chapter 2, the emergence of the *disc jockey (DJ)* in radio formats provided a local connection, because they related to listeners through references to community events and people. Stations used DJs for local promotions, such as appearances at merchant outlets, which in turn generated more interest in radio advertising among local merchants. In this manner many radio stations survived the transition from network programs and national advertisers to personalized formats supported primarily by local businesses.

Radio Dayparts

Radio stations cater not only to the demographics of listeners, but to their *psychographics,* or lifestyles, as well. As radio programming changed from a mass audience appeal to one geared to specific demographics, the broadcast day also changed. During the golden age of radio, prime time was during the evening hours, when people sat down after the workday to listen to the radio. In the 1950s television took over the evening programming, and radio had to design programming that met listeners' needs during different times of day. This need saw rise to the *daypart*—segments or divisions of the broadcast day used for programming and research purposes.

Audience research indicated that people tended to listen most to the radio during the early-morning and late-afternoon hours, as entertainment and information while going to and from work or school. Thus, prime time for radio became drive time—those hours when people were in their cars commuting to work and back. These dayparts, generally the 6 to 10 A.M. *morning drive time* and the 3 to 7 P.M. *afternoon drive time,* offered radio the largest audiences. But the morning daypart also offered a significant audience in the home: People getting up and ready for the day tended to tune in to their favorite radio station for music, news, and weather.

The broadcast day is segmented by radio dayparts as follows:

- Morning drive time: 6 to 10 A.M.
- Midday: 10 A.M. to 3 P.M.
- Afternoon drive time: 3 to 7 P.M.
- Evening: 7 P.M. to 12 A.M.
- Overnight: 12 to 6 A.M.

Radio formats broadcast during each daypart are designed to be compatible with listener lifestyles. Drive-time dayparts contain more information and personalities than do other dayparts. During the *midday,* for example, stations tend to play more music as people listen while they work at the office or at home. During the evening, radio audiences are generally younger, as much of the adult audience is watching television. As a result, many stations offer music specials such as live concerts to appeal to these evening listeners.

Of course, some radio stations offer alternatives to music in all-news and news/talk formats. These formats usually vary by daypart as well. The drive-time dayparts offer a mix of information, including news and weather, and the other dayparts generally focus more on talk shows. Talk radio is particularly popular during the *overnight daypart,* when many listeners turn to radio for company during graveyard shifts or sleepless nights. Again, the programming strategy is compatibility: Formats vary according to listener lifestyles and activities.

THE RADIO CLOCK

As noted, contemporary radio programming is based on a formula of broadcasting: a music or information format that reflects the style of a station throughout the day. The format is usually programmed through a radio *clock* or *hot clock*—a breakdown of each program hour into segments according to a formula. Within each daypart the clock, also referred to as a *format wheel,* displays the particular rotation of musical selections and information elements of the format. Figure 8.1 shows a sample radio clock for a music format.

Note that the clock begins with a *station identification (ID),* as required by the FCC, consisting of the station's call letters and city of license. Often stations will follow this legal ID with their logo such as "Y-95," "K-Lite," or "Z-Rock." The logo can be used before or after the legal ID, but not between the legal call letters (KOY or KKLT) and the city of license.

The music elements of the format are then laid out in quarter-hour segments. Generally, music formats will include a music sweep of three or more songs followed by a *stop set*—a break from the music

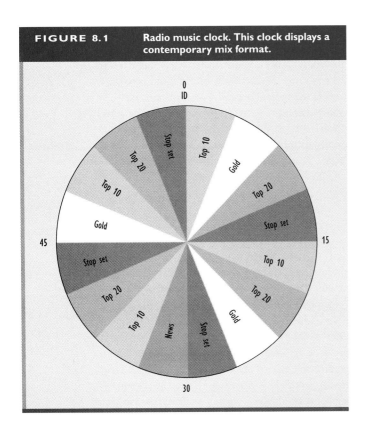

FIGURE 8.1 Radio music clock. This clock displays a contemporary mix format.

for commercials and other information elements, such as station promotional announcements and weather. The music generally is categorized within the specific format. For example, in a Top 40 format, the first segment of the clock might include an A (top ten) song, a B (top twenty) song, and a C (recent hit) song. Top ten hits are often placed in heavy rotation, which means they are played numerous times throughout the broadcast day. They may be repeated within a given daypart, for example, whereas other songs may be played only once a day.

Contemporary Radio Formats

Current radio formats can be categorized into two very general types: information and music. The information category primarily includes news/talk formats but can also represent such specialty formats as all-sports and children's programming, often referred to as *niche formats*. A wide variety of music formats is available in contemporary radio. The variety increases as the radio-listening audience continues to fragment into more-specific demographics.

Adult Contemporary

One of the most popular radio formats of the 1990s was *adult contemporary (AC)*. According to Arbitron (a national media research firm), AC accounted for the largest format share of the national radio audience in 1998. It is probably the broadest music format, because it can encompass light rock, oldies rock, soft rock, and adult rock. The common element is the demographic to which the format is designed to appeal: adults. Generally, this includes listeners ranging in age from 25 to 54. As the U.S. population (particularly baby boomers, born between 1945 and 1964) ages, many radio stations have focused on the older segment, ages 35 to 54. This is considered the "money demographic," because its members tend to make the most product purchases, as they have the most disposable income. In terms of gender, AC formats tend to attract more women than men.

The adult contemporary format is characterized by popular hits; some versions include hits from the 1960s, but others are restricted to more-recent hits from the seventies and eighties. In the soft rock or light rock versions of AC, the music is carefully selected to exclude hits that are too heavy or loud. Some of the newer versions of AC, designed for a younger demographic, include selected contemporary hits along with oldies but exclude rap and heavy metal. This version of AC is sometimes described as a "mix" format.

The role of the personality, or DJ, in the AC format varies. In mainstream AC formats, personalities play a larger role than in AC–soft rock, where music is the main attraction for listeners. Personalities do play a more important part in morning dayparts, providing information and humor, but generally remain more mellow than DJs found in Top 40 formats. For the most part, they are adults communicating with adults.

Top 40/Contemporary-Hit Radio

Top 40 is one of the oldest radio formats. As stated in chapter 2, which explores the origins of this format in the 1950s, it was patterned after the jukebox on which people tended to play the same songs over and over. The *Top 40 format* programs the current hits played in a rotation and repeated throughout the day. More recently, the format has evolved into *contemporary-hit radio (CHR)*, which tends to appeal to a younger demographic (12 to 34) than did traditional Top 40 (18 to 44).

The CHR sound is quite different from adult contemporary. It moves quickly from element to element with absolutely no dead air (as opposed to the slower pace of classical or fine arts stations). The music is upbeat and current with high-energy personalities doing humorous routines between songs. Listeners are often included in the antics through call-in participation. The idea is to grab the attention of young listeners and hold it as long as possible. This is not easy, because the target audiences for CHR, teens and young adults, tend to punch to another station if they do not hear their favorite hits right away or if they become bored with the radio personality. Top 40/CHR was the fourth most popular music format in 1998, but its audience shares had declined as much as 40 percent since the late 1980s. Part of the reason for this decline is the fragmentation of rock-and-roll music, which now ranges from heavy metal to rap, and one is not likely to hear all of the categories on one Top 40 station.

Many traditional Top 40 stations have abandoned the format in search of older demographics that have more advertiser appeal. During the early 1990s, the number of stations programming Top 40/CHR dropped by about half.

AOR/Classic Rock

Album-oriented rock (AOR) grew out of the Top 40 format in the late 1960s. In some ways it was a rebellion against Top 40 in that it avoided chart hits in favor of longer album cuts by popular artists. It reflected the rebellious youth movement of the 1960s and 1970s. It brought with it the *music sweep*—an uninterrupted series of songs—and a less structured, more laid-back announcing style. Initially, AOR appealed to a young adult audience, but it was primarily male, the other end of the spectrum from AC and Top 40. During the 1970s AOR grew in popularity as Top 40 waned.

During the 1980s, however, AOR began to lose younger listeners as Top 40/CHR regained popularity because of the emergence of MTV. Younger demographics could no longer relate as well to standard AOR artists like the Doors, the Grateful Dead, and the Moody Blues. So a splinter of AOR, called *classic rock*, became new competition for AC stations in the 25-to-54 demographic. The format features hits of the past, but with a harder edge than the AC formats. Typical artists in the classic-rock format include Bob Seger, ZZ Top, Bruce Springsteen, and the Rolling Stones.

Country

By market share, *country* was the second-most-popular radio format in the early 1990s. By station numbers, however, this format was found on more stations than any other. By 1994 country was the number one format, with the largest listener share in radio markets nationwide, but the format declined to number three by 1998. Country, formerly called country and western, is one of the oldest radio formats. WSM in Nashville has been playing country music on what became the Grand Ole Opry since the 1920s. Country-music stations have been common in rural areas for years, but more recently the format has grown in popularity in metropolitan areas across the country.

The demographic for country music is one of the broadest in radio; it claims listeners of all ages—virtually from the cradle to the grave. The format has now splintered into several segments including traditional, progressive, and current hits. In most major radio markets, there are now two and even three country stations. Stations promote themselves as "rebel country," "young country," and "hot country." Phoenix, Arizona, historically a strong country-radio market, was dominated by KNIX-FM for years, but two more FM country stations have been pulling shares of the country audience. KNIX also provided a more traditional country format that is now distributed across the country via satellite network by ABC.

Urban Contemporary

Urban contemporary (UC) is the descendant of a number of older formats, including rhythm and blues (R&B), soul, and disco. It often is associated with ethnic minority audiences, although listeners for this format are diverse. Indeed, African American and Hispanic American listeners are well represented in the UC demographic, whose growth in popularity reflects the growth of these U.S. populations. During the early 1990s, UC showed one of the strongest growth patterns among all radio formats, with the second-largest share of the radio audience in the top twenty-five markets by 1996.

Musically, UC formats include a variety of styles ranging from rap to classic R&B cuts. The consistent element in the format seems to be the rhythm of the music. The roots of the UC format—soul and disco music—focused on dance music, and that focus has carried over. Although there are dance variations of the CHR format, listeners consistently tune to UC stations for danceable music. Personality styles are appropriate to the music: upbeat, high-energy deliveries much like those found in CHR radio.

Easy Listening

The easy-listening format's history goes back to the golden age of radio. While network radio affiliates were programming situation comedies and soap operas, some independent stations were playing "good music," which during that period usually meant instrumental music performed live by studio orchestras.

After the transition of network programs from radio to TV, "good music" became "beautiful music," a format that featured smooth, orchestral arrangements of popular melodies. During the 1950s and 1960s, AM radio operators applied for FM licenses to keep the frequencies away from the competition. Many of these FM stations were automated (that is, operated by automatic tape players) to save expenses and were programmed with beautiful-music formats. By 1970 music services provided hours of this type of music prerecorded for automated stations. One of the largest at the time was Shulke Stereo Radio Productions. Because the format sounded rather like a Muzak service one might hear in reception areas or elevators, it came to be known as "elevator music." Today many radio stations are automated through computers.

More recently, the format has been categorized as *easy listening*. Part of the reason for the name change is the change in the music mix. Current easy-listening formats are a mix of instrumental and vocal music. The format historically has appealed to an older demographic, but now that group includes baby boomers who grew up with rock music and don't want to hear large orchestral arrangements of their favorite popular songs. They want to hear the originals. Consequently, easy-listening music mixes include vocals by John Denver, Neil Diamond, Barbra Streisand, Celine Dion, and even the Beatles. The key element in the format is still mellow music designed to complement work or relaxation activities.

OTHER MUSIC FORMATS

A large majority of the more than twelve thousand U.S. radio stations program one of the music formats just described, but other music formats exist that appeal to even narrower demographics, or *niches,* in the radio marketplace. These include classical/fine arts, alternative, jazz, Spanish, religious, New Age, and Big Band and they attract a smaller but often very loyal group of listeners. Some are supported by advertisers seeking a specific target audience; others are supported by listeners through donations made directly to the station.

Most of the radio stations that program classical, jazz, or a combination of the two are noncommercial or public stations. They are licensed to nonprofit organizations such as colleges and universities. They do not sell commercials, but are supported by their licensee institutions as well as by corporate grants and listener donations (see chapter 2). In many radio markets, audiences for these formats are too small to attract sufficient advertiser support, but a few successful commercial stations with jazz or classical formats exist in large radio markets such as New York or Los Angeles.

One of the fastest-growing niche formats today is Spanish. The format offers ethnic music, such as Mexican/Ranchera, along with Spanish-language announcers. The listener demand for this format has grown along with the expanding Hispanic population in the United States, particularly in the Southwest. In the fall of 1992, a Spanish-language station was ranked number one in the Los Angeles radio market for the first time.

Religious stations have occupied a small share of radio markets across the country for years. Most offer a combination of Christian talk and music. Recently, Christian rock has emerged as one of the most popular types of religious music along with the traditional favorite, gospel. Religious formats are generally on small AM stations that are supported by listener donations. As of the end of the 1990s, more than two thousand radio stations in the country broadcast at least some religious or gospel programming every week.

Alternative-music formats also have been categorized as modern rock or new rock. The term *alternative* refers to an alternative to popular (CHR/Top 40) formats, focusing on new music and artists who are on the cutting edge of contemporary music. Programmers in this format specifically avoid hits and, in fact, remove songs from their rotations if they become mainstream hits on CHR stations.

On opposite ends of the format spectrum, adult standards and adult alternative are two more niche options. *Adult standards* primarily feature original recordings of big bands from the 1940s and 1950s. Logically, the format appeals to an older demographic—50 and over. *Adult alternative,* on the other hand, offers listeners a mix of jazz, blues, and light

News, News/Talk Radio Shares Jump

Birch Scarborough Research was best known for radio ratings, so Birch was especially interested in what effect the 1991 Persian Gulf War had on radio listening. Overall, radio did very well during the crisis: Approximately 18 percent of the U.S. public first found out about the war through radio, doubling the listening level in the 7 P.M. to midnight daypart when the news broke.

The top ten markets (New York, Los Angeles, Chicago, Philadelphia, San Francisco, Boston, Detroit, Dallas, Washington, and Houston) were examined to determine where the radios were tuned. Ratings for the month of January were compared with the previous two-month Birch Radio report (November–December 1990). Not surprisingly, the share for all-news stations went up in all markets except one, with an average gain of approximately 66 percent, as shown in Figure 8.2. News/talk (as differentiated from all-news) stations also showed a large increase—approximately 36 percent compared with their overall share from the previous two-month report, radio listening (as measured by the persons-using-radio [PUR] figure), however, did not show any significant decreases or increases in the ten markets. Therefore, radio listeners were adjusting their listening habits, looking for the latest news rather than increasing their radio listening.

Source: From E. Cohen, *How America Found Out About the Gulf War*, Coral Springs, FL: Birch Scarborough Research (1991).

rock selections, most of which are instrumental. It is essentially a splinter of beautiful music designed to provide a musical escape from the pressures of the workaday world. Adult alternative is designed to appeal to the 25-to-54 demographic as an alternative to AC formats.

Music formats continue to splinter or fragment as radio becomes more competitive. Just as adult contemporary splintered into oldies, soft rock, and light rock, country has splintered. Fragmentation and specialization will continue to be major trends in the radio industry.

As national radio services become more readily available on satellite and cable, broadcast radio stations will need to focus even more on localism.

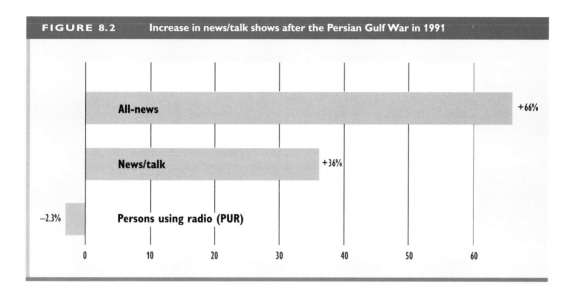

FIGURE 8.2 Increase in news/talk shows after the Persian Gulf War in 1991

INFORMATION: NEWS/TALK

Information formats include all-news, all-talk, and a combination of the two. The *news/talk format* has been the only growth format for AM radio stations. In fact, in large radio markets, it has been the savior of AM. With the exception of a few niche music formats such as those just described, AM has not been able to compete with FM. In large radio markets across the country, news/talk formats rank consistently among the top ten stations. In 1998 Interep ranked news/talk as the most popular radio format in the United States. Three news or talk stations were among the top ten stations in New York; in the Los Angeles and Chicago markets, news/talk stations were among the top ten stations as well. The format attracts a very broad audience, particularly for coverage of breaking news events such as the 1991 Persian Gulf War (see figure 8.2).

In 1988 three hundred radio stations across the country programmed a news/talk format. By 1998, the number had grown to more than nine hundred. Part of the reason for this dramatic growth was the popularity of nationally syndicated talk shows such as *The Rush Limbaugh Show* and *Imus in the Morning*. Much of the growth has been in small markets, because of relatively inexpensive delivery of national talk shows by satellite.

Most information formats are a combination of news and talk. Generally, the morning dayparts focus on news, sports, and weather during the early segment and then move into talk programming for the transition into the midday daypart. Talk hosts interview guests and take call-in questions and comments from listeners. Usually, there are regular news updates on the hour and half-hour. Many news/talk stations are affiliates of radio networks as a source for national and international news.

LOCAL PROGRAMMING

Local radio stations hire *program directors* to design station formats. In large markets radio stations may have both a program director and a music director. In smaller markets the program director also functions as a music director. Usually, the program director is responsible for the overall sound of the station: news, public affairs, and other community programming in addition to music. A music director primarily is responsible for the station's music library and playlist. Some radio stations have full-time news directors, but many smaller stations have air personalities do newscasts as part of their regular shifts.

A number of factors must be considered in programming the local station. First, program directors must conduct a *market study* to assess the competition (the other radio stations in the market). They must determine what formats are represented in the market and what niches are not being served.

Who Dominates Talk Radio?

Talk-radio programs have been criticized for being dominated by politically conservative male hosts who attract the same type of listener. In 1995 the most popular radio talk-show host was Rush Limbaugh, who claimed 37 percent of the national radio talk-show audience. His program was heard in more than five hundred radio markets across the country. The other nine radio talk-show hosts of the top ten programs in 1995 were also males, including G. Gordon Liddy, Bob Grant, Tom Leykis, Oliver North, and Michael Reagan. Who were their listeners?

A 1995 survey by Adams Research showed that 60 percent of talk-radio listeners nationally were men, but only 38 percent were Republicans. Nearly all (92 percent) were registered voters, 39 percent had college degrees, and 27 percent were more than sixty years old. But by 1998 the radio talk landscape had changed. An audience survey by *Talker* magazine indicated that a woman had taken the talk-show host crown from Rush Limbaugh. Dr. Laura Schlessinger's show was ranked number one, with Rush Limbaugh number three, behind Howard Stern.

The talk-radio guru of younger listeners in the 1990s was "shock jock" Howard Stern. Though his controversial program was heard on stations in only twenty-four radio markets in 1995, nine were among the top ten markets (largest cities) in the country. By 1998 Howard Stern's syndicated morning show was more popular than Limbaugh's. Stern's ratings were the highest among listeners between 18 and 34 years old.

Source: Data from *Broadcasting & Cable* (October 2, 1995 and March 18, 1996) and *Talker* (April 1998).

Should they try to defend a market share with an existing format or try something new that might take listeners from another format?

Once program directors have made a format decision, the clock must be designed. As described earlier, they must designate music sweeps, stop sets, and the role of the air personality. A *music playlist*, often generated by a computer that is programmed to follow the rotation of the clock must also be compiled. As new music comes into the station, it is added to the music library and playlist by the program/music director.

Once a format is in place and functioning, two ongoing processes are necessary for the success of the station: research and promotion. Audience research is conducted by radio stations to determine listeners, music preferences, and personality popularity. Radio program directors must be able to interpret the statistics produced by market research, which help position the station to appeal the target audience or demographic. The format must be evaluated continually. Stations subscribe to national ratings services (see chapter 11) and conduct their own telephone research to determine listener taste for new music. Program/music directors monitor music charts published in trade journals to track CD and record sales around the country. Among the most prominent of these trade journals are *Billboard, Radio and Records,* and *Gavin*.

The other ongoing process is the promotion of the format. Most radio markets are very competitive; listeners have many stations from which to choose. Very simply, stations constantly have to remind listeners of their call letters, logos representing their formats, and their on-air personalities. Many radio stations employ full-time promotion directors to handle this job. Promotional campaigns include public appearances by station personalities, contests, live concerts, community events, advertising on television, billboards, and in newspapers, and, of course, on-air promotion on the radio station itself.

Sources of Programming

Radio stations use programming from a number of sources beyond that produced locally by the station.

Local radio programming comes primarily from the station's music library. Stations that program popular music formats obtain music service from major record (CD) companies; that is, new music is sent to stations free of charge by these companies. The recording and radio industries have a close relationship—they need each other. The record companies need radio stations for exposure of their artists and music; the radio stations, in turn, receive the music programming they need to fill their clock hours. The stations pay for the music through annual copyright fees to music licensing agencies such as ASCAP and BMI. The fees generally amount to far less than the cost of buying each new CD or album that comes into the station.

Some niche formats require that stations buy music for their libraries. Classical music, for example, is generally not distributed to stations on a promotional basis. The market for this type of music is much smaller than that for popular music, so record companies are not as willing to send product without charge. Oldies are not as readily available as are current hits. Often, oldies stations have to build a library of past hits simply by purchasing them.

Radio stations also obtain programming from syndication services and networks. Syndication services tend to provide individual programs; radio network services range from individual programs to entire music formats. Generally, a radio station purchases prerecorded programs from a syndication service while contracting for live program feeds from a network. In the 1990s, however, the distinction between syndication and network distribution blurred, as many companies called "music syndication services" in the past became radio networks. Music services that were distributed on large reels of tape in the past are now distributed by satellite.

In 1993 Infinity Broadcasting Company, which had been syndicating the Howard Stern morning radio show to several stations in large markets, purchased the Unistar Radio Network. Through this purchase a syndicator became a radio network operator. Today, Westwood One Radio Networks distribute these 24-hour radio formats by satellite-to-station affiliates around the country. On a barter basis, station affiliates receive a satellite feed of music and personalities from a large market like New York or Los Angeles. The stations have the opportunity to "localize" the format with local station promotional announcements, local news breaks, public service announcments (PSAs), and commercials.

Affiliates of Westwood One, and other music networks like it, sacrifice some localization because they lack local personalities. But the cost savings outweigh this and is often the major factor in choosing a radio network service. Local stations save substantial personnel costs because they don't need program/music directors or on-air personalities. The difference in operating costs has been critical for many stations in smaller markets. The music networks have been particularly valuable to small AM stations that have been all but squeezed out of their markets by FM competition. Some stop sets contain a slot for network commercials as well as local commercials. The networks obtain revenue from the sale of commercial spots to national advertisers and do not charge a fee for the programming; the stations obtain the programming without a cash exchange, but trade commercial time to the networks.

After the emergence of TV in the early 1950s, radio network services dwindled to the point that they were providing only newscasts and sporting-event coverage. But in the 1980s and 1990s, radio networks showed a growth pattern. As noted, a wide variety of radio programming now is available from networks, ranging from 24-hour music and information formats to features like ABC's *Paul Harvey News*. One of the most notable trends in radio syndication has been the growth of talk radio. Some radio managers credit the Larry King and Rush Limbaugh shows, and others like them, with helping create a "renaissance for radio." This renaissance, based on strong radio personalities, brings with it new listeners and national advertisers.

One of the radio network casualties of the 1990s was the Mutual Broadcasting System (MBS). Started in 1934 as a source of programming for a group of independent radio stations, Mutual provided competition for NBC and CBS during the 1930s and 1940s (see chapter 1). When the major networks added television programming in the 1940s and 1950s, Mutual remained a radio-only network. It survived through numerous ownership changes until 1999, when its last owner, CBS, pulled the plug.

Three major factors contributed to a growth in radio syndication: increased popularity of talk radio, wider access to relatively inexpensive national programs, and more-efficient delivery of programming by satellite. With more than half of all U.S. radio stations having lost money in the 1990s, the lower cost of satellite network and syndicated programming has made it very attractive despite the lack of localism. In addition to talk programming, the traditional radio networks—ABC, CBS, NBC—offer their affiliates a mix of news and features. These program services are designed to complement various music formats.

Diversified Radio Network: ABC

In 1996 there were fourteen radio networks operating in the United States. According to RADAR 54, a network radio audience survey conducted by Statistical Research Inc., ABC was the number one radio network in 1996, with a national audience share of 45.2 percent. ABC's competition came primarily from Westwood One, which offered a series of satellite music networks along with the NBC, Mutual, and CNN radio networks. Westwood One had merged with Infinity Broadcasting in 1993, and then Infinity was absorbed by CBS/Westinghouse in 1996. This brought the Westwood One, NBC, Mutual, CNN, and CBS radio networks together under the control of one company. By the end of the 1990s, Mutual had ceased operation and ABC was the only major radio network not under the control of CBS/Infinity. ABC radio networks offer seven network services:

1. **ABC Contemporary** is designed for CHR and urban contemporary stations; offers newscasts, one-minute newsbriefs, and comedy bits.
2. **ABC Direction** provides newscasts, sports reports, and features for listeners in the 25-to-54 demographic.
3. **ABC Entertainment** is also designed for the 25-to-54 demographic and features *American Country Countdown*.
4. **ABC-FM** serves listeners in the 18-to-34 bracket with youth-oriented newscasts and sports reports.
5. **ABC Information** offers headlines, news, sports, business reports, and commentary.
6. **ABC Rock Radio** is targeted to listeners in the 15-to-24 demographic and is designed for AOR stations.
7. **ABC Talkradio** offers daily, three-hour talk segments featuring personalities Deborah Norville and Tom Snyder.

From *Broadcasting & Cable* (March 11, 1996).

Radio Programming Issues

Indecency

As program directors struggle to compete in today's complex media marketplace, they face a variety of issues that affect daily programming decisions. There is constant pressure to increase audience share and, in turn, increase advertising revenues. Program directors must be creative to stay ahead of the competition. At times, the quest for larger shares has been at odds with the public-interest responsibilities of radio stations as licensed by the Federal Communications Commission.

One of the most controversial issues of the 1990s was indecent programming, with the emergence of radio "shock jocks." The most renowned shock jock on contemporary radio is Howard Stern, a morning personality on WXRK-FM in New York City. Stern is well known for his language, which many consider offensive and, in fact, indecent. In spite of the fact that the Communications Act of 1934 prohibits "censorship over radio communications or signals transmitted by any radio station," the FCC does regulate indecent programming. Indecency on the airwaves, according to FCC regulations, is "descriptions of sexual organs or excretory functions that are patently offensive when measured by contemporary community standards."

The U.S. courts have made a distinction between indecency and obscenity: indecency is protected speech under the First Amendment; obscenity is not. Under these rulings indecent programming is legal so long as it is aired when children are not likely to be in the audience. The FCC established a "safe harbor" to allow indecent programs on radio (and TV) during the hours between 10 P.M. and 6 A.M. after a 1995 court decision supporting the FCC's indecency policy.

Problems in radio have arisen because most shock jocks are on the air during morning drive time, when audiences are largest.

The licensee of WXRK-FM was fined nearly $2 million by the FCC for indecent programming. As of 1995 this was the largest fine ever levied on a broadcast licensee by the FCC. The station licensee, Infinity Broadcasting, appealed the fine on the grounds that indecency regulation by the FCC is unconstitutional. Infinity finally paid the fine to ensure FCC approval of its purchase of additional radio stations. In the meantime Howard Stern and others like him present a dilemma for radio programmers. Even though this type of radio personality is found on a very small percentage of the stations in the country, many broadcasters feel that the FCC should not regulate broadcast content in any way. So although they do not endorse shock jock programming, they generally support Infinity Broadcasting in its appeal against FCC content regulation.

CONSOLIDATION

Another factor affecting radio programming is ownership limits. For most of its history, the FCC allowed an individual or group to own no more than 21 U.S. broadcast properties (7 AM, 7 FM, and 7 TV). In the 1980s, under the deregulation banner, the limits were raised to 12 each. Then in the 1990s, the FCC raised the limits again to 20 AM, 20 FM, and 12 TV. In 1996 Congress passed legislation removing national limits on radio and TV station ownership (see chapter 5). Perhaps more important, the limits on broadcast properties held in one market changed: The FCC now allows ownership of up to 8 radio stations in one market, 5 in one service. The ownership of 2 or more stations of the same service (AM, FM, or TV) is referred to as a *duopoly*.

The FCC and Congress raised radio station ownership limits and allowed duopolies in reaction to the radio marketplace of the early 1990s. As noted, by that time, more than half of the radio stations in the country were losing money. Part of the problem was a struggling economy, but another significant factor was the saturation of many markets with radio signals. With so many radio stations competing for fragmenting audiences, many simply could not attract enough advertiser support to pay the bills, let alone make a profit. In authorizing duopolies within markets, the FCC made it possible for station owners to consolidate the operation of two or more stations and reduce expenses. For example, sales for two FMs and two AMs could be handled out of one office. Programming could be consolidated for several stations, as could general business management. In this way stations that were on the verge of going dark (going off the air) could be saved and their signals maintained in the market.

The expanded ownership limits are controversial because they decrease diversity of station ownership in a particular market. Critics have asked what this would mean for program diversity and localism. Historically, the FCC has supported a more-is-better philosophy in broadcast regulation. Essentially, the FCC promoted more media outlets to ensure a marketplace of ideas for the listener or viewer. But in the 1990s, economic realities tempered that philosophy. FCC commissioners realized that many markets simply could not support dozens of radio stations. To maintain the current number of radio signals in medium and small markets across the country, it was apparent that group owners must be allowed to consolidate station operations.

What does consolidation mean for radio programming in today's market? Let's consider the station purchase by Capital Cities/ABC in the Atlanta market. ABC had owned an AM-FM combination in Atlanta for some time. Both stations offer country music, but the FM is contemporary country and the AM airs a more traditional country format from a satellite network. ABC took the opportunity offered by the FCC to simply buy its competition. It bought the other major FM country station in the market. The FM stations still compete for contemporary country listeners, but their operations (and expenses) are consolidated. The AM-FM-FM combination allows ABC to dominate the country music format in Atlanta with a combined share of 13.8, a very healthy share in a large radio market.

In the ABC duopoly, format diversity was not affected, because the country format of the third station was not changed. What this duopoly means to the marketplace of ideas in the Atlanta radio market remains to be seen. Most radio stations in large markets are operated by group owners whose commitments to localism vary. Consolidation allows more stations to be controlled by fewer group owners. As they consolidate the operations of these stations, localism may, indeed, suffer. Now that the national cap on radio station ownership has been lifted, there has been a flurry of station-buying activity and corporate mergers. In 1999 the largest radio group in the country was created by a merger of AMFM and Clear Channel, which now operates 830 radio stations. Among the largest concentrations of radio stations in one market is Phoenix, Arizona, where Clear Channel owns or operates five FM and three AM stations.

New Technology and the Future

Other chapters discuss the point that broadcasting and cable are technology-driven businesses. In fact, new technologies are developing so quickly that policy and programming cannot keep up. This certainly has been the case in cable television and it applies to radio programming as well.

The effect of new technology on AM radio stations was evident during the 1980s when the FCC authorized AM stereo but did not designate a standard system. As a result, a number of incompatible AM stereo systems emerged at stations around the country, leaving consumers confused as to which receiver to purchase for AM stations in their markets. By the 1990s the Motorola C-Quam system was recognized as the standard for the most part, but relatively few AM stations have converted to stereo and instead have turned to niche formats, many talk-oriented, to survive in the radio marketplace. In this case, programming ultimately was more important to the future of AM radio than was new technology.

Digital technology (described in detail in chapter 13), on the other hand, will have a significant effect on the future of radio broadcasting. *Digital audio broadcasting (DAB)* will offer listeners CD-quality audio over the air. Although DAB can currently be distributed nationally by satellite and on cable systems, radio broadcasters are concerned that it might be detrimental to local stations, because digital broadcasting requires more spectrum space than standard AM and even FM channels provide. The problem with DAB has been where to place the service on the broadcast spectrum. If local radio stations cannot provide DAB, will the national music services put them out of business?

New compression technology allows radio broadcasters to compress more signal information into a smaller space. This technology may allow local radio stations to broadcast CD-quality digital audio. Obviously, this will enhance the quality of FM signals, but could have an even more significant effect on AM stations. Indeed, AM stations may be able to compete with FM music formats, something they have not been able to do in the recent past.

In addition, digital technology makes possible such services as the *radio broadcast data system (RBDS)*, which allows radio stations to code their formats as adult contemporary, country, or news-talk and be displayed as such on the receiver. Radio receivers show the data on a small display built into the unit. Listeners can code in "country," for example, and the receiver will seek all of the country stations in the market. The RBDS provides another advantage to stations by displaying call letters when tuned to a particular frequency. The system can even be used as a mini-billboard for advertising in data form.

In the future, radio programming will be influenced by the continued development of digital technology, the growth of satellite networks, and the consolidation of radio station ownership. As audiences continue to fragment, research will become even more important, as stations compete for niches in the marketplace. Radio station operators will need to find ways to make their stations stand out in saturated markets. They will need to position themselves with a niche format and find creative ways to promote the station to potential listeners. Many stations distribute live programs on the Internet. As national radio services become more readily available on satellite and cable audio, broadcast stations will need to focus even more on localism. This will be the challenge of radio programming in the twenty-first century.

SUMMARY

▶ After the advent of television in the late 1940s, radio network affiliates gradually lost network programming to TV. It was obvious that radio had to change or die. Radio evolved from a mass medium to a personal, mobile, and localized medium geared to segments of the audience. Formats were developed to appeal to specific demographics. Among the first of these was Top 40, developed in the 1950s along with rock-and-roll music. Whereas most radio stations program music formats, some offer all-news or all-talk or a combination of both. Radio stations today cater not only to audience demographics, but to psychographics, or lifestyles, as well.

▶ As radio changed as a result of television growth, the broadcast day changed with it. After TV took over the evening prime-time hours, radio stations offered programming that met listener needs during other periods. Prime time for radio became drive time—those hours when people were in their cars commuting to work and back. During these dayparts and others throughout the day, radio formats are designed to be compatible with listener lifestyles.

▶ Radio formats usually are programmed through a radio clock, which shows a breakdown of each program hour into segments according to a formula. Within each daypart, the clock displays a particular rotation of music and information elements. The elements are generally arranged in quarter-hour segments, with stop sets for commercials and other information.

▶ Contemporary radio formats can be categorized into two very general types—music and information. Music formats include Top 40, adult contemporary, AOR/classic rock, country, and easy listening. The majority of the U.S. stations program one of these music formats. Other formats appeal to even narrower niches in the radio marketplace, including classical/fine arts, alternative, jazz, Spanish, religious, New Age, and big band. Classical/fine arts and jazz are often found on public or noncommercial radio stations. Many AM stations program news/talk formats as an alternative to FM music formats.

▶ Radio stations hire program directors to design station formats. In large markets radio stations may have both a program director and a music director. The program director is responsible for the overall sound of the station, and the music director is responsible for the music library and playlist. Once a format is in place, the program director is responsible for two ongoing processes: research and promotion. Audience and music research data help the program director position the station in the marketplace. Promotion includes contests, live concerts, and public appearances by station personalities.

▶ Radio programming sources include record companies, syndication companies, and networks. Most music for the station library comes from record companies on a promotional basis. Some music must be purchased, as in the case of classical and oldies. Stations pay for the use of the music through annual copyright fees to music licensing agencies such as ASCAP and BMI. Radio stations also receive programming from syndication services and networks. National talk programs became very popular in the 1990s.

▶ Radio programming issues today include indecency regulation by the FCC, group ownership and duopolies, and the effects of new technology. In the future, radio programming will be influenced by the continued development of digital technology, the growth of satellite networks, and the consolidation of station ownership. As audiences continue to fragment, research will become even more important. As national radio services become more readily available on satellite and the Internet, broadcast stations will need to focus even more on localism.

INFOTRAC COLLEGE EDITION EXERCISES

In the twenty-first century, radio programming will be distributed by satellite and on the Internet as well as broadcast, but the key to success will remain in localism.

8.1 You read in chapter 8 about the growth of radio syndication in the 1900s. In an article on radio syndication in *Broadcasting & Cable* (August 30, 1999), John Merli suggests that the golden age of radio may be 1999.

  a. What is the basis for his suggestion?

  b. The owner of the United Stations Radio Network is probably more famous for music on television than on radio. Who is he?

8.2 You read in chapter 8 that the largest radio group in the United States in 1999 was Clear Channel. In an article in *Broadcasting & Cable* (October 11, 1999), Elizabeth Rathbun reports that with 830 stations, Clear Channel can reach an audience previously enjoyed only by network television.

  a. How does Clear Channel compare with ABC Television in terms of listeners versus viewers?

  b. What is the dollar value of the merger between Clear Channel and AMFM?

WEB WATCH

Here is a list of a few URLs (Internet addresses) for some of the organizations or corporations discussed in this chapter. Please explore these Web sites and follow the links to learn more about the complex business of the electronic media. Add your descriptions and your own favorite sites at the end of the list. Please keep in mind that the dynamic nature of the Internet allows sites to come and go but also allows organizations to update information about themselves very quickly.

Address	Description
http://www.chancellormedia.com	
http://www.westwoodone.com	
http://www.npr.org/	
http://www.clearchannel.com/	
http://www.rronline.com/	
http://www.radiostation.com/	
http://www.radioinfo.com/radiosearch.shtml	
http://www.icr.org/radio/radiolog.htm	
Other favorite sites:	

CHAPTER 9

BROADCAST

TELEVISION

PROGRAMMING

Television programming is the bait needed to attract the audience to the station.

It has been said that people listen to radio *stations*, but they watch television *programs*. You probably have a favorite radio station, or several favorites, that you listen to regularly. But do you have a favorite television station? Probably not. That's because we tend to think of TV in terms of programs rather than of the stations or channels on which they are delivered. Television viewers, especially soap opera viewers, become very attached to their favorite programs and don't want to miss them. So it is not surprising that viewers complain when program times change, let alone when programs change channels.

In 1994 television viewers in Phoenix, Arizona, experienced some dramatic changes in network and station affiliations. Because of a corporate buyout, the local CBS affiliate became a Fox affiliate. The former Fox affiliate then picked up ABC through another corporate deal. This left the former ABC affiliate and an independent station competing for CBS. The group owner of the independent station also owned some CBS affiliates in other markets, so the Phoenix independent became a CBS affiliate. The apparent loser in the network shuffle was the former ABC affiliate, which was left to survive as an independent. Phoenix was not alone in these network changes; more than three dozen stations in seventeen television markets were affected at about the same time.

The major concern expressed by television viewers as these changes unfolded was about programs: "Where will I be able to see my favorite sitcom, soap opera, or news program?" In other words, viewers didn't care what channel or station delivered their favorite programs so long as they could watch them. But for TV stations, changes in network affiliations have major effects on their survival as businesses. For them, the channel is the **product shelf** needed to attract TV program audiences, which are sold to advertisers.

This chapter describes how programming content is created in video formats and examines the programming process that networks, television stations, and cable systems use to fill their limited shelf space. Programming as product is also discussed. Distribution and pricing, from programming wholesaler to programming retailer, from local broadcast station to the cable system, and from the proprietary uses to the home video store, are examined.

The components of the video industry discussed in the chapter are diagrammed in figure 9.1. Note that television stations, cable systems, and video rental outlets are considered *retailers* because they provide services directly to consumers. Networks, syndicators, and videocassette distributors are *wholesalers* because they supply retailers with the product. Production companies, or manufacturers, create and distribute the product to wholesalers. Advertisers buy the audiences provided by programming and constitute a major source of revenues; audiences are *consumers* of the programming. The Federal Communications Commission (FCC) is responsible for the regulation of the broadcast/cable industry, including some regulation of the programming of the station or cable system.

The Programming Department

The programming department in the local television station is often headed by a program director (although this job is being eliminated at many stations to save money, and the duties of the program director are spread to other station executives). Programming departments often are composed of (1) operation units that schedule programming and operate the switching equipment that actually puts the content on the air; (2) production units that produce, stage, write, and direct programs; and (3) acquisition areas that manage libraries of product and maintain records of programming contracts, episode use, scheduling, and audience ratings. (Some smaller stations also place news operations and promotion activities within the programming department.) The programming department budgets are often the largest in the station, because many of the station employees are part of the programming department and the station's greatest expenses are in the acquisition of programming rights.

The program director, operations manager, news director, promotion manager, and production manager are all involved in the selection of programming and, therefore, are directly responsible for the broadcast content. As such, they must protect the station license by understanding and complying with the programming rules and regulations of the FCC as well as other laws that affect content of the electronic media. These laws, discussed in chapters 4 and 5, cover such areas as copyright, accuracy, fairness, balance, and privacy.

In addition, those directly involved in the programming decisions are on the front lines when generating profits for the station. They must attract the largest possible audiences by selecting and

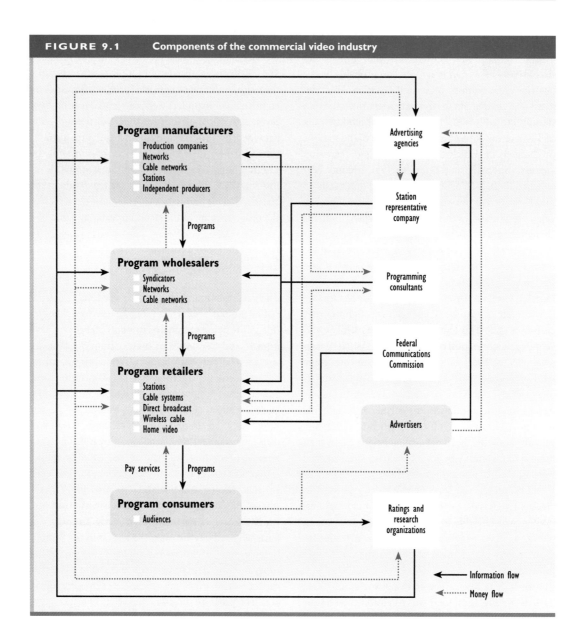

FIGURE 9.1 Components of the commercial video industry

scheduling the content that audiences want as well as promoting or advertising the programming to the potential viewers and listeners.

The television programmer, or programming team, has the rather heavy responsibility for selecting and scheduling programs for the station. This is an awesome responsibility, because millions of dollars often are at stake in the programming process, even at the local level. These executives' decisions are critical to the station's success and bottom line, so they must carefully consider product available, rating track records, competition in the market, and audience viewing behavior.

The programmers' specific tasks include the following:

1. Research the market, the potential audience, and the competition

2. Select programs to fill available time slots

3. Schedule programs to attract and hold the largest audiences

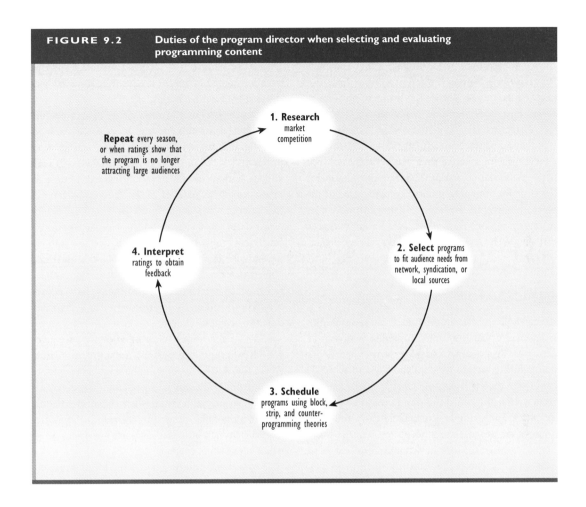

FIGURE 9.2 Duties of the program director when selecting and evaluating programming content

4. Interpret ratings information to obtain feedback concerning the success of the programming and scheduling

These tasks are repeated as new programming schedules are prepared for each season (see figure 9.2).

TELEVISION AUDIENCES

To perform their tasks effectively, television programmers must know their audiences. Audience research in the early 1990s indicated that average television viewing time had increased among adults over the past twenty years. By the early 1990s, the average household had the television set on more than seven hours per day.

In the 1990s, with many more channel choices available through cable and satellites, the percentage of heavy TV viewers (more than three hours per day) in the audience increased. Most heavy TV viewers were found among socially disadvantaged people, although those with more education and income have also increased their average viewing time. Also, fewer people classify themselves as TV nonviewers. In terms of gender, women tend to watch more TV than do men, although men have increased TV viewing time slightly over the past two decades.

PROGRAMMING ON THE PRODUCT SHELF

The goal of the television programmer is to create the largest possible audience of a specific demographic type for each time period. The large audience, reflected by the ratings, can then be sold to the advertiser. The only way that the television station can create the audience is by providing programming that the public is interested in watching. With only one channel, the television station has only one chance in each time period to attract that audience. Programming time is limited and, once lost, can never be recaptured. The station therefore must make every effort to use the time effectively.

Programs are the vehicles used by television stations and networks to attract and sell audiences to advertisers. In a sense they are brightly colored packages on the store shelf. The television channel is the shelf for arranging products or programs in the most attractive manner. Television programmers often refer to programs as **product**, even though what is actually sold by stations and networks is the audience.

Television Dayparts

In scheduling television programs, the programmer must ensure that they are compatible with viewer activities. Programmers consider what viewers are doing during various times of the day and schedule programs that will complement those activities. Hence, the programming day is segmented into **dayparts**. For example, during the early-morning hours of the weekday, most people are getting up and preparing to go to work or school. From television programs, they generally want information—news, weather, sports, and features—but they want to be entertained as well. The networks offer audiences a good mix of information and entertainment in *Today* and *Good Morning America*. This is the first television daypart: *early morning* (6 to 9 A.M.). Some programmers break out a separate daypart called *early-early morning* (6 to 7 A.M.).

The next television daypart is *daytime* (9 A.M. to 3 P.M.), during which programs are geared primarily to those viewers who stay home during the day. The networks tend to program talk shows, game shows, and soap operas, and independent stations often schedule movies and off-net syndication, or sitcoms that have already been broadcast in prime time by the networks. Again, programmers want to create a schedule that is compatible with home viewer activities during the day. Although many women work outside the home, the home viewing audience is still primarily women.

The *early fringe* daypart follows daytime at 3 P.M., when children begin coming home from school and adults from work. Television programming becomes more fragmented to serve a more diverse audience. A variety of children's programs, talk shows such as *Oprah*, and more off-net programming such as situation comedies or action-adventure series air over the next two to three hours.

Depending on time zones across the country, the period between 5 and 7 P.M. or 6 and 8 P.M. contains the programming window for local and network news programs. In the larger television markets, network affiliate stations usually program two half-hour local newscasts split by network news. The last half-hour of this daypart is known as *access*, because it leads into *prime time* (7 to 10 P.M. or 8 to 11 P.M., depending on time zone). This is a critical time, because stations depend on access programming to maintain and build audiences for prime time. Independent stations tend to schedule entertainment programs, particularly situation comedies, during this period as alternatives to network and local news.

By far the largest viewing audiences for television programs are present during prime time. Obviously, during the evening hours most people are home from work and are available for television viewing, so this is the time when networks schedule their most popular programs. (Scheduling strategies for prime time and other dayparts are discussed in detail later in this chapter.)

Late fringe follows prime time and generally begins with local news on network affiliates or syndicated programs on independents. After the news the battle for late-night audiences begins with network talk programs such as *The Tonight Show* and *The Late Show with David Letterman*. Of course, there are alternatives for late-night viewers, including sitcoms, movies, talk, and action-adventure on independent stations and cable channels. The last television daypart, *overnight*, offers more of the same, but audiences drop off significantly after midnight.

Because different types of people are available to watch television in the various dayparts, the programmer schedules a variety of programs to appeal to the different audiences. Program-appeal research, first conducted in the early days of television, has determined that people of different demographics are likely to watch different types of programs. In general, that research has shown that age, gender, educational level, and income range are all factors that can help predict the type of television programming desired. Because the age, and to some extent gender and

Programming Appeals

In the early days of television, Harrison Summers, professor at Ohio State University, conducted considerable research on the reasons that certain demographic groups of viewers select certain types of television programs. Lawrence Lichty and Joseph Ripley later described the program viewing process in terms of *major audience appeals*—motivations that would drive the individual to select one type of program over another:

- Conflict/competition
- Comedy
- Sex appeal/personality
- Human interest
- Information

Generally, each program has several appeals built into it. For example, a dramatic program with a lot of conflict will often have an element of comic relief. Producers often use several of the audience appeals to broaden the audience that will be attracted to the program. Comedy is often called the universal appeal. Situation comedies on television often draw the largest audiences in all of the demographic groups.

Source: From Lawrence Lichty and J. M. Ripley, *American Broadcasting: Introduction and Analysis: Readings*, Madison, WI: College Printing and Publishing (1969), reprinted in S. T. Eastman, *Broadcast/Cable Programming: Strategies and Practices*, 4th ed. Belmont, CA: Wadsworth, (1993).

educational level, can vary with the daypart, it is wise for the programmer to study the demographic composition of each daypart and to schedule programs that are known to appeal to the available audiences.

Program Types

The types of television programs that we take for granted today were developed, for the most part, by network radio in the 1930s and 1940s. Radio programmers developed the situation comedy, the variety program, the drama, and the soap opera. These programs made the transition from network radio to network television during the late 1940s and 1950s (see chapters 1 and 2).

Programs exist to provide entertainment and information. Although many programs have elements of both, we generally include news and public affairs programs as information. News programs, which have become very important to the television industry, are examined in chapter 10. This chapter focuses primarily on entertainment programming.

The entertainment program types that we are all familiar with include situation comedies, action-adventure programs, dramatic programs, soap operas, game shows, and reality programs. (Reality programs are a fairly recent phenomenon in entertainment television and are closely related to news programs. Examples of successful reality programs are *Rescue 911* and *Cops*.) Other categories of entertainment programs that may be included are children's programs, talk shows, sporting events, westerns, docudramas, miniseries, and of course movies.

During the first two decades of network television programming, prime-time schedules were dominated by action-adventure, movies, situation comedies, general drama, and variety programs. More recently, variety programs have been relegated to the late-night spot and, in the 1990s, newsmagazines and reality programs emerged in prime-time network schedules. Table 9.1 shows network prime-time programming. Note that there are more situation comedies on the air than any other program type; they have always attracted the largest prime-time audiences, with the exception, of course, of special events covered by television.

Program Sources

The television station programmer has three sources from which the programming, or product, of the station can be selected. The sources can include the network (for a network-affiliated station), a program syndicator, and local production.

TABLE 9.1	Network prime-time program types
TYPE OF PROGRAM	QUANTITY
Suspense and mystery	5
General drama	24
Situation comedy	51
Adventure	1
Feature films	7
Variety	2
All programs	201

Source: Nielsen Media Research National Audience Demographics Report (December 1996), reprinted in *Broadcasting & Cable Yearbook*, New Providence, NJ: R. R. Bowker (1998).

For most television stations, networks are the major source of entertainment programming. There are a number of advantages to being affiliated with a network, the major one being that most of the station's schedule (about 70 percent) is provided by the network without cost. In exchange for airing network commercials, stations are given rights to broadcast network programming. In addition, they are given commercial time within or adjacent to the heavily viewed network program, which they can sell to local or national spot advertisers. The adjacent commercial time is often sold at a higher cost to the advertiser than commercial time near local or syndicated programs. Another way that the affiliated station is paid by the network for the use of its time is through *compensation*, or a direct payment, usually a share of the advertising revenues from network programs.

Another major advantage of network affiliation is access to high-quality programs. Television stations generally do not have the budgets to produce their own action-adventure or situation comedy programs, so station programmers must obtain these programs from networks or program syndicators.

An additional advantage is that network programs still have the highest audience shares during prime-time viewing. When a program develops a national time slot and reputation, affiliates have the advantage over independent stations in their markets during this period, because larger portions of the audience will be attracted to the network program.

The traditional full-service television networks—ABC, CBS, and NBC—have provided programming to affiliated stations in all of the dayparts for much of the history of television. The Fox network emerged as a major ratings contender in the short period that it has been in operation. In the early spring of 1995, Warner Brothers and United Paramount both started television networks to provide prime-time material to stations.

The Public Broadcasting Service (PBS) is a national television network that provides programming to public broadcast stations in the United States. It operates differently from the commercial networks in that public stations must pay PBS for much of the programming that they use rather than receive compensation.

A second major source of television programming is *syndication*, a distribution method in which television stations buy the rights to broadcast programs or product directly from the program production company. There are two types of syndicated programs: off-network and first-run. *Off-network programs* are aired on a major network for a few seasons and then sold to stations as reruns. The popular CBS series *M*A*S*H* has been syndicated as an off-net program for years. Other off-net series from the early days of television that are still popular include *I Love Lucy* and *The Andy Griffith Show*. These programs are said to be *evergreen*—to have the ability to attract audiences over a long period of time. No matter how old they are or how often they have appeared on television, they always seem able to find an audience.

First-run programs are produced for distribution directly to television stations; they do not air on a network first. Examples of first-run syndicated programs include *Star Trek: The Next Generation* and *Wheel of Fortune*. Syndication is discussed in more detail later in this chapter.

A third source of television programming for stations is local production. This source usually provides the smallest percentage of the station program schedule. Local production in television stations consists almost exclusively of news and public affairs programs. News has become a profit center for many television stations, and stations have been expanding this type of programming since the mid-1990s.

SCHEDULING STRATEGIES

The primary responsibility of television programmers is to attract and to retain program audiences. They use a variety of programming strategies designed to achieve *audience flow*—the movement of viewers from one program to the next. Audience flow can occur

ABC 2000—MILLENNIUM COVERAGE

Television is often at its best when covering breaking news. Certainly major news events over the past fifty years have attracted huge audiences, as the American public collectively turned to television when history was in the making—for information, to grieve, or to celebrate. Nearly every American of the baby-boomer generation can tell you where he or she was when television broke the news of the assassinations of President John F. Kennedy, Martin Luther King Jr., and Robert Kennedy; man's first steps on the moon and the *Apollo 13* drama are clearly burned into their memories. Equally ingrained in the minds of the generation that came of age in the eighties are the moments they learned of the deaths of John Lennon, Princess Diana, and Kurt Cobain, and the images of student protesters in Tiananmen Square, the Challenger explosion, the fall of the Berlin Wall, and the Scud missiles over Baghdad.

Most of television's defining moments have been brought to us on the spur of the moment, as network news departments scrambled to make sense out of breaking stories. But television also covers planned events as *specials*. Many of those special programs have attracted the largest audiences in the history of the medium. Any Super Bowl broadcast, for example, is almost guaranteed to garner the highest ratings of the season, which is why advertisers pay so much for that airtime and spend millions developing clever new commercials to debut. Other rating bonanzas have included the miniseries *Roots*, the television premiere of *Gone With the Wind*, coverage of the O. J. Simpson trial, and the Barbara Walters interview of Monica Lewinsky.

There have also been television series that through extensive promotion have captured the attention of the mass audience. The last episode of *M*A*S*H*, the birth of Little Ricky on *I Love Lucy*, and the "Who shot J.R.?" episode of *Dallas*, and the *Seinfeld* swan song were all ratings blockbusters. But more often high ratings come from specials, which is why there are so many annual awards programs, holiday specials, and sports spectaculars.

On Friday, December 31, 1999, ABC presented an "unprecedented global telecast" with a seventy-two-city satellite linkup for live, twenty-four-hour coverage of the turn of the century (some say millennium) around the world. The program, hosted by Peter Jennings, included live feeds of festivities provided by broadcast networks in Australia, Japan, Israel, Egypt, Germany, France, Great Britain, and Canada.

Although the program probably benefited from dire predictions of the end of life as we knew it because of Y2K problems, the number of viewers tuning in to ABC affiliates was phenomenal. An estimated 175 million Americans (nearly 70 percent of the population) watched some part of the ABC program. In Phoenix, affiliate KNXV-TV chalked up ratings consistently double those of the nearest competitors that day. The $11 million that ABC spent to produce the program paid off, at least in terms of ratings.

---

from one program to the next on the same channel or from one channel to another, sometimes called *in-flow* or *out-flow*.

Audience flow is more difficult to achieve today than it was in the early years of television. Before the advent of remote-control devices, programmers could take advantage of a phenomenon called **channel inertia**, wherein viewers were reluctant to get up and change the channel unless the next program was particularly unappealing. Channel inertia still exists to a degree, but with remote-control devices, viewers are much more likely to surf through the channels during breaks between programs.

There are six major strategies that the television programmer might use to develop and to hold audiences for the station: block, strip, counter, power, mood, and cyclical programming.

One strategy used to achieve audience flow is **block programming**, or scheduling of a series or block of similar programs, such as situation comedies, one after another. The major networks schedule blocks of programs during prime time. Note in table 9.2 that programs are stacked one on another throughout the evening. NBC has been particularly successful in achieving audience flow with its Thursday night line-up including *Friends, Jesse,* and *Frasier*. An audience is established with the first situation comedy and maintained through the block. In fact, audience shares increased for the last two programs as a result of in-flow from other channels.

TABLE 9.2	Prime-time block programming (fall 1999 season)	
MONDAY	**CBS**	
8:00 P.M.	King of Queens	
8:30 P.M.	Ladies Man	
9:00 P.M.	Raymond	
TUESDAY	**ABC**	**NBC**
8:00 P.M.	Spin City	Just Shoot Me
8:30 P.M.	Oh Grow Up	3rd Rock from the Sun
9:00 P.M.	Dharma & Greg	Will & Grace
9:30 P.M.	Sports Night	
WEDNESDAY	**ABC**	**NBC**
8:00 P.M.	Two Guys and a Girl	
8:30 P.M.	It's Like, You Know	
9:00 P.M.	Drew Carey	The West Wing
9:30 P.M.	Norm	The West Wing (cont.)
10:00 P.M.		Law & Order
10:30 P.M.		Law & Order (cont.)
THURSDAY	**ABC**	
8:00 P.M.	Friends	
8:30 P.M.	Jesse	
9:00 P.M.	Frasier	
9:30 P.M.	Stark Raving Mad	

When new programs are introduced to a network schedule, programmers often use a block strategy called *hammocking*, wherein the new program is scheduled between two popular programs to take advantage of audience flow. The intention is for the audience established by the first popular program to flow through the new program into the second popular one. NBC used hammocking when it scheduled a series called *Jesse* between established winners *Friends* and *Frasier*.

The opposite of hammocking is *tentpoling*. In this block strategy, programmers schedule a strong, established program between two weaker ones. Again, they hope to take advantage of audience flow into and out of the strong program. The tentpole strategy was applied by ABC with its popular series *Dharma & Greg*. The program was scheduled at 9:00 P.M. on Tuesday nights between two weaker programs.

Strip programming, or *stripping*, is used during other dayparts. Also known as *horizontal programming*, this strategy schedules the same program at the same time every weekday (that is, horizontally across a weekly program chart). This is the strategy used for scheduling soap operas, talk shows, and game shows, as programmers hope to establish viewing habits in audiences. Again, audience flow is the programming goal, but a flow from day to day is achieved. This strategy can be so effective that viewers actually schedule their activities around program times.

Checkerboarding, another form of strip programming, involves scheduling horizontally across the week at the same time by program type: a series of situation comedies or action-adventures. One might program a "comedy hour" every weeknight at 7 P.M. with different programs each night, hoping to build viewing habits. Viewers would know that, whatever else

TABLE 9.3 Counterprogramming on a television station

TIME	CHANNEL 3	CHANNEL 5	CHANNEL 8
5:30 P.M.	Network news	Network news	**Seinfeld**
6:00 P.M.	Local news	Local news	**Coach**
6:30 P.M.	Hard Copy	**Jeopardy**	Inside Edition

At 5:30 and 6:00 P.M., the comedies *Seinfeld* and *Coach* on channel 8 provide counterprogramming to the network and local news programs on channels 3 and 5. At 6:30 the quiz program *Jeopardy* on channel 5 is counterprogramming to the tabloid news programs *Hard Copy* on channel 3 and *Inside Edition* on channel 8. (Examples of counterprogramming are in bold type.)

might be on other channels, they can find comedy on the station every night at 7.

Stripping also is used by local television stations. Independent stations often schedule programs at the same time every day to counterprogram network affiliates in their market.

Counterprogramming is the strategy of scheduling a program that will attract an audience different from that of the competition. A prime example of this technique is scheduling a situation comedy against network or local news on an affiliate station (see table 9.3). Another example would be scheduling local news against the last hour of network prime-time programming. In both examples independent station programmers hope to attract viewers with alternatives to network programs.

Stunting, often used as a counterprogramming strategy, is scheduling onetime programs such as music specials or sporting events to attract viewers from the competition. For example, a network might schedule a Garth Brooks country music special against a popular prime-time program on another network. Of course, the special does not build regular audiences for the time period, but it does pull viewers away from the competition temporarily.

Power programming is the opposite of counterprogramming. In this strategy, also known as *challenge programming*, programmers compete head-to-head with similar program types. If the competition has scheduled a situation comedy in a particular time period, they schedule another one against it, hoping to draw viewers away. This is common during network prime-time schedules. Networks often schedule comedies, movies, or miniseries against each other to compete for prime-time audiences. Local stations also use power programming in scheduling syndicated programs, such as *Oprah* and *Rosie*.

Television stations often schedule local news programs in a power-programming situation or in direct competition with one another. The reason for this is that certain time periods have become traditional news slots. Scheduling local news before and after the network news feed develops audience flow by providing a block of programs and therefore greater audiences. Scheduling local news following the network prime-time programming is also common.

Mood programming, or *narrowcasting*, is a strategy wherein all of the programs on the station (or the cable channel) are designed to appeal to the same audience. It is as though the entire schedule of the station were designed as a long block of programs. This strategy is usually found on a cable network and provides an entire schedule of the same type of programming. For example, ESPN provides only sports programs, CNN provides only news, and Nickelodeon serves the needs of a children's audience. Some cable channels now have gone beyond narrowcasting to what might be called *slivercasting*—providing programs of a single type to a very small demographic audience. The Weather Channel or Court TV are examples of slivercasting. The audiences at any one time are usually small; over time, however, many different people are likely to tune in to the channel for specific information or entertainment. Many advertisers are interested in reaching the very specific audiences that these channels can generate.

With the expanding number of independent television stations, low-power stations, direct-broadcast television services, and cable television channels, there is an ever growing need for programs, or product, to fill the massive amount of programming time. Programming is expensive to produce, and good programming that will attract an audience is even more expensive. The television industry therefore subscribes to the *parsimony principle,* which states that resources should be used as efficiently as possible. To use the money spent on expensive television programs in the best way to gain the largest possible return on the investment, the television programmer should provide different audiences the opportunity to view the program in several different time slots on the schedule. In theory, a program that costs $100,000 for one broadcast will cost only $50,000 per broadcast if shown twice, or $25,000 per run if shown four times on the station. The parsimony principle accounts for the many *reruns* that we see on television: Each station wants to get as much mileage out of each programming dollar as possible.

Cyclical programming, like a Top 40 radio station, repeats programming over and over throughout the day. Examples of cable networks using this systematic scheduling of reruns are CNN Headline News and MTV. The theory is that the audience changes over time, and those who saw the program in one time period will not be watching it in the next time period.

Network Programming

The traditional television networks have offered a full schedule of programming to affiliated stations for many years. In the morning hours, the programming includes fare such as *Good Morning America* and *The Today Show.* Daytime programming consists mostly of game shows and soap operas. Prime-time programming includes primarily sitcoms, action-adventure, reality programs, and movies. Late-night shows include *The Tonight Show* with Jay Leno and *The Late Show with David Letterman.* In addition, each network has a large news operation and a sports department and provides a variety of specials. With the exception of many of the prime-time programs, the network produces and owns the programming that it distributes to its affiliates. To obtain programs for prime time, the traditional networks buy the rights to broadcast entertainment programs from major Hollywood production companies. The rights usually allow the network to broadcast the program twice. The process of selecting the program series that will be broadcast each season is a long and complex one.

Each year hundreds of producers and writers present, or *pitch,* ideas for new prime-time television series to the network or to a production studio. Generally, only ideas from established television writers are considered. The major production studios then pitch to the networks what they consider the best ideas. Each network has identified weak spots in its programming schedule and is often receptive to programming ideas that will bolster the weak time periods. Often the network will hire a writer to develop a script, which is then presented to a focus group for *concept testing.* If concept testing indicates that the idea is viable, a pilot program is ordered. A *pilot* is a single, custom-produced television program that gives the audience a sample of the style and writing of the prospective series. The setting and characters are introduced, and a typical story line is presented. The pilot is then tested with an audience in a theater setting. Audience members can indicate acceptance or disapproval of any part of the program with a computer-controlled handset. Good audience reaction can ensure that the program goes into production; a bad reaction might spur rewriting or recasting—the program might even be scrapped.

Because television production costs are high, many times the pilot for a series, even those not good enough to find a time slot on the network schedule, becomes a "movie of the week" on the network. A pilot that tests well usually becomes an episode of the scheduled series.

If the network finds a spot on the schedule for a new series, it usually orders only five or six episodes to be produced. If the series doesn't generate the expected audiences, no more programs will be made, and the network will schedule something else for the time slot. If the audiences are large during the first one or two broadcasts, more episodes will be ordered. If the program does well in ratings, it will stay on the network schedule, many episodes will be produced, and the program may eventually go into syndication.

If the program does poorly, or *bombs,* in the ratings, it will soon be replaced on the network schedule. Programs that attract a moderate rating may be replaced in the *second season*—the time (usually in late January) when the networks schedule prime-time replacement series for those programs that have been canceled because of low ratings. In the early days of television, the network ordered thirty-nine episodes of television series. Thirteen of the episodes were repeated during the summer months,

GAME SHOW FRENZY

Popular network television programs tend to follow trends. When a new network program type attracts high ratings, similar programs pop up on other networks seemingly overnight. Each network is eager to provide the programming that the mass audience wants.

The first of the television network popular genres was the adult western. In the late 1950s, the popularity of *Gunsmoke* kicked off a trend of dozens of network westerns. By the end of the decade, there were thirty westerns in prime time, constituting seven out of ten of the most popular programs. Then came the big-money game shows, and the westerns rode off into the sunset. But the game show era was short-lived—to generate larger audiences, the producers "fixed" the quizzes. The government investigated, and the genre abruptly ended.

Other program genres soon followed. The early 1960s gave rise to the "idiot sitcom" with *The Beverly Hillbillies* and *Gomer Pyle*. The "relevance" era rounded out the 1960s, with such programs as *Mod Squad* and later *All in the Family*. The mid-1970s gave us a "fantasy" era with *Fantasy Island, Happy Days,* and *Charlie's Angels*. The 1980s brought the prime-time soaps like *Dallas* and *Real People*. The 1990s was the era of "news magazines" and "reality" à la Fox. The turn of the century brought network TV programming trends full circle: The game show era of the late 1950s was resurrected in force.

The original game show era was stillborn, ending in scandal with the revelations of dishonesty in the most popular network programs of the time—*The $64,000 Question, The $64,000 Challenge,* and *Twenty-one*. It would seem that the game show was gone from prime time forever, but then came a new generation of viewers and ABC's *Who Wants to Be a Millionaire?*

The program first aired in August 1999 on a nightly basis for two weeks. By the end of its summer run, it had doubled its ratings from a 3.9 to an 8.7. ABC quickly saw the advantage of programming *Millionaire* during the regular season, to ratings of 10s and 12s. The other networks noticed the high ratings, too. NBC dusted-off the forty-year-old *Twenty-one,* and Fox contracted with Dick Clark Productions to produce *Greed: The Multimillion Dollar Challenge*. Even cable was bitten by the game show bug: Comedy Central started running *Win Ben Stein's Money*. The prime-time programming genre of the early 2000s appears to be the game show.

---

when audiences were smaller. As the cost of program production went up, the number of episodes ordered went down. With shorter program contractual commitments, the networks now premiere the new television season in September each year and offer a second season in February to replace the programs that are marginal in ratings.

Both the networks and the production companies want success for the television series that are developed. But audiences are fickle, and what brings high ratings one year may not attract a large audience the next. In fact, the failure rate for new prime-time television programs is about 75 percent. To try to beat the odds, proposed network programs are tested and retested, schedules are revised and refined, and successful programming formulas from the past are resurrected.

Networks have used spinoffs in the past to ensure audiences for their program schedules. A *spinoff* is a new program series that uses established characters from an older series in a new situation. For example, the popular *Cheers* of the 1980s provided the character for the successful 1990s spinoff series *Frasier*. A decade earlier the series *All in the Family* gave birth to the later series *Maude* and *The Jeffersons*.

In addition to using established characters as the basis for new series, networks and production companies often rely on the talents of producers, writers, and directors with established track records of network prime-time production. Because of this practice, it is often very difficult for new talent, with new ideas, to break into prime-time television production.

THE BUSINESS OF SYNDICATION

The program syndication industry has grown immensely in recent years with the introduction of electronic-media marketing and delivery systems. Today broadcast stations, home video, corporate video, and cable and satellite distribution all provide profitable means of distributing syndicated video material.

Wheel of Fortune hostesses from around the world gather with Merv Griffin for the 20th anniversary of the popular syndicated program. Stations and cable networks rely on syndication as a source of programming.

Traditional broadcast syndication is the front line; it is the profitable business. First conceived during early radio as a means of filling out the broadcast schedule, syndication has become a stable of programming for all kinds of electronic-media distribution systems. The amount of airtime a broadcast station devotes to the syndicated program is staggering. Consider, for example, the traditional commercial television stations—each acquires its programming from three primary sources: the network (if they are affiliated), the syndicator, and/or local production. If a station is affiliated with a network, it acquires about 30 percent of its daily schedule (6 to 8 hours per day) from the syndication market, including audience favorites such as *Wheel of Fortune, Jeopardy,* or *Oprah*. If a station is an independent (not affiliated with a network), it will purchase about 90 percent of its programming (18 to 24 hours per day) from program syndicators. Even the affiliate at 6 to 8 hours per day purchases approximately 2,500 hours of programming per year.

Remember, the audience turns to an electronic-media program as a source of entertainment or information. The audience has little loyalty to a program distributor or producer, be it a broadcast network, cable network, or local station. With perhaps the exception of a local news personality, the audience turns to a channel for a program. The channels (stations and cable networks) turn to syndication as an inexpensive program product source. Many people recognize the names of the top syndication suppliers: Lorimar-Telepictures, Warner Brothers Television, the Walt Disney Company (Buena Vista), Columbia-Embassy, Twentieth Century Fox, Paramount Television, Universal Television, MGM/United Artists, MTM, New World Pictures, Orion Pictures, Viacom, and King World. Some creators also own syndication firms, for example Mary Tyler Moore (MTM) and Stephen J. Cannell. These producers provide almost 50 percent of prime-time television programs and almost all of the syndicated programming.

Program syndicators are basically program wholesalers. *Syndication* can be defined as the uniting of two or more program developers, under central management, to create, market, and distribute video programs on a national market level.

Syndicators provide most of the television content we see today on broadcast and cable. It is the syndicator, wholesaler, and/or creator who owns the copyrights to the programs and then contractually rents the program rights to the station or cable network for telecast.

Syndicators who are involved in creating program material for distribution through channels other than a network have a large task before them. First, they must research the program marketplace and then analyze the competition and program versus market potential. Assuming the research contributes positive encouragement, their next task is restructuring the financial investment and the budget to create the program. Investors and corporate managers must be sold on the program's potential success. Contracts are drawn up at this stage of the process, then there is the creation of the program itself. Once the program is finished, or "in the can," it still must be packaged, promoted, and distributed and sold. Without these last elements, even award-winning creativity can be of no avail.

Nationwide the dollar amount of syndication is astonishing. King World Productions has reported nearly a 50 percent return on its investment. Barter programs alone passed the $1 billion mark. The top syndication shows see a 97 percent national penetration rate without a network or cable affiliation. These figures are achieved via direct, station-by-station, market-by-market, system-by-system contract sales.

Syndication is a complex industry of program creation, supply, and distribution. The producers, major studios, and independents create and distribute programs for a profit. They sell and distribute product to networks (both broadcast and cable), and they sell to stations and to videocassette distributors. The stations purchase the programs to attract an audience which, in turn, is sold to advertisers. Even in the videocassette rental business, advertising is making its way. There are numerous channels through which a syndicator may distribute the program. The cassette distributors sell or rent their product for home viewing.

The control and supply of syndication programming itself involves program and financial strategy. It is a phenomenon called *windowing*—releasing a program in multiple distribution channels at different times. In other words, a firm may sell the same program product, at different times, and at a different price, to different buyers. For example, a theatrical release will first appear in the theater—then overseas theaters, home video, overseas home video,

second-run cable, and syndication to local stations. The options are many and varied. The hierarchy of distribution generally depends on audience interest and the rate at which that interest declines. The audience is generally willing to pay $7.50 to see a new theatrical release; however, as interest declines, that same movie will later appear on cable or direct-broadcast pay-per-view for $2.99, and so on down through the various distribution channels.

The foreign or overseas window is becoming increasingly important, and international syndication is growing. Programs from the United States dominate the global market, becoming so prevalent that some countries have passed importation and quota laws to encourage program creation from within their own borders. Canada, for example, mandates 60 percent Canadian content—a regulation passed to stem the tide of importation and to enhance the Canadian national culture. Other countries have adopted similar laws. Despite the efforts of other governments, the popular demand for U.S. programming abroad continues to grow. The foreign market accounts for approximately 50 percent of the sales for films, programs, and major market producers.

The commerce of program and syndicated materials is fiercely competitive. Major producers first target the networks. Despite their continuing decline in audience share, the networks are still effective distribution systems, and success as a network run can lead to continued success in syndication. *Off network* syndication refers to programs that have first been aired on the networks. The ideal off-network series has at least 130 episodes to allow programming flexibility at the individual market level. A program does not have to air on the network to be successful in syndication, however. Today's reality and talk-show programs have not aired on the networks. *Oprah, Rosie, Sally,* and *Judge Judy* are marketed directly to the station through syndication.

The sale of syndication on a station-by-station basis is referred to as *first-run*. These are original programs created without the network in mind that make their first appearance in syndication. They too need multiple episodes to be successful.

The sale of the program, whether off-network or first-run, requires a contract with each individual station in each market. The contract itself contains several common elements that are important to understand, including barter, cash, license period, exclusivity, length of run, reruns, delivery, promotion, and copyrights. Each of these is negotiated and becomes a part of the individualized, station-market contract.

Barter means the exchange of program material for commercial station spot time. It pertains to the way a program is sold and perhaps created. In the selling of a program, the syndicator may give it to the station at a lesser price in exchange for commercial avails during its showing. In other words, the station pays less for the program but gives up some commercial time slots, which the syndicator may in turn sell. If the barter arrangement is made before the program is sold to stations, the presale of time within the program is one way the syndicator may defray the costs of investment. *Cash* simply means that barter is not involved, and programs are exchanged for money.

The cash and/or barter elements of the contract are obviously critical, and it is not merely a matter of the cost and popularity of the program, but of estimating revenues that can be obtained for it within the syndication contract. Barter affects the number of minutes that are available for sale and promotion. For example, let's say there are five minutes and thirty seconds of commercial time available for the station to sell within a specific program, and the sales department predicts a 90 percent sellout rate. This means five minutes, or ten 30-second units, are available for sale. Ten units times the average spot rate of $1,000 equals $10,000. Subtract any applicable agency commissions and the unit cost of the syndicated program, and what remains is the estimated profit-per-telecast dollar figure.

The *license period*, generally one to three years, covers the time within which the station has approval to air the program. The syndicator is likely to favor the shorter period, because it provides the opportunity to renegotiate based on program success.

Exclusivity means the station purchasing the program will be the only one in the marketplace with the rights to air it. It is a part of the negotiation process, because the syndication firm's representative can go from station to station within the same market, trying to get competitive offers for the exclusive rights to a popular program.

Length of run refers to the number of episodes within the series that are available for syndication. The number of episodes is important, because it affects the way in which the program may be scheduled. Strip programming, for example, requires a minimum of 130 episodes. Because there are exactly 260 workdays per year, a program with 130 episodes can run Monday through Friday for six months and, if one assumes a one-year contract allowing for two runs of the program, the series fits nicely into the schedule. Some programs have a long length of run, such as *I Love Lucy*, *M*A*S*H*, and *Seinfeld*. These programs have holding power not only because of their popularity, but because the length of the series itself is conducive to forming viewing habits.

As stated, *reruns* are simply the number of times a station is allowed to run the program within the contract period. Both the length of run and number of reruns affect exactly how a station may program the series. Some syndication packages are short and allow minimal reruns. These are marketed as blockbusters or special programs, which can be scheduled and promoted during a ratings period.

Delivery concerns the method of sending the program to the station. It could be sent by satellite each day, or it could be mailed to the station on videotape.

Promotion is simply what the syndicator and/or station will do to hype the program. The syndicator may produce a promotional package to be used by the station. The station itself may contract to do specific promotions on behalf of the program.

Copyright concerns what the station can do with the program. For example, the issues of program alteration and duplication are considered. What duplication is permitted? How may the program be edited to produce promotion materials? These are questions of program rights. The syndicator will relinquish enough rights to increase promotion and marketing efforts, but will retain major duplication and alteration rights for future syndication efforts. Remember, the syndicator, not the station, holds the copyrights.

Programming Costs

As noted earlier, the programming schedule on the television channel could be considered a product *shelf* on which programs are displayed for the viewer. The purpose of the schedule is to attract audiences, which can then be sold to advertisers. Advertisers buy time within or between programs—the commercial *avails*—in which to air their commercials. Account executives, who sell the time to advertisers, know that the product shelf contains a limited number of avails. Hence, it is in the station's or network's best interest to maximize the value of the avails sold. This is done by building audiences that advertisers want to buy; the larger the audience, the more the avail is worth in advertising revenue.

Programming costs must be balanced against revenues, with the ultimate goal of making a profit for the station or network. For the local television station, programming is available from networks and syndicators or it can be produced locally. For the network affiliate, network programs are obtained without cash payments, but these programs are not free, because affiliate stations give up commercial avails. The networks sell a certain number of minutes each hour to national advertisers. In some cases, affiliates share in these revenues through compensation—they receive a portion of network revenues for each minute sold. In addition, stations usually have about two minutes each network hour to sell directly to advertisers during station breaks.

The networks have been reducing compensation to affiliates over the past few years, primarily because of decreased audience shares for network programs. For the average network affiliate, compensation provides less than 10 percent of its advertising revenue. Still, many affiliates consider compensation important enough to lobby against projected network cuts, although network compensation may eventually be eliminated entirely.

This means that all network programs would be provided on a barter basis—a simple trade of program for commercial time. In this arrangement, already in effect for some network programs, the networks and stations share the avails within a program or hour.

Barter is also common in the acquisition of first-run syndicated programs, which are distributed directly to stations without airing on a network first. As stated, the rights to the programs can be obtained on a cash, barter, or cash-plus-barter basis. The cash-plus-barter arrangement involves both a trade of commercial time for programming and a cash payment per episode. *The Cosby Show* was syndicated on a cash-plus-barter basis; stations paid a fee for each episode and gave one minute of time within the program to the syndicators.

Although cash-plus-barter syndication has been successful for some of the most popular off-network programs, such as *The Cosby Show,* in some cases the cost is too high for many television stations. In 1992 MTM Productions tried to syndicate the CBS series *Evening Shade* on a cash-plus-barter basis. Few stations were willing to take the series on those terms, and it was eventually taken off the syndication market and was subsequently sold to the Family Channel cable network.

Costs for syndicated programs vary widely, depending on market size and, of course, the program itself. *Seinfeld,* one of the most successful situation comedies of all time, became available for syndication in 1994. Sold on a cash-plus-barter basis, the series went on the market in Chicago for $110,000 per episode, in New York for $125,000, and in Los Angeles for $130,000. These were the lowest, or *floor,* prices from which stations were expected to start bidding. In addition, stations would give up one minute of barter time in each episode. Certainly, these prices were on the top end of the syndication cost scale, but *Seinfeld* provides an example of the expense of off-network programs.

The average network affiliate station (in the top fifty markets) allocates about 25 percent of its budget to programming costs, and the average independent spends nearly 40 percent. This reflects the difference in program sources for the two types of television stations. As noted, the independent must purchase, barter, or produce virtually all of its programming, and the affiliate receives the majority of its programming from the network without cash outlay.

CABLE PROGRAMMING

The cable operator faces an entirely different task in programming than does the radio or television station programmer. With the exception of a cable local-origination channel, the cable programmer does not place specific programs into a schedule to attract the largest audience in any given time period. Instead of selecting programs for a single channel, the cable programmer chooses entire channels to fill the vast shelf space of the modern cable television system. The goal of the programmer is to provide a diversity of channel choices that will encourage subscribers to buy cable service—a task more difficult than one might imagine.

Most cable television systems provide subscribers with between fifty and a hundred channels. Even though the cable programmer has many channels to fill, the available information is vast. In addition, federal law, FCC regulation, and city franchises dictate specific channels that must be carried. In other cases, the programmer must pay on a per-subscriber basis to have the rights to carry the programming of the cable network. If the audience for a specific channel is small, as is often the case, the cable operator must consider the value of the program in generating new subscribers compared with the cost of providing the channel. The programmer also wants to limit *churn,* or disconnects, caused by dissatisfaction on the part of the subscriber. Often the cable operator bundles different channels and then markets them to the customer at tiered prices (see table 9.4). Ideally, the bundles are packaged to encourage the subscriber to pay higher rates to obtain desired programs.

Although this section addresses programming a cable television system, the same considerations apply to the newer multichannel electronic-media systems. Selecting programming for wireless cable (MMDS), *satellite master-antenna television (SMATV),* or *direct-broadcast satellite (DBS)* systems is about the same as for a cable television system.

To fill the system shelf space, the cable programmer selects channels from five sources: broadcast stations, superstations, pay-cable networks, basic-cable networks, and local-origination/access programs.

As discussed in earlier chapters, the first cable television systems, then called *community antenna television (CATV) systems,* were established in the late 1940s to pick up the faint signals of broadcast television stations and distribute them into homes of subscribers. Carriage of broadcast television stations—to provide improved signal quality and a greater choice of programs—was, and still is, an important function of the cable system. Nearly 50 percent of the viewing by audiences in cable homes is of network-affiliated broadcast television stations.

Must Carry

A variety of rules have been established over the years concerning the relationships between the broadcast television station and the cable television system. Many of these rules stipulated which broadcast channels the cable system was required to carry. For many years the FCC's **must-carry rule** dictated which stations the cable system was required to offer on the system. The Cable Television Consumer Protection Act of 1992 provided a set of rules for carriage of the local broadcast station on the cable system that are still in effect today (see chapter 5 for details). In essence, this law allowed the local station either to be considered a "must-carry" and therefore guaranteed a channel position on the system, or to give the right of retransmission consent. In granting **retransmission consent,** the station gives the cable system the right to carry it in exchange for cash, additional channel space, or other valuable considerations.

In many markets broadcast channels constitute 20 percent or more of the entire shelf space of the cable television system. They must be provided, however, because of regulation or economic necessity. Viewers have proven that they want the programming of the

TABLE 9.4 Tiering of monthly cable programs and services

TYPE OF PROGRAM SERVICE	RATE
Limited basic	$ 8.02
Expanded service	13.73
Standard service	21.75
Expanded plus	1.48
Classic	23.23

PREMIUM SERVICES

Premium channels and expanded plus customers must have at least the standard level of service to subscribe to the following:

HBO	$ 10.43
Showtime	8.52
TMC	8.52
Cinemax	7.57
Music Choice	0.50
Total TV	3.25

MISCELLANEOUS CHARGES

A la carte	
AMC	$ 0.74
Showtime	0.74
TNN	0.74
TNT	0.74
Converter	1.45
Additional outlet	3.19
Remote control	0.41

broadcast station as part of the cable service to which they subscribe.

SUPERSTATIONS

A *superstation* is an independent broadcast television station licensed by the FCC to serve a local market. Because its unique programming is of interest to a wider audience, however, its signal is distributed via satellite by a common carrier to cable television systems across the country.

WTBS, channel 17 in Atlanta, was the first superstation. In 1976 Ted Turner arranged to have the signal of his independent television station placed on the satellite transponder that carried the signal of HBO. Cable systems, looking for something new to offer their audiences, quickly arranged to carry the channel. The cable operator paid the common carrier for satellite delivery of the signal to the cable system head-end. The superstation benefited by selling advertising time made more valuable because of the increased national audience. Ted Turner's Atlanta Braves quickly became a hometown team to many who could, for the first time, watch their games on WTBS.

Other superstations have since come into existence, including WGN from Chicago, WWOR and WPIX from New York, WSBK from Boston, KTVT from Dallas, and KTLA from Los Angeles. Some have since disappeared as superstations. To carry a superstation, the cable operator must pay a few cents per month per subscriber to the common carrier, as well as pay copyright fees for programming. Because of the costly copyright payments, most cable television systems offer only one or two superstations.

PAY NETWORKS

Sometimes called *premium channels,* these networks offer movies, sports, and special events to the cable subscriber for an additional fee. Most of these services offer a variety of programs for a monthly price, but some provide *pay-per-view (PPV)* through interactive, or addressable, technology, enabling the subscriber to select and pay for individual programs rather than subscribe on a monthly basis.

HBO was the first satellite-delivered national pay-movie service. It started in 1972 and delivered recently released theatrical movies to the home. As it expanded its programming day, HBO began producing made-for-cable movies, episodic series, sports, and specials.

Other pay cable network services include Showtime, Starz!, and Encore. Many cable systems offer several premium channels in hopes that the subscribers find them different enough to warrant subscribing to all.

The cable television system collects a separate payment from the subscriber for each of the premium channels used. The cable operator passes on a portion of that fee to the pay network for the cost of programming and operations. The remainder is kept by the system operator and forms an important revenue stream for the cable television system.

BASIC CABLE

With cable's *broadband* technology, which allows multiple channels to be sent over the same wire, many different satellite-delivered cable television networks have been developed. Starting in the early 1980s, cable system operators began searching for something different from the off-air broadcast channels to offer their subscribers. Although pay movie services and superstations came first, cable networks were not far behind. Most cable networks were designed to offer a specific type of programming: One might offer all sports programming, another might provide an all-country-music format, and a third might provide an entire channel of news programs.

Since the early 1980s, the number of basic cable networks from which the cable operator can select has grown to more than a hundred. New cable networks seem to spring up overnight, while others die because of lack of carriage on enough cable systems to attract advertiser support. The early networks, such as MTV, ESPN, and CNN, seem to have found a permanent position on most of the nation's cable systems. These networks have the potential of reaching the homes of nearly 60 million subscribers.

Although most cable networks design programming schedules to appeal to a specific audience, several operate like the traditional broadcast networks and program by daypart to a broad audience. These, including TNT, TBS, the Family Channel, and USA Network, show old movies and off-network reruns during most of their programming day.

Other cable networks specialize their schedules. Nickelodeon and Nick at Nite provide programs for children and families. Univision and Telemundo seek out the Spanish-speaking audience. The Discovery Channel, the Learning Channel, and the History

A Short History of HBO

Chuck Dolan, one of the pioneers of cable television, and publishing giant Time, Inc., became partners in Sterling Communications during the mid-1960s. The company was struggling to build a cable television system in lower Manhattan long before cable was viewed as a profitable business. Because Time was beginning to have doubts about cable—as well as about television in general—Dolan was under some pressure to make the business pay. He developed an idea called the Green Channel that would provide sporting events and movies to cable customers for an additional fee. That channel, which was soon renamed Home Box Office (HBO), signed on the air with its first programming in 1972. It had fewer than 350 subscribers on a single cable television system in Wilkes-Barre, Pennsylvania, when the first programming was beamed by microwave from New York City.

HBO grew quickly, however, because of the management decision to include local cable television systems in the revenues generated by subscribers. The more subscriptions sold, the more money the local cable operator was able to keep. In addition, HBO offered movies, sports, and specials that could not be found on broadcast television stations. Soon microwave links were carrying sports from Madison Square Garden to many cable systems in the Northeast. The real surge in subscribers occurred in 1975 when, under the leadership of Jerry Levin, HBO leased a transponder on the new RCA *SATCOM I* to distribute four hours of HBO programming each evening. He then arranged for satellite dish manufacturer Scientific-Atlanta to reduce the price on its $100,000 antenna to be affordable to the cable operator, on the promise that many new orders would soon be coming from the cable industry. HBO redefined the cable television industry with satellite distribution of its signal to systems across the nation.

FCC regulations limited movies that could be shown on a pay system to those that were less than two or more than ten years old. HBO went to court, and in 1977 the FCC was forced to drop its rules (which had favored the broadcaster at the expense of the cable operator). HBO was free to buy the rights to show additional movies and sporting events, but the rule changes brought competition. ESPN, CNN, CBS Cable, and Satellite News Service were all established. Direct competitors in the pay field included Showtime, the Movie Channel, and Spotlight. HBO started a second service called Cinemax. By 1981, less than ten years after its beginning, HBO, under the programming genius of Michael Fuchs, became a major profit center for Time, Inc. Finding programming, however, was becoming increasingly difficult.

Realizing that it would have to create its own programming, HBO began producing television specials as early as 1975, but its audiences wanted more first-run movies. Hollywood just wasn't turning out enough product to satisfy the needs of the competing pay-television services. In 1982 HBO partnered with Columbia Studios and CBS to form the new motion picture studio, Tri-Star. In addition, it bought rights from other movie studios for product long before it was produced. Some movies turned out to be bombs, but many others were hits acquired for amazingly low prices.

Under the leadership of Michael Fuchs and Chairman Frank Biondi, the very profitable HBO became synonymous for everything that was new and exciting about cable. Then subscribers began to drop out and profits leveled as a new competitor had entered the field: The videocassette recorder and the video store that allowed the viewer to rent any number of movies and play them when convenient. Soon those that had reinvented pay television were out of jobs at HBO, and a new generation took over the reins. Competition from home video has now been joined by direct-broadcast satellite and pay-per-view. HBO is still the healthiest of the pay services, however, with 18 million subscribers, as compared with Showtime's 13 million, but the glory days appear to be over.

Source: From G. Mair, *Inside HBO*, New York: Dodd Mead (1988).

Channel all provide educational programs. Those, as well as Arts & Entertainment (A&E), provide cable network programming similar to that of public television stations.

In addition, there are movie channels such as American Movie Classics, Bravo, and Galavision for Spanish-language audiences. MTV, VH-1, and the Nashville Network provide music programming. There

A Short History of CNN

Ted Turner was left with a billboard company in Atlanta, Georgia, after his father's death in 1963. Turner was twenty-five years old at the time. By 1970 he had parlayed the billboard business into a multimillion-dollar operation. He used the profits to buy a money-losing UHF television station in Atlanta and another in Charlotte, North Carolina. Within two years both stations were operating in the black by broadcasting reruns of such family favorites at *Gilligan's Island, Leave It to Beaver,* and *The Andy Griffith Show,* along with old movies and Atlanta Braves games. Turner began distributing the signal of his now-popular channel 17 by microwave to the new cable television systems that were springing up in the valleys of the southeastern United States.

When Western Union launched its first domestic satellite in 1974, Turner was told that it would cost a million dollars a year to have his station's signal broadcast from "the bird." He was reported to have exclaimed: "You mean that I can reach the whole country for only a million dollars?"

Home Box Office became the first cable-satellite network when it went on *SATCOM I* in 1975. But Turner put WTCU (later WTBS) on the satellite in 1976 and called his station, which was delivered to cable television systems, a "superstation." He soon bought the Atlanta Braves so that he could have access to all of their games. The Braves thus became the hometown team in small cable markets from Montana to Maine, thanks to the communication satellite and superstation channel 17.

As cable television systems picked up the superstation, Turner was able to sell advertising time at national rates, and the profits for Turner Broadcasting soared.

By 1978 Ted Turner began talking about starting a second satellite-delivered service, one that would provide news twenty-four hours a day. He went to the cable industry to finance his new venture but found little support for his radical idea. The Charlotte television station was sold to Westinghouse Broadcasting to help raise the money needed to establish the Cable News Network (CNN).

With no competition, CNN was in the black by 1985, just five years after its launch. Looking for other challenges, Turner attempted a hostile takeover of CBS. Although he failed, the challenge cost CBS a large sum of money in stock acquisition that subsequently led to downsizing the award-winning news department of what became known as the Tiffany network, because of the money spent to make its operation first-class.

Not one to sit still for long, Turner went from his failure to acquire CBS to the successful buyout of the movie studio MGM/UA. The rights to the film library gave Turner enough product to establish the successful channel TNT (Turner Network Television). But the debt load of the acquisition was more than the combined profits of WTBS and CNN could handle. To save his empire, Turner took on partners from the cable television industry and Time, Inc. In the process he lost majority control of the Turner Broadcast System, and the freewheeling decision process took on a definite corporate attitude.

Source: From H. Whittemore, *CNN: The Inside Story,* Boston: Little, Brown (1990).

---

are also sports networks (ESPN), shopping networks (QVC and Home Shopping I and II), and religious networks (Trinity Broadcasting and the Eternal Word Television Network).

News and informational networks have proliferated on cable television. One of the earliest was Cable News Network (CNN), established in 1980 by Ted Turner as a companion to his superstation, WTBS. CNN has been very successful and now provides twenty-four-hour news service in many countries around the world, as well as a news feed to many U.S. broadcast television stations.

CNN has been joined by CNN Headline News, CNN International, the Weather Channel, the Travel Channel, C-SPAN (Cable-Satellite Public Affairs Network), C-SPAN II, CNBC, MSNBC, and Court TV. There are plenty of choices for the viewer who wants to know what is happening in the world, in the U.S. House of Representatives (C-SPAN), or even in the British Parliament.

Local-Origination and Access Channels

Programming produced or provided locally by the cable operator is called *local-origination (LO) programming*. Programming produced for a cable channel by a local citizens' group, a community organization, an educational institution, or a government organization is called *access programming*. Local-origination and access channels are often mandated in the franchise that licenses the cable system to operate in a city. The city can, and often does, require that the cable operator provide a public-access channel, an education-access channel, and a government-access channel as one of many conditions for receiving the right to operate a cable television system. Some of the channels exist in name only, but in other communities the LO and access channels are very active. About 50 percent of the cable systems provide regular programming on LO or access channels. Most of the systems with active local programming are in the larger cities or in college communities, where there is more interest in local television programming.

Local Origination

The local-origination channel is operated by the cable television system operator. Its programming can range from a few hours per week to a twenty-four-hour programming service and from the simplest programming to full-blown sports remote coverage. Some LO programming is provided as a service to the community by the cable operator, but much is designed for local advertiser support.

Some cable systems provide their own local news inserts in CNN programming as local origination; others arrange with local broadcasters to provide full news service either as local inserts or on separate channels. A few cable operators provide well-funded, professionally produced regional news services. *News 12,* serving subscribers of Cablevision on Long Island, is a good example of a successful regional cable news service.

In a few markets, cable operators have established regional pay sports networks. These operators obtain the broadcast rights to local and regional high school and college football and basketball games. The games are produced using remote television production vans, some of which are even equipped with slow-motion units for stop-action instant replay. The games are usually provided to subscribers on a premium channel that must be purchased separately.

In a handful of communities, the cable television operator has established comprehensive local-origination services. Systems in such cities as Rochester, New York; Toledo, Ohio; and Eureka, California, have even purchased the rights to off-network television series to program on a local-origination channel.

A few cable *multisystem owners (MSOs)* have produced local-origination programming that has been distributed to other systems owned by the company.

Nearly all cable operators provide one or more text channels. These can be used to give the subscriber the cable channel lineup, to provide weather information, to serve as a community bulletin board, or even to provide a classified advertising service to the local community.

Access Channels

Access channels are provided by the cable television service for free to the public or to an education or government entity for their use. Often these are called the *PEG channels*—an acronym for *public, educational,* and *government*. A fourth access channel is *lease access,* which is provided by the cable company to a commercial business for a fee. The common attribute of all access channels is that the cable operator provides the channels for others to program and exercises very little control over programming content. It should be noted, however, that there are some rules concerning the content of cable programming. Hate language and pornography, for example, are usually banned.

The public-access channel is provided so that a citizen can have an electronic-media soapbox, or forum, from which to present his or her opinion. The cable operator often provides a community-access center that offers television station production equipment as well as trained staff or volunteers. Often the community-access center facilities are used to produce entire television programs that will be shown over the system. The facility is usually provided free on a first-come, first-served basis. Cable operators and city governments are often nervous about the content of programs that are cablecast over the public-access channel. Some programs feature extreme points of view; others broadcast nudity into the homes of subscribers.

Government- and education-access channels are often used to provide gavel-to-gavel coverage of public meetings. Many cities offer live cablecasts of city council meetings or meetings of the planning and zoning committees. School districts often provide coverage of school board meetings. Some educational-access channels are used for the delivery of instructional material; others offer the opportunity for students to produce television programs that will be seen by an audience.

Access programming often does not draw large enough audiences, or have enough programs, to justify a full-time channel for each of the PEG access services. Many cable systems have provided a hybrid community-access channel that includes public, educational, governmental, and leased-access programming on a single channel.

SUMMARY

▶ Audiences listen to radio stations but select television programs to watch.

▶ Programs are the product of the broadcast station, but the program is used to generate an audience that is sold to the advertiser.

▶ The television programmer (1) researches the market and the competition to determine what to put on the air, (2) selects the program to meet the needs or wants of the audience, (3) uses programming strategy in scheduling the programming, and (4) interprets the ratings to obtain feedback as to the success of the programming efforts.

▶ The television broadcast day is divided into dayparts. Each daypart has its own audience demographic and requires its own type of programs.

▶ Program sources for the broadcast television station include the following: (1) the network (for network-affiliated television stations), (2) the syndication market for both first-run and off-network television programs, and (3) local production, of which news is probably the most important program.

▶ To attract and hold the largest possible audience of the desired demographic type, the television station programmer uses a variety of programming strategies. Although less effective now that the remote control is so common, most programmers use blocking, strip, counter-, mood, and cyclical programming when developing the program schedule.

▶ Cable television systems have different programming goals than do broadcast stations. The cable operator is not interested in which program will attract the largest audience in a competitive situation, but must select the channels of programming that will increase penetration and limit churn.

▶ The cable television operator selects channels of programs to fill the channel shelf space from the following: (1) local broadcast television stations, (2) superstations, (3) pay, or premium, satellite-delivered cable networks, (4) basic satellite-delivered networks, and (5) local-origination and access channels.

InfoTrac College Edition Exercises

People listen to radio stations, but they watch television programs. We tend to think of TV in terms of programs rather than of the stations or channels on which they are delivered.

9.1 You read in chapter 9 about television program syndication. Use InfoTrac to locate the article "Syndication in the Eye of Justice" by Joe Schlosser in *Broadcasting & Cable* (January 3, 2000) and answer the following questions.

   a. Why did the Justice Department investigate the merger of media giants Viacom and CBS? How does that relate to television syndication?

   b. Name several popular syndicated television programs owned by Viacom-CBS.

9.2 You read in chapter 9 about network television program production. Use InfoTrac to locate the article "A Mouse In-house" in *Broadcasting & Cable* (November 29, 1999) and answer the following questions.

   a. The consolidation of ABC's prime-time entertainment division and Disney's TV studios—Buena Vista Television Group—was the first of its kind in Hollywood. Why were Hollywood producers concerned about the merger?

   b. ABC Television President Pat Fili-Krushel said that there were a number of reasons for the consolidation of the studios and the network. Name and briefly discuss one of those reasons.

WEB WATCH

Here is a list of a few URLs (Internet addresses) for some of the organizations or corporations discussed in this chapter. Please explore these Web sites and follow the links to learn more about the complex business of the electronic media. Add your descriptions and your own favorite sites at the end of the list. Please keep in mind that the dynamic nature of the Internet allows sites to come and go but also allows organizations to update information about themselves very quickly.

Address	Description
http://www.nbc.com/	
http://www.abc.go.com/	
http://www.cbs.com/now/section/o,1636,100-311,00.shtml	
http://www.fox.com/frameset.html	
http://www.viacom.com/	
http://www.disney.go.com/	
http://www.universalstudios.com/tv/	
http://www.paramount.com/	
http://www.emmyonline.org/	
Other favorite sites:	

CHAPTER 10

News and Information Programming

News and information programming have become the business of news and information.

Defining News

"News exists in the minds of men," Wilbur Schramm wrote a few decades ago. If he were writing today, he'd define news as existing in the minds of people—people interacting with people, people interacting with their physical surroundings, and people trying to reconstruct the framework of events affecting people. Schramm's theoretical definition of journalism transcends both the electronic and print media to reconstruct the framework of events affecting people that has been the traditional news mission. It focuses on people and the communication of information. Its tools are writing and aural/visual presentation. Reports are based on journalistic values of objectivity, fairness, accuracy, and responsibility. The definition of news typically includes values of *consequence,* which affect people's lives; *interest,* which present unknown or unusual facts; *timeliness,* which are new and current; *proximity,* which have more timeliness and consequence to local interest; and *prominence,* which relate to people whose names we recognize—events surrounding a public figure make more news than those same events surrounding the average person. Finally, and much to the distaste of the traditional newsperson, news has recently acquired another value—*entertainment.* We address this phenomenon later in this chapter.

Today we are at a pivotal point in the evolution of electronic journalism. Definitions and values are under attack as the number and profitability of news and information programs are growing. The purpose of this chapter is to briefly look at the growth of electronic journalism in radio and television. This includes discussion of regulatory and marketplace constraints, basic newsroom and a news worker's responsibilities, news-gathering sources, newscast programming variables, and audience-perceived trends in news and information.

Network News

Electronic journalism has both its theoretical and practical roots in print and radio, which defined the practices used in television and succeeding electronic media delivery systems.

The radio news program as such scarcely existed before 1937. Earlier chapters talked about KDKA's election coverage, President Coolidge's 1923 network speech, President Roosevelt's Fireside Chats, and public debate over platforms of Roosevelt's New Deal or the European "phony war." These illustrate the style of information programming that existed in the early days of the radio. There were daily newscasts, but not as we know them. The local stations and networks carried both news and commentary, and a few stations, owned by newspapers, broadcast bulletins and information in hopes of stimulating newspaper sales. Commentary is the closest forerunner to what we envision today as news. H. V. Kaltenborn and Lowell Thomas were perhaps the most well-known commentators. The press-radio war restricted the rapid growth of radio news to bulletins and forced radio's information programming of the thirties to emphasize coverage of special events. As William L. Shirer put it in *Twentieth Century: The Nightmare Years* (1976), he and Edward R. Murrow were busy in Europe, "putting kids' choirs on the air for . . . Columbia's American School of the Air."

The growing possibility of U.S. involvement in World War II and the ideological debates that surrounded the New Deal fostered growth of news and increased coverage of world events. During the late 1930s, both CBS and NBC distributed guidelines for their news divisions that Paul White said required commentators to "elucidate and illuminate the news of common knowledge and to point out the facts on both sides" (1947). These network organizations were small at first, with personnel primarily from print and wire service organizations. As World War II unfolded, however, the job descriptions of "commentators" were transformed to those of "correspondents" so the networks could keep pace with the influx of news. Writing in *History in Sound,* William S. Paley, president of CBS, said that "radio news grew up with World War II." During this war radio reporters filled the airwaves with the tragedy of conflict. Radio and television news stations and networks today continue to emphasize news gathering, on-the-spot factual reporting, and eyewitness event coverage—all within the roundup format developed by the networks during World War II.

Edward R. Murrow began his career in journalism in 1935 as a director of talks and education for CBS. His fame as a leading journalist began with his broadcasts from 1940s London as Hitler unleashed his blitzkrieg of nightly bombings over England. Americans, many not yet convinced of the Axis threat, heard for the first time the broadcast sounds of a

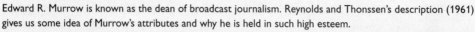
Edward R. Murrow

Edward R. Murrow is known as the dean of broadcast journalism. Reynolds and Thonssen's description (1961) gives us some idea of Murrow's attributes and why he is held in such high esteem.

In Murrow's *This is London* series of broadcasts from 1939–1945 . . . Murrow, then in his thirties, earned an international reputation as a reporter of the happenings of World War II. His job was to report the hard news of the war: This he did with penetrating insight and not a little eloquence. Without distortion of value, however, he also paused on occasion to take stock of man's enduring pleasures. Although the ugliness of war was everywhere in view, he could yet report on the beauty of a flower in early spring. . . . With the democratic order in peril, he spoke with renewed enthusiasm of individual dignity. . . .

Murrow's basic creed seems rooted in his deep-seated belief in the dignity of the individual—a dignity which entails the individual's right to hold and express beliefs, to formulate and fulfill his own plan of living, to assume a personal responsibility in the use of his rights, and to recognize his obligation to the social units of which he is a member.

Murrow prepares his talks with an eye to a news formula . . . the network has not editorial views; neither is it disposed to advocate. . . . In short, its news analysts are expected not to air editorial positions. . . . Murrow believed that the reporter had no right to use the microphone "as a privileged platform from which to advocate action."

Murrow's wartime reports from London testified eloquently to the effectiveness of his reportorial prescription. . . . [He] captured the spirit of a moment in history and translated it into a language that could be understood by people distantly removed from the scene. . . . Without sensationalism, he gave urgency and dramatic impetus to ideas and happenings which were the common concern of people. . . . He reported both on events and on the climate of opinion in which the news occurred. . . . It included such things as laughter, jokes, gripes, hopes, aspirations, fears and quarrels of men and women in the less privileged class as well as of those in positions of responsibility. It included a recognition of what people believed to be true, whether it was or not. . . . Vignettes of distinguished men are common in Murrow's reports.

The distinction of Murrow's style derives in no small part from his selection of materials. He chooses his subject matter with good taste. He ignores . . . the tawdry details of the current scandals and the petty doings of headline hunters. He respects personalities. He achieves effectiveness without sensationalism or abuse. With intelligent discernment he senses the course of historical events and reports the facts that seem to make the biggest difference in the lives of men.

From O. T. Reynolds and L. Thonssen, "The Reporter as Orator: Edward R. Murrow," in Loren Reid (ed.), *American Address: Studies in Honor of Albert Craig Baird*, Columbia: University of Missouri Press (1961). Reprinted by permission.

city at war. Murrow's signature opening—"This is London"—was followed by vivid, sensitive descriptions of the bombs' disastrous effects on the city and its people.

After the war Murrow continued to work as a radio news broadcaster for CBS. When a coast-to-coast system of coaxial cable and microwave towers made it possible, *See It Now* debuted in 1951 as the first national TV news broadcast with Murrow as host.

During the early 1950s, strained relations with Russia had helped create increasing fears about communism in the United States. The hysteria reached a peak during Senate and House hearings, when intellectuals and artists—many from the acting and writing professions—were summoned to Washington to answer questions about their own and their friends' political beliefs. In the Senate much of the witch-hunting activity was spearheaded by Sen. Joseph McCarthy, whose reckless denunciations helped end many careers and caused several suicides. Fed up with McCarthy's terrorist tactics, Murrow invited him to appear on *See It Now* in 1954. McCarthy's unbalanced demeanor and comments helped end his own career. The power of television—and the integrity of Ed Murrow—were demonstrated to the nation simultaneously.

Murrow left television, whose power he had helped create, in the early 1960s to head the U.S. Information Agency for the Kennedy administration. He died of lung cancer in 1965.

THE TELEVISION DOCUMENTARY

A. William Bluem describes the idealism of the era of the television documentary—noting that at some point between the end of World War II and the start of the 1950s, television evolved, taking with it the news practices from radio. As it did so, it emerged from a stage of technical experimentation and became a major communications force. Discussing our documentary heritage, Bluem maintains that the challenges of television documentary during the fifties posed some questions familiar today. The documentaries Bluem mentions became the foundation for what we now call "TV news magazines," seen nightly on one network or another.

The documentary idealism of the fifties and sixties is reflected in statements from John Grierson, the father of documentary. Grierson suggested that the function of documentary was "to make drama from life . . . to make observation a little richer than it was by creative interpretive accounts of actuality." The power of the documentary according to Grierson was "making drama of our daily lives and poetry from our problems."

Where does "art" become distinct from "journalism"? What are the differences between "objective" and "subjective" accounts of life? And when is the documentary idea and spirit to be separated from mere social propaganda in which the authentic is subservient to partisan and untruthful reconstruction?

So long as fiction filmmakers were content to rely upon costume and makeup, studio settings, and the art of the performer, the direction of documentary was clear: It would go to the natural location, it would seek the natural man, it would represent nothing that did not exist in truth, and it would attempt to fashion its story only from these records.

But when the fiction film was inspired to leave the studio and move to the streets, the forms blended once more.

In American television news documentary history, there were three programs of prominence: *CBS Reports*, the NBC *White Paper*, and ABC *Close-up*.

CBS Reports, an idea conceived by Frank Stanton, had its roots in Edward R. Murrow and Fred Friendly's *See It Now*. In this documentary experienced journalists, believing that television was an instrument of transportation, applied film in a way it had not heretofore been used on television—to relate a news story.

See It Now built its largest audiences with its deliberate choice of those social conflicts that define the course of a free society and thereby were of compelling interest to most Americans. There were not added dramatic values of sound effects, mood music, camera manipulation, or stylized editing to sustain these programs as documentary in the best traditions.

Source: From A. W. Bluem, *Documentary in American Television*, New York: Hastings House Publishers (1965).

The decade following World War II was one of transition for radio (see chapter 8) and it was catalytic for television. In 1950 only 9 percent of U.S. homes had a television; ten years later this figure had grown to 87 percent. News grew with the industry to include national and international features on both radio and network television. The fifties, the era of the cold war, inspired special reports and classical documentaries, and some broadcasts influenced the course of national events. The coverage of the Army-McCarthy hearings and United Nations debates, for example, produced riveting exchanges and swayed public opinion. Documentary programs added dimension and depth to the audience's ability to understand the events close to home and around the globe.

News programming and the network documentary activities grew slowly but steadily during the fifties and sixties. At the network level, CBS, NBC, and later ABC expanded their international and national news operations and established foreign bureaus. Eventually, the networks employed twenty-five to thirty reporters in their New York headquarters, another fifty in their Washington, D.C., bureau, and a dozen or so in bureaus in Chicago, Atlanta, London, Paris, Bonn, Moscow, Tokyo, and Rome. Later the Vietnam War brought an expansion at bureaus in Los Angeles, San Francisco, and Tokyo. The budget for *CBS News* during the latter part of the sixties was approximately $45 million per year. At this time the *CBS Evening News with Walter Cronkite* featured approximately thirty news items—twenty written in-house by Cronkite or one of several writers, the remainder filmed or taped stories from bureaus all over the world. Simultaneously, the NBC budget for news and sports was around $100 million.

Local News

In the postwar era, small-market radio news struggled with unprofitable and low-budget news operations. It remained oriented toward national and international news supplied from the wire services. Local news was secondary on radio. Television stations offering news tended toward local coverage, but only for the early-evening news. They were not profitable programs, and many of the low-budget stations were only "rip and read," wherein the newscaster simply read copy directly from the wire service printout. Some even used the local newspaper as a source of daily information.

If a station was large enough to have a budget, most of it went into the purchase of one of the wire services: Associated Press (AP) or United Press International (UPI). News departments were small—frequently with fewer than three people. TV news departments consisted of an anchor who doubled as reporter, and a reporter who may also have worked as the news director, executive, or a station manager. In television's larger markets, the reporter likely doubled as writer and videographer. These shops were called "one-man bands." News was not a high priority in the station's program lineup, and few watched. It was produced as a service that, along with public-affairs programming, was monitored by the Federal Communications Commission (FCC) as part of the station's commitment to its community.

Local TV news programs grew slowly, especially during the fifties and sixties, and were overshadowed by increasingly popular network news. By the early seventies, however, local news too was attracting a sizable segment of the audience and was starting to be recognized as a marketable commodity. This realization made the business of local electronic-media news fiercely competitive. The local news around the country became the centerpiece of a station's program lineup. Today local newscasts are no longer a mere public service but are programmed into profitable time blocks adjacent to the network news. In these blocks the networks dominate international and national coverage, and local affiliates cover their communities. It was a balanced partnership for the viewers.

Deregulation and Local News

The transition in news from an unprofitable public service to a profitable commodity paralleled, ironically, the FCC's deregulatory efforts. Prior to deregulation the stations promised a tabulated amount of news and public-affairs programming in their applications and were monitored by the FCC at the time of license renewal, when theoretically a station's promise was matched against its performance. A part of the license renewal actually called for this comparison. News was a part of the regular schedule, as were the public-affairs programs, which were in-studio, live-on-tape, round-table discussions with community leaders and newsmakers. Unfortunately, public-affairs programs were scheduled in what was caustically referred to as "the Sunday morning ghetto," a time period that had so few viewers it could not be sold to advertisers.

Deregulation drastically altered news and public-affairs programming on radio. Initial research conducted in 1985 by the Radio and Television News Directors Association (RTNDA) reported that following deregulation there was no immediate change in the actual amount of news or public-affairs programming stations offered. That same research, however, concluded that 17 percent of the radio stations had reduced news staffs. More-recent studies have pointed to more-dramatic changes. News and public affairs have virtually been dropped by many radio stations, which saved money in the process. With the exception of all-news stations, the local radio industry clearly has deemphasized news and public affairs in favor of more-profitable music formats.

Coinciding with the deregulation of radio and the deemphasis of radio news was the increasing interest in television news. This interest grew during the 1950s and 1960s because of increasing ownership of TV sets and, by the 1970s, news was considered a marketable commodity. Today the local news is the most profitable program on most major-market TV stations. It is promotable, it is profitable, and it often constitutes three to six hours of a station's daily program schedule.

So today's news is different—it is profitable. It is America's primary source of news (see figure 10.1). Radio news has moved to the all-news and talk formats, with only remnants of news left on music formats. Today news comes in different forms based on marketplace demand. There is a change in substance and format. Formats include the traditional hard news, as seen primarily on the networks and affiliated

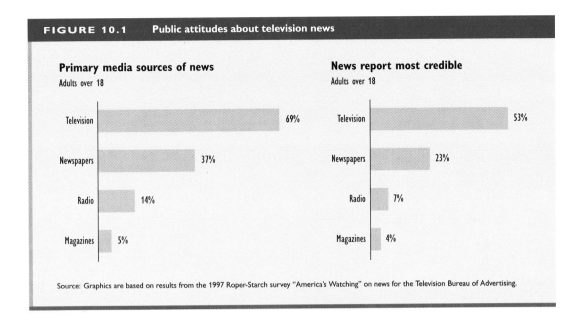

FIGURE 10.1 Public attitudes about television news

Source: Graphics are based on results from the 1997 Roper-Starch survey "America's Watching" on news for the Television Bureau of Advertising.

stations; news magazine and documentaries, such as *60 Minutes, 20/20,* and *Prime Time,* produced by the networks; and talk shows, produced by local stations and networks, which include such programs as *Meet the Press* and *Washington Week in Review* and contain solid news content. Syndicated tabloids (sometimes calling themselves news magazines because of their format similarities) are a new addition to the lineup. These include programs such as *Inside Edition* and *Hard Copy;* shows produced by cable such as *Rivera Live;* live sports; and a host of information-related genre that contain what media critics are beginning to call "infotainment."

As news magazines of all descriptions flood the schedule, the line between journalism and tabloid is difficult to accept for some traditional newspeople. Asked how an audience is to tell the difference, veteran newsman Bernard Shaw remarked, "The terms *tabloid* and *journalism* were contradictions in terms."

Technology: Reshaping the News

New technology is not only shaping the news-gathering profession, it is reshaping news program options. These options can clearly be seen as news on the Internet and differing corporations working together to provide specialized news services.

News is seen on the Web sites of every radio and television news station. Immediately after the nightly news, a station and/or network's Web address appears on-screen. Often it is promoted within a story: "For additional information on this story and others, visit our Web site at . . ." Web sites are not only increasing in number, they are becoming essential for both the news-gathering and the news promotions process.

Not surprisingly, one of the leaders in the Internet broadcast news services is CNN. Its service, CNN Interactive, is now typical of that offered by the other networks—ABC, CBS, and NBC. It provides the Internet surfer updated news twenty-four hours a day. This all-day service, launched in 1995, was staffed by forty-five people who provide news organized topically much like you see it in broadcast: national, international, sports, politics, business, and technology.

Television news and programming executives are learning to profit from Internet activities. Most news and information sites are offered free. Researching the specific information one may desire, however, often necessitates weaving your way through advertising and promotional banners offered to regular advertisers as an "added value" for purchasing time on the station.

Joint operations, such as Microsoft's and NBC's MSNBC venture, are another twenty-four-hour news service comparatively new to the marketplace. This is similar to other specialized news services, such as CNBC and the *Wall Street Journal* specializing in financial reporting. Dialing in to the MSNBC Web site yields the day's top stories, business, sports, technology, opinions, and the Microsoft Network. "Get the best news on the Web," touts MSNBC as it welcomes the online user.

The reshaping of news is a trend that students of journalism cannot ignore.

The Newsroom

Television newsrooms across the nation are as different as management personalities, budget resources, and markets. They range from one-person operations to departments with sophisticated equipment and budgets in the millions of dollars. Despite the differences in management, finances, and markets, all news departments face a similar challenge—defining news and covering it. This challenge is met with reasonably similar organizational structures.

The daily news decisions in all organizations are made by *gatekeepers*—those who control the information flow of any news item; they decide which stories are covered, how they are covered, and what is going to air. They include news directors, producers, assignment editors, writers, and almost everyone who has input as a story progresses. From an idea or event to its final production and transmission in story form, these people form the basic organization of the newsroom.

The News Director

On the local level, the *news director* represents station management within the news department; in a major market, the person may be a vice president in charge of news. In years past, the news director reported to the program director, but today's news directors are part of the top management team: They sit on decision councils, report to the station's general manager or the chief executive officer, and interact with other department managers, such as sales, promotions, engineering, and programming. The news director oversees day-to-day news coverage, conceives and develops specials, trains and hires staff, and handles all aspects of the budget. This includes supervising personnel, ensuring competent performance, and developing long-range plans. People reporting to the news director include producers, reporters, talent, and a host of technical staff. The position not only necessitates management skills and good news judgment, it demands a healthy respect for public relations as well as news, because it is often the news director who, along with the anchors and talent, will represent the station within the community in regard to its information and overall news practices.

The newsroom organization chart (figure 10.2) reflects the growing importance of consultants and anchor talent, who, along with the news director, have the ear of the station manager. The news consultants' research plays an important role in news programming decisions. The talent, likely the more experienced newspeople on staff, are often elevated to this management decision level because of their position and the investment stations have in anchor personnel.

In the syndicated shows, the counterpart of the news director is the show's creator and/or executive producer. These individuals develop organizations not unlike those in local stations, but assume information-gathering duties and responsibilities directed only toward program syndication. Although syndicators likely will be based in office suites or in production houses as opposed to a station's newsroom, these executives nevertheless determine the general direction of the program and are responsible for the overall organization. They too work in the company of other department heads and chief executive officers.

Producers and Assignment Editors

The *producer,* the person in charge, may be assigned to specific programs or, on larger documentary and magazine-style programs, may supervise a program segment. In radio they are called *editors* and work in shifts developing the overall sound of the newscast. In both media they organize news content, determine a story lineup, establish time limits on stories and program segments, and direct the overall presentations. The producer's daily meeting is a review of the day's events and pertinent news. The producer's job is one of dealing with constant changes—in content, direction, the reporting of stories, times, and the ever changing event itself.

Working closely with the producer is the *assignment editor (AE),* whose job is to know everything and cover everything. They maintain a "futures file" that contains information on significant stories for each day

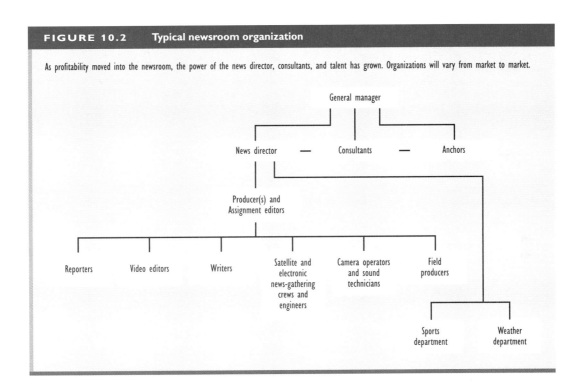

FIGURE 10.2 Typical newsroom organization

As profitability moved into the newsroom, the power of the news director, consultants, and talent has grown. Organizations will vary from market to market.

Before moving to NBC, *Later Today* co-host Jodi Applegate anchored *Good Morning Arizona* for KTVK-TV in Phoenix.

News director and video editor (left) select videotaped segments for a nightly newscast.

The director (center) calls the shots during a live newscast. The technical director (right) pushes the buttons that put the pictures on the air.

of the month. Every day, they receive hundreds of press releases and story ideas from within and from outside the newsroom; each must be read, analyzed, assigned, filed, or tossed in the recycling bin. Assignment editors view many hours of video press releases, satellite and network feeds, and all-news source materials. The assignment editor chooses the reporter for the selected stories and may stay in touch with the reporter and field crews as the story unfolds, coordinating live shots. In general, the assignment editor's task is to be ahead of all the daily events, assign reporters to cover those newsworthy events, and coordinate the coverage.

REPORTERS AND FIELD PRODUCERS

The *reporter* is on the front line, on location at the event. After receiving the assignment, he or she researches and gathers the information, writes the story, and then directs the story's content, production, and presentation. Reporters at larger-market stations often specialize in politics, police, government, science, sports, health, or whatever *beats,* or assignments, are appropriate to audience interest. These assignments are a matter of definition within the marketplace and the audience's need to know.

The field producer, a position of growing importance with today's technology (especially within the local market), acts as a reporter's producer. *Field producers,* whatever the setting, work directly with the reporter and assignment editors, conducting the initial research and preliminary interviews with subject experts. Once the story's general thrust has been established, they coordinate the technology of the field process. In some operations field producers may actually videotape the interviews, although they will not be seen on camera. Rather, the reporter or producer takes the raw material from the field producer and directs the story. Field producers are sometimes station employees who supervise remote units. Often, however, they are independent freelance contractors who can be hired by news directors at considerable savings. Although this position is important to the traditional newsroom, it is of particular significance to the syndicated "news reality" program genre.

ANCHORS AND TALENT

The *anchors* and *talent* are the key individuals of a station's news presentations. On radio we recognize their voices; on television we identify with personalities and invite them into our homes to deliver the news. Most talent have experience as reporters. Although talents' jobs are varied, most stations expect their anchors to be seasoned newspeople. Nevertheless, a few talent are little more than readers who look good in front of the camera, friendly types who happily read what a supporting staff of reporters, writers, and producers have prepared. Major-market stations work to communicate the value of such a personality to their audiences by putting anchors in the field as well as in front of the studio desk. They want anchors to be involved with the community.

Support Staff

There are a host of other positions within the typical newsroom. Dozens of people contribute to the information we see an anchor deliver. Radio, like TV, employs anchors, reporters, and assignment editors. In addition, both media have desk assistants and associate producers, who pull and organize wire service copy and police, fire, and other emergency services transmissions, and log traffic, weather, features, and network feeds coming into the station. Television additionally employs videographers, camerapeople, editors, news production coordinators, computer graphic artists, engineers, field and satellite engineers, and a studio production crew. Behind every anchor, in both small and major markets, are a lot of people working to deliver the news.

Sources: AP and UPI

Regardless of their size, all of the major-market stations, syndicated operations, and most of the small markets subscribe to one or all of several wire services that are the informational foundations of many newsrooms. The primary services have been the Associated Press and the United Press International. They function around-the-clock, feeding subscribers a continuous flow of news and information, usually in printed form.

The largest of the organizations, the Associated Press was founded in 1848 as a not-for-profit cooperative news print service. Today it offers a range of services to all AP-franchised media. The AP gathers its news from an established network of national and international bureaus, which feeds the national headquarters in New York. There are approximately 230 AP bureaus supported by a workforce of nearly three thousand full-time reporters, editors, and photographers. In addition to its bureaus' reports, the AP also receives input from its member broadcaster stations and newspapers. A reporter may file a story with the AP from his or her station. If the story is cleared through the bureau and is broadcast over the wire, the AP will pay the reporter a small fee. Because fourteen hundred newspapers and six thousand broadcast stations contribute to the AP, in addition to its employees, the range and amount of information available is vast.

The Associated Press has grown over the years and is the dominant wire service. It networks far more bureaus, reporters, and news-gathering sources than its competitors or other supplementary wire services. As the electronic media have diversified their news operations, the AP has maintained a competitive edge by diversifying its services in tandem. Today AP digital video is on the horizon.

In 1907 the United Press (UP) was founded by a group of small newspapers, because competing newspapers could not subscribe to the AP. The UP and the International News Service joined forces in 1958 to form United Press International. As with the AP, UPI was initially conceived as a newspaper service that, when broadcasting evolved, also made its services available to the new media. During the midfifties UPI and the AP were similar in size and extremely competitive. Both were primarily newspress wires but also offered radio services. The significant difference between UPI and the AP at that point was that UPI was a contract service that had to be paid for and the AP, as noted earlier, was a member cooperative. In the competition between UPI and the AP, however, UPI has suffered considerable losses. In the early 1990s, UPI was all but out of the business, having suffered heavy financial losses; but its purchase by AGI (a holding company) put wind back in its sails and is currently restoring its competitiveness.

These two wire services remain the backbone of most electronic-media news operations. AP/UPI wires feed hard news stories, five-to-ten-minute summaries, headline summaries, business and stock information, sports news and features, commentary, analyses, and weather information. They also feed audio and photographs and are moving into digital video. The receiving stations utilize the information in its various forms in a number of different fashions, depending on their commitment to news and their adopted formats. The music stations are likely to read only the headlines—not wanting to disturb the audience flow in music. The news/talk stations likely subscribe to all services and have the news staff rewrite wire service copy, localizing the information for their own communities, as well as develop news information from the wires and cover the events of the community. Utilization is determined by the function of the station, the news programs within the station lineup, and the specific program format.

Television stations and some of the larger radio news stations may have the budget to draw from additional news sources. These include supplemental wire services and databases as well as network suppliers, regional or national bureaus, corporate exchange services, photo and video suppliers, and satellite news gathering. There are as many news services as a budget can afford.

Specialized Wire and Computer Services

The two major wire services are supplemented by a host of specialized and foreign wire services. Reuters, Agence France-Presse, and Canadian Press and Broadcast News are among the larger supplemental services. Reuters is a British news service that focuses its news gathering on financial information. Reuters is currently making a strong attempt to expand significantly into the U.S. market and become competitive with the AP and UPI. Although at first glance the French and Canadian organizations may seem specialized, they do provide differing perspectives on international events. Larger newspapers also offer supplemental wires to broadcast stations, including the *New York Times* News Services, the *Los Angeles Times* Syndicate, the Chicago *Tribune*, and the Gannett News Service.

The Internet has introduced new database services available to the newsroom that at times look a lot like wire services. The *Wall Street Journal*/Dow Jones service, for example, provides financial news and information. The onslaught of computer information retrieval services such as NEXIS, DIALOG, and library search engines are also playing a role in investigatory journalism, as these programs provide new accessible sources of quick information. They offer hundreds of databases ranging from the scientific and legal to newspapers and newswires. NEXIS is, to some, the database of choice because of its size and full-text (as opposed to summarized) information. DIALOG has some full-service text, but it offers primarily bibliographic searches.

Network Feeds

Network-affiliated stations have access to regular network-program feeds, coverage of breaking stories, and delayed electronic feeds. Regular program feeds include the evening news, specials, and regularly scheduled programs. These may be accepted by the local station and/or used as an information source. Breaking news coverage gives the local station the opportunity to capitalize on the network resources—obviously better positioned to provide coverage of global current events—for major stories. A *delayed electronic feed (DEF)* is simply a network feed containing outtakes from network news or stories that never made the air. These, in addition to the network newscast itself, can be edited and used by local stations within their news blocks, usually in the late-evening and early-morning news. They are used to recap and present news from the early-evening news that a viewer may have missed. Viewers who watch a lot of news will notice a great deal of repetition and recutting as stories evolve and are repeated in succeeding time slots beginning with the early-evening news. If there are continuing developments within the story, it can begin all over again.

News Bureaus

National and regional bureaus are organized, particularly by corporations owning stations in the multiple markets, as an important adjunct of news-gathering operations. The larger stations often have bureaus in the nation's capital, perhaps in the state capital, and in other metropolitan locations. These bureaus are generally small and include a reporter, a videographer, and often an intern. The metropolitan bureau covers a community defined by a local station's coverage area, then distributes reports of interest to other corporate-related stations. Duties and audiences are multiplied by differing stations within the corporate conglomerate; in other words, if a corporation owns more than one station, the bureau may have the responsibility of covering Washington, D.C., news for more than one market. A Washington, D.C., bureau

DIALOG

> DIALOG and NEXIS are full-text and bibliographic-text retrieval services. At present they exist only in the larger-market newsrooms where investigatory journalism opportunities exist. DIALOG's self-description gives some idea of its range and extent. It claims to offer the advantage of having a newspaper morgue or the *Readers' Guide to Periodical Literature* at the user's fingertips:
>
>> Contained in the DIALOG information collection are millions of documents drawn from more sources than any other online service—from scientific and technical literature—from full-text trade journals, newspapers, and news wires. There's more too . . . details on millions of chemical substances, information on patents issued worldwide, demographic data, and company financial statistics. Compiled from the world's most respected and authoritative information sources, DIALOG can provide you instant answers from just one source.
>
> DIALOG information bridges a huge span of subject areas ranging from news and travel to scientific research and patents. Major areas of focus are:
>
> - agriculture, food, and nutrition
> - biosciences and technology
> - business news and information
> - chemistry
> - company information and financial data
> - computers and software
> - energy and environment
> - engineering
> - government and public affairs
> - industry analysis
> - law
> - medicine, health care, and drug information
> - news and complete-text publications
> - patents, trademarks, and copyrights
> - people, books, and consumer news
> - physical science and technology
> - social sciences and humanities
> - travel
>
> Source: Courtesy of Knight-Ridder Information.

industry info

covers the workings of representatives and senators; reports of legislative importance are fed daily back to the home stations.

Satellite News Gathering

Satellite news gathering (SNG) has played a major role in expanding local news-gathering opportunities. In effect, it has taken news control from traditional network and wire services. The local station today is no longer dependent solely on its own news-gathering wire services, network feeds, or DEFs. With satellite and *electronic news gathering (ENG)* technology, a local station's news producer may appear to send a reporter anywhere in the world for a same-day story. In reality the producer is having a freelance independent reporter cover events and relay them back to the station. This technology has resulted in the formation of both formal and informal SNG groups, which have experimented with arranging distant coverage to local stations, which in turn trade and share their coverage with other stations.

One of the first such experiments occurred in 1979, when the Bonneville International Corporation sent station reporters from KSL-TV in Salt Lake City to cover events in Israel. The coverage provided via SNG was then shared with remote locations throughout the State of Utah. Westinghouse stations were also among the first to exchange information via SNG.

Today, Conus Communications, Inc., is one of the major players in the development of SNG distribution technology. Conus is organized to share coverage for member stations across the nation. Thus, Conus has become a point of connection, linking member stations in various parts of the country. The Conus member simply calls Conus control and is connected with another member in the area where an event is occurring. A reporter is then sent to the scene, the story is written, edited (in the field), bounced off the satellite via Conus, and fed back to the requesting station. Conus has no bureaus; it merely calls on member stations, which exist in almost every market in the United States, to act as news-gathering units and uses satellite technology to relay stories from one station to another.

Closely related to the function of SNG is the station's own reporting efforts within its local coverage area. The technology of electronic news gathering, the microwave, and sometimes the satellite make it possible for a reporter in the field to transmit live (through a remote van) back to the station. Using satellite these same reports and/or reporters can be transmitted to other stations in other markets.

The flexibility of SNG and ENG not only has extended a station's presence on the scene of a news event and its ability to reach out to those events—it's also a good promotion deal. The traditional newsperson would view the practice with distaste, but it is not uncommon for a news van to be on the scene of a news event when the shot has little relevance to the story—a night shot, for example, in front of a darkened building where news occurred earlier in the day. This technology also accommodates promotions tours, featuring the remote truck and anchors, that often take place during sweeps week. There is little of news relevance to the tour—it is simply a promotion to gain attention and increase the number of viewers.

CABLE NEWS NETWORK

In a discussion of news and news sources, one cannot ignore the impact of the Cable News Network (CNN) on local and national news. When Ted Turner launched his all-news network in June 1980, ABC, CBS, and NBC dominated the news ratings. When the major networks began to lose their ratings share, however, because of audience fragmentation and lulls in the economy that forced them to cut back at their headquarters and on their bureaus, CNN rapidly expanded. Those who had scoffed at Turner's venture no longer did so. ABC and Westinghouse tried to counter CNN with SNC—the Satellite News Channel—but this organization was purchased by Turner and transformed into CNN II, now CNN Headline News.

CNN's advantage was that while the networks were downsizing and becoming dependent on fewer bureaus and smaller staffs, CNN was taking advantage of new technology and expanding. SNG technology had made it possible for Turner to organize the country's first twenty-four-hour news service. It was different from the traditional networks and Conus in that it enlisted several over-the-air stations, as well as its own staff, bureaus, and freelance reporters in news coverage. Member stations, freelance reporters, and CNN staff reporters feed into the central system in Atlanta, at which point the information is downloaded in the CNN "pit" for distribution on one of several CNN programs, its cable networks, or simply affiliate service feeds.

The impact CNN has had on national and international news organizations was clearly evident

CNN newsroom.

during the coverage of the Persian Gulf War (1991) and the crisis in Moscow (1993). While the NBC, CBS, and ABC networks were asking former World War II and Vietnam-era correspondents how it was to cover those wars, CNN already had reporters on location providing riveting around-the-clock coverage. CNN was simply the network with the greatest resources. Today CNN's worldwide services are seen in 142 countries and 60 million households. A distant second in this global market is the BBC World News Services, seen in 83 countries and in 20 million households.

CNN provides member stations with a variety of network information services, which are contracted or bartered with the stations. The Newsource service provides multiple feeds throughout the day. Newsource materials consist of hard news, national and international news, sports, weather, and features. The CNN Live Breaking Coverage service provides primarily independent stations the opportunity to promote themselves as network affiliates—they too can be on the front lines of breaking news events. With coverage contracted through CNN, breaking stories are fed to stations in their entirety. Independent stations may also subscribe to the total network thirty-minute news service, which provides them a full-service news operation with minimal investment. Additional services, including CNN Excerpting, Headline News Excerpting, NewsBeam, and the CNN Library also provide news-editing and program options that augment a station's overall news programming. The CNN services, which were originally targeted to the independent stations, are picked up by many network affiliates who use them to augment network and local coverage.

VIDEO NEWS RELEASES

New technology not only bolstered the growth of CNN and new methods of video delivery between stations, but it also aided in the growth of video news release and news insert syndication industries. Most traditional news executives claim they rarely use the VNR or the syndication news sources. However, Medialink and companies that produce and track such usage claim that nearly 75 percent of all stations use video news releases regularly, and many of them also see syndication as a means of holding down their news budgets and promotions during a ratings sweep.

A *video news release (VNR)* is basically a press release on videotape. A public relations firm, corporation, or anyone who has something to say or sell can create a VNR and send it to the local stations in hopes that it will be aired. Such press material has heretofore been limited to the printed press releases, which fit comfortably into the newsroom assignments system. The VNR, however, is different—it is visual. Corporations and public relations firms producing news stories deliver them to stations via satellite for the stations' unrestricted use and can do so at low cost. Costs of satellite delivery are generally paid by the corporate client, which uses VNRs to network a newsworthy story. The subject and usage trends in VNR are becoming increasingly diversified, because it is easy to produce a message and purchase satellite time for distribution to local stations.

The local station's reluctance to use VNR-produced materials can be traced to traditional journalistic values. The uneasy relationship revolves around the news department's journalism ethics: The station wants control over the content of the video it airs, and not having such control is seen as a breach of public confidence. The news department sees itself as in a position of public trust—it has a responsibility to present a true and objective picture. Lacking editorial control over the content of a VNR therefore creates suspicion within the newsroom, the newsroom staff believing that, because corporate public relations people produce VNRs, they must have a point of view, a product to push, a hidden motive.

Today's effective VNR producers have responded to the newsrooms' concerns by adopting newsroom standards while maintaining what they hope to be an effective means of communication. They provide VNRs much like written press releases. For example, information is provided in different forms, thus allowing the newsroom to exercise editorial content control. Scripts, B-roll video (picture and sound without a reporter), natural sound, and narrative tracks are provided separately, allowing the stations to screen the material for accuracy and to localize the VNR using an in-house reporter to do his or her own standup report. Thus the station gains editorial control over the finished package, and the corporation sees its video on the air. An effective VNR producer leaves advertising and promotion messages to their respective corporate departments, producing news release materials that can be utilized cost-effectively by the stations while allowing them editorial control over the content.

The mistrust between stations' news departments and the corporate world's public relations departments will likely continue. The fact is, however, that the VNR has become an important part of the corporate public relations approach, and more and more stations are using VNRs as source materials, especially because news budgets are tight. Hollywood movie producers are perhaps the most sophisticated creators of VNRs. Designed to introduce a new motion picture, these VNRs include interviews with the stars, clips from the film, and tours featuring talent appearances that are tied to local activities and personal appearances. All require separately produced video news releases that are seen throughout the nation in a variety of configurations on networks, local stations, and syndicated programs alike.

News Inserts

In 1965 the first syndicated television news inserts appeared. A *news insert* is simply a short (60- to 90-second) feature story. The first series was Joe Carcione's *The Green Grocer*. What Carcione's producer, Dave Meblin of Mighty Minutes, offered was a daily 90-second news report about fruits and vegetables. Carcione started in San Francisco at KRON-TV, providing this feature material from the local farmer's market. He simply produced a report focusing on the best buys of the day and provided some cooking ideas. It was a straight video shoot of talent on location. The series ran daily on 133 stations for thirteen years.

From *The Green Grocer*, the production of news inserts and news syndicated materials has grown. In the mideighties N-I-W-S and Newslink arose, marketing fifteen to twenty features that specialized in consumer affairs and medicine. These and other major news syndicators appeared quickly—and some disappeared almost as quickly, including Newslink, its longevity shortened by financing, marketing, and acceptance difficulties. Dave Meblin, syndicator for Joe Carcione, indicated that it took *The Green Grocer* almost two years to turn a profit and that "there are some news inserts that have taken nine years to get into the black." The major players in the news insert business on the air today include N-I-W-S; Pinnacle Communication, producing *Lifeplanning*; HMS Communications, producing *Dr. Dean Edell*; SPR: News Source, Inc., featuring Leroy Powell; the *Wall Street Journal Report on Television*; *Dr. Red Duke*; and Steve Crowley, CPA, producing the *Money Pro Report*.

News directors use the syndication inserts as features. They deal with them as a syndication business, just as program directors provide entertainment syndication, although in news the syndicated product is first-run and produced specifically to be utilized within the local news format. What the audience gets in a news syndication package are homey, quick bits of information that transcend markets and are created to evoke a response such as "I didn't know that!" or to perhaps send viewers to a sponsoring client where they can pick up a brochure for further details. The medical doctors, Red Duke and Dean Edell, have practical health information. Steve Crowley's *Money Pro Report* is designed to give viewers the latest in financial news and relative advice. The station buys the inserts on an exclusive basis for their *ADI (area of dominant influence)*, and inserts are sold as a continuing service or as promotional packages often used in a ratings sweep. The inserts themselves come to the stations via satellite or as tape with scripts, separate sound, and narrative tracks, allowing the station opportunities for localization and content control. The larger syndication companies also produce promotional materials that allow sales and audience interaction with the syndicated materials.

The usage of VNR and news insert material can complement today's newsroom, which is trying to streamline. It is more cost-effective to purchase the syndicated material than it is to send a reporter out into the field to do the same feature report. Thus inserts can free a reporter to cover more-pressing hard news.

News Programming Strategies

Today's news programs have changed significantly over the years, not only in the amount of news delivered, in technology, and in various delivery systems, but also in the way the information is presented and programmed. Networks are no longer dominant, and satellites have given stations and producers global reach. News sources are rich and varied far beyond network and wire service sources.

This chapter has so far discussed the evolution of news, news-gathering sources, SNG technology, the typical newsroom organization, and news sources. It has also touched on gatekeepers and their individual responsibilities in establishing the lineup of a day's newscast. Besides a program lineup, what other elements of news programming can be manipulated to attract the largest market share? This section explores those elements: content treatment, program promotions, the lead in and out for the program itself, the consultants' research, and the talent.

CONTENT TREATMENT

At first glance the manipulation of news program content may seem questionable. A program's lineup is determined by the priority of events in the day; a fair and truthful story is going to look and sound the same regardless of differences in its approach—*isn't it?* Not necessarily.

The content treatment of news today is determined by management—station management, network management, and the syndicated news program creators. Management orientation and philosophy are the primary factors directing content treatment. For example, a network affiliate's news and a religious broadcast station's news could be different, as could a syndicated television news magazine creator who sees news and information differently than do the networks. The MTV network programs environmental news differently than does the Discovery Channel. A hard rock radio station's news sounds a lot different than that of the all-news station or the adult contemporary station. These perspectives represent a management orientation, a philosophy, and the niche in the genre of informational programming that controls the direction of information content and production. Management directs the news program itself to complement a format and accomplish a program objective: traditional hard news, tabloid, or "infotainment." However dedicated the traditionalist is in programming a mainstream news operation, one cannot deny the existence of alternative content treatment even on the network affiliate and independent stations.

Management control may be more subtle than the mere dictation of a format or an approach. There are station owners, conservative and liberal, whose operations portray the character of the ownership/ management. One manager may be concerned enough about the news program's profitability and ratings to permit experimental forms of presentation not approved by the traditional news manager. A conservative manager whose station policy is to preempt a network program because "it features too much sex" is not likely to look favorably on a reporter or producer who schedules a porn star for an interview on the noon news. A liberal manager, likewise, may influence the coverage of a rural environmental issue—and the content will vary from what it would be if handled by a local conservative owner. This is not an issue of political conservatism versus liberalism nor management interference in the newsroom; it is the news manager's assessment of management's orientation that more subtly guides content direction.

PROMOTIONS

The importance of promotions and programming is discussed in chapter 9. News programs conduct not only promotions to boost ratings, but also promotions focusing on image, public relations, advertising, marketing, and even merchandising. It is a monumental task that plays a key role in determining the image of a local station's news acceptance. Many stations, however, budget more for the latest equipment than they do for promotions. The promotional spot is often the first one cut to make room for other commercial spots.

News promotion really involves the complete packaging of the station's news image, much the same as a marketer would package any product for which he or she hopes to attract a buyer or audience. The overall design of a good promotions campaign should create "a positive and significant relationship between the station and the public," according to Promax, the Broadcast Promotions Association. In the case of news, that relationship historically has promoted interplay between the anchors and the audience, and the reporters and the audience. With the anchors' salaries taking up a significant portion of the newsroom budget, some news directors attempted to deemphasize the talent connection and reemphasize the product and service in their promotions and programming. This was the CNN strategy: Its salaries were not as hefty as at the other networks; presentation was deemed secondary to information. Research has shown, however, that people watch TV news not only for the news itself, but because of the presenters. Effective campaigns pushing this anchor/viewer relationship include spot campaigns, public appearances, kits and contests, personality identification campaigns, billboard image and spot, newspaper, and advertising. Today news itself is also merchandized—the audience can call an 800-number to acquire copies of stories, scripts, or video for a reasonable price. Many stations also invite feedback from viewers via e-mail and Web sites.

LEADS—IN AND OUT

As discussed in chapter 9, the success or failure of a show can often be attributed to the programs that precede and follow it. The audience interested in news and information flows from one informational program or block to another. The current success of *Oprah* as it leads in many major market news blocks illustrates the point. The lead need not necessarily be of the talk, information, or news genre. Game shows such as *Wheel of Fortune* and *Jeopardy!* attract the affluent working woman and currently provide the bridge into and out of a news block. If the first program is successful, acquiring good ratings, one can count on a number of viewers to flow to the program that follows. A good program buy like this reinforces the news. A program that falls on its face leading into the news does not offer such support.

CONSULTANT RESEARCH

News consultants have made a significant impact in the development of news since the midsixties. Today it is almost impossible to program a competitive news operation without research input. A consultant evaluates audience reaction to news and presentation, and then, based on this analysis, content and presentation are manipulated to enhance ratings.

Consultants provide research data for local managers faced with decisions about programming, audience development, and especially news, because it is the locally created program that reflects the overall station image. They are seeking that ever changing and elusive audience. Consultants are constantly assessing and reassessing the needs of a station and its competition.

The three primary news consultants are Frank N. Magid Associates, McHugh-Hoffman, and Audience Research and Development (AR&D). These are among the largest; they have been in business the longest and are full-service research firms. All are business-concept driven, specializing in audience and behavioral research.

These three companies started out in the sixties and seventies as news consultants, but as the industry diversified, so did the consultants. Each issue of the *Broadcasting & Cable Yearbook* lists a growing number of research firms. Don Fitzpatrick and Associates deals in talent and management placement; Reimer and Associates, the Broadcast Image Group, and Feedback Unlimited work in producer training. Research consultants today provide services not only for local stations, but also for program producers, developers, syndicators, and networks, and they operate around the globe.

Consultants are criticized for "boilerplating" their research—they ask the same questions in every market. As a result, it is not surprising that their reports are similar and that stations, over time, start to all look alike. Some contend that consultants have produced a sameness because they represent only one station in each market; they take similar ideas from market to market. They are criticized when a popular talent is fired by the station because of unfavorable research results. Managers have no doubt made mistakes in their decision making by relying too heavily on research. They have produced actor anchors. Consultants also produce fear in the newsroom, because the research is largely proprietary—"for management eyes only"—a situation that generates rumor and criticism. When one of these firms visits the newsroom, the atmosphere turns into one of paranoia; many ask, "Whose job will be gone tomorrow after the research report has been turned over to management?" There is a reason such research is so closely guarded: The information was acquired to provide the station with a competitive edge, so a leak of such information to competitors could lessen its impact. Besides being proprietary, the research is very expensive, and stations are not eager to have their investments in these reports become the talk of the newsroom or of the town.

Despite the criticism, consultants continue to play an important role in audience research and program development. They have made a difference in the image of local news, having produced a faster-paced entertainment presentation and fierce competition. Certainly many consultant reports have taken stations

The Consultant's Report

Research from a consultant's report is guarded. The consulting contracts contain strong language about the confidentiality of the findings. The general layout of research design is not unlike other public audience research reports. The methodology is not complex, and one can access similar information studies in many research and university settings.

The reports exist to assess the attitudes and opinions toward television in a given market. The study begins with a statement of the problem, including clear identification of client and market designations as well as general questions serving as a basis for the study. Methodological explanations and sample base data include information on sampling techniques, the number of interviews, and tabulation of statistical information. The development of the questionnaire is critical. Structured questions provide statistical information, and open-ended questions allow a freedom of response from the audience. Questions revolve around the basic problem statement. To help management understand the statistical data, descriptive interpretations also provide information on how the data was analyzed and interpreted.

The organization of the report follows a basic outline with detailed questions as reflected in specific station concerns, such as:

I. Attitudes and Opinions Toward Television in . . .
II. Early Evening Viewership
III. Late Evening Viewership
    Recent viewership
    Late evening news for the station
    Audience loyalty
IV. News Station Preference
V. News Presentation Preferences
VI. Weather and Sports
VII. Personalities

It is the Personalities section that strikes fear in the hearts of newsroom personnel—it can be extremely complimentary or devastating. For example, a reporter/anchor may be described as "widely recognized," "perceived to be serious," or "competent and friendly" or not so positively described: "displaying little personality" or "not particularly authoritative." Other descriptors are added by the consultant as he or she interprets the data gathered relative to personalities. There are few limits. The real decision on the use of such information remains in the hands of responsible management. Research functions only to provide a data source and input. Taken with other factors and used responsibly, it assists in the decision-making process.

---

from the bottom of the ratings scale to the top. In fact, consultants have created such an awareness of the value of research that some larger stations have developed their own research and marketing departments. These departments work with the consultants and conduct research and organize input of all kinds from the array of sources available to the stations (see chapter 11).

Nowhere is the consultants' research more visible to the audience than in their work with news talent. Research firms not only survey a talent's effectiveness, but often act as headhunters, talent coaches, and agents.

Talent

Perhaps one of the most difficult adjustments for students entering the profession of electronic journalism is the concept that talent—reporters and anchors—are a product commodity. These people are moved, made up, managed, and manipulated to fulfill ratings expectations and to bolster a station's image. News talent has become one of the most important variables in the news ratings war. Anchorpeople are the centerpieces of the news presentations. The traditionalist wants such people to have experience and sound news judgment and be seasoned professionals. Some managers, however, seeking higher ratings, hire people based on appearance alone, believing that presentation skills are more important than sound news judgment.

Consultants have played an important role in elevating (or reducing) the anchor talent to the position of news commodity. They have also taken part in the hiring shuffle. The earliest consultant researchers tested on-air talent. Their questionnaires for the local market started with identification questions assessing attitudes and opinions toward a station's talent, asking audiences who they recognized and who they did not. Viewers were asked to recall the names of those air personalities they recognized and identify their function as well as the station on which they appeared. From these identification questions, generalized descriptive questions were asked about what the audience thought about the personalities: what kind of people they were, how well they worked on television, whether their physical appearance and other characteristics were appealing. The descriptive questions led to a personality test where each personality was rated on a five-point scale. The researchers then ranked the talent and composed tables and descriptions of a talent's value for the station management's perusal. In the hands of the station's news director and manager, these research reports have been used to position, train, hire, and fire talent—all attempts to improve the station's image and increase its ratings.

Critical Trends in News and Information Media

We have already seen the trends that have affected the traditional newsroom: new technology, the expanded news-gathering sources, and marketplace trends that lead to the success and profitability of news. In our examination of news, however, it is important to briefly describe the electronic-media news trends as seen through the eyes of the consumer. Consumers are aware of the technology; they are demanding more and speedier information and entertainment.

News is going through a period of unprecedented change in the eyes of both the consumer and the professional. Competition and technology have left professional standards rudderless and open to new debate. We are a society engulfed in a communications revolution—the information society. "As a nation we have fallen in love with the concept of communication," according to Ries and Trout (1986). They were writing more than a decade ago and describing the business of advertising, but the revolution has also had an effect on news. Research and marketplace news, according to critics, has created a shift from those traditional news values, discussed earlier in this chapter, to selling and entertainment news. News that once provided an audience with *needed* information, now works to fulfill what a social scientist tells a news director that the audience *wants*.

Not too many years ago, the consumer had an easy task in deciding what was news. News came from newspapers, magazines, radio, and television. The news was almost unquestionably objective and truthful. It was easy to tell the difference between the *National Enquirer* and the *New York Times;* in television there were no tabloids, only Walter Cronkite—the most trusted man in America—and the *CBS Evening News*. In the past decade, however, the information era has exploded upon us and much of the information had become eroded.

Everyone is moving quickly, expecting decisions just as rapidly. Our families, culture, and society have changed forever because of this information revolution. To the consumer, this electronic superhighway means easy, speedy access that is user-friendly and cost-effective. To the professional it means an added responsibility to communicate fair and accurate information. As Edward R. Murrow (1958) said, "This instrument can teach, it can illuminate and it can inspire."

Trends in news have not all been information access driven. The values of entertainment in news are attracting an audience, but one of the more disturbing criticisms of news recently has been a growing trend of public distrust. The questionable actions of people

EDWARD R. MURROW: A BROADCASTER TALKS TO HIS COLLEAGUES

This just might do nobody any good. At the end of this discourse a few people may accuse this reporter of fouling his own comfortable nest; and your organization may be accused of having given hospitality to heretical and even dangerous thoughts.

But the elaborate structure of networks, advertising agencies, and sponsors will not be shaken or altered. It is my desire, if not my duty, to try to talk to you journeymen with some candor about what is happening to radio and television in this generous and capacious land.

Our history will be what we make it. And if there are any historians about fifty or a hundred years from now, and there should be preserved the kinescopes for one week of all three networks, they will there find recorded in black-and-white, or color, evidence of decadence, escapism, and insulation from the realities of the world in which we live. I invite your attention to the television schedules of all networks between the hours of 8 and 11 P.M. eastern time. Here you will find only fleeting and spasmodic reference to the fact that this nation is in mortal danger.

So far as radio—that most satisfying and rewarding instrument—is concerned, the diagnosis of its difficulties is rather easy. And obviously I speak only of news and information. In order to progress it need only go backward—to the time when singing commercials were not allowed on news reports, when there was no middle commercial in a fifteen-minute news report; when radio was rather proud, alert, and fast. I recently asked a network official, "Why this great rash of five-minute news reports (including three commercials) on weekends?" He replied: "Because that seems to be the only thing we can sell."

One of the minor tragedies of television news and information is that the networks will not even defend their vital interests.

DOLLARS VS. DUTY

One of the basic troubles with radio and television news is that both instruments have grown up as an incompatible combination of show business, advertising, and news. Each of the three is a rather bizarre and demanding profession. And when you get all three under one roof, the dust never settles. The top management of the networks, with a few notable exceptions, has been trained in advertising, research, sales, or show business. But by the nature of the corporate structure, they also make the final and crucial decisions having to do with news and public affairs.

It may be that the present system, with no modifications and no experiments, can survive. Perhaps the money-making machine has some kind of built-in perpetual motion, but I do not think so.

I do not advocate that we turn television into a twenty-seven-inch wailing wall, where longhairs constantly moan about the state of our culture and our defense. But I would just like to see it reflect occasionally the hard, unyielding realities of the world in which we live. I would like to see it done inside the existing framework, and I would like to see the doing of it redound to the credit of those who finance and program it.

This instrument can teach, it can illuminate; yes, and it can even inspire. But it can do so only to the extent that humans are determined to use it to those ends. Otherwise it is merely wires and lights in a box. There is a great and perhaps decisive battle to be fought against ignorance, intolerance, and indifference. This weapon of television could be useful.

Stonewall Jackson, who knew something about the use of weapons, is reported to have said: "When war comes, you must draw the sword and throw away the scabbard." The trouble with television is that it is rusting in the scabbard during a battle for survival.

Source: From "A Broadcaster Talks to His Colleagues," by Edward R. Murrow (1958), speech delivered at the annual meeting of the Radio and Television News Directors Association. Reprinted by permission of the RTNDA. All rights reserved.

JOHN CHANCELLOR

Broadcast journalism is the newest chapel in the cathedral—only about fifty years old. In those earliest of days, men like Paul White set standards of conduct, performance and ethics which are still with us today. Because of people like Paul White, broadcast journalism got off to a good start. Those early settlers cleared the land and built the roads, they knew they were building for an important future. They insisted on excellence. And when the television age began, the standards which had been set for radio became the standards of television news. . . . The concern for honest journalism, which was there at the beginning, shaped the history of this craft.

When I think of the early settlers I wonder what they would think of the world of broadcast journalism today? I don't think they would be surprised by the changes and advances brought about by technology. . . . But the understanding they showed half a century ago of the scope and the reach of radio indicates to me that . . . their main interests would be in what messages are being carried by the technology. They were interested in content, the meaning of words and the accuracy of the information. . . . News, documentaries and special events were something that were considered necessary and worthwhile. But those were the days before television news became a profit center. . . . [when news] turned out to be a gold mine that changed things. It changed some basic attitudes about the work.

Thirty years ago . . . [people] didn't go into journalism for money or fame. You went into it because it was fun, exciting and possibly because it was socially useful. That is still true today . . . the overwhelming percentage of American journalists will not get rich or famous and most sensible young journalists know that. They are not in it for the money. But when the broadcast news operations at the networks and the stations became profit centers this understanding, that journalism was white collar work at blue collar wages, this understanding began to change . . . now alas we have added glamour and big bucks and those are hard to resist . . . and this is changing the way people think about their jobs. . . .

Broadcast journalists in any medium concentrate on their own communities and what's going on down the street or around the block . . . the best of them become identified with their own communities and rarely leave them. Today, in television, the words *community* or *town* are too often replaced by the word *market*. Local anchors, people and reporters have become rootless vagabonds searching always for the larger and more lucrative market . . . [today] the emphasis is not so much on what an anchor or reporter has learned but how it is said.

[Profits have] changed the dynamics of the business. It has created news programs which have little news, but lots of glitter, it has emphasized competition—which is probably good—and performance instead of content which is unquestionably bad.

[Challenges] can only be overcome if we remember that our business is journalism, that our primary mission is to serve the public and that we have fifty years of tradition to back us up. There are standards and we know what they are.

Source: From a speech by John Chancellor (1983) delivered before the Radio and Television News Directors Association. Reprinted by permission of the RTNDA. All rights reserved.

involved in Watergate and Vietnam initiated a general mistrust of our country's institutions—including the news media. But in the competitive scramble for ratings, the electronic-news practitioners have taken little time for retrospection. The proliferation of programs, syndicated news, magazine shows, and "infotainment" has forced traditional news into an accelerated pace of competitive entertainment and, in this process, values have changed. News programs have become specialized to fit the demands of different formats. Professionally, the traditional broadcast managers decry the injection of the tabloid program into the genre of news and information programming. They decry the "infotainment" approach. Ironically, however, they have also been purveyors of the entertainment values in news, adopting the entertainment elements of protagonist and antagonist, conflict, pacing, talent personality, visual imaging, and graphic/information manipulation. These program values translate into audience values and are mixed within the news program itself, as well as in the programs that surround the early-evening and late news. The result is an audience that makes little distinction among news genre.

Along with entertainment, one of most striking news values today is that of "firsts." *Scooping* (being first) has long been a competitive value in journalism, but today it has taken on new connotations—first to the story, first to make an impression, first in the ratings. The report must immediately focus the attention of the community and the station's management. The preference is for a dramatic first as opposed to a healthy competitive scoop. The spectacle of O. J. Simpson being chased down the Los Angeles freeway by almost a dozen police cars and the ensuing media circus is an illustration of this point. Television professionals have always placed a heavy emphasis on the value of the lead (or the first) story: It is the most important story of the day, according to most professionals. But because of increased competition, the lead is now defined as the story most likely to grasp audience attention and win the overnight ratings score. This resulted in the criticism of stories featuring crime, violence, and sex leading the newscasts. This perception not only attracts the audience but leads to negative criticism. The lead story influences audience perception of the overall newscast—a perception easily verified by the asking: Is the most dramatic video narrative the most important story of the day? Or does it merely lead to pull a rating?

These values are quite different from those discussed at the beginning of this chapter and have led to headlines across the nation, illustrating this growing trend and the distrust. The headline creating the biggest sensation of the decade was "NBC Apologizes to GM"(see figure 10.3). The story concerned an episode of *Dateline,* an NBC program, in which a fiery crash of a General Motors pickup truck was shown as video documentary evidence of the poor safety record of the trucks. After admitting that the incident was staged, NBC settled out of court, apologized, fired a few of the top staff from its *Dateline* crew, and deepened the audience's suspicion of journalistic practices.

In the spring of 1998, CNN ran a story alleging that United States troops had used nerve gas on U.S. defectors during the Vietnam War. The story attracted international headlines, but it was false. Two producers were fired, a senior producer resigned, and a reporter was reprimanded. The headlines retracting the story were smaller, but thankfully the retraction was headlined and not buried on page 2. CNN was forced to apologize, which it did for the "serious faults" in its story.

Speaking on the issue of trusting the press, former president George Bush commented, "The national press . . . are unaccountable and if they lose the confidence of the American people . . . they'll [be] undermining the confidence that we have in how we

FIGURE 10.3 Cartoon commenting on staged documentary evidence

Reprinted by permission of Tribune Media Services.

get our news" (Van Der Werf 1998). Such instances of poor journalism and errors in judgment fuel the public's mistrust of the news and the news process.

Today's audience is becoming increasingly aware of the news process. Hardly a day goes by without accusations that the media overemphasize, present a pro- or anti-administration bias, and talk about what is wrong in the world rather than what is right. Critics and consumers are complaining in increasing numbers that news is slanted, distorted, negative, pessimistic, cynical, and depressing. Political figures have even accused the media of being "out to get" them, and some TV personalities have used the media to effect a personal political career agenda.

The subject of news criticism is a sensitive one within the industry. It is closely aligned with the traditional journalists' definition of news, news priorities, audience perceptions of news, and news ethics. In fact, by equating controversy and violence with importance, many journalists seem unable to conceive of news that does not give preference to conflict and emotional leads. Critics indicate that the audience feels the media place too much emphasis on crime, violence, and the negative. The perceptions of the audience, however, are not always the same as the perceptions of news practitioners. Whereas news departments have readily adapted to meet the competition of the entertainment tabloids, they have been slow to recognize the more serious democratic criticism of the news product—a criticism, reported recently, placing the believability of news reporters on a par with politicians.

Another critical issue related to the news process is that of "ambush journalism." The paparazzi get most of the blame for this type of tabloid information, but the techniques are not limited to a few overzealous photographers. The death of Princess Diana of Britain focused attention on the paparazzi, but the actions of these so-called journalists has long been a concern of celebrities. Jacqueline Kennedy-Onassis finally won a restraining order against one such photographer who was ordered by the courts to stay away from her. More recently, Paul Reiser and Michael J. Fox appeared in 1998 Judiciary Committee investigations to provide personal accounts. Fox told of how "tabloid photographers" had "bribed local officials to infiltrate [his] wedding...cornered his 85-year-old grandparents and pumped them for information,...posed as medical personnel at a hospital to get pictures of a newborn son, and pretended to be mourners and snuck into the wake at my mother's home." Barbra Streisand wrote, "I feel as if I am a prisoner in my own home." Sharon Stone complained that she felt it was not safe for her to walk down the street. Tom Hanks had a similar complaint, noting that "ordinary family events," were often frustrated by the actions of photographers. As a result of these pestering techniques, Congress is considering bills that would make it a federal crime to endanger anyone's safety just to get a picture or a story to sell. Previous legislative efforts have been promoted and failed, but the issue is of growing concern. It should not be an issue only of safety, but also of journalistic integrity. An industry that cannot govern itself will be regulated.

Despite the recent attention on the paparazzi, the techniques of ambush journalism are not limited to them. Reporters and videographers who hide in bushes or rush into offices with cameras blazing to confront a wrongdoer are placing safety at risk—not only of the crew, reporter, and subject, but of those surrounding the event as well. We've all seen the subject who's been so angry as to attack the reporter. This is the headline that will attract viewers to the evening news because it is real-life drama. But is drama news?

Perhaps one of the more interesting criticisms brought upon news in the adoption of dramatic news values is the backlash created by the television "news star." These are the anchors who read out loud and look good on the air. Unfortunately, many of today's anchorpeople are selected not for their journalistic experience, but for values of image. According to David Halberstam, Pulitzer Prize winner for his coverage of news in Vietnam, "The celebration of journalism [anchors] has generated a public backlash and deepened distrust of the media" (Kammer 1998). Under contract, the "news doctors" (consultants Magid, McHugh-Hoffman, and others) are tutoring anchors to present themselves in a friendly, warm posture, that is, one received and identified by the home viewer as authoritative. If a reporter or anchor does not score well in these personality profiles, his or her career is generally altered.

There is good news amid the criticism of news. Most people still believe that journalists do get the story right. Most polls will agree that stories reported in the media are accurate. The reporters may seem more arrogant, cynical, and excited, but television is still considered a credible medium (see figure 10.1). So, although some news is favorable overall, it is worth noting that audience trust is eroding. One can only speculate that if the profession is losing the trust of the audience, will the ratings follow?

SUMMARY

▶ News and electronic-media information processing have become important components of our culture. On a social level, the new media are a powerful means of mass communication—a means of rapid-fire delivery of information across nations and our globe. On an industrial economic level, a television station's news department accounts for a significant portion of the station's revenue flow. News has become profitable, and, as a result, programs featuring differing treatments of news and information have become popular offerings on many different delivery channels. On a professional level, the practitioners find themselves in a continuing debate over changing values.

▶ News exists in the minds of people—people interacting with people, people interacting with their physical surroundings, and people trying to reconstruct the framework of events that affect people.

▶ Electronic journalism has both its theoretical and professional roots within the practices of print and radio.

▶ Radio news grew up with World War II.

▶ The decade following the war was one of stabilization and growth for radio and television news. The fifties was an era of classical network television documentary film.

▶ Local radio and television news were originally seen as community services by the stations and their markets. They were trusted to be fair and objective; criticism was rare.

▶ Deregulation of radio and television led to a significant decline in public-affairs informational-style programming, a drastic reduction in radio news, and slight reductions in television news programs.

▶ Local news has become both the most expensive and the most profitable department in most major-market station operations.

▶ Local news departments in stations across the nation are similar in organization. The daily news is determined by people referred to as gatekeepers.

▶ News sources have expanded greatly over the years. The Associated Press (AP) and United Press International (UPI) remain newsrooms' basic sources, but today these are supplemented by many different news wire services, computer databases, network feeds, news bureaus, and satellite news-gathering (SNG) opportunities.

▶ CNN's advantage was that while the networks were downsizing and becoming dependent upon fewer and smaller bureaus, it was expanding and taking advantage of new technology.

▶ A video news release (VNR) is a press release on video.

▶ A syndicated news insert is an independently produced news story sold to stations using a market-by-market syndicated approach. Such features typically focus on practical health, finance, or living information that transcends markets.

▶ News programming is more than determining the lineup of different stories occurring within the day. Content treatment in news programming includes management orientation, promotions, leads, consultants, and talent.

▶ Consultants have had a significant impact in the development of local news. They are business-concept driven and specialize in audience and behavioral research. They provide audience data to managers faced with decisions relating to programming, audience development, and news.

▶ Trends in news and information programs include new technology, expanded sources, profitability, and a proliferation of new programs in the genre.

▶ Consumer-perceived trends in news information include a growing distrust of the media.

InfoTrac College Edition Exercises

News programming has long been a public-interest responsibility for television stations; today it is also a profit center. The potential for profit has contributed to the blurring of the line between news and entertainment programming.

10.1 The traditional news directors don't always care much for "infotainment" news programming. Use InfoTrac to locate articles in *Broadcasting & Cable* and other periodicals that present the differing positions on this trend in news.

   a. What is causing this trend?

   b. Why is it profitable?

   c. What are the critics saying?

   d. What is the audience doing?

10.2 Use InfoTrac to find an article about television news and answer the following questions:

   a. How does traditional news differ from tabloid news?

   b. What are the prime-time news magazines? Find an article about each.

   c. Name two news source databases excluding those mentioned in the text.

10.3 Use InfoTrac to find an article about the laws recently passed in California that restrict paparazzi freedoms.

   a. What limits were imposed?

   b. Are other states considering legislation? Is Congress?

WEB WATCH

Here is a list of a few URLs (Internet addresses) for some of the organizations or corporations discussed in this chapter. Please explore these Web sites and follow the links to learn more about the complex business of the electronic media. Add your descriptions and your own favorite sites at the end of the list. Please keep in mind that the dynamic nature of the Internet allows sites to come and go but also allows organizations to update information about themselves very quickly.

Address	Description
http://www.cnn.com/	
http://www.foxnews.com/	
http://www.cbs.com/	
http://www.msnbc.com/news/	
http://www.rtnda.org/	
http://www.poynter.org/	
http://www.spj.org/	
http://www.upi.com/	
http://www.ap.org/	
http://www.abcnews.com/	
http://www.digxpr.com/conus.html	
http://www.medialinkworldwide.com/	
Other favorite sites:	

CHAPTER 11

AUDIENCE ANALYSIS AND MARKETING

Because the audience is often considered the product of broadcast stations and cable systems, audience research is essential to the electronic media.

esearch. For college students, just the mention of the word may conjure up images of working all night to finish that most dreaded of all research projects—the term paper. For many students the term paper is the only experience they have had with research, and, too often, it was a negative experience. But research is very much a part of the electronic media and it actually can be interesting and even fun! Audience research is essential to broadcast and cable programmers and advertisers. In fact, everyone who aspires to work in the broadcasting and cable media should have a basic background in research.

This chapter examines the audience analysis process and its application to broadcast and cable programming and marketing, including a look at basic research techniques, ratings services, sampling, and data gathering, as well as at the interpretation, utilization, and criticism of research.

UNDERSTANDING RESEARCH

The *Merriam-Webster Dictionary* defines research as (1) "careful or diligent search (2) studious and critical inquiry and examination . . . aimed at the discovery and interpretation of facts." We conduct research because we seek information that is not available without a search of some kind. If we are assigned a term paper on a historical figure, we search for information in the library. But what if we are assigned the task of determining an estimate of the audience for a particular television program or a radio station? These two research assignments have different purposes and require different research techniques.

There are a number of ways to categorize research. One way is to categorize research as either qualitative or quantitative, referring to the degree of measurement precision used. *Qualitative research* usually involves observation of behavior by a researcher or even participation in a behavior or lifestyle. The researcher then reports what is observed without quantifiable data or information, such as numbers, that can be verified or replicated. *Quantitative research*, in contrast, usually results in specific data that can be tested for verification. Quantitative research generally is considered to be more reliable than is qualitative research.

Another way to categorize research is by the technique used. These categories generally include (1) historical-critical, (2) experimental, (3) content analysis, and (4) survey. A term paper assignment provides an example of the *historical-critical research* technique. We could examine historical documents to gather information on a particular figure, although college students are not likely to do *primary research* on the original documents themselves. They are more likely to do *secondary research* by reading books and articles written by someone else about the historical figure.

The second technique, *experimental research*, generally involves a test of an effect as a result of a variable. This type of research is conducted in relation to effects of the electronic media, such as the many studies done to test the effects of violent television on children. Usually, the studies involve two groups, an experimental group and a control group. The experimental group views a program containing violence, whereas the control group views a program with no violence. Children from both groups are then observed in play situations to determine differences in aggressive behavior. Observed differences in aggressive behavior between the two groups often are attributed to the variable, violence in programming.

The third research technique, *content analysis*, is, as the name suggests, a study of the content or message of a program. Through content analysis, we can examine the messages in a television program or commercial and isolate stereotypes of women or ethnic minorities, for example. Or we might choose to examine the use of graphics in television news programs. This technique does not address effects or audience behavior patterns.

The fourth technique, *survey research*, is most often conducted for and utilized by the electronic media. Here we gather information through a survey and describe the information collected. We do not attempt to manipulate behavior or effects as in the case of experimental research. In fact, we make every effort to be as objective as possible so as not to *bias* the information gathered. In the electronic media, we gather information on the viewing and listening behavior of the audience.

Why are we interested in the viewing and listening behaviors of audiences? Because audiences are the *products* of broadcast stations and cable systems. As businesses, they sell audiences to advertisers. As explored in previous chapters, broadcast stations do not sell their programs directly to the general public; they provide programming at no charge to the public and sell their audiences to advertisers. Cable systems have the advantage of both selling audiences to advertisers and selling programming directly to subscribers for a monthly fee.

To sell audiences to advertisers, we need to know as much as possible about our product. Certainly, we need to estimate the size of an audience for a TV program or a radio station. Audience estimates generally are referred to as *ratings*. We also want to know the makeup of the audience: males versus females and the various age groups represented. These categories of audiences are **demographics**. Many products are designed for, or *targeted to,* a particular demographic, such as women ages 18 to 34. The advertiser for the product therefore seeks media that reach that demographic. Advertising is targeted to a demographic through a particular television program or radio format.

Advertisers also are interested in the **psychographics** of audiences—the values or lifestyles of viewers and listeners. These include such characteristics as purchasing behavior, self-concept, and leisure interests. An advertiser might target women in their early twenties who drive foreign cars, for example. The fact that she drives a foreign car gives the advertiser psychographic information about a potential buyer. She also might be a single parent who prefers health food and belongs to a fitness club. All of these factors provide valuable information to advertisers in defining a target audience for a product.

Early Research Companies

National surveys of radio audiences began in 1929, when Archibald Crossley initiated telephone surveys of network radio listeners. Crossley's was the only research service for radio ratings until the mid-1930s, when another company, C. E. Hooper, began offering its own radio survey reports. The Hooper method of telephone surveying (discussed later in this chapter) proved superior to that of Crossley, who went out of business by 1946. C. E. Hooper took over network ratings and added local program surveys.

In the meantime another company had developed an alternative method of gathering listener data. In the late 1930s, the A. C. Nielsen Media Research Company developed the *audimeter,* an electronic meter attached to radio sets that monitored set use, listening time, and radio channels tuned. In 1950, when television viewing measurement started, Nielsen eliminated its competition by buying out Hooper. Today the electronic media ratings business is dominated by two companies, Nielsen and Arbitron.

Whereas Arbitron focused on ratings reports for local television and radio markets around the country, Nielsen specialized in national network ratings. When Arbitron dropped its local TV market reports in 1994, that business was left to Nielsen, which produces the *Nielsen Television Index,* a ratings report of national television audiences. It also produces overnight ratings for network television programs based on electronic meters in a number of large cities. Nielsen's local market television report, the *Nielsen Station Index,* is produced four times a year. The company offers a number of specialized audience research services as well, such as a report on syndicated television programs. Recall that syndicated programs are those sold or leased directly to television stations, not to networks (see chapter 9). Another report, *NSI Plus,* focuses on *audience flow,* the movement of viewers from one program to the next. The report also provides information on audiences viewing the same program daily or weekly.

Arbitron specializes in local market reports and, until the early 1990s, produced ratings for both radio and television. In 1993 Arbitron announced its intention to end its local TV market reports and specialize in radio ratings. The company, which began in the 1950s as the American Research Bureau, surveys more than 250 radio markets in the country at least once a year. The ratings reports for radio are based on a *diary,* or weekly listening log (discussed later in this chapter), kept by a sample of radio listeners in each market. Arbitron local television market reports were based on the information recorded in diaries and by electronic meters attached to television sets. Until the mid-1990s Arbitron surveyed more than two hundred TV markets for local reports published four times a year. Arbitron also offers specialized research services such as Arbitrends II, a computer software program, that allows station programmers to analyze audience data through their own personal computers. The utilization of audience research in programming decisions is examined in more detail later in this chapter.

Arbitron also developed the Network Cable MediaWatch service, which monitored commercials run on major cable networks, including CNN, ESPN, MTV, TNT, and Lifetime. The service identified commercials on the cable networks and provided a daily monitoring system to track commercial runs. This function also was left to Nielsen when Arbitron dropped TV ratings services in 1994.

Audience research companies charge radio and television stations subscription fees for their services. The fees range from thousands to hundreds of thousands of dollars annually, depending on the size of the market. Ratings services have become very expensive, but they are essential to the electronic-media business because advertisers demand some evidence of listening and viewing audiences.

Arbitron and Nielsen have had some competition from upstarts over the years, but none has been able to overcome their dominance. For example, in 1978 Birch Radio took on Arbitron in the radio ratings business with a telephone survey service. It offered less expensive surveys, but the company could not maintain enough clients to compete with Arbitron as radio station profits dwindled by the 1990s, and it went out of business in 1991. A former Birch product, the *Scarborough Report,* an annual report of radio listener lifestyles, is now available through Arbitron. The report provides audience profiles showing viewer/listener psychographics. Consumer patterns are correlated with radio listening and television viewing habits.

There are some smaller, more specialized companies that offer audience research. For example, the *Media Audit,* produced through the Radio Research Consortium, surveys radio listeners about consumer habits, attitudes, and values. The report offers information on audience income and education that allows radio stations to develop listener profiles to market to potential advertisers.

Defining Market

To conduct a radio or television audience survey for a local market, the boundaries of the market or area to be surveyed must first be defined. We might assume that a local market is simply the city or county in which radio and television stations are located. In reality, however, stations often are heard or viewed outside of

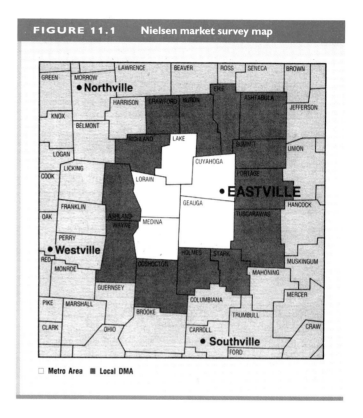

FIGURE 11.1 Nielsen market survey map

those boundaries. For local markets Arbitron developed the *area of dominant influence (ADI),* defined as "the geographic area that reflects those counties in which the dominant share of television viewing is to home market stations." Nielsen uses a similar survey area referred to as a *designated market area (DMA).* Note the Nielsen map in figure 11.1 showing a DMA for a local television market.

The primary survey area used by Arbitron is the *metro survey area (MSA),* defined as "the core of the market . . . [it] usually has the highest concentration of population in the market." Each of the survey areas reflects a different demographic makeup; one may be more suited to an advertiser's needs than another. The ADI was developed by Arbitron for local television markets, but it is still shown in radio reports for comparison of radio and television markets.

For radio markets, advertisers generally focus on the metro survey areas. In many cases the signals of radio stations are more concentrated than those of television stations. Hence, radio listeners are more concentrated in the metro areas, so radio advertisers

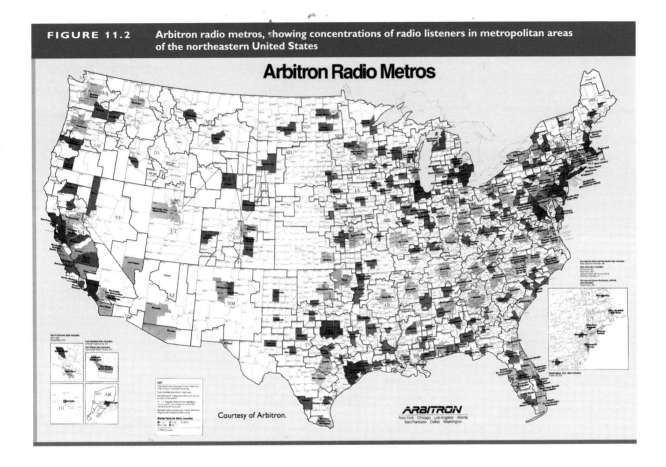

FIGURE 11.2 Arbitron radio metros, showing concentrations of radio listeners in metropolitan areas of the northeastern United States

usually buy commercials based on metro area audience estimates. Figure 11.2 shows an Arbitron radio metros map. In contrast, a cable TV market is defined by the number of subscribers to a particular system, not by geographic boundaries.

THE RESEARCH SAMPLE

How can we determine who is listening to a radio station or viewing a television program? Obviously, we can't interview every member of the television audience to find out what programs they watched last night. Instead, we rely on a *sample* of the population—a small group to survey about their listening and viewing behaviors. Based on this sample of audience behavior, we generalize to the population and obtain audience estimates. The figures obtained are estimates of the actual audience and are subject to *sampling error*, which reflects the probability that an estimate is the same as the actual audience. Research companies try to keep the error within 5 percent, which means a 95 percent chance that the estimate is correct. Can a small portion of the population give a good picture of national television viewing behavior? It can, if the sample is representative of the population.

To generalize viewing or listening behavior from a sample to the population, we must use a *random sample*. This means that every member of the population, or universe, has an equal chance of being selected for the sample. To draw a random sample, we first designate the population we wish to sample. Will it be the entire population of the United States or the population of a particular city or county? Once the population is designated, a *sampling frame* is determined. This is a list of every member of the selected population. A commonly used sampling frame is the telephone directory. In most communities virtually every person has a telephone, and most are listed in the directory. A percentage of telephone numbers are unlisted, however, so that must be taken into consideration. To obtain a sample of the unlisted numbers, a computer can be programmed to dial numbers on a random basis.

There are numerous methods for selecting a random sample from a sampling frame. One is to use a numbering system in which every nth (tenth or twentieth) name on the list is selected. This helps reduce bias in the selection of names from the list. Another method is to use a list of random numbers generated by a computer or from a table of random numbers. Again, the purpose is to draw a sample that represents all demographics in the population as accurately as possible. The sample should reflect the percentages of males and females as well as the various ages and ethnic groups in the population. If the sample is drawn correctly, it is possible to generalize the population to obtain audience estimates.

How large should a sample be? Simple statistics can answer this question. One might assume that larger samples will result in more-accurate or more-reliable estimates. This is true to a limited extent. Statisticians have shown that the level of *reliability* (degree of accuracy) of a sample increases as the square of the sample size. If we begin with a sample of 100 names or units, for example, we cannot double the reliability of the sample by doubling the number to 200. To double the reliability, we would have to increase the sample size to 10,000 (100^2). So, as the size of a sample increases, the degree of reliability increases, but in smaller increments to the point that they become essentially meaningless. We eventually would reach the point where the sample size would be too expensive to justify the small increase in reliability.

How large a sample would we need to estimate national audiences for network television programs? Nielsen uses a sample of about five thousand homes to estimate network television audiences. That sample is used to generalize audience behavior to a U.S. population of about 99 million television households.

Data Gathering

Once we have selected the audience sample, how do we gather the information or data needed? There are several methods available, including diaries, telephone surveys, electronic meters, and personal interviews.

Diaries

Diaries, written logs of viewing or listening patterns, are distributed to individuals selected for a sample. They are asked to keep track of TV programs watched or radio stations listened to for one week. As an incentive to keep the diary, individuals in the sample are offered a small cash payment, usually one to five dollars. Ratings companies generally contact individuals by phone to ask for participation and then send the diary by mail. Figure 11.3 shows a typical radio diary.

The sampling unit for television viewing usually is the *television household (TVHH)*, a unit that originates from a time when U.S. families had one television set and watched as a household. Today U.S. households commonly have multiple sets, so a diary is kept for each set and is then combined to produce

FIGURE 11.3 Arbitron radio diary page

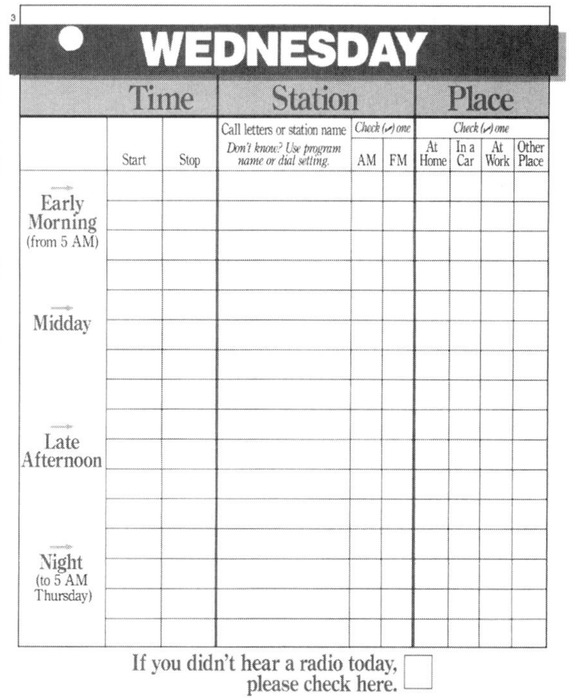

Radio listeners use this form to keep a log of their listening for one week.

TVHH audience patterns. For radio, the sampling unit is the individual, so each member of a household is asked to record his or her own radio listening for the week. Of course, the reliability of the diaries depends on accurate recording of viewing or listening by the individuals involved.

Diaries allow ratings companies to gather a significant amount of information, including the times receivers were in use and who was watching or listening, as well as demographic information on the individuals in each household. Diaries also provide documentation of viewing or listening habits that can be stored and referred to at a later time.

Telephone Surveys

There are two methods of telephone interviewing: coincidental and recall. The **coincidental method** requires respondents to identify programs or stations they were watching or listening to when they received the phone call. This survey method is quite accurate, because respondents don't have to rely heavily on memory to list programs watched at an earlier time. This is the problem with the **recall method** (originated by the Crossley company): Respondents are asked to recall the programs or stations they listened to or watched over the past twenty-four hours. In this method, of course, the accuracy of the information depends on the memory of the respondent.

Electronic Meters

An alternative to surveys by mail and telephone is the *electronic meter*. In this method a meter is attached to television sets in sample households. The first Nielsen audimeters just recorded when the set was on and to what channel it was tuned. The advantages of the meters are that they eliminate human error in logging viewing patterns and that the information is available immediately. The meters are connected through telephone lines to central computers for information retrieval. The disadvantage is that the meters do not provide demographic information on those watching programs. Ratings companies rely on supplementary diaries for that information.

In the late 1980s, the *peoplemeter* was introduced. This electronic meter is preprogrammed with demographic characteristics for each member of a television household. The meter has a button to be pushed by each family member when he or she watches television. They punch in when they begin watching and punch out when they finish. In this manner the meter records not only time and program watched, but also who was watching. The peoplemeter eliminates the need for a supplementary diary for demographic information. It is interactive in the sense that it requires that viewers take action before watching TV.

Another improvement on the current peoplemeter is the *pocket peoplemeter*, developed by Arbitron in the early 1990s, which can be carried by the television viewer like a beeper in a pocket or on a belt. It automatically picks up a code in the soundtrack of each television program viewed. The device also detects codes broadcast by radio stations. Those selected for audience samples are provided with the device along with a modem with which to send the data to the research company. The pocket peoplemeter eliminates human error in audience sampling, but requires that a code be included in all television and radio programming.

Ratings companies have developed a *passive peoplemeter* that eliminates the need for viewer interaction. The meter is programmed with the facial features of each member of the family and scans the room and recognizes family members watching television. Such a device, if functioning perfectly, would eliminate inaccuracies in ratings caused by human error (forgetting to punch in). But there is some question about whether families would want such a "big brother" device in their homes.

The peoplemeter offers other advantages beyond accurate viewing logs. It also tracks VCR usage such as programs that were recorded and viewed at a later time. This is known as *time shifting*. It also detects the viewer practice of zipping, or fast-forwarding, through commercials in a taped program, an obvious concern to advertisers.

Personal Interviews

Although some electronic media research is conducted using personal interviews, this method is not commonly used for audience research, primarily because it is very expensive and people today are less willing to invite interviewers into their homes. The personal interview does provide the opportunity for human interaction and probing questions, but these advantages are outweighed by the cost factor.

In-House and Consultant Research

Many radio and TV stations conduct in-house research to supplement the market research available from Arbitron and Nielsen. One example of this is *call-out research*, whereby members of the station staff call listeners to survey music preferences. Listener names can be drawn from station lists of request callers or contest winners; these are known as *active listeners*. Names also can be drawn at random from telephone directories for a list of potential or *passive listeners*. Station staff attempt to determine the popularity of songs played in the station music mix. How long should a song remain in the rotation? What new songs should be included? At what point do listeners grow tired of a hit song? Could a decline in listenership be a result of the music mix or is it more related to the on-air personality?

Another type of in-house research technique utilized by radio and TV stations is the *focus group*, wherein ten to twelve potential listeners or viewers are brought together to discuss their opinions of a particular station. A researcher conducts the session and attempts to elicit information on station programming and image. Are there certain programs or personalities that irritate members of the group? What elements would cause one to tune out? Again, the purpose of the research is to determine the *why* of audience trends that are revealed by the ratings books.

Focus groups can be followed up by *perceptual surveys*, which are in-depth telephone interviews of targeted listeners or viewers. The purpose of the survey is to verify information gathered in focus groups. An image problem with a radio personality or a television anchor can be verified through a larger telephone sample. Stations generally hire professional research companies to conduct perceptual surveys.

Another audience research technique used by radio and TV programmers is *auditorium testing* (also referred to as *theater testing*), wherein a sample of potential listeners or viewers is recruited by telephone and gathered in a theater setting, where they are shown television pilot programs or listen to a variety of music cuts. After viewing or listening, they are divided into groups and questioned about their responses to program materials. Professional researchers attempt to determine reasons for positive and negative reactions to programs or music.

Samples of viewers can be expanded through a technique called *cable testing*, in which a cable channel is used to access viewers, rather than showing pilot programs in a theater or auditorium. A sample of viewers is selected from a subscriber list and contacted to participate in the testing. This method allows for a more representative sample and enables those participating to view programs in a home setting. Cable testing is generally thought to provide better data on viewer preferences than does auditorium testing. Because the cost of producing new television programs is so high, this type of research is particularly important to major production companies.

In addition to ratings, in-house research, and program testing, radio and television stations employ a variety of *consultants* to address specific programming or management problems. One of the largest electronic media consultants is Frank N. Magid Associates. The firm offers a number of research services, but tends to focus on news talent evaluation (see chapter 10). Television stations are particularly interested in audience reaction to news anchors, because they play a major part in station image. Consultants like Magid, sometimes called *news doctors*, sample viewers for opinions of news personnel and provide advice based on the results. They critique factors ranging from anchor hairstyles and dress to studio set colors and computer graphics. News consultants have been criticized for contributing to a trend away from hard news or bad news on television newscasts and toward lifestyle features and "infotainment." Critics contend that news consultants tend to make television newscasts look and sound alike.

In radio, consultants provide input into programming decisions such as whether to change formats, revise a music playlist, or use a satellite or syndicated programming service. Consultants are also used in other areas such as station business practices, promotion, technical design, and financing.

Data Analysis: Ratings and Shares

When audience sample surveys are completed, the information from diaries and peoplemeters is input into the computers of Nielsen and Arbitron. Diaries mailed in are reviewed by staff to determine which are usable. Those that are incomplete, filled out incorrectly (wrong days), and without demographic information on families or individuals are rejected. Arbitron and Nielsen anticipate that a portion of the sample diaries

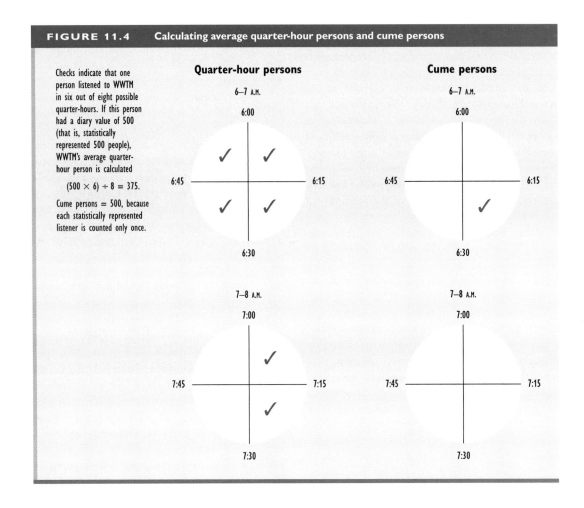

FIGURE 11.4 Calculating average quarter-hour persons and cume persons

returned will be unusable and inflate the sample size to account for them. Those diaries that are used in the ratings reports are referred to as *in-tab diaries*.

As noted earlier, ratings reports provide stations and advertisers with audience estimates based on samples. For radio, the estimates include persons estimates, ratings, and shares. *Persons estimates* reflect the estimated number of persons listening to a radio station at a given time. These estimates include average quarter-hour persons and cume persons.

Arbitron (1991) defines *average quarter-hour persons* as "the average number of persons estimated to have listened to a station during any quarter-hour in a time period." *Cume* persons are defined as "the estimated number of *different* people who listened to a station for a minimum of five minutes in a quarter-hour within a reported daypart [emphasis added]." The difference between the two estimates is that cume persons (short for *cumulative*) represent the unduplicated audience; they are counted only once during a survey period (one week). In contrast, average quarter-hour persons are duplicated; they can be counted up to four times each listening hour.

The cume persons estimate gives the **reach** of a station or a program—the total number of different people who heard or watched a specific program or a commercial. This is rather like newspaper circulation, which reflects the number of different households that receive the paper. Figure 11.4 shows the relationship between average quarter-hour persons and cume persons.

Ratings and shares reflect percentages of the population and of the radio audience. A *rating* is the percentage of the total population listening to a radio station at a given time. A *share* is the percentage of the radio audience listening to a radio station at a given

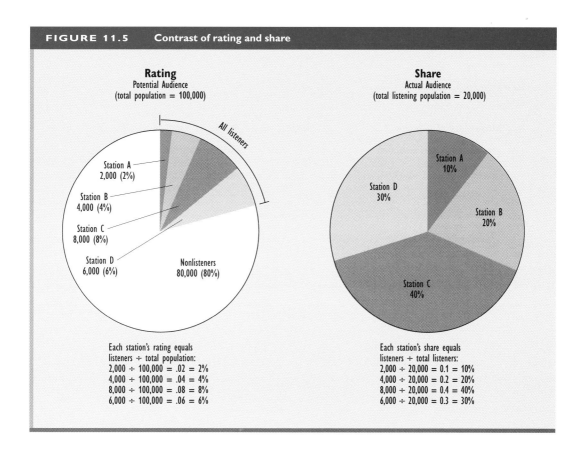

FIGURE 11.5   Contrast of rating and share

time. Thus, the rating is a percentage of the *potential* audience and the share is a percentage of the *actual* audience. Figure 11.5 shows the relationship between a rating and a share.

For television the reports include four estimates: homes using television, rating, share, and cumulative audience (cume or reach). As noted earlier, television ratings reports still use TVHH as the sampling unit even though most households have multiple sets. The number of television households is used as the population for the calculation of television ratings, but the number of households using television is used to calculate shares. Nielsen defines *homes using television (HUT)* as "the number of U.S. TV households using their television sets during the average minute of each 15-minute time period." In television reports, cume/reach is shown for both households and persons.

Research companies report both the rating and the share for a radio station or a television program, because they represent different yet equally important parts of the population. As shown in figure 11.5, the share pie excludes that portion of the population that was not listening to radio at all at a given time. Hence, the share is always larger than the rating, because the pie is smaller. A particular station may show a large share of the listening audience, but the total radio audience might have been quite small at that point in the rating period. A station with a small share of a large audience might have more persons listening than one with a large share of a small audience.

Shares do not provide an indication of the absolute size of a station's audience. Consider the illustration from Arbitron in which a station share of 15 percent in the morning actually represents a greater number of listeners than a 20 percent share at night (see figure 11.6).

The rating is an indication of how the station did in terms of total population or potential audience. In the preceding example, the total morning radio-listening audience represented only 25 percent of the population of the market. All of the stations' ratings

> **FIGURE 11.6   Ratings and shares**
>
> Total metro population = 100,000
>
> Total listeners for *all stations:*
>   Morning listeners = 25,000
>   Night listeners = 15,000
>
> *Rating* = total listeners ÷ total population
>   Morning rating = 25% (25,000 ÷ 100,000)
>   Night rating = 15% (15,000 ÷ 100,000)
>
> Total listeners for *one station:*
>   Morning listeners = 3,750
>   Night listeners = 3,000
>
> *Share* = station's listeners ÷ total listeners
>   Morning share = 15% (3,750 ÷ 25,000)
>   Night share = 20% (3,000 ÷ 15,000)

together equal 25 percent. Each station's share represents a percentage of that piece (25 percent) of the population pie. Ratings and shares together give the advertiser and the programmer a more accurate picture of the audience.

Research Reports

Radio: Arbitron

A typical ratings report, known in the broadcasting/cable industry as "the book," will help contrast an Arbitron local radio book with a Nielsen local television book. *Arbitron Radio Market Reports* for larger markets are published four times a year. The reports begin with a map showing the market surveyed, including the ADI and metro survey areas. As mentioned, the ADI survey area was developed for television markets but is still included in radio market reports, because it is useful for advertisers in comparing radio markets with television markets.

Following the map, information on the population of the survey area and the station facilities is listed. This usually includes population estimates for men and women in the area broken out by age groups. Station call letters are listed along with frequencies and power outputs of each facility. Most radio reports provide a *metro market profile* containing demographic and socioeconomic characteristics of the population. Here household incomes, education levels, occupations, and car ownership are listed to give advertisers more information on potential buyers for their products.

The audience estimates provided by the radio book include ratings, shares, and persons using radio. These estimates are generally categorized by dayparts and demographics. *Dayparts* are segments of the radio broadcast day as designated by listener usage (see chapter 8). Radio listening patterns vary by dayparts, such as morning drive and midday, rather than by programs, so the audience estimates displayed in a radio book are so listed.

The first estimates in an Arbitron radio book are *metro audience trends,* comparing shares, persons, and cume ratings for the five most recent metro survey periods. The numbers allow advertisers and programmers to see station audience trends over the past year. An advertiser might wonder how consistent audiences for a particular station have been over time. A station might have a particularly good book for one rating period, but could not maintain its listeners over time. Obviously, advertisers and programmers both are interested in listener consistency on radio stations.

A typical audience estimate page from an Arbitron radio report is shown in table 11.1. The first column contains estimates for the morning-drive daypart, Monday through Friday. The station with the

TABLE 11.1 Sample information from *Arbitron Radio Market Report*

Target Audience (persons 25–49) (demographic group targeted or sought by these stations)

STATIONS	Monday–Friday 6 A.M.–10 A.M.				Monday–Friday 10 A.M.–3 P.M.				Monday–Friday 3 P.M.–7 P.M.				Monday–Friday 7 P.M.–Midnight				Weekend 10 A.M.–7 P.M.			
	AQH (00)	CUME (00)	AQH RTG	AQH SHR	AQH (00)	CUME (00)	AQH RTG	AQH SHR	AQH (00)	CUME (00)	AQH RTG	AQH SHR	AQH (00)	CUME (00)	AQH RTG	AQH SHR	AQH (00)	CUME (00)	AQH RTG	AQH SHR
KCWW METRO	6	47	.1	1.2	5	52	.1	.2	8	53	.1	.5					28			
TSA	6	47			5	52			8	53							28			
KDKB METRO	154	860	1.8	6.4	155	767	1.8	6.6	101	874	1.2	5.9	15	263	.2	2.9	82	710	.9	6.2
TSA	168	920			175	844			113	988			15	263			97	761		
KESZ METRO	88	545	1.0	3.6	117	445	1.4	4.9	78	504	.9	4.5	21	263	.2	4.1	63	393	.7	4.8
TSA	88	556			117	453			79	513			21	263			63	393		
KFYI METRO	71	310	.8	2.9	87	338	1.0	3.7	56	327	.6	3.2	23	170	.3	4.5	11	108	.1	.8
TSA	76	323			101	389			85	360			41	190			22	141		
KISP METRO	21	57	.2	.9	14	73	.2	.6	14	66	.2	.8	9	43	.1	1.8	8	59	.1	.6
TSA	21	57			14	73			14	66			9	43			8	59		
KKFR METRO	41	344	.5	1.7	31	387	.4	1.3	33	486	.4	1.9	17	241	.2	3.3	39	341	.5	3.0
TSA	46	396			41	438			36	449			27	315			43	399		
KKLT METRO	154	752	1.8	6.4	183	828	2.1	7.7	115	829	1.3	6.7	33	386	.4	6.5	81	497	.9	6.1
TSA	170	817			207	938			125	920			34	406			82	525		
KMEO METRO	1	24			2	39		.1	2	33		.1	1	30		.2	7	16	.1	.5
TSA	1	24			2	39			2	33			1	30			7	16		
KMLE METRO	146	858	1.7	6.1	160	664	1.9	6.8	137	794	1.6	7.9	45	469	.5	8.8	153	842	1.8	11.6
TSA	157	980			191	780			156	875			46	477			171	925		
KMXX METRO	67	494	.8	2.8	79	472	.9	3.3	48	427	.6	2.8	8	141	.1	1.6	26	356	.3	2.0
TSA	68	501			80	492			50	435			10	161			26	356		
KNIX METRO	325	1351	3.8	13.5	295	1258	3.4	12.5	184	1235	2.1	10.7	35	578	.4	6.8	179	1044	2.1	13.6
TSA	342	1478			320	1470			211	1432			47	727			210	1207		
KONC METRO	24	152	.3	1.0	22	88	.3	.9	12	111	.1	.7	8	65	.1	1.6	7	93	.1	.5
TSA	24	152			22	88			12	111			8	65			7	93		
KOOL METRO	27	153	.3	1.1	26	144	.3	1.1	28	210	.3	1.6	4	113		.8	24	140	.3	1.8
TSA	27	153			28	164			29	217			4	113			24	140		
KOOL-FM METRO	108	771	1.3	4.5	128	697	1.5	5.4	112	931	1.3	6.5	37	366	.4	7.2	68	674	.8	5.1
TSA	114	809			136	739			122	1000			42	414			69	693		

Courtesy of Arbitron.

most listeners in the 25-to-49 demographic during this daypart is KNIX-FM, with 32,500 persons listening in the metro area during an average quarter-hour. During the week (Monday through Friday), Arbitron estimated that 135,100 cume or *different* persons listened at least once to KNIX during this daypart. The rating and share figures show that, on the average, 3.8 percent of the population (persons 25 to 49) and 13.5 percent of the radio audience were listening to KNIX during this daypart. The other dayparts are listed across the page in the other columns.

These four audience estimates are then displayed on later pages of the report for different demographics or target audiences: first for all persons, then for men, women, and teens. Following the section on **target audience**, the Arbitron radio report has a *specific* *audience* section, where all of the demographic groups are contrasted for each station. Average quarter-hour and cume estimates are shown for both metro and total survey areas by daypart. These estimates allow the advertiser or programmer to isolate the demographic strength of each station in the market. One station might be popular among men 18 to 34, but not appeal to women in that age group. This audience information could be critically important to a particular advertiser.

The Arbitron radio report further breaks down audience estimates into various dayparts and demographics: persons, ratings, and shares are analyzed on an hour-by-hour basis. These figures help pinpoint the specific hour within a daypart in which the most listeners in a target audience can be reached. The book also gives information on *listening locations,* including at home, in the car, and other locations, and estimates for *exclusive audience*—persons who listened to only one radio station during a particular daypart. Finally, the radio book provides estimates on overnight, ethnic, and ADI target audiences.

Television: Nielsen

A television ratings book, for example the *Nielsen Station Index,* is different from the radio book in several ways. Beyond the obvious difference of television versus radio, the television book is based on both diaries and electronic meters. The television book uses households and persons as sampling units, and the time periods or dayparts measured are different from those in radio. Actually, the television book provides audience estimates for dayparts and individual programs. (See chapters 8 and 9 for the differences in radio and television programming.)

The television book begins with a map showing the survey areas used for a particular market. Nielsen uses the DMA and consolidated metro area as survey areas. Demographic characteristics and television station facilities for the market are listed along with other market information. Among the audience estimates are *time period estimates.* Figure 11.7 shows a sample page from a Nielsen local television market report.

Station ratings and shares are shown for DMA households. Then *DMA ratings* are displayed for the various demographics in the market, including daypart estimates for persons, men, and women of various age groups, as well as for teens and children.

The Nielsen television book, like the Arbitron radio book, further breaks down audience estimates by daypart and time period. About the last half of the book is devoted to *program averages,* where audience estimates are provided for each program aired during the broadcast week. Beginning with 6 A.M. Monday, programs for each station in the market are listed with ratings, shares, and persons estimates. A *program title index* is provided at the back of the book, listing programs alphabetically with the station and time aired for each. This allows advertisers to locate and evaluate specific programs in each television market.

Applying the Research

Once information on radio and TV audiences has been gathered and reported in ratings books, how is it applied to programming and advertising decisions? Such decisions generally involve a great deal of money, whether a programmer is purchasing programs for a station or an advertiser is purchasing time for commercials. Good utilization of audience research is essential to both programmers and advertisers.

Radio

In radio, programmers are obviously concerned with the position of their respective station in the market. Radio books will tell where the station ranked among all stations in the market on the average. But remember that radio audiences are very fragmented, and the key to success is demographics. The station format is designed to appeal to a particular segment of the audience, primarily males or females of a certain age group. The overall market ranking may not be particularly useful in programming the station: It may be ranked tenth in the market for the entire broadcast week for all listeners ages 12 and over, but the more important statistic is its ranking for listeners in its demographic. For example, what is the station's audience share for men, ages 25 to 49?

In the Arbitron ratings book, audience estimates for the target demographic determine how effective the format has been in attracting that group of listeners. One measure of this is *time spent listening (TSL),* an estimate of the time that members of the radio audience listen to a station during a specific time

CHAPTER 11 AUDIENCE ANALYSIS AND MARKETING 245

FIGURE 11.7 Sample of a Nielsen report

Courtesy of Nielsen Media Research.

TABLE 11.2 — Time spent listening (TSL)

	QUARTER-HOURS IN TIME PERIOD	×	AVERAGE IN AUDIENCE	=	GROSS QUARTER-HOURS OF LISTENING	÷	CUME PERSONS	=	TSL (QUARTER-HOURS)
Station 1	100	×	14,200	=	1,420,000	÷	102,400	=	13.9
Station 2	100	×	14,300	=	1,430,000	÷	61,000	=	23.4
Station 3	100	×	19,800	=	1,980,000	÷	108,800	=	18.2

period (such as Monday through Friday). This estimate is usually reported in quarter-hours and can be calculated for total audiences and for specific demographics. Table 11.2 shows the calculation of time spent listening for three hypothetical stations.

Such TSL estimates can help determine format effectiveness by calculating the *efficiency of target audience*, a formula that uses TSL estimates to determine how target audiences compare with total audiences. To use the earlier example, how long do men ages 25 to 49 (target audience) listen to the station during the week as compared with all listeners? Dividing the target-audience TSL by that of the total audience gives an efficiency rating. That rating, compared with those of the competition, gives a good indication of the station's success (or failure) in attracting its target audience.

The station may also need to determine how consistent the audience has been throughout the broadcast day and week. How do its audience shares for the morning- and afternoon-drive dayparts compare? Audience estimates help determine *audience recycling*—the extent to which morning listeners tune in again in the afternoon. This information can be related to programming components such as music mix, personalities, promotional and commercial breaks, and dayparts. They might discover a decline in audience share over time in a particular daypart.

Television

On a national level, the *Nielsen Television Index* provides programmers with audience estimates for various program types. For a specific period of time (one week, for example), Nielsen provides national audience estimates for demographic groups and program types, such as dramas and situation comedies. This information is useful to television programmers in purchasing and scheduling network programs.

The television networks depend on Nielsen for their national audience estimates. Nielsen provides audience estimates on a quarter-hour basis for network prime-time programming. Audience shares are shown as well as average ratings by the quarter-hour. Again, network programming decisions are largely based on these national audience estimates.

Sales and Marketing Applications

Beyond the programming function, audience research is essential to the sales and marketing functions in broadcast stations and cable systems. As noted, the products of these organizations are audiences. Once we attract an audience with programming, how do we then use audience research to promote and sell our product?

Units of time—*spots*—are priced and sold to advertisers, who use the time to promote their products and services to radio and TV audiences (see chapter 7). Audience estimates are used by station sales staffs and advertisers to make buying decisions. In buying television time, for example, advertisers usually consider *gross ratings points (GRP)*, calculated by multiplying the number of spots (commercials) purchased in a particular program by the average rating for that program.

Advertisers also have a market guideline called *cost per point (CPP)* to use in planning the purchase of television spots in a given market. Generally, advertising agencies and media buyers establish CPP estimates based on the buying histories of local markets. These figures allow advertisers to project the cost of buying a certain number of ratings points in each market planned.

The cost of radio or TV commercials also can be translated into numbers of persons or homes reached through the calculation of *cost per thousand (CPM)*—the cost of reaching one thousand persons or homes through spots placed on radio or TV stations; CPM is calculated by dividing the cost of a spot schedule (total commercials purchased) by the audience estimated for the programs or dayparts in which the spots will run. Ratings books provide audience estimates in the form of *gross impressions* (average persons) in radio and TV, or average homes in the case of TV only. The equation in figure 11.8 shows an example of a CPM calculation. The CPM figure can be used to compare the cost of one medium with another. Cost per thousand can be calculated for radio, TV, newspapers, magazines, or any medium. In designing a media-buying campaign for a particular product or service, CPM is a valuable tool.

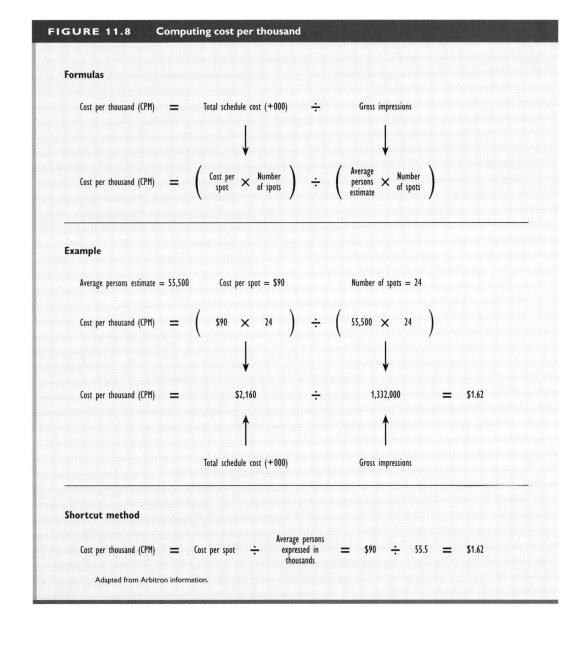

FIGURE 11.8 Computing cost per thousand

FIGURE 11.9 Computing frequency

Formulas

Gross impressions = Average persons × Spot schedule

Frequency = Gross impressions ÷ Reach (Cume)

Example

Reach (cume) = 44,600   Average persons = 10,900   Spot schedule = 12

Gross impressions = 10,900 × 12 = 130,800

Frequency = 130,800 ÷ 44,600 = 2.9

Media buyers are concerned with two other audience estimates in formulating buying plans: reach and frequency. Recall that reach is the number of different households or persons that are exposed to (watched or listened to) a particular spot schedule. This number is expressed in the ratings books as *cume persons*. Cume persons give us the number of *different* persons who watched or listened to the commercial spots throughout the schedule.

Advertisers also want to know how many times viewers or listeners were exposed to their spots. This figure is referred to as *frequency*. It is an estimate of the average number of times a person saw or heard a commercial during a specific period of time. Reach and frequency estimates generally represent a one-week period, because diaries are kept for one week. Monthly reach and frequency figures can be obtained only through metered markets, where households are monitored continuously. Reach and frequency for a one-month schedule can be simulated in diary-only markets by generalizing one-week estimates to the total number of spots scheduled for the entire month. The equation shown in figure 11.9 provides an example of the calculation of frequency based on gross impressions and cume persons.

When broadcast or cable spots are actually sold to advertisers by account executives, prices are broken out on *rate cards* (see chapter 7). Prices vary according to a number of variables, but primarily by daypart and/or program. The difference in price is supported by audience estimates. In radio, morning drive (usually 6 to 10 A.M.) is the most expensive time to buy, because this daypart delivers the most listeners on average. In television, prime time (evening hours) generally is the most expensive time to buy. Again, the size of the audience is the variable. Account executives need the book to justify their rate cards to advertisers.

Audience research is an integral part of broadcast and cable sales and marketing. Audience estimates are used to market stations to potential advertisers. A good example of this process is found in the marketing plan of KBSG-FM in Seattle. Their promotional material states that the format of the station—oldies—is targeted to "an even balance of men and women

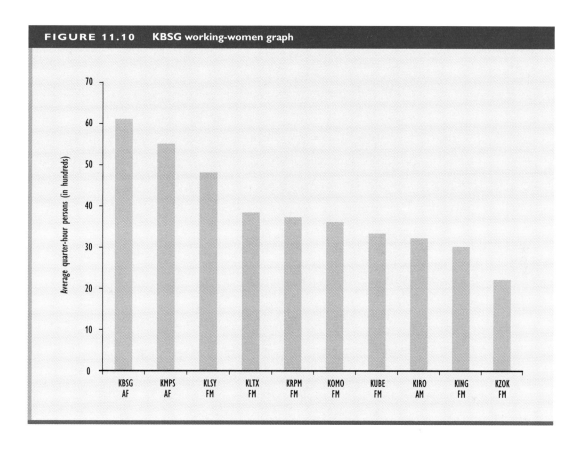

FIGURE 11.10 KBSG working-women graph

ranging in ages from 25 to 54 with particular emphasis between 35 and 44." They use the *Scarborough Report* to produce audience estimates that show station rankings for that demographic. The bar graph in figure 11.10 shows contrasting average-quarter-hour-persons rankings for working women ages 25 to 54. Note that KBSG is ranked number one in the Seattle market with just over 6,000 average quarter-hour persons in that demographic. This means that during the entire broadcast week, about 6,000 women between the ages of 25 and 54 who work full-time listened to KBSG in an average quarter-hour.

Another KBSG marketing tool, as shown in figure 11.11, displays three demographic categories and the probability that KBSG will reach those listeners. Note that KBSG's listeners are 54 percent more likely to have a household income in excess of $75,000 than the average radio listener in the market.

The *Scarborough Report* provides stations with demographic and psychographic information. For example, KBSG can focus audience estimates to show station listening among women 35 to 44 who have shopped in a particular department store in the Seattle area. This information is very useful in marketing the station to the department store and to other merchants located in the same shopping center.

The research director of a television station in Phoenix used *Scarborough Report* estimates to design a TV spot schedule for a concrete company. The company, which produces concrete walls for homes, was advertising in newspapers, magazines, and on radio, but not on TV. The research director profiled the potential customer for the product: upper-income adults 25 to 54 who had done construction or remodeling in the past year. She then outlined a schedule of TV programs on the station that reached that target audience. The company revised its advertising budget to spend 60 percent on television spots.

Some major-market stations have developed their own marketing departments, ranging from a single individual to a multiperson operation, such as that of Fisher Broadcasting's KMA. The one-person departments gather research data from various sources and organize them as sales and programming support.

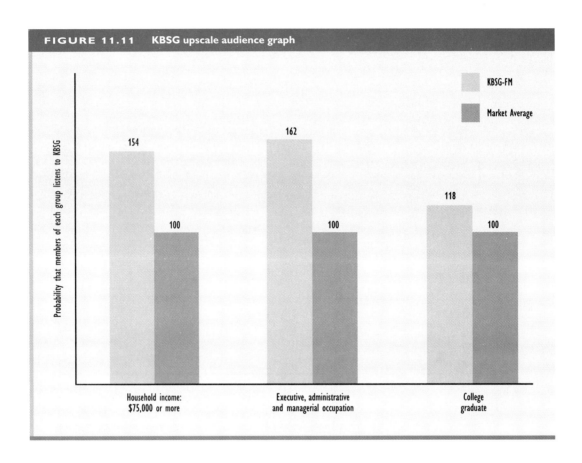

FIGURE 11.11 KBSG upscale audience graph

The larger operations, such as that of KMA, actually contract and conduct their own research. KMA conducts regular *market audits* throughout the year and correlates "chief shopper"—the person who does the most shopping for the household—retail buying with station audiences. These psychographic data are used primarily by the station sales department, but also are made available to retailers free of charge.

Criticism of Audience Research

The importance of audience research to the successful operation of broadcast stations and cable systems has been established. Still, there are many critics of the ratings and the role they play in programming, sales, and marketing decisions.

There is continuing concern about the sampling process that is the basis of audience research. Because it is impossible to survey the entire population of the United States about their listening and viewing habits, researchers must depend on samples of the audience. Critics contend that samples are too small to be truly representative of the population and that some groups within the population are underrepresented. Some producers of programs designed primarily for black audiences suggest that Nielsen samples do not provide an accurate picture of black viewing. Based on the 1999 U.S. Census projections, about 13 percent of the population is African American. Producers of programming targeted to blacks claim that the proportion is actually closer to 18 percent. This underrepresentation, they suggest, has an adverse effect on sales and marketing of their programs.

Cable networks (TNT, ESPN, USA, and Nickelodeon) have expressed concern about the ratings for their programs. Advertising agencies have been using quarter-hour estimates for cable network audiences, but cable operators claim that quarter-hour samples for their programs are much too small and result in a large sampling error. The Cablevision Advertising Bureau suggests that, in some cases, the sampling error in a cable network rating is as high as 140 percent. Cable networks have proposed that advertising agencies use Nielsen estimates for cable programs on a weekly basis, arguing that this method will provide a larger audience sample and reduce the sampling error.

Even if one accepts a sample as representative of a population, there are still potential problems with the information gathered. The diary method depends on those filling out the diary to provide accurate information. Errors certainly occur due to faulty memory or carelessness. Nielsen and Arbitron indicate that they account for these errors in their samples. In the meter method, the potential for human error is less, but there may be technical problems with equipment. Still, human error is a factor with peoplemeters, because viewers must punch in and out.

Even the stations that benefit from the ratings can play a role in biasing audience estimates. Those that subscribe to ratings services are expected to be consistent with station promotions and contests throughout the year. And though stations are not allowed to make reference to ratings periods on the air, Arbitron authorized an experiment in the Atlanta market in which radio stations encouraged listeners to participate in an Arbitron sample if selected. The campaign was an attempt to increase the response rate for Arbitron radio samples. Critics of the experiment argued that the on-air announcements would bias the audience estimates.

In-house research provided by station marketing departments is criticized not only for methodological flaws, but also for a lack of objectivity. Critics contend that this type of research can be self-serving and misleading and warn that in-house research data should be viewed in a comparative context.

Audience research is not a perfect process. In spite of criticism of the ratings, there appears to be no viable alternative at this time. Broadcast and cable advertisers demand some evidence of audiences before buying commercial time; audience research provides that evidence. Certainly, research companies have improved their methodologies and technologies in recent years, but so long as sampling is used, there will be sampling error. The challenge is to reduce sampling error as much as possible. Audience research will become even more important as the media marketplace expands and more channels are available to the consumer, and the demand for better audience measurement will continue to grow.

Summary

▶ Research is essential to the electronic media. The primary type of research conducted by and for these media is the survey of audience behavior. Audiences are the products of broadcast stations and cable systems. As businesses, they sell audiences to advertisers. To effectively sell their audiences, these organizations need to know as much as possible about their product. Through audience surveys, estimates of listening and viewing—ratings—are produced. These ratings portray audiences for radio and television programs in terms of demographics, such as age and gender. Advertisers also are interested in the psychographics, or lifestyles of the audience.

▶ The two major research companies in the country today are Arbitron and Nielsen.

▶ Nielsen specializes in television ratings on the national and local levels; Arbitron specializes in radio ratings.

▶ Using random samples drawn from telephone directories in local markets, research companies survey audiences using several data gathering methods: diaries, telephone surveys, and electronic meters.

▶ A rating is the percentage of the TV households or population tuned to a particular program or station.

▶ A share is the percentage of homes using television (HUT) or persons using radio tuned to a particular program or station.

▶ Stations also conduct additional audience research through call-outs, focus groups, and consultants to supplement ratings books.

▶ Audience estimates also are used to market stations to potential advertisers targeting specific demographics.

▶ Critics contend that audience samples are too small to be truly representative of the population.

▶ Audience research will become even more important as the media marketplace expands and more channels are available to the consumer.

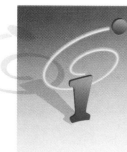

InfoTrac College Edition Exercises

The audience is the product of broadcast stations and cable systems, so audience research is essential.

11.1 Chapter 11 focuses on audience research for commercial broadcasting and cable, but noncommercial broadcasting needs audience research also. Read the report on the history of ratings in public television in the *Journal of Broadcasting & Electronic Media* 42 (Fall 1998) and answer the following questions.

   a. Why did public television programmers and managers start paying attention to audience ratings in the 1980s?

   b. What role did the Corporation for Public Broadcasting (CPB) play in the increased use of ratings in public television?

11.2 Nielsen Media Research was sold to a foreign company in 1999. Read the article on Nielsen in *Broadcasting & Cable* (August 23, 1999) and answer the following questions.

   a. What foreign publishing company purchased Nielsen?

   b. Nielsen offers a service called NetRatings, Inc. What does this service provide?

WEB WATCH

Here is a list of a few URLs (Internet addresses) for some of the organizations or corporations discussed in this chapter. Please explore these Web sites and follow the links to learn more about the complex business of the electronic media. Add your descriptions and your own favorite sites at the end of the list. Please keep in mind that the dynamic nature of the Internet allows sites to come and go but also allows organizations to update information about themselves very quickly.

Address	Description
http://www.arbitron.com/	
http://www.acnielsen.com/	
http://www.asiresearch.com/	
http://www.sriresearch.com/	
http://www.adtalk.com/	
http://www.demographics.com/	
http://www.nielsenmedia.com/	
http://www.roper.com/	
http://www.srds.com/	
Other favorite sites:	

CHAPTER 12

CORPORATE AND INSTRUCTIONAL ELECTRONIC MEDIA

Private television delivers instructional, training, and marketing messages to select audiences. Operators in this control room remotely control cameras in classrooms and studios to feed instructional television signals to students at remote sites.

We normally think of radio, television, and cable as sources of information and entertainment for a mass audience. They are an important part of what we call "the media," but they can also be used for the distribution of information to smaller, private audiences. Over the past two decades, video, especially, has been adopted by the corporate world as a medium for delivering internal and external communication messages to employees and customers. In addition, radio and television have a long history of delivering educational and instructional material to students, both in and out of the classroom. This chapter explores the uses of electronic media in education—called *distance learning*—and in the corporate world. Students preparing for a role in the media might consider the employment possibilities in distance learning or in corporate communications. Large corporations generally pay better salaries and have better benefits packages for starting employees than do traditional commercial radio or television stations.

In corporate and instructional video, which we can call **nonbroadcast video**, the tools, techniques, and basic skills used to create programs are the same as those in broadcast video. The major differences between broadcast and nonbroadcast video are primarily in the audiences served and in the purposes, or objectives, of the communication message.

Corporate Video

The term *corporate video* is a fairly recent addition to the communications lexicon. It is interchangeable with *industrial video, private television, nonbroadcast television,* or *organizational video.* All of these terms are used to describe a video, or television system, that is used for private communications—either externally between a corporation and its customers or internally between management and employees. Generally, corporate video messages are not designed for the mass audience watching television at home, although some corporations are now producing *infomercials* (program-length sales pitches for the company's products or services) to reach this mass audience via television stations or cable systems.

Although corporate video is relatively new, the concept of using the media for private communications dates back to the late 1920s, when some major manufacturing firms, such as the Ford Motor

Many corporate video departments are moving to nonlinear digital video-editing systems.

Company and United States Steel Corporation, used the new technology of 16mm motion picture film to document the wonders of the assembly line. Copies of these films were then distributed for public relations purposes. Within a few years, many of the more affluent centralized school districts around the country had invested in 16mm film projectors. An organization named Modern Talking Picture, Incorporated, was developed to expedite the distribution of industrial films. By the 1950s a school district could order free films from a catalog of more than two thousand titles on a wide variety of topics. Most were produced by major manufacturing corporations to enhance their organizations' image. Other films provided an instructional view of the industry.

A few of the larger corporations that were early producers of films about their businesses established internal corporate video departments in the late 1950s and 1960s. Smaller organizations were slow to enter corporate video, because the only television equipment then available—the same used in broadcast—remained much too expensive for most. It wasn't until after the introduction of the inexpensive helical scan videocassette recorder in the early 1970s that corporate video departments became cost-effective.

Because the corporate video department does not generally broadcast a signal over the airwaves, the Federal Communications Commission (FCC) does not require that expensive, high-quality broadcast television production equipment be used. Since the 1970s, television equipment manufacturers have developed a less expensive line of professional equipment that is sold to cable

TABLE 12.1   Comparison of selected analog videotape formats

FORMAT NAME	TAPE SIZE	SCANNING FORMAT	PHYSICAL FORMAT	HISTORIC USE	EQUIPMENT COST	VIDEO QUALITY	EASE OF EDIT	SHIPPING COST
2" Quad	2"	Quadruplex	Open reel	Broadcast (1956–1980)	Expensive	Good	Difficult	High
1" Type C	1"	Helical	Open reel	Broadcast (1910–1990)	Expensive	Excellent	Difficult	High
U-matic	¾"	Helical	Cassette	Educational, corporate, newsgathering (1970–1990)	Inexpensive	Good	Easy	Moderate
VHS	½"	Helical	Small cassette	Home video, educational/corporate distribution (1980–present)	Cheap	Acceptable	Easy	Low
Betacam	½"	Helical	Small cassette	Newsgathering, corporate production (1985–present)	Expensive	Excellent	Easy	Low
8mm/Hi-8	8mm	Helical	Tiny cassette	Home use (1990s)	Inexpensive	Acceptable	Easy	Low
S-VHS	½"	Helical	Small cassette	Corporate production (1990s)	Inexpensive	Good	Easy	Low

television systems, educational institutions, and corporate video production units. It is somewhat more expensive and of better quality than that sold for home-recording use, but it costs much less than equipment required for over-the-air broadcast applications. Table 12.1 shows characteristics of several analog tape formats.

The development of the ½-inch VHS videotape system provided the ideal format for distribution of corporate video material. The playback machines have become very inexpensive and nearly every business, as well as most U.S. homes, has access to one. In addition, the blank cassettes can be purchased for less than $2 each. A corporate video department can purchase a VHS cassette, have the program duplicated on it, and ship the videocassette package to the user for less than the average cost of a paperback book. Advances in technology and decreases in cost have made corporate video an attractive communication tool for many large and medium-sized businesses and corporations. New uses for corporate video are being developed daily.

Corporate Programming Goals

The purpose of the video unit in the corporation is to obtain, produce, and/or distribute information in a video or television format. That information or programming can be, for example, an interactive instructional message encoded on videodisc, or it can be an infomercial or public service announcement aired over a television station. It can be a public relations message shown over a local cable television system or a newscast designed to reach employees watching in a company cafeteria. Content can include

Many corporations provide company information or training on video. These students are watching a live instructional program originating from a distant university classroom.

something as simple as a video record that documents a procedure or an event that the organization would like to keep in pictorial form in the archives.

Generally, the content of the television program is designed to serve either the internal or external communication needs of the organization. Internal needs can include training programs for employees, state-of-the-company messages from the president or CEO, or motivational material in the form of a company video newsletter. External needs can be videotaped programs sent to commercial television stations as *video news releases,* played in retail stores as point-of-purchase sales pitches, or sent directly to the potential customer. Nissan, for example, will allow you to "test-drive" a new Maxima by videocassette. Salton/MAXIM Housewares packages a VHS videocassette of consumer directions in the box with its cappuccino maker.

Television programs are usually produced to either inform or entertain. This does not mean that an informational program cannot be entertaining or that an entertainment program cannot contain information. An effective program usually contains elements of both information and entertainment.

Different formats and types of programs are produced to serve different communication needs within the corporation. Corporate program design begins with the client. In the case of corporate video, the client is usually a division within the organization. Corporate clients might include sales and marketing divisions, human resources departments, or manufacturing areas. Generally, the client department has particular goals and objectives that must be met by the video program. The goals should state what the program is to accomplish and identify the target audience. Because the client will want accountability for the money spent designing and producing the corporate video program, the program objectives should be written in a way that can be easily measured (see table 12.2). A measurable objective is stated in precise language, covers one subject, and invites an observable change in behavior, attitude, or information on the part of the viewer. An example objective is: "The new employee, immediately after viewing the program, should be able to list ten categories of benefits that the

TABLE 12.2   **Sample objectives for a corporate employee training program**

The new station employee (tested immediately after viewing this training program with 95 percent accuracy) should be able to identify and describe the functions of the following equipment:

1. oscilloscope
2. vector scope
3. production switcher
4. VU meter
5. line monitor

company provides to him/her." This objective is stated in a way that can be measured. The new employee can be tested immediately after viewing the program to determine if the appropriate knowledge has been retained. If a number of employees are tested and shown to have retained the information, the effectiveness and validity of the video program are confirmed. Determining effectiveness of the program is important in the corporate system, which does not have access to audience ratings.

In those programs designed to reach the public, the corporation must be especially concerned that its philosophy, or image, is correctly portrayed. A large corporation that spends millions on commercial television advertising to develop an image will usually want the same image to come through in its public relations video programs. The image of the organization reflected in the video program should be consistent with the organization's other communications.

Although the goal of providing an effective internal and external communication system is very important, the overriding concern of many corporations is cost-effectiveness. Corporate video can be an expensive business, even when the technical systems are equipped with lower-end professional cameras and recorders. The cost of equipment, personnel, and distribution systems must be offset by the instructional advantages of the combination of sight, sound, color, and motion that video can provide. Like commercial television, the corporate video production process includes compromise. Although more money, more time, and better equipment allow the message to be strengthened, the corporate video program must be measured not only in terms of effectiveness, but also in terms of cost to obtain that effectiveness.

Budgetary considerations will influence the style of the corporate video program. Using the proper style is important to the complete understanding of the message, but an increased budget does not necessarily buy the correct style of program or increased understanding of the message. The most appropriate style of a video program conveys the message in the least complex manner. A simple program, done well, is usually far superior to a complex one done poorly.

The corporate video producer must remember that the potential audience for his or her program is an expert in television production who has watched thousands of hours of network programs. Not only must the corporate video production match the technical quality of what the viewer sees on television at home, but, more important, it must match network clarity of story line.

Internal Communications

The more-typical uses of corporate video programs are those that meet the needs of an internal audience composed of the corporation's own employees. The communication needs of this audience generally are job orientation, job training, information, or motivation.

Orientation

One of the most important times in the relationship between the employee and the employer is during the orientation period, when the values of the organization are transmitted to its new member. The new employee will want to do a good job: His or her mind is open and will easily accept first impressions. The manner in which the corporation presents information about the job will form powerful first impressions, and information presented in the video can be carefully designed to ingrain the corporate philosophy in the employee's mind.

The new employee will want to learn certain things about the organization, such as the company rules and the available benefits. Videotaped information can be repeated for each new employee as well as for long-term employees who wish to review their benefits.

In addition to the low cost of multiple presentations, there is an advantage to consistency of information: Each employee receives exactly the same information presented in the same manner each time the tape is shown; with control of the playback unit, the individual can structure the pace to suit his or her own needs; and the message can be rewound and repeated as many times as necessary for understanding.

Training

In a world where information and knowledge are presented at an ever-increasing pace, duties within an organization must also change so that it can remain competitive. Employees must constantly upgrade their skills to remain employed. To make the best use of its

TABLE 12.3 Partial list of training programs for AlliedSignal Aerospace customers

COURSE NUMBER	SUBJECT	LENGTH (IN MINUTES)	PRICE
TPV-011	**TPE331 PAB Fuel Nozzle Maintenance**	32	$65
	Describes the procedures used to properly maintain TPE331-piloted airblast fuel nozzles. Maintenance practices include: checking the nozzles on-the-wing, installing a new nozzle and manifold assembly, establishing a zero-time baseline, and checking nozzles as an assembly on a flowbench.		
TPV-015	**P3 Rolling Diaphragm Replacement on Woodward Fuel Control Units**	18	$65
	This program is also available in Spanish: TPV-015(S).		
	Provides step-by-step procedures necessary to replace the TPE331 P3 rolling diaphragm on Woodward Fuel Control Units—based on Garrett Alert Service Bulletin TPE331-A73-0203 and Woodward Service Bulletin WG-64042.		
TPV-020	**Agriculture's Answer: The Garrett TPE331 Engine**	19	$65
	Overviews the maintenance requirements for Garrett TPE331-powered agriculture aircraft. Reviews the engine logbook and maintenance manual. Discusses special concerns of AG operators and provides AlliedSignal service and support information.		

human resource investment, the corporation must provide continued training for employees and must train each new employee for specific duties.

Although the training function is key to the long-term success of the company, training efforts are expensive, and education departments are often the first to feel any corporate budget cuts. But corporate video can be a cost-effective method of providing necessary employee training, offering the advantages of consistency of content and repeatability just discussed.

Many corporations conduct training with the aid of corporate video. AlliedSignal Aerospace, for example, produces and sells video training tapes to instruct field service personnel on the newest techniques in aircraft engine maintenance (see table 12.3). The U.S. Army, Navy, and Air Force have long used video to instruct recruits on the proper way to perform specific tasks necessary to operate high-tech armaments. The Honeywell Corporation uses interactive video training materials to instruct users in the operation of its very sophisticated electronic industrial control system.

INFORMATION

Corporate video can be used to provide standardized information to employees to help them understand the organization's philosophy and their role in helping the company reach its goals. Recognition of outstanding employees presented in a video format can positively affect others in the organization by developing pride, loyalty, and a sense of belonging. Many corporations use video to enable the CEO to address all members of the organization simultaneously, even those in remote locations. Timely information provided by an authority figure can help quell damaging rumors that are inevitable in any large organization.

Samaritan Health Services, of Phoenix, Arizona, developed the video memo format to send timely information directly from the president of the organization to each employee. AlliedSignal sent each of its employees a VHS tape on which the president of the corporation explained the organization's new goals and described the adoption of a new total-quality management program to be instituted in the company's facilities worldwide. The Arizona State University used video to comply with a board of regents directive that each employee receive training in identifying and handling cases of sexual harassment as well as other legal issues.

MOTIVATION

Most employees will work harder for an organization if they feel that they are an important part of its society. Motivation can be inspired, in part, by involving the employee in the organization. The history and traditions of the company should be communicated to employees so that they feel a part of the organization's social fabric. It is important for employees to know and to identify with other members of the organization, especially in large and diverse corporations where the typical employee will have little opportunity to come in contact with those outside his or her own work unit. Video can be used effectively to explain the roles of different departments within the organization or to introduce the officers to the employees. As the employee learns the roles and players within the organization, he or she is better able to understand his or her own job function. An employee will be a better-motivated worker in an organization that is not just a faceless bureaucracy. Video can convey the faces of the organization and explain its goals in human terms.

One of the first duties of many newly established corporate video departments is to produce a news program for employees. Most corporations that support video departments are now producing such programs. The formats generally follow a news magazine style, with four or five in-depth stories in a fifteen- to twenty-minute program produced and delivered monthly to employees. Topics range from safety tips and descriptions of new products to interviews with the employee of the month and messages from the CEO. The Samaritan Health Services video news magazine included a "people feature" each month whose goal was to have as many of the employees on-camera as possible. People like to watch themselves and their friends on television and will watch management messages if there is a chance that they will see themselves on the program.

DOCUMENTATION

Many corporations develop and manufacture products of a technical nature. Often in the engineering process, the corporate video department is called on to record or document product tests. The video, as well as computer data, then is examined to gain insight into how the product reacts under certain conditions. In regulated industries such as aerospace, this data is necessary to obtain certification for certain products before they can be offered for sale. AlliedSignal Aerospace maintains corporate video departments in its facilities in Torrance, California, and in Phoenix, Arizona, specifically to document engineering tests that

On-location filming of video newscast for U-Haul Corporation. Many corporations use company newscasts to keep employees informed of company activities.

are conducted in the development of new aircraft engines and control systems.

Other documentation services that the corporate video department might provide include recording and archival storage of milestones in the company's history. For example, video recordings of stockholder meetings, groundbreakings for new facilities, and proclamations by company officials might all be saved for posterity by the corporate video department.

EXTERNAL COMMUNICATIONS

External communications are designed to reach a variety of audiences outside the corporate organization, including customers, potential customers, the broadcast media, industry regulators, and the general public. Generally, these messages are sales or image related, but in recent years many corporations have been using corporate video to train or to educate customers in the proper use of the product. Most of the video material produced is distributed on videocassette, but some organizations are using multimedia, wherein a videodisc or CD-ROM is coupled to a personal computer to allow interaction between the user and the program. Some large franchise organizations have even developed their own satellite networks to connect each outlet with the company's home office. The Chrysler Corporation enables the technicians in its Jeep dealerships to view training videos via a satellite network. The Salt River Project utility company uses video to train employees in safety procedures.

SALES INFORMATION

In today's marketing-oriented world, corporate survival often means the constant development of new products, or at least the refinement of established ones. The introduction of those new products is usually entrusted to the marketing expertise of a national advertising agency and the universal reach of the mass media. Both wholesale and retail sales staffs, however, must understand new products and product features. Traditionally, sales staffs are brought together at annual sales conventions to gain new product knowledge. A less costly method of providing product information uses video. Some corporations use video teleconferencing to bring the hype and excitement of the sales conference to sales personnel located across the country. The video format enables the presentation of detailed information and can show the product in action. Video can be an inexpensive method of providing accurate information about new products to salespersons located in remote areas.

PRODUCT DEMONSTRATIONS

Generally, these programs are directed to the potential buyer. They include point-of-purchase programs, point-of-information programs, product service information, and infomercials. All use television techniques to demonstrate the advantages of the product.

Major department stores often use television monitors to show continuously looping videotaped fashion features or demonstrations of kitchen appliances near where the displayed items can be purchased. The Broadway Southwest stores placed monitors next to the escalators to show a somewhat captive audience the merchandise that can be found on the next floor. Such *point-of-purchase (POP) programs* are designed to entice the shopper to make an impulse buy of the item being displayed.

Similar to POP programs but usually more subtle in the sales pitch, *point-of-information (POI) programs* provide information on how to accomplish a particular task. Monitors for POI programs might be found in a home improvement store, such as Home Depot, showing how to install a three-way lightswitch or how to design and install a backyard sprinkler system. Supplies needed for the job are located next to the monitor; often those supplies have been manufactured by the corporation that produced the instructional program and made it available to the store.

In our visually oriented society, the complex instructions necessary to repair, or even operate, some consumer devices is better demonstrated to the end user than explained in reams of directions, schematics, and operating instructions. Answering this need, *product-service information (PSI)* on video has gained popularity in recent years. The Ford Motor Company has long used video to show auto mechanics certain complex repairs on a Ford automobile. Now VHS videocassettes demonstrating how to actually assemble or operate the device are being included in the packaging of the product in lieu of the printed operation manual. With 80 percent of U.S. homes having access to one or more videotape players, the adage *a picture is worth a thousand words* has never been more true in the marketing arena.

Infomercials—program-length sales pitches—have become a common marketing device given the proliferation of cable channels and independent television stations since the early 1980s. Programs fifteen to thirty minutes in length are designed to provide interesting information interspersed with series of commercials for a product that is closely related to the topic of the show. For example, a program that at first glance is about medicine, health, and eye care is really a very long commercial for sunglasses. Although most corporations large enough to support a corporate video department do not rely on infomercials to market their products, the techniques used to produce the programs are the same. Infomercials are often produced by small independent production companies on inexpensive or S-VHS video formats. The company marketing the product then buys inexpensive time slots on local television stations or cable systems.

Public Relations

One of the major uses of corporate video is for *public relations (PR)*. The corporation is interested in putting its best foot forward when presenting itself to the public, the potential customers, and the stockholders. In the current climate of investigative reporting, it is especially important that a company have a carefully planned method of disseminating positive information about itself. In large corporations the PR departments have grown tremendously; now even smaller organizations without the resources to employ full-time PR departments usually retain the services of a consulting firm. The corporation must be able to tell its side of any story that reaches the mass media.

Video is one of the most effective tools that the PR department can use to get its message to the public. Video is often used for high-quality presentations to boards of directors, civic organization meetings, governmental boards, classrooms, and citizens' groups. Other programs are developed for presentation to mass audiences. For years large corporations have been producing television programs that extol the virtues of the company. Some of these can still be seen on cable systems and on small-market television stations in off-hours. These programs generally describe a particular industry and, in doing so, prominently display the specific corporate identification or logo of the company that produced the video.

News Releases

Corporate public relations departments have always churned out news releases to get the corporation's name in the news. The news releases often concern a new product, but they can also describe promotions, expansions, or other business matters. Occasionally, the press release is issued in response to a breaking news story about the corporation. A recent development in this arena is the *video news release (VNR)*, which performs the same function. As introduced in chapter 10, a VNR is a thirty- to ninety-second positive news story about the company, completely packaged on a video format and distributed to news departments in television stations.

Although most television news directors state that their stations would never air a VNR, many stations select footage and sound bites from the VNR and present the story as one that the news department has developed. Usually, the corporation is happy to have any part of the story broadcast and is therefore satisfied that the VNR has served its purpose.

Stations in smaller markets, where news personnel and resources are more limited, often broadcast more VNRs than do stations in larger markets. Because some video news releases are eventually broadcast, many corporations are now producing them. News directors might find several dozen VNRs crossing their desks over the course of a week. Some companies also provide a service that, for a fee, will distribute the corporation's VNR via satellite to television stations across the country. The obvious advantage is the almost instant access to every broadcast station simultaneously.

Corporate Video Departments

The organizational structures of corporate video departments are as varied as the companies they serve. Some are part of larger training and educational departments, others belong to PR divisions, and still others report directly to the organization's president or CEO. Some departments create very professional programs that use computer-generated graphics, professional actors, and the highest-quality production equipment. Others make due with antiquated television cameras and recorders and rely on employees to serve as talent. Some produce all programs in-house, using the facilities that the corporation owns, whereas others develop the program script in-house but rely on commercial production houses to convert the script to an actual video program. The structure and resources of the corporate video department should be determined by the complexity and number of the programs it is required to produce. Usually, however, the corporate video department consists of only one or two employees with a very tight operating budget, and inexpensive and dated production equipment.

The Future of Corporate Video

As corporate video reaches maturity, some trends point to new directions in the business. First, although more corporations are becoming interested in video communications, they are reluctant to invest in even the low-end video production equipment needed to furnish a full-service television studio and post-production facility. Second, the number of personnel hired to staff a corporate video facility is shrinking. The bygone era of specialization that allowed producers to produce, editors to edit, and videographers to shoot pictures is over. The people who are hired today are expected to be generalists with the necessary skills to create the program from script to screen.

Third, the tools of the trade are increasingly computer based. Nonlinear desktop video is replacing the videotape editors, special-effects amplifiers, and tape machines of the past. One producer sitting at a computer will be able to generate, sequence, and edit much of the material that will make up the video program of the future. Fourth, corporations will rely more on freelance talent hired to create specific images and communication messages than on full-time employees. Corporate video production will become a cottage industry dominated by freelance producers operating out of their homes. Fifth, there will be an ever increasing demand for electronic messages. With many employees moving to computerized workstations, a geographic dispersal of facilities to locations worldwide, interactive video teleconferencing, and computer-based multimedia information systems, there will be a need for employees with computer and telecommunication skills as well as for people who can write and create effective messages. The video producers of the future will need to be computer literate.

Distance Learning

The use of video for educational purposes has a long and interesting history: It could be called the granddaddy of corporate video and the father of public broadcasting. Its precursor was called *instructional television*.

History of Instructional Television

In 1917, several years before the advent of commercial broadcasting, the University of Wisconsin was operating an experimental radio station. That station, 9XM (later WHA), still serves the Madison area along with sister stations WHA-TV and WERN, an FM outlet. Many colleges and universities established radio stations as laboratories for electrical engineering departments, but few stayed on the air during the Depression years. Several land-grant colleges did view radio as a way to deliver agricultural instruction to farmers. In the 1930s the Universities of Wisconsin, Michigan, Kansas, and Minnesota all broadcast educational radio programs to the citizens of their respective states. The Ohio School of the Air was established by the Ohio State Board of Education and broadcast over the superpower commercial station WLW in Cincinnati. Although educational radio had its supporters, it never received the financial support needed to reach its potential as a means of distribution for instruction. Neither the Federal Communications Commission nor its predecessor the Federal Radio Commission acted to reserve channels for educational AM radio.

Like radio, television was an object of interest for many university electrical engineering departments. The State University of Iowa is often given credit for establishing the first educational television station in the early 1930s. Once the technology was developed, it was only natural for the university to turn to professors to provide content to send over the new communication system. Professors did in front of the camera what they knew how to do—they taught. The presentation of the content did not take advantage of the medium, and the programs were often very boring, consisting of only a "talking head" in front of a gray drape.

Although few universities had the money to build their own television stations, many provided instructional programs over the local commercial stations that were established in the late 1940s and early 1950s. The University of Michigan provided one of the first telecourses over Detroit station WWJ-TV. Western Reserve University (now Case Western Reserve) in Cleveland was the first university to offer a televised course for college credit. Soon many commercial television stations were offering college courses; called "Sunrise Semester" programs, they were broadcast in the early-morning hours before the commercial schedules began.

Major financial support for the early experiments in educational television came primarily from the Ford Foundation, which dumped hundreds of millions of dollars into various instructional television programs during the 1950s and 1960s. The foundation created two funds for educational television. One was to provide formal in-school instruction via television during the cold war era, when the United States was thought to be losing the space race. The then Soviet Union was able to gain the upper hand by launching a satellite before the United States did. Television was to bring the scientific and mathematical wisdom of master teachers into the rural classrooms of the nation. The second fund was to provide continuing education for adults.

In the area of instructional television, several interesting projects were undertaken. For example, the Chicago Board of Education established station WTTW (Window to the World) to provide all the junior college courses necessary to obtain an associate of arts degree. A student in the station's coverage area could obtain all the credits necessary simply by watching classes on television and passing the appropriate tests.

The fund also provided the money to establish several instructional *closed-circuit television (CCTV) systems*. In cooperation with AT&T and the Ford Foundation, the Washington County Board of Education constructed a wired distribution system that connected classrooms in the county with production studios in Hagerstown, Maryland. Six studios allowed live master teachers to reach county students on six different channels at the same time.

South Carolina established an in-school closed-circuit distribution system to reach every classroom in the state, and CCTV was used as the primary instructional delivery system in an experiment in American Samoa. Both were funded by the Ford Foundation. Probably the most interesting and costly of the foundation's experiments was the Midwest Program on Airborne Television Instruction (MPATI). Long before communication satellites became possible, engineers knew that a television signal broadcast from a high altitude would cover a greater area than one broadcast from closer to the ground. MPATI engineers stripped the interior out of two four-motor propeller-driven commercial airliners and installed two television transmitters in each, along with associated power systems and videotape recorders. The airplanes then flew in a figure-eight pattern over central Indiana, broadcasting instructional programs to classrooms in eight surrounding states. Bad weather, tricky engineering problems, and huge fuel and maintenance costs made the MPATI system unreliable and very expensive. Although it quickly became apparent that the costs far exceeded the benefits, the system continued for several years. Even after the airplanes were grounded, the instructional videotapes were distributed to educational television stations by the Great Plains National Instructional Television Library in Lincoln, Nebraska.

The early instructional CCTV systems served as demonstration projects and encouraged many school districts to jump on the *instructional television (ITV)* bandwagon. By the early 1960s, more than five hundred CCTV systems were estimated to be operating in school systems. The equipment for many of these was funded by a variety of federal government sources. One system was constructed by the Hancock County Schools in Weirton, West Virginia, with the aid of funding from a local bond issue and the National Defense Education Act. In this system master antenna television distribution systems allowed television receivers in each classroom of the seventeen school buildings to be connected via cable to a central studio. In Pittsburgh, most of the in-school programs were broadcast to classrooms from the educational television station WQED-TV, but some were produced locally with the broadcast-quality production equipment that the school district owned. Although this closed-circuit system was similar to many operated by school districts, several attributes allowed it to stand out as pioneer. First, it was well funded, operating with a full-time staff of five as well as the assistance of about fifty high-school student interns. Second, it had a close working relationship with the local cable television system, which allowed one of its three channels to be delivered to the homes that subscribed to cable; thus the district was operating an educational-access channel almost ten years before the FCC designated such a channel. With the aid of production support, the system covered most community events and provided to the community more than forty hours of scheduled, locally originated programs each week. Third, the system offered one of the first rudimentary dial-access video retrieval systems in the country: A teacher could choose a videotape, film program, or instructional unit from a large catalogue of those housed in the system library, and within five minutes of receiving the phone order, the material would be presented on a television receiver assigned to a particular classroom. Although the system was manual, a proliferation of telephones, cable channels, television receivers, and videotape and film playback equipment created a system that allowed convenient customized use of instructional video materials in the classroom.

Television for instruction was also used in higher education. College courses for credit were telecast to adults who were not part of the regular college environment. Closed-circuit television was later used for direct instruction in the college classroom; it was seen as a way to provide enough class sections to meet the needs of the baby boomers who inundated the higher-education system in the 1960s and 1970s. By the early 1980s, instructional CCTV was out of favor, but some colleges and universities were again searching for ways to deliver instruction to off-campus students. The adult learner and the nontraditional student were becoming more significant to university administrations. Instructional broadcasting had come full circle as the concept of distance learning took hold at the university level.

Many universities had long provided correspondence courses for military personnel and others who lived too far from campus to attend as regular full-time students. As the nation's demographics aged and technological advances demanded continuing training, universities looked for ways to serve new markets. Telecommunication technologies were viewed as a way of providing direct instruction to the distance learner. Broadcasts on public and commercial television stations could no longer meet the needs of distance learning, because they could air only one lesson at a time, the student could not respond because the channels were not interactive, and, most important, broadcast time had become extremely expensive. Educators experimented with other communication technologies.

Some universities put distance-learning classes together via telephone. With an audio bridge to connect the students' telephones, the professor taught the class over the phone on a party line. A major disadvantage of this system was the students' inability to see examples the instructor might offer. Other systems used computer networks to connect students and teachers on a real-time basis, but this system required that each student have not only access to a terminal but also the knowledge to use it effectively.

Most early distance-learning systems were developed around television technology so that the advantages of sight, sound, color, and motion might all be used to assist in the learning process. Inexpensive professional television cameras were mounted on movable pan-and-tilt heads in a college classroom. The light level of the room was increased slightly so the

A studio/classroom, the site of traditional classroom instruction that is also broadcast to remote locations. The camera at left images the instructor. The camera above the blackboard is positioned to broadcast graphics placed on the desk; TV monitors allow students in the classroom to see the same graphics. Individual microphones allow questions from students in class to be heard by viewers at remote locations.

camera could function better, and a microphone was provided for the instructor and students. Classroom monitors showed the students pictures from the cameras or from ancillary sources such as videotape players, film chains, or computers. The audio and video signals were also fed from a control room to a distribution system, which could be one of several types: closed-circuit via coaxial cable to another classroom or to a more distant student via a private channel on a cable television system; by an *instructional television fixed service (ITFS)*—a low-power, private broadcast system; or to a satellite uplink for distribution to a national or international student population. A telephone link allowed students to ask questions of the instructor.

During the 1970s the high-maintenance instructional CCTV systems that delivered direct instruction to the classroom disappeared. They were replaced by inexpensive portable videocassette recorders that the classroom teacher could use on an occasional basis to provide an enrichment program for the class. The VCR took the place of the 16mm film projector, and teachers began assembling personal libraries of videotaped materials. The inclusion of the VCR in the classroom returned control of the media to the teacher.

School television studios increased in numbers during this time because of the development of inexpensive equipment and the willingness of cable television operators to donate production equipment to educational institutions. The function of the studio became a training facility for high-school students learning television production techniques. Many of these studios were connected to a cable television educational-access channel. The programming found on these channels often consisted of live coverage of board of education meetings, school sporting events, and student-produced newscasts. These programs were more attuned to the public relations goals of the school rather than to its instructional mission.

During the 1980s the personal computer (PC) was touted as the instructional aid that would turn around the faltering U.S. public education system. The instructional television equipment of the 1950s and 1960s disappeared into closets to make room for the computer terminals of the 1980s and 1990s. Video is again becoming an important instructional tool in the classroom, but the twenty-first century version, displayed on the student's terminal, allows the student to interact with the program by selecting specific clips of information to meet the particular objectives of the lesson, and includes graphics, vivid color, fast motion, and phenomenal sound effects. In the new classroom system, the **multimedia** hardware is contained entirely in the PC and is controlled totally by the student. Its software is produced, tested, and marketed by an educational resource firm. Yesterday's textbook publisher will distribute its product on CD-ROM today and on the Internet or a computer chip tomorrow.

The communication satellite system offers a technology that might improve the educational programs of the public schools. The U.S. Department of Education has provided funding through its Star Schools Program to establish educational telecommunication partnerships needed to deliver instructional programs to schools. For example, Northern Arizona University currently provides direct instruction in Spanish to grade-school students over the satellite system. Oklahoma State University offers credit courses in German to high-school students via satellite.

DISTANCE-LEARNING TECHNOLOGIES

Although many public television stations continue to broadcast instructional television programs into elementary classrooms, in most cases they are meant to serve as enrichment, not direct instruction. Instructional television is not going to replace the classroom teacher in the public schools.

DISTANCE LEARNING TODAY

Arizona State University operates a distance-learning system that has been used as a prototype for many recent systems. It was established in the early 1980s to deliver engineering instruction to the many high-tech telecommunication manufacturing companies located in the Phoenix, Arizona, area. Companies such as Honeywell, Motorola, and Intel needed their engineers to continually upgrade their skills, because knowledge was expanding so quickly in that area. Rather than send them back to school, the companies asked for the instruction to be delivered to their facilities. The university responded by establishing a four-channel ITFS system that allowed the simultaneous delivery of four classes to subscribing companies located within

ITFS control room, Arizona State University. Operators control the television signals coming from two adjacent studio/classrooms.

about twenty miles of the university. Because the low-power broadcast system operates in the 2,500 MHz range, an antenna and converter are needed at each receiving site to enable the signal to be shown on a regular television set. An on-site telephone makes the system interactive, and a daily courier service delivers printed materials and picks up homework and exams from the sites. Four studio/classrooms, complete with control rooms, three remote-controlled cameras each, monitors, computers, and associated audio and video control equipment allow everything that is said or done in the classroom to be sent to the remote sites.

The Arizona State University distance-learning system now serves about four thousand students per year with more than eighty different courses each semester. In addition to courses delivered over the ITFS system, noninteractive classes are provided over cable television and the local PBS station. Interactive courses are also offered, via cable, wireless cable, microwave (two-way video), and satellite. Arizona State University is also a member of National Technological University, a consortium of universities that provides a complete engineering curriculum to a national student base by selecting specific classes from more than forty universities and delivering them to students via communication satellite. Noncredit seminars are also delivered via satellite to corporations.

Many distance-learning courses are offered over the Internet. The Web-based courses are asynchronous in nature. A student with an Internet connection can interact with a professor and other students on individual schedules.

Over the past two decades, universities have developed distance-learning departments to expand the schools beyond their campuses. Many were originally developed to distribute engineering or business classes, but schools now use distance learning to distribute coursework in many other disciplines.

The Public Broadcasting Service (PBS) has developed its Adult Learning Service to provide distance-learning opportunities through the public broadcasting channels. National Technological

University distributes engineering programming to students via satellite, and the Mind Extension University offers college classes for credit over cable television systems. The Open University, based in Great Britain, offers college courses in many countries, including the United States, via the Web.

With the continuing development of new electronic delivery systems, it is expected that more and more nontraditional students will choose to obtain part, if not all, of their college education through distance learning. The classroom of the future might be the student's own living room, and the instruction may be delivered through a for-profit corporation rather than a public or private university.

SUMMARY

▶ Many corporations are now using video produced in their own studios to communicate with their employees and customers as well as the general public.

▶ Corporate uses of video can be either internal or external. Internal uses include employee training, employee information, and employee motivational programs, as well as documentation of company events or engineering tests. External communication is directed to persons outside of the organization. Most external communication is either used for sales and marketing purposes or directed to the general public to enhance the image of the corporation.

▶ Many corporations are now including VHS videotapes in their product packaging to serve as the operations manual for instructing consumers in the proper use of the product.

▶ As television production equipment becomes digital and computer-based, it is expected that more corporations will use desktop video to provide both internal and external visual communications.

▶ Television used for direct instruction by schools, colleges, and universities has a long history. Although most school-owned television facilities are no longer used for sending direct instruction into the classroom, many public stations still offer some classroom instruction directed to the public schools. Most of this material is now used as classroom enrichment, rather than as an integral part of the curriculum.

▶ Many colleges and universities have developed distance-learning systems to deliver classroom instruction to the nontraditional student who lives or works beyond the geographic bounds of the university campus. A variety of delivery systems, including instructional television fixed services (ITFS), coaxial cable, telephone lines, satellite, and the Internet, will allow the student of the future to telecommute to class.

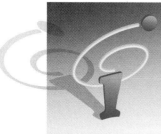

INFOTRAC COLLEGE EDITION EXERCISES

Most of us think of broadcast or cable programming when we think of television, but television (and video) has long been used to deliver instruction and other non-broadcast visual messages.

12.1 Using InfoTrac to access *Educational Technology News* or *Technology & Learning*, find examples of new ways of using communication technology in distance learning.

    a. Find similar articles on distance learning via the Internet in computer journals such as *PC User, PC Week,* and *PC/Computing.*

    b. Based on your reading of articles found through InfoTrac, what do you think the future of distance learning is?

    c. Based on your readings, describe the "desktop university" of the future.

12.2 Use InfoTrac to access articles in *Information Technology and Libraries*.

    a. Compare today's access to information to that of twenty years ago.

    b. Find articles about the use of video or computer technology used to train employees.

WEB WATCH

Here is a list of a few URLs (Internet addresses) for some of the organizations or corporations discussed in this chapter. Please explore these Web sites and follow the links to learn more about the complex business of the electronic media. Add your descriptions and your own favorite sites at the end of the list. Please keep in mind that the dynamic nature of the Internet allows sites to come and go but also allows organizations to update information about themselves very quickly.

Address	Description
http://www.itva.org/	
http://www.itfs.org/	
http://www.dlt.asu.edu	
http://www.pbs.org	
http://www.gwu.edu/~cats/	
http://www.gwu.edu/%7Egwutv/	
http://www.petersons.com/dlearn/	
http://www.usdla.org/	
http://www.dlnetwork.com	
Other favorite sites:	

CHAPTER 13

Audio and Video Systems

The future of radio and television certainly lies in digital technology and computer applications.

lmost all of us have become accustomed to coming home after a hard day at school or work and turning on the TV set. We can instantly see and hear news events as they happen, taking place around the corner or in distant parts of the globe. We can watch dramatic programs or comedy. Advertisements for new products are available day and night and are ballyhooed with all the glitz and glamour of the opening of a Broadway musical. If nothing on broadcast radio, television, or on the multichannel cable system suits our fancy, there is always the cassette we can slip into the VCR to watch a favorite movie.

How does it work? From where do the sound and pictures come? How can we see and hear them? A generation ago the public was amazed that a machine could be devised that would allow us to watch movies in our living rooms. A generation before that, people were awestruck over the ability to hear concerts in the home or see images move on the motion picture screen. Television, which today is commonplace worldwide and taken very much for granted, was the equivalent of the eighth wonder of the world to our grandparents. And the technology that makes it all possible is considerably more high-tech today than it was at its inception just seventy years ago.

Video control room. This area requires complex electronic equipment to select, process, store, and manipulate the audio and video signals required to produce a television program.

Tools of the Trade

This chapter describes the inner workings of audio and video technology so that we can understand the tools that telecommunication industries use to entertain and to inform us. The radio and television transmission chains are broken down into categories of electronic equipment and described in terms of function.

These categories (see table 13.1) include the following:

- *Encoding equipment,* used as a transducer to change light and sound energy into an electronic signal; the primary instruments in this category are the microphone and the video camera, although the computer has recently been developed as a device to create audio and video signals

- *Electronic manipulation equipment,* including audio control boards, digital video effects generators, special-effects generators, video switchers, time base correctors, and monitoring instruments such as waveform monitors and vectorscopes, as well as speakers and equalizers

- *Information storage equipment,* including audiotape recorders, photographic film projectors, videotape recorders, videocassette recorders, optical laser disc players, and audio and video computer servers

- *Signal transmission equipment,* including the radio or television transmitter, microwave systems, coaxial cable, fiber optics, and satellite communication systems

- *Presentation or display equipment,* including the radio tuner, speakers, television receiver, the video monitor, and the video projector

The technical operation of the equipment used in the radio or television transmission chain is complex and based extensively on physics theories. It is assumed here that most students in the broadcasting business will not go on to become electronics engineers, so the

TABLE 13.1	Categories of video equipment			
ENCODING	**MANIPULATION**	**STORAGE**	**TRANSMISSION**	**DECODING/DISPLAY**
Video camera	Video switcher	Videotape recorder	Broadcast transmitter	Television receiver
Character generator	Special-effects generator	Videocassette recorder	Antenna	Video monitor
Computer graphics system	Digital video effects generator	Videodisc recorder	Modulator	Video projection system
	Distribution amplifier	Still-frame recorder	Coaxial cable	
	Time base corrector	Film projector	Fiber-optic cable	
	Waveform monitor	Computer servers	Microwave	
	Vectorscope		Communication satellite	
	Edit controller			

information in this chapter is greatly simplified but detailed enough to allow the novice to understand how electronic communication systems operate.

After examining how electrons in microphones, television cameras, monitors, speakers, and tape recorders work to form sound and pictures that can be sent through the air, this chapter describes the electronic tools used to create programs. This chapter is but a preview of the fascinating process used to create messages in a broadcast format.

Properties of Sound

Sound. It is a form of physical energy that is virtually always present. From the deafening roar of a jet engine to the almost inaudible sound of air moving in a room, sound is a consistent element in our lives. Total silence is rare; our ears are constantly reacting to a variety of sounds. Much of the sound that we hear every day comes to us through radio and television. Through these media we are bombarded with words, sounds, and music designed to inform, entertain, or persuade us. We are sophisticated media consumers who take for granted a vast array of programming available in full color with stereo sound, listening to our favorite music on compact discs and on the radio at the touch of a button. But how are the voices and music we hear on radio and television produced and transmitted?

Sound, like many other forms of physical energy, is created through vibration—a vibration must be set in motion to create sound. For example, when someone speaks, air is forced from the lungs through the vocal chords. Because the vocal chords are stretched tightly across the throat, they vibrate when air is forced between them. This vibration or pattern can then be transferred through the mouth to the air. This occurs because air is elastic—that is, it will compress and expand if enough physical pressure is applied.

This property of air allows sound to be carried outward from the source of vibration in wave patterns, similar to the ripples created when a stone is thrown into a pond. When the stone enters the water, it sets in motion an energy pattern that can be seen as waves radiating outward in all directions from the source. This rhythmic wave pattern is called *oscillation.* Sound also travels as oscillations like the ripples in water. Sound requires a medium in which to travel—it will travel through air, water, and some solids, but it cannot travel through space, because in a vacuum there is nothing to transfer the vibration pattern. Even physical objects can transfer sound if enough energy is applied to them: Sound will travel through the wall of a room if it is loud enough; that is, if it has enough strength to cause the wall to vibrate and transfer the energy to the air in the next room.

FREQUENCY AND AMPLITUDE

Sound waves travel in patterns or cycles. One complete cycle of the wave includes a positive and a negative phase, as illustrated in figure 13.1. Every sound has a unique pattern that differs from other sounds in frequency and strength. The *frequency* is the number of cycles the wave goes through in one second, hence the term *cycles per second*, more commonly referred to in the electronic media as *Hertz (Hz)*, in honor of German scientist Heinrich Hertz, who discovered radio waves.

Frequency determines the *pitch* of a sound: As the frequency of the wave increases, so does the pitch. This can be most easily demonstrated by the notes on any stringed musical instrument, such as the piano. The lower notes on the piano are generated by longer, thicker wires as they are struck by the hammers connected to the keys. The longer wires vibrate more slowly than the shorter wires, so they sound lower in pitch. The lower notes also generate longer *wavelengths*, measured by the distance from the beginning of one cycle to the beginning of the next. Frequency (and pitch) are inversely related to wavelength: The longer the wavelength, the lower the pitch; shorter waves vibrate faster, producing a higher pitch.

Sound waves also vary in *amplitude,* or strength. The human ear perceives amplitude as loudness, shown by changes in the height of the peaks in a sound wave (see figure 13.1). As more energy is applied to the vibration creating a sound wave, the amplitude or loudness will increase. As you turn up the volume control on your boom box or on the sound system in your car, the amplification or power is increased and the music sounds louder.

As shown in figure 13.2, the human ear is capable of hearing sounds that range in frequency from about 20 Hz (cycles per second) to about 16,000 Hz (Alten 1999). When a sound wave penetrates the ear, it strikes the ear drum, a membrane similar to a drum head. The ear drum picks up the unique vibration of a particular wave and transfers it to the brain. In this process the physical energy of the sound wave is changed or *transduced* to the electric energy of a brain wave. We perceive this energy as sound. It is very important that the ear drum transfer the pattern of the sound wave as accurately as possible so that none of the unique information carried in that wave will be lost.

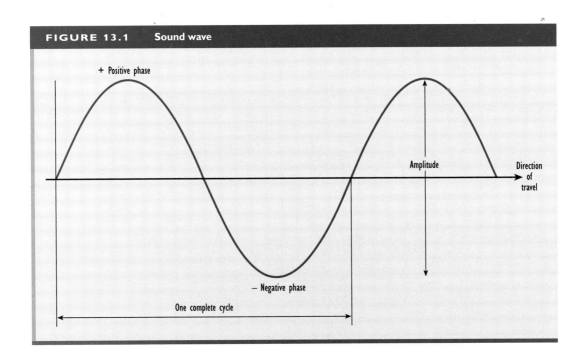

FIGURE 13.1   **Sound wave**

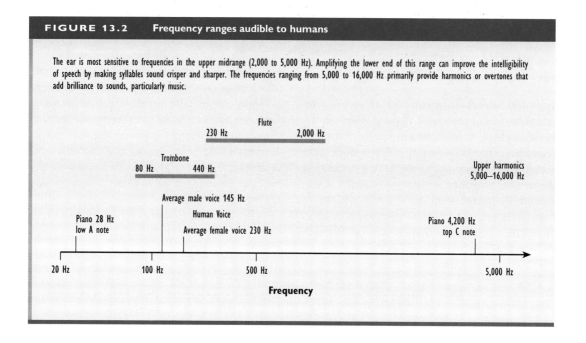

FIGURE 13.2 Frequency ranges audible to humans

THE SOUND PATH

The *sound path* is simply a vehicle for carrying sound information from one point to another. Sound waves become the audio signal in radio and television programs. The path that sound takes from the point of origination—the studio—to the point of destination—the receiver—begins with the sound of a radio announcer's voice, which is created by vibration of the vocal cords. As the announcer reads news stories from a script, the sound of a voice is produced in the studio. Anyone sitting in the studio with the announcer would hear the voice because their eardrums would pick up the physical pattern and transfer it to their brains. But how can the announcer's voice travel from the studio to radio receivers in homes or in cars? The path traveled involves five steps and a variety of audio equipment (see table 13.2).

TABLE 13.2 Categories of audio equipment

GENERATION/ ENCODING	PROCESSING/ MANIPULATION	STORAGE	DISTRIBUTION/ TRANSMISSION	RECEPTION/ DECODING/ DISPLAY
Microphone	Audio control console	Turntable	Broadcast transmitter	Radio receiver
Computer	Distribution amplifier	Audiotape recorder	Twisted-pair wire	Amplifier/ speaker
Computer/ MIDI	Equalizer	Audiocart recorder	Coaxial cable	
	Edit controller	Audiocassette recorder	Fiber-optic cable	
		CD player	Microwave	
		Computer servers	Communication satellite	

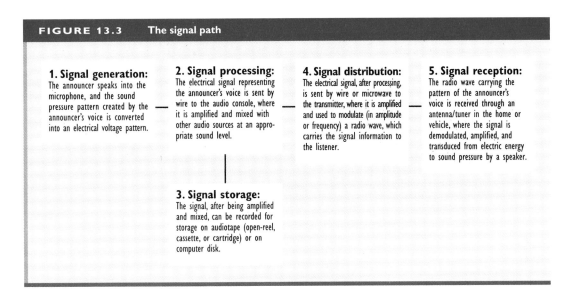

FIGURE 13.3 The signal path

1. Signal generation: The announcer speaks into the microphone, and the sound pressure pattern created by the announcer's voice is converted into an electrical voltage pattern.

2. Signal processing: The electrical signal representing the announcer's voice is sent by wire to the audio console, where it is amplified and mixed with other audio sources at an appropriate sound level.

3. Signal storage: The signal, after being amplified and mixed, can be recorded for storage on audiotape (open-reel, cassette, or cartridge) or on computer disk.

4. Signal distribution: The electrical signal, after processing, is sent by wire or microwave to the transmitter, where it is amplified and used to modulate (in amplitude or frequency) a radio wave, which carries the signal information to the listener.

5. Signal reception: The radio wave carrying the pattern of the announcer's voice is received through an antenna/tuner in the home or vehicle, where the signal is demodulated, amplified, and transduced from electric energy to sound pressure by a speaker.

Signal Generation

The first step in this process is *signal generation* (see figure 13.3). To prepare the announcer's voice for radio broadcasting, the sound wave must be converted or transduced into another form: electricity. Just as the human hearing system converts sound into electrical brain waves, the microphone converts sound into electrical voltage patterns. The microphone, or mic, must maintain the unique pattern of the sound wave to transmit it accurately. The microphone, then, is the first in a series of transducers that the announcer's voice encounters in the sound path of radio broadcasting.

Microphones accomplish their functions as transducers through a variety of structures. The three basic types of microphones are the moving-coil, the ribbon, and the condenser (or capacitor) (see figure 13.4).

The *moving-coil microphone* has a coil of wire suspended in a magnetic field. Attached to the coil is a diaphragm similar in function to the human ear drum. As sound pressure strikes the diaphragm, the coil moves with the pattern of the wave. This movement within the magnetic field generates a weak electric current. The form of energy is changed, but the pattern of the wave remains the same.

The *ribbon microphone* has a thin strip of metal, rather than a coil, suspended in a magnetic field. The metal ribbon vibrates when struck by the pressure of a sound wave and, again, the movement within the magnetic field creates a weak electric current. The ribbon mic is more sensitive to sound than is the moving-coil type. It also tends to emphasize the bass frequencies of sounds, especially as you move closer to it; this is called the ***proximity effect***. Hence, it tends to provide a fuller, warmer sound, but it also is more susceptible to damage from air pressure or sudden changes in volume in a sound wave.

The third type of microphone is the *condenser* or *capacitor*. It is different from the first two types of mics because it is electrical; the first two are mechanical. The condenser mic requires a power supply; it has to be plugged in because it uses two plates, one of which carries an electric charge, rather than a moving coil or a ribbon. The charged plate, or capacitor, moves in response to sound pressure entering the microphone. The variation of its distance from the back plate, which does not move, produces a variation in the electric charge. This variation mirrors the pattern introduced by the sound wave. Again, sound energy has been transduced into electric energy.

Condenser microphones are inherently more sensitive than both the moving-coil and the ribbon types and are very common. These mics use a small battery to charge the capacitor instead of an external power supply. Their smaller size and more durable construction allow for their use in a variety of on-location applications.

Microphones also vary in type according to directional characteristics. Each microphone has a specific ***pickup pattern*** in which sound can be "heard"

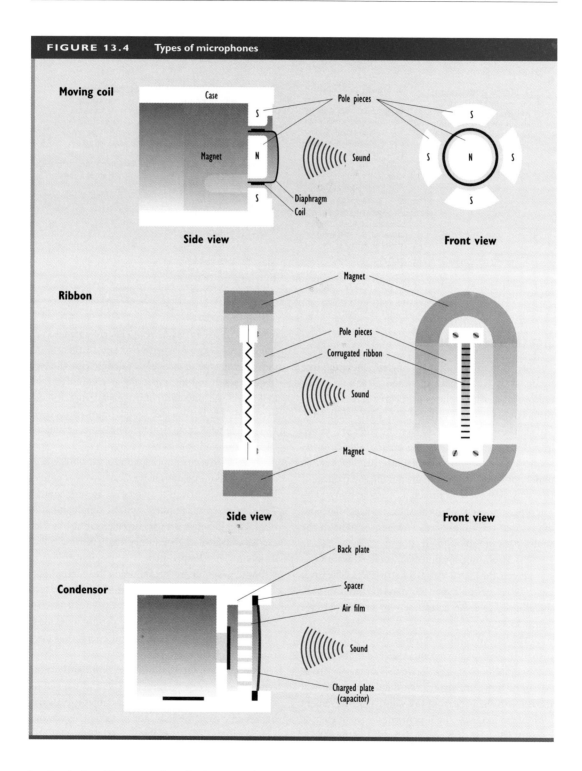

FIGURE 13.4 Types of microphones

by the device. There are three basic microphone patterns: omnidirectional, bidirectional, and unidirectional or cardioid (see figure 13.5).

Omnidirectional microphones, as the term suggests, pick up sound equally from all sides. Most moving-coil mics have omnidirectional patterns. They

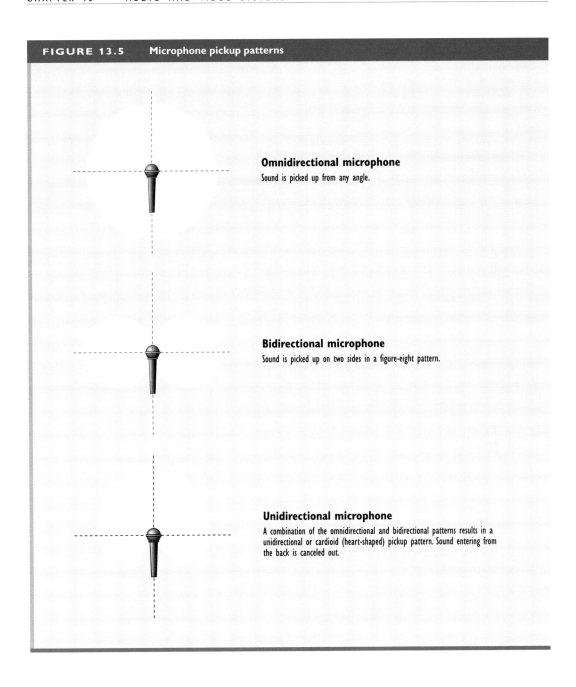

FIGURE 13.5 Microphone pickup patterns

Omnidirectional microphone
Sound is picked up from any angle.

Bidirectional microphone
Sound is picked up on two sides in a figure-eight pattern.

Unidirectional microphone
A combination of the omnidirectional and bidirectional patterns results in a unidirectional or cardioid (heart-shaped) pickup pattern. Sound entering from the back is canceled out.

are versatile because they will pick up sounds from many directions, but they will also pick up unwanted noise when used on-location. These microphones will not isolate noise from the desired primary sounds. Still, this pattern is common in very small electret condenser microphones used in television applications. These microphones, called *lavalieres,* are clipped to the lapel or collar of a newscaster on-camera. In a controlled studio environment, they are unobtrusive and pick up the newscaster's voice very well.

The other two types of microphones, bidirectional and cardioid, allow more control of unwanted noise. In the *bidirectional* microphone, the pickup pattern is shaped like a figure-eight.

The *unidirectional* or *cardioid* pickup pattern is probably the most common in contemporary audio applications. It picks up sound primarily from one side in a heart-shaped pattern, hence the term *cardioid*. When picking up an announcer's voice on-location, the cardioid mic will emphasize the voice and minimize surrounding noises. Because of this control feature, this type of mic is most often used on-location, where noise can be a problem. An example of this type is the *shotgun* microphone often used for live coverage of sporting events.

Signal Processing

The second step in the sound path is *signal processing* or *mixing*. After physical sound energy was produced in the studio or on-location (the announcer's voice), it was transduced into electric energy so that the information could be processed electrically before being sent to the home or stored on tape. The signal-processing step occurs in the control room, where electronic equipment shapes and mixes the signals.

In the control room, the electrical signal representing the announcer's voice is sent by wire to the *audio console,* which controls the audio levels, or loudness, of each audio source. The announcer's voice is one audio source; others might include music on compact disc, cassette, or cartridge tape; other voices on tape or live via telephone, microwave link, or satellite; and sound effects produced live or on tape.

The audio console has three basic functions: input, output, and monitor. In the input function, each of the audio sources comes into the console and is assigned to a channel with its own volume control, or a *potentiometer* (commonly abbreviated *pot*). In older consoles these were rotary knobs, but contemporary consoles have vertical sliding controls called *faders* or *sliders*. Many audio production computer programs have slider icons on-screen that can be adjusted with a mouse.

Sources are amplified by the console, and the sound levels are mixed or set at the appropriate loudness level by an audio console operator. A single announcer's voice reading news is relatively simple to control, but when other audio sources are introduced the process becomes more complicated. Music can be added underneath the announcer's voice at a lower level. Other announcers' voices can be originated live or from tape and must be mixed to match the live announcer's voice level. The console operator is

Audio console in contemporary radio station.

creating a package of sounds at appropriate, matching volumes to send out to listeners.

The second function of the audio console is output or routing. Each channel must be amplified and routed by the console to various destinations. If the operator is mixing a live program, the signals are sent to the transmitter, where they are broadcast on a radio wave. If the program is being taped for later broadcast, the console can send the signal to a variety of tape machines for recording. Or it can do both at once.

The third function of the audio console is monitoring. To mix audio signals, the operator must be able to hear them in the control room, so the console feeds an amplified audio signal to studio monitor speakers. Console operators listen to the audio levels through these speakers or through headphones while talking to radio listeners on a microphone. They also monitor the program visually by watching *VU meters* (short for *volume-unit*) that show volume levels on a scale.

There are two general types of audio consoles: broadcast and production. The broadcast type is found in the on-air booth or control room and is used by music disc jockeys (DJs) to mix their voices with the other audio components of a radio program—music, commercials, listener call-ins, and promotional announcements; these originate from other audio sources in the control room, such as tape machines, compact disc players, telephones, and remote microwave links (a wireless connection between an announcer on-location and the on-air studio).

Each of the audio sources is assigned to a console channel, where volume levels can be controlled. The DJ continuously mixes the audio levels as the live radio program is being broadcast. In this situation the disc jockey is functioning as both operator and talent—a situation often referred to as a "combo" operation.

In other situations, such as at a talk radio station, the announcer sits at a microphone in the studio while an operator sits in the control room at the audio console. Because radio talk-show hosts are too busy managing listener calls to operate the audio console at the same time, they need a separate operator to control the levels of the various audio signals that make up the talk program.

The audio production console performs different functions from those of the broadcast console, because it is designed to meet the more complicated audio demands of the production of commercials and promotional announcements. Whereas the broadcast console generally has 8 to 12 channels, the production console has 16 to 24, which allows the input of many more audio sources, such as multiple microphones and a wide variety of music sources. The production console can send many signals to multitrack tape recorders, which assign each audio signal to a separate channel on the tape head (see figure 13.6). Thus, instead of having all of the signals assigned to only two channels—either left or right—as in a normal stereo signal, each of the sixteen or even twenty-four signals is recorded on a separate tape track. This allows the operator to mix the tracks after they have been recorded. This process is used in most music recording, where instrumental and vocal tracks are recorded separately and then combined in the final mix. Through a process called *overdubbing,* a recording can be begun with music tracks to be followed by vocal or announcing tracks at a later time. Examples are radio commercials with an announcer reading copy over a music background, or bed (referred to as a *voice-over*).

Audio signals can be processed electronically in other ways to achieve special effects. A graphic equalizer allows the operator to alter the announcer's voice by emphasizing the bass or treble frequencies. Echo or reverberation can be added to enrich the sound of the voice. The speed of the voice or other sounds can also be altered to achieve desired effects.

Audio Signal Storage Equipment

The third step in the sound path is *signal storage,* which takes place in the audio control room. The control room contains a number of other transducers. If the radio newscast must be recorded on audiotape to play back later, another transducer—the magnetic tape head—is used. This electromagnet converts electrical information to a magnetic pattern on the surface of audiotape. Once the announcer's voice is converted to electric energy, or voltage, the pattern can be directed to the tape head and stored by reproducing the pattern in the metal filings on the surface of the tape. When the tape is played back, the magnetic pattern is read by the playback head of the tape machine and converted back into voltage. This process occurs in cassette, cartridge, and open-reel tape recorders and on a hard drive of a computer.

All of the tape recorders in audio control rooms have three basic parts: the tape transport, the magnetic heads, and the record and playback amplifiers. The function of the *tape transport* is to move tape across the magnetic heads so that a signal pattern can be recorded or played back. This process must be performed at a constant speed so that there is no loss or distortion of the information contained in the signal pattern.

In the *magnetic head* mechanism, there are generally three heads: erase, recording, and playback. The tape runs across the heads in this order while recording, so any signal patterns that were stored on the tape earlier are erased by the first head. That clears the tape so that new signal patterns can be placed in the metal filings by the recording head, which takes the electrical signal originally created by the microphone and re-creates it magnetically on the tape surface. This signal pattern can then later be read from the tape by the third, playback, head. Its function is the reverse of that of the recording head: It converts the magnetic pattern back into electricity or voltage.

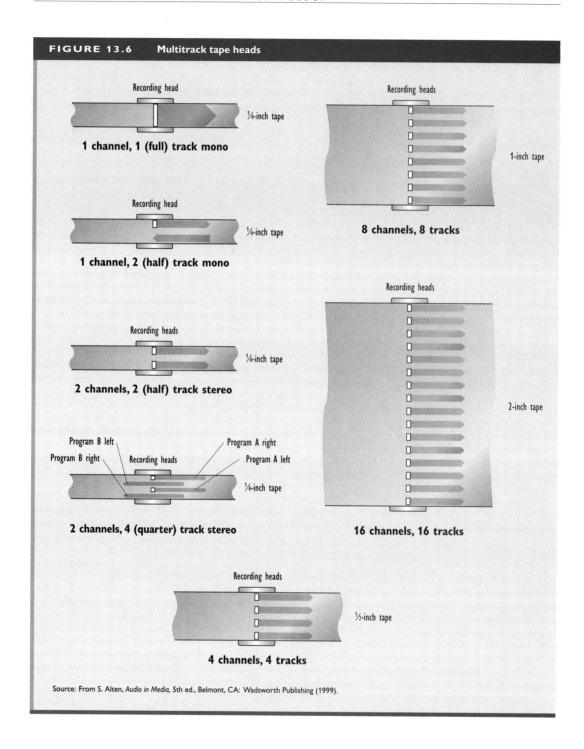

FIGURE 13.6 Multitrack tape heads

Source: From S. Alten, *Audio in Media*, 5th ed., Belmont, CA: Wadsworth Publishing (1999).

The third component of the tape recorder consists of the record and playback amplifiers. The *record amplifier* takes the audio signals coming into the tape recorder, varies the strength, and feeds them to the recording head. The *playback amplifier* takes the signals coming from the playback head, varies the strength, and feeds them to the output of the recorder. From there the signals can be routed back to the audio console, to another tape recorder, or to the transmitter to be broadcast to the listener.

Analog Versus Digital Transduction

So far we have been describing transducers that utilize *analog* technology—converting one form of energy into another by producing a copy, or analogy, of the original sound. Accurate reproduction requires an exact image of the announcer's voice in the studio so that it will sound the same when it emerges from the receiver. But producing an analog image of the original introduces a certain amount of *noise*, or error, in the transfer process. And the copies are subject to decay with time: Phonograph records become scratched and can wear out with use; audiotape also loses quality over time as magnetic patterns in the metal filings break down (commonly referred to as *tape dropout*).

These problems have been eliminated with digital technology, which uses an entirely different method of transducing energy. Rather than producing an analog of a sound, *digital* technology translates each part of the original signal into digits or numbers by *sampling* the sound (see figure 13.7). These digits are then coded into *binary* form using various combinations of 0's and 1's (or off and on).

The binary code is then read or translated back into electric energy by a laser beam. The result is an exact duplicate of the original. The most familiar example of this technology is the compact disc (CD). The sound quality of the CD is clearly superior to that of the phonograph record, because there is virtually no noise in the signal. Compact discs don't wear out like phonograph records and tapes, because there is no friction or wear involved as the laser reads the information on the disc; CD players do occasionally skip like turntables due to surface damage and also need regular realignment of the laser beam.

Digital technology also has been developed recently for tape recorders. Digital audiotape (DAT) is now available for both audio professionals and consumers. It applies the same functions to magnetic tape that are found in the compact disc and records as well as plays digital signals. Very small DAT machines record sound in digital stereo on a cassette tape the size of a large postage stamp.

Digital audiotape recorders sample the frequency response of the signal being recorded a

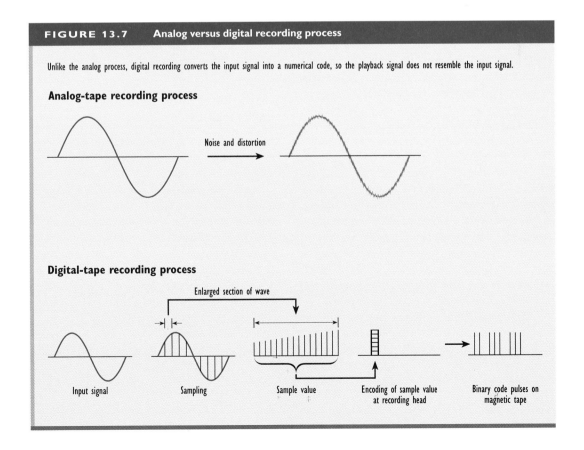

FIGURE 13.7 Analog versus digital recording process

Unlike the analog process, digital recording converts the input signal into a numerical code, so the playback signal does not resemble the input signal.

number of times every second. These sampling intervals generally vary from forty to fifty times per second. Because the samples represent only the frequencies of the signal and not those of noise introduced at other frequencies, digital audio recording is nearly noise-free. On the other hand, DAT does not reproduce very high frequencies (20,000 Hz) as well as does analog recording.

Many radio stations are now using digital technology, replacing tape-cartridge, or cart, machines with ones that use computer disks for storage. These digital cart systems store audio content such as commercials, promotional announcements, and musical selections directly on floppy disks. They can be cued automatically and played back on demand.

After the announcer's voice has been converted into electric energy, it is sent into the control room for processing, where it is mixed with other audio signals (music on tape for the introduction of the newscast, other news reports on tape or called in live on the telephone, and commercials on tape) in the right order and at the proper levels to produce a good radio newscast.

Signal Transmission or Distribution

The fourth step in the sound path is *signal distribution,* which is accomplished through wireless (broadcast) or wired (cable) distribution or a combination of both. In wireless distribution the vehicle on which the radio program is carried is the *radio wave.*

Radio energy is electromagnetic energy; that is, it consists of both electric and magnetic energy. Like other forms of energy, radio energy is generated through vibration or oscillation. In this case electromagnetic waves are generated by feeding an alternating electric current into a radio antenna. The alternating current, like that used in homes, creates oscillations that alternate between positive and negative phases. These waves radiate outward from the antenna in all directions in the same manner that sound waves radiate from their source (see chapter 14).

Radio waves differ from sound waves in a number of ways. First, because radio waves are electromagnetic (rather than physical like sound), they do not require a medium in which to travel. Sound waves will travel through air or water but not through space; radio waves will travel through air, water, *and* space.

Second, although radio waves are undetectable to the human senses, they are present virtually everywhere on the earth. As are sound waves, radio wave frequency is measured by the number of cycles per second (Hz) completed. Sound energy oscillates up to about 20,000 Hz (or 20 KHz), but radio energy oscillates much faster. Waves at the lower end of the radio portion of the electromagnetic spectrum oscillate at more than 500,000 Hz, or 500 KHz.

Earlier we described the process of transduction—changing one form of energy to another. We noted that it is very important to maintain the pattern of an energy wave when transducing it. When we changed the news announcer's voice from sound to electric energy, the unique pattern of the wave was maintained. In the control room, we amplified the electric energy and mixed it with other audio sources, but we did not change the critical information or pattern.

The radio program or signal must now be moved to the transmitter via wire or microwave. The transmitter will amplify the signal, generate a radio wave, and modulate or reshape the wave so that it matches the sound pattern of the announcer's voice.

Signal Reception, Decoding, and Presentation

The final step in the sound path from the studio to the home or car is *signal reception,* which is a reversal of the first part of the path.

The radio signal is in the atmosphere (and in the ground if it is an AM signal), so another antenna is needed on the receiving end of the path to pick up the signal. AM radio signals can be picked up more easily than can the signals of FM radio and television due to the strength of their ground waves. For good reception FM and TV signals generally require a whip antenna like those on cars or the rabbit-ears antenna configurations found on portable TV sets.

The antennas are connected to radio receivers that can be tuned to the specific frequency being generated

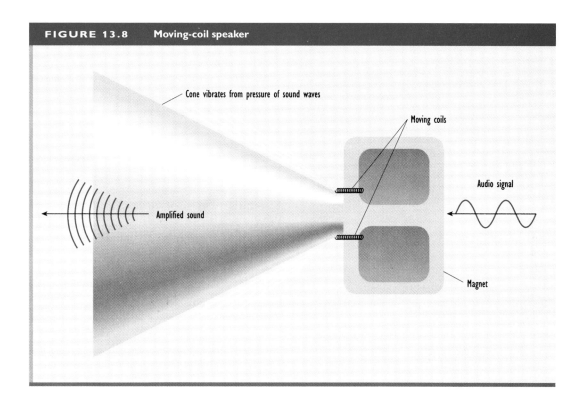

FIGURE 13.8 Moving-coil speaker

by the transmitter. By tuning a receiver to a particular frequency, the radio wave that carries the newscast is detected in the atmosphere. The process that started with the announcer's voice in the studio must be reversed: The audio signal must be separated from the radio wave by *demodulation*—eliminating the radio wave. What is left is the analog of the electrical signal created with the microphone in the studio.

Ideally, the characteristics that make the radio announcer's voice unique have been retained. But the audio signal has been weakened by its travel through the first three steps in the sound path, so it must be amplified. Amplification in a receiver, like that of a transmitter, is measured in *watts*. A receiver may have 100, 200, or even more watts of power to allow listeners to "pump up the volume."

Now that the audio signal has been received, demodulated, and amplified, it must be converted back into sound energy so that people can hear it. This function needs another transducer: the speaker (see figure 13.8). Like the microphone, it converts one form of energy into another. The speaker completes the sound path by converting electrical waves into sound waves audible to humans.

Like the microphone, the most common type of speaker uses a moving coil to convert energy, but the speaker works in the opposite direction from the microphone. The moving coil moves in reaction to electric current fed to it from the receiver-amplifier and is attached to a cone that, in turn, moves air. As the cone vibrates, it creates a sound pressure wave in the surrounding air. The sound wave pattern is the analog of the pattern created in the studio with the announcer's voice, and listeners hear the announcer's voice in their homes or cars much as they would hear it if they were sitting in the studio.

The miracle of this process is that it is virtually instantaneous. Because electric and radio energy travel at the speed of light (186,000 miles per second on the earth's surface), the announcer's voice reaches listeners at home at virtually the same time it reaches a listener in the studio.

Digital audio workstation. With this editing facility, audio signals are converted from analog to digital and displayed on the computer screen.

DIGITAL BROADCASTING

Radio receivers now utilize digital technology to improve signal tuning. The frequencies of radio stations are shown on a *liquid-crystal display (LCD)* in the same manner as in a digital watch. The stations tuned are indicated in actual numbers, such as 98.3 MHz, rather than as a point on the radio dial. *Digital audio broadcasting (DAB)* will allow radio stations to broadcast CD-quality sound in the near future. Currently, there is not enough spectrum space in each AM and FM frequency assignment to allow digital broadcasting: There is simply too much information, once the sound is sampled and converted to binary form, to fit into each radio station's space on the spectrum. Digital audio broadcasting will require a reallocation of frequency space by the Federal Communications Commission (FCC) for distribution on AM or FM. Currently transmitted by satellite or cable, when DAB is more widely available to listeners it will bring with it a variety of new audio services, such as *format finding,* a function that would allow the receiver to automatically find certain formats in a particular radio market. Each station will send an identifying digital code with its broadcast signal that will cue the receiver that it is a jazz or progressive rock station.

THE TELEVISION SOUND PATH

Now that we have traced the sound path for radio, let's consider the path of the audio signal in television. The steps in the television sound path are essentially the same as those for radio, but in TV the audio signal must match a video signal. Historically, sound played a secondary role to the visual aspect of television, but with the advent of stereo TV there is much more focus on sound quality today.

As in radio, the broadcast begins with the creation of sound in the television studio. Because the television newscaster is seen as well as heard, the microphone must not be a distraction in front of the newscaster's face. With its omnidirectional pattern, a small lapel-mounted lavaliere mic will pick up the newscaster's voice very well yet remain almost invisible on-camera.

Other microphones commonly used in television are handheld and shotgun mics. The handheld mic is usually a moving-coil or condenser *electret* type with a unidirectional pickup pattern. These mics are often used by TV reporters while doing reports on-location. They are held in the hand at chest height so as not to distract from the reporter's delivery. The unidirectional pattern allows the reporter to control surrounding, or *ambient,* noise on the scene of the report.

The shotgun microphone, as noted earlier in this chapter, has a directional pickup pattern that allows it to literally be pointed at a sound source. This factor is useful in television applications, because the shotgun can be placed at a distance from the sound source out of camera range. This mic can be used to pick up a reporter's voice or that of a person being interviewed on-location or in a TV studio.

As stated, the basic function of microphones is to convert sound energy to electric energy. They provide an audio signal from the studio or from the scene of a news story or remote production to be sent to the control room, where an audio console performs the same functions as it does in the radio sound path: It amplifies and mixes the signal with other audio sources. In the television control room, these other sources come from video- and audiotape as well as live audio feeds via telephone, microwave, and satellite. Once television audio has been mixed and processed, it is sent along with the video signal to the transmitter. When a signal is received in the home, the audio is separated from the video. In monaural TV receivers, the audio signal is amplified and sent to a very small speaker mounted to the side of or below the screen. The speakers convert the audio signal back into sound energy that viewers can hear as they see the video reproduced on the screen.

Now that stereo sound is more common in television programming, more attention is being paid to the audio portion of the signal. This is particularly true of the music programming offered by cable channels such as MTV and VH-1. On television music channels, the audio portion of the program has become the most important part of the signal. In music videos the video content provides visual support for the audio content which is, of course, the music.

Partly as a result of increased viewer demand for better sound, stereo receivers now offer viewers the option of two small speakers mounted in the set or audio output from the receiver that can be connected to a stereo sound system for more volume and frequency response. In fact, monitors that provide video only can be incorporated into modular audio systems that reproduce sound for both radio and television. In these systems high-quality speakers with bass, midrange, and treble frequency response complete the television sound path. But for the sound path to result in high-quality audio, a high-quality stereo-audio signal must be generated in the first step.

Many audio control rooms now feature at least one *digital audio workstation (DAW)* that incorporates computer technology with disk recording. Audio is recorded on computer disk rather than on magnetic tape that is pulled across a head mechanism. With these systems sound can be recorded directly onto the computer disk in digital form. Sound that has been recorded earlier in analog form can be converted and stored on disk as well. Once the audio information is stored on disk, it can be manipulated just as are other types of digital information.

Audio workstations convert all audio signals into digital data that can be displayed visually on the computer screen. This allows graphical editing of the audio material stored on the computer disk. To edit, the editor simply locates the sound, such as a cough or other unwanted noise, isolates it as one would text on a word processor, and gives the computer command to eliminate it. The change or edit is made digitally and then converted back to audio so there is no opportunity for a pop or a click to be introduced. These are *nondestructive edits* as well, meaning that the sounds that are removed are not destroyed. They are simply stored until the editor decides they aren't needed anymore. If the editor decides to put a sound back into the program, that can easily be accomplished through a restore command.

Digital audio can be manipulated by computer in other ways as well. Audio can be compressed without affecting the speed of the sounds heard. For example, if an announcer reads copy for a thirty-second radio commercial that is actually thirty-two seconds long, it could be corrected by reading again or by speeding up the tape. Speeding up the tape would affect the sound of the announcer's voice, because the pitch would change. But with digital audio, the commercial could be compressed into thirty seconds without affecting the sound of the announcer's voice in any way.

PROPERTIES OF VISION

The human eye is a complex receptor of a narrow portion of the *electromagnetic spectrum*—the range of all naturally occurring vibrations or oscillations of energy arranged by frequency (see figure 13.9). At the lower end of the spectrum, the vibrations are slower; at the upper end, they are faster. The visible portion of the electromagnetic spectrum includes only those oscillations in the range of 10^{15} cycles per second. Energy that oscillates more slowly (such as radio waves) or faster (such as X-rays) are invisible to the human eye.

Objects are visible as light waves are reflected off them, are concentrated, are focused on the eye's retina, and are detected by the rods and cones in the eye. That information is transmitted to the brain, where it is interpreted as an object that can be recognized as a car, a house, or a tree. The color information in the natural picture is dependent on the wavelength of the light energy detected by our eyes. The lower wavelengths are seen as blue, the higher ones as red. A combination of equal portions of all the various wavelengths of light is called white, and the lack of light reflected back to the eye is black.

In addition to detecting a variety of wavelengths of the electromagnetic spectrum with the eyes and interpreting those patterns as objects with the brain, the human being has the capacity to remember for an instant the "picture" that the eyes have received. This effect is called *persistence of vision,* and it can be tested by staring at the print on the page for a few seconds

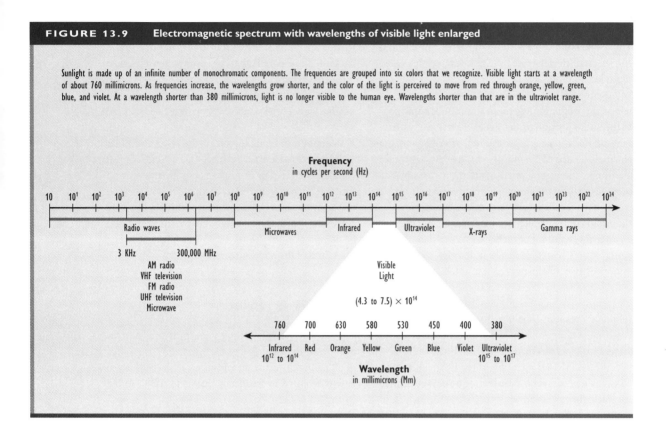

FIGURE 13.9 Electromagnetic spectrum with wavelengths of visible light enlarged

Sunlight is made up of an infinite number of monochromatic components. The frequencies are grouped into six colors that we recognize. Visible light starts at a wavelength of about 760 millimicrons. As frequencies increase, the wavelengths grow shorter, and the color of the light is perceived to move from red through orange, yellow, green, blue, and violet. At a wavelength shorter than 380 millimicrons, light is no longer visible to the human eye. Wavelengths shorter than that are in the ultraviolet range.

and then staring at the white space on the page. Most people probably will see a negative image of the print for an instant after they look away. You may have also experienced the effects of persistence of vision if you have inadvertently looked into a bright light for a second and then looked away; you probably remember seeing a black spot in your vision where the bright light had been. We also have the ability to mentally connect the slight differences between a new picture that we see and the one that remains in our persistence of vision. This ability to remember the picture for an instant is necessary for us to see a continuous picture on a television or motion picture screen.

At the turn of the century, Thomas Edison capitalized on the persistence-of-vision effect to develop the motion picture camera and projector. The electronic motion picture, or television, relies on the same series of individual still pictures and the persistence-of-vision effect, but the technology required to create the effect is considerably more complex.

Television, like all modern communication media, is based on the principle of electron flow. Certain elements—such as, gold, silver, copper, aluminum, and lead—have free electrons in the outer shell of the atomic structure (see figure 13.10). These elements, called *conductors,* have electrons that are easily dislodged from one atom and moved readily to a neighboring atom. Other materials, called *insulators,* do not have free electrons that can be easily dislodged. Some materials, called *semiconductors,* have properties that under some conditions allow free flow of electrons and under other conditions inhibit their flow.

If a copper wire is passed very quickly through a magnetic field, free electrons in the copper will move from one atom to another. By continuing that movement at a high rate of speed, an electric current is created. If the *polarity,* or direction of the current flow, is changed sixty times per second, we have created the sixty-cycle alternating current (AC) that is used to power the lights in our homes as well as our television receivers. The movement of electrons and the ability to control the direction and amount of that flow is the basis upon which all the electronic media operate: That same current, boosted con-

CHAPTER 13 AUDIO AND VIDEO SYSTEMS

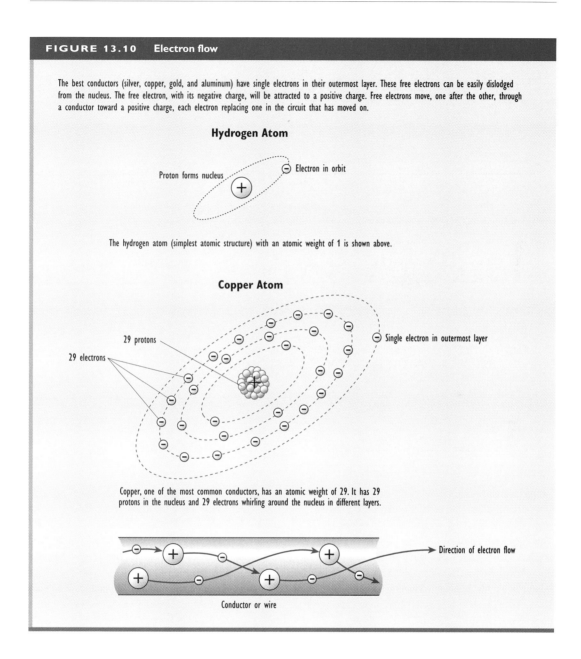

FIGURE 13.10 Electron flow

The best conductors (silver, copper, gold, and aluminum) have single electrons in their outermost layer. These free electrons can be easily dislodged from the nucleus. The free electron, with its negative charge, will be attracted to a positive charge. Free electrons move, one after the other, through a conductor toward a positive charge, each electron replacing one in the circuit that has moved on.

Hydrogen Atom

The hydrogen atom (simplest atomic structure) with an atomic weight of 1 is shown above.

Copper Atom

Copper, one of the most common conductors, has an atomic weight of 29. It has 29 protons in the nucleus and 29 electrons whirling around the nucleus in different layers.

siderably in frequency and in change of polarity and power, becomes the basis of the radio or television carrier signal that allows picture and sound to be broadcast from the station to your home.

Normally, we would examine the video process starting with origination and ending with the audience. For purposes of ease of understanding, however, we will instead consider the five categories of video equipment starting with display in the home.

DECODING TECHNOLOGY

THE TELEVISION RECEIVER

The final link in the television chain—the *receiver*—is the easiest place to start the description of how everything works. The most obvious part of the television receiver is the picture tube. If the case were to be removed from around a television receiver or a

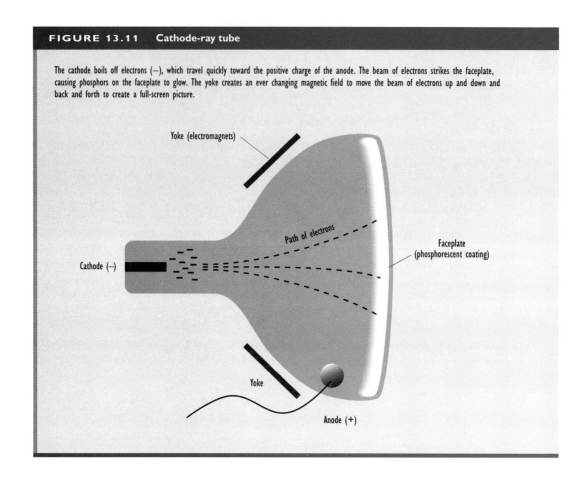

FIGURE 13.11 Cathode-ray tube

The cathode boils off electrons (−), which travel quickly toward the positive charge of the anode. The beam of electrons strikes the faceplate, causing phosphors on the faceplate to glow. The yoke creates an ever changing magnetic field to move the beam of electrons up and down and back and forth to create a full-screen picture.

computer monitor, the picture tube would look like a large glass bottle lying on its side (see figure 13.11). This is called a *cathode-ray tube (CRT)*. The semiflat bottom—the screen—is coated on the inside with a phosphorescent substance that glows when bombarded by electrons. Most color receivers have tubes that are coated with thousands of dots of red, blue, and green phosphors that glow when hit by electrons.

At the neck of the tube is a *cathode,* or small piece of metal, that gives off electrons when heated by an electric current; this is also called an *electron gun.* Electrons have a negative charge and can be attracted by a positive electric charge. If a very strong, positive electric charge, or *anode,* is placed near the faceplate of the tube, the cloud of electrons near the cathode will hurtle toward it. A stream of electrons traveling in the vacuum of the tube from the cathode toward the anode will strike the phosphorous coating on the inside of the tube, causing it to glow brightly. A single bright spot on the screen is not very useful, so the beam of electrons must be made to move up and down and from side to side to cover the screen. The process, called *scanning,* is created by a variable electromagnetic field that is applied around the neck of the tube by the yoke. The stream of negative electrons inside the tube can be bent by the charges in the magnetic field surrounding the tube. A device in the television station called a *sync generator* creates a series of horizontal and vertical pulses that are transmitted to the television receiver as well as to the cameras in the station. The pulses control the amount of magnetism needed to deflect the beam up or down and right or left. The beam of electrons can be used to paint a picture on the faceplate of the CRT. The older *National Television System Committee (NTSC)* standard calls for 525 of these lines to be drawn across the screen at the rate of 30 times per second. Because of persistence of vision, our eyes blend each of these 30 frames (or pictures) per second into a continuous flicker-free blank screen. Different standards are used for digital television.

To create a picture on the blank screen, the level of intensity at which the electrons strike the screen must be modulated or changed. In the dark parts of the picture, the beam does not strike the phosphorous; therefore, no light energy is created. In the bright spots, the beam of electrons strikes with great intensity. The combination of bright and dark spots, with the intermediate shades of gray, creates the highlights and shadows that we interpret as a black-and-white picture.

The color television receiver operates in basically the same way as the black-and-white one, except that there are three electron guns in the tube, each firing a beam of electrons toward the faceplate (see figure 13.12). Near the faceplate is a metal screen with thousands of tiny holes called the *shadow mask*. Its purpose is to allow only electrons from the red gun to hit the red dots of phosphors on the screen, the green gun to hit only green phosphors, and the blue gun to hit only blue phosphors. Some tubes have stripes of colored phosphors deposited on the screen rather than dots, but the effect is the same.

Using the *additive color* theory, white light is created by adding together the proper amounts of red, green, and blue light (see figure 13.13). Any color of light can be created by mixing the three primary colors of light in the right proportions. If the red and green guns in the color tube are firing electrons in the right proportions and the blue gun is blanked (turned off), the result is that red and green phosphors on the screen glow, and our eyes perceive yellow. Because the NTSC color television signal must be compatible with black-and-white television sets, the person watching the black-and-white set would see a light gray or white image in place of the yellow one displayed on the color set. To achieve compatibility between black-and-white and color receivers, the television signal contains a *luminance* component, which determines the brightness or darkness of each element in the picture, and a *chrominance* component, which determines both hue and saturation of each element in the color picture. *Hue* determines the particular wavelength of the spectrum, which is perceived as red, green, blue, or a combination of two or more of the primary colors. Incorrect hue information will determine if flesh tones on the screen appear greenish or purplish rather than natural. *Saturation* determines the purity of the color; an impure red color, for example, appears pink on the screen.

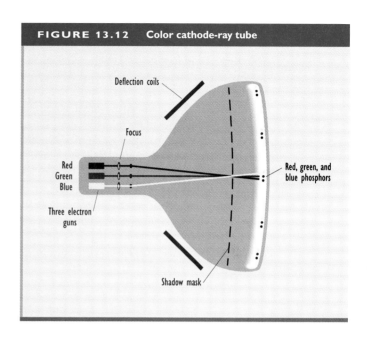

FIGURE 13.12   Color cathode-ray tube

NEW DISPLAY TECHNOLOGIES

Most television receivers operate based on standards developed by the National Television System Committee back in the 1940s, but there have been many improvements in the technology since then. Some

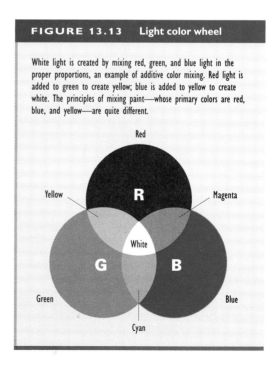

FIGURE 13.13   Light color wheel

White light is created by mixing red, green, and blue light in the proper proportions, an example of additive color mixing. Red light is added to green to create yellow; blue is added to yellow to create white. The principles of mixing paint—whose primary colors are red, blue, and yellow—are quite different.

recent developments that have significantly changed the way that we display visual information include projection television systems, liquid-crystal display systems, high-definition television systems, and advanced television systems.

Projection television systems use three high-output light tubes (one for each of the primary colors) rather than one direct-view tube. The light from each of the tubes is concentrated by an optical lens system onto a flat rear- or front-projection screen. The screen then reflects the light waves back to the eye. The advantage of the projection system is that the picture can be greatly enlarged, which allows larger audiences to view the video program in a theater setting and allows the home viewer the luxury of a theater-size picture in the living room.

Liquid-crystal display screens use technology similar to that of a digital wristwatch or cell phone display and may well be the large, flat-screen display technology of the future. Applying a small amount of current to each tiny dot, or *pixel*, on the screen changes that spot from light to dark or to a specific color. Because each pixel on the screen has a specific address, a computer chip is required to direct the small amount of electric current to each pixel at the appropriate time to display a picture. Although there are advantages in small, flat-display screens for portability, the technology has not yet developed to the point that allows hanging a huge LCD screen on the wall of a media room for a low cost, and the small liquid-crystal screen produces a very flat picture that lacks detail.

High-definition television (HDTV) and its cousin **advanced television (ATV)** are just around the corner for most people in the United States and are already being used in other countries. Each technical system promises much improved picture definition, but the basis for that improvement remains the same in all systems. The addition of more scanning lines will mean better picture quality and the ability to enlarge the screen without loss of picture detail. The HDTV systems offer more than twice as many scanning lines as the conventional NTSC television system and a much wider screen, using the 16 × 9 motion picture **aspect ratio** rather than the traditional 4 × 3 TV system (see figure 13.14). With the addition of more scanning lines and a wider screen, the amount of information that must be transmitted to the television set increases, and more of the electromagnetic spectrum will be required to carry the information for each television channel.

The FCC has mandated that all television stations broadcast in a digital advanced television (ATV) format by 2003. The move to new standards has given the television receiver manufacturers the opportunity to design new display devices. These "smart" television sets incorporate computer chips that allow the receiver to display video from a number of different transmission formats, as well as to perform other communication tasks now the purview of the PC. Some major manufacturers are marketing large-screen (3 foot by 5 foot), flat display devices that can be hung on the wall. Although the price of these ATV display devices is currently quite high ($3,000 to $6,000) when compared

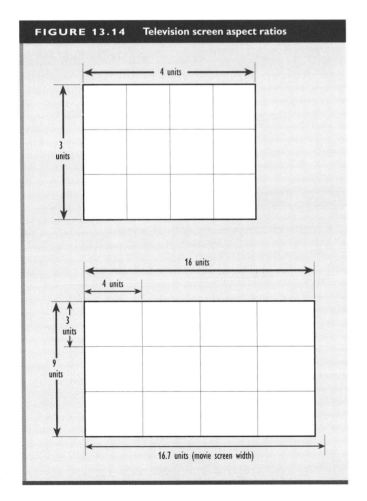

FIGURE 13.14 Television screen aspect ratios

with older STV (standard television) receivers, the price will drop as manufacturing volume increases.

The FCC has also stated that all broadcasters cease transmission on the old NTSC system by 2006. Viewers who still have analog television receivers will probably have to buy a set-top converter to enable the old television sets to pick up the new digital signals.

With the new digital standards, the opportunity exists to reinvent the television display device into something very different from the glowing box in the corner of the living room. The television display devices of the future may well allow three-dimensional pictures to be projected into our living spaces through the use of holography.

Encoding Technology

The Video Camera

The television receiver is the final link in the television transmission chain; the *first* link is the *video camera*. The camera is an electrical transducer: It gathers light energy and changes it into electric energy. For our purposes, the camera can be divided into three parts: the optical system, the pickup device, and the monitoring system.

The optical system includes the lens, which gathers and concentrates light waves. Most cameras come equipped with a variable-focal-length lens called a *zoom lens,* which allows the camera operator to select a wide or panoramic shot, a medium shot, or a close-up without physically moving the camera.

In addition to the control that allows the focal length of the lens to be changed, another adjustment allows the image to be focused sharply on the faceplate of the pickup device. A change in focus is required each time the camera is moved in relationship to the object being photographed. A third adjustment on the lens controls the aperture, or iris. This device can be adjusted to allow varying amounts of light energy through the optic system to strike the faceplate of the pickup device. The standard scale that indicates the amount of light that is allowed through the lens is called the *f-stop.* The larger the *f*-stop number, the smaller the aperture. In bright, sunny conditions, the aperture must be stopped

Lightweight portable video camera used to cover sporting events. This type of TV camera can be held on the operator's shoulder for portable work or, with an attached monitor as shown here, mounted on a tripod and operated in the same fashion as a studio camera.

down to $f/16$ or $f/22$ (which means the aperture is smaller) to keep too much light from washing out the picture. When shooting indoors the lens is opened up to $f/5.6$ or $f/4$ so that more of the limited light available can be focused on the pickup device. Most lenses are now equipped with *servo-zoom mechanisms* that allow smooth zoom motion, automatic focusing, and instantaneous automatic iris adjustment to compensate for changing light conditions as the camera pans or moves from one view to the next.

An additional optical system needed for three-tube color cameras is called a ***beam splitter*** (see figure 13.15). The purpose of this system is to break the light energy down into component parts of red, green, and blue and to direct each into the appropriate pickup device. For example, light of the red wavelength is focused on a mirror that allows part of the light to pass through and part to be reflected. A series of these

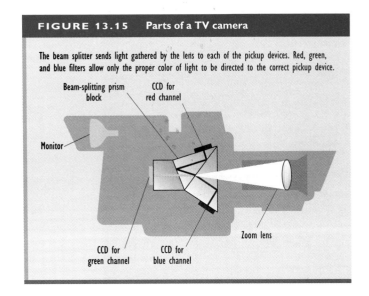

FIGURE 13.15 Parts of a TV camera

The beam splitter sends light gathered by the lens to each of the pickup devices. Red, green, and blue filters allow only the proper color of light to be directed to the correct pickup device.

mirrors is used to channel a portion of the light that enters the camera to each of three pickup devices. This system allows the light reflected into the camera to be broken into the three primary colors with a separate pickup device for each color of light.

There are also color cameras that use only one pickup device. Cameras of this type are common for home use, but because the resolution of the picture is not high, they haven't found much acceptance in the broadcast arena.

The second part of the camera is the pickup device. Almost all cameras use solid-state *charge-coupled devices (CCDs)* to convert light energy to electric current.

Light energy is changed into electric energy by a chip (see figure 13.16). The image is focused on a photosensitive layer on the chip. The solid-state sensor

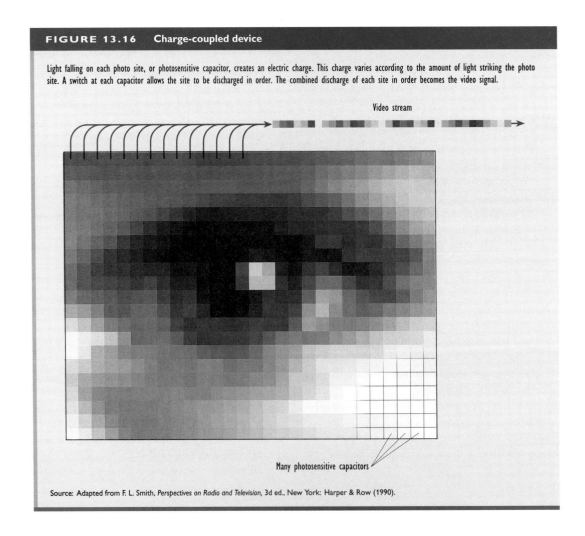

FIGURE 13.16 Charge-coupled device

Light falling on each photo site, or photosensitive capacitor, creates an electric charge. This charge varies according to the amount of light striking the photo site. A switch at each capacitor allows the site to be discharged in order. The combined discharge of each site in order becomes the video signal.

Source: Adapted from F. L. Smith, *Perspectives on Radio and Television*, 3d ed., New York: Harper & Row (1990).

is composed of hundreds of thousands of pixels, or separate photodiodes that are picture receptors. Each pixel has a specific address that a computer chip reads to obtain the picture information. By reading each pixel in order, the computer chip acts in the same manner as the scanning beam of electrons found in the cathode-ray tube.

After the video signal is obtained from the pickup device, it is amplified and processed to make it strong enough to be recorded or transmitted. Signals from the red, green, and blue pickup devices are also combined in an encoder to create a composite video signal that contains both luminance and chrominance elements that can appear as a color picture on a color television receiver.

The third element in the video camera is the monitoring system. This is simply a small video monitor mounted permanently to the back of the camera, allowing the camera operator to compose or frame the picture, as well as to focus it. Generally, the monitor is black-and-white even though the camera is color. Portable cameras used for electronic news gathering often have a tiny electronic viewfinder that is seen through a magnifying lens as the operator holds the camera on his or her shoulder.

COMPUTER-GENERATED VIDEO

Even though most video signals are created initially by a television camera, the signal can also be synthesized by computer. Most of the words in the titles or credits for a television program or the graphics used in commercials are composed at the keyboard of an advanced video graphics word processor called a *character generator*. These devices are digital in operation, so their output must be converted to analog for insertion into composite video. The high-end character generators allow the creation of very complex graphic designs using many different fonts for lettering and hundreds of color combinations for shading both the lettering and the backgrounds. Graphic designs can be stored in memory or on a magnetic disk after composition and called up for later use in the program. Most character generators allow the graphics to be animated, or to move on the screen. With the wider acceptance of digital technology, computer graphic generators of the future may not be limited to flat graphic designs, but may in some instances take the place of the camera in the creation of a natural-appearing video picture.

SIGNAL MANIPULATION

The composite video signal from a single television camera can be fed directly into the transmitter for transmission to the home receiver, but generally the signal must be monitored, processed, or even stored before it is sent into the home.

THE VIDEO PRODUCTION SWITCHER

Perhaps the most basic unit of the manipulation process is the *video switcher,* which allows the video signal from two or more video cameras to be selected for editing or transmission. When using several cameras in a television studio, the signal from one camera can be put on the air while the cameraperson is framing and focusing the next picture in a sequence. The simplest video switcher consists of a row of buttons called a *bus*. Each button corresponds to the output of an individual video camera, a character generator, a videotape recorder, or **black**. Although the simple video switcher would allow the selection of one video source from a variety of cameras, recorders, and other signal generators, the effect on the home screen would not be very pleasing unless some additional features were added to the device. For example, the picture on the TV set would roll vertically each time a new picture was selected unless all of the picture sources were locked together and driven by one sync generator.

The signals coming into the switcher must also be timed to arrive in the proper *color phase.* Because there is resistance to electron flow in any wire, each foot of cable between the picture source and the switcher causes delay in the time that it takes for the signal to arrive. It is likely that each cable is of a different length; therefore, each signal is likely to be delayed by differing amounts. The effect of this on the screen is that flesh tones from two properly balanced cameras will appear greenish in a shot from one camera and purplish in a shot from another. Timing of the signals at the switcher is necessary to match color from different sources.

Most production switchers are anything but simple. Many have rows and rows of duplicate buttons for each of dozens of separate video sources. This allows transitions and special effects to be used between shots. The simplest transitions are the cut, the fade, and the dissolve. The **cut** or *take* is an instantaneous change in picture from one camera to another, accomplished by pushing the button corresponding to the desired camera on the live bank of the switcher. It is used to show the viewer new information in the same scene in the program. The *fade*, a slow transition from black to a picture or from a picture to black, usually tells the viewer that a new program is starting or that the program has finished, as in "fade to black." To create a fade on the switcher, two banks of duplicate buttons are required. A fader bar that deactivates one row of buttons is slowly moved while activating the preset output of the second row.

A third transition, called a *dissolve*, is accomplished in the same manner, except a second camera is selected on the second bank instead of black. The effect is one picture slowly being replaced by another. This transitional device is normally used to indicate a change in scene, in location, or in time in the program.

Most production switchers are equipped with a **special-effects generator (SEG)**, which allows more manipulation of the video signal. For example, a fourth transitional device called a *wipe* is possible with an SEG. The effect of this transition is that a diamond, a circle, or almost any other geometric shape is used as a figure in which a new image can be expanded to replace the previous image on the screen.

Another effect made possible by SEG circuitry is the *key*, used most often to insert graphics or titles at the beginning of a program and credits at the end. The lettering of the title is cut into the original picture yielding a combination of a background picture with lettering. A *chroma key* operates through the SEG in a similar manner but allows the operator to delete all of one primary color from the original picture and to replace it with elements of a second picture. The most common use of this effect is on the news, when the weatherperson stands in front of a blue background. All of the blue signal is deleted and replaced with a picture of a weather map, a radar screen, or a satellite photo, making it appear as though the weatherperson is standing before a huge map. He or she knows where to point by watching the combined image on a monitor located out of range of the camera shot. Most on-air talent learn quickly to avoid clothing that was near primary blue in color, because the chroma key would drop out the color of the clothing as well as the background. An anchorperson wearing blue might lose significant body parts in the resulting chroma-key picture!

Digital video effects allow the television picture to be compressed, shrunk, expanded, rotated, flipped, and otherwise distorted in hundreds of ways. The digital effect has found a home as an attention-getter in locally produced commercials: Who hasn't seen a tiny picture spinning and flipping out of the center of the screen to form the image of the local celebrity used-car salesperson pushing the latest values in cheap transportation?

Monitoring Systems

Although not strictly among the equipment to manipulate the signal, monitoring systems are a necessary part of the process. Engineering personnel and production people both must monitor the video signals closely to obtain the best possible results from the television equipment. Both rely on video monitors to display the picture before it is transmitted to the home or recorded on videotape. Usually, there is a video monitor on each piece of originating equipment

A technician evaluates picture composition and quality on a video monitor.

and duplicate monitors in each production control room in the station as well as in the master control area. The camera operators, the director, the technical director, the talent, and the producer all rely on the video monitors to compose and frame the shots as well as to determine which picture is on the air and which shot should be switched to next. Most of the source monitors are black-and-white, but the *line monitor* and the *preview monitors* connected to the production switcher show color pictures. Technicians use video monitors to cue tape to the start point as well as to evaluate the picture quality.

Technicians must also adjust the video signal to the proper levels needed for transmission. To do this they must be able to rely on some standard reference of quality and adjust the signal to that reference. The *waveform monitor,* the standard test instrument that allows the video signal to be displayed graphically, creates a visual representation of the synchronization signals as well as luminance information in one frame of the video signal. Brightness information can be read from a scale on the face of the monitor. Optimum brightness for the whitest parts of the picture should create a waveform, or graph, that peaks at 100 percent on the scale. If the peak is not that high, the iris in the camera lens can be opened a stop to allow more light to fall on the pickup chips, more lighting instruments can be turned on to raise the light level on the set, or the camera amplifier can be adjusted to create a brighter picture. Contrast in the picture is created by adjusting the black level to a point where the darkest part of the picture registers slightly above 7.5 percent on the waveform scale.

The *vectorscope* provides the engineering staff with a graphic display of the color information in the video signal. Signal phasing as well as the saturation and hue of each color in the picture can be determined with the vectorscope. The circular scale of the vectorscope displays the color burst (which tells the TV set that the picture should be in color rather than black-and-white); and each of the primary colors—red, green, and blue; as well as the secondary colors—magenta (red and blue combined), cyan (blue and green combined), and yellow (red and green combined). White light is displayed in the center of the screen, red light toward the top, and blue on the right side. The farther the graphic image moves from the center of the display, the more saturated the color. The vectorscope can allow the measurement of chrominance information in much the same way that the waveform monitor measures luminance information.

PROCESSING AMPLIFIERS

Most video cameras have controls that allow the operator to adjust the various components of the signal. If the cameras are located in the studio or in a remote truck, the *camera control units* are located together at an engineering location. The technician can watch the video monitors as well as the waveform monitor and vectorscope for each camera. By remotely adjusting the iris in the camera lens and by adjusting the black level, the engineer can match the studio cameras so that the picture from one camera looks the same technically as the picture from every other camera. In the studio this is a routine operation, but on an outdoor sports remote multiple cameras and ever changing light levels can pose a challenge for the video quality engineer.

Many programs, news stories, and commercials are videotaped on-location with one camera. The videotaped information is then returned to the station for *postproduction*—the editing of the separate videotaped segments into one continuous story. The signals from the videotape playback and record machines needed in the editing process are generally routed through *time-base correctors,* whose function is to lock the *videotape recorder (VTR)* or the VCR to the station sync generator so that images from one tape player can be mixed with another for fade and dissolve effects (this process is called *AB-roll editing*). The time-base corrector also contains a processing amplifier, or *proc amp,* that allows the signal from the tape player to be manipulated in a variety of ways. The *gain* (signal strength) and *pedestal* (black level) can be adjusted as well as the hue and the amount of color in the picture. If careful attention is paid to the waveform monitor and vectorscope, each videotaped picture can be matched to the next even if shot at different times and under varied lighting conditions. The controls can also be used creatively to enhance a mood or atmosphere by subtly adjusting the color of the shot. Most time-base correctors have provision for some limited special effects, such as *freeze-frame,* wherein all motion in the shot is stopped.

Edit Controllers

Most video production today is recorded on videotape. Only news, sports, and an occasional special is broadcast live. Much of the videotape recorded for a television program is shot out of sequence and must be edited to form a complete television program. Without directly manipulating the video signal, the *edit controller* allows the videotape recorder to be brought under the systematic and precision control needed in the editing process.

Videotape editing systems are either on-line or off-line. *Off-line editing* systems generally consist of two small-format videotape recorders with logic circuits connected to a simple control unit through which the operator can locate and cue audio or video segments. The second VTR then records the selected scenes in the desired order. The recording can be made either in an assemble mode or in an insert mode. In the *assemble mode,* each subsequent shot is simply tacked on to the end of the preceding shot. Although the process is more time-consuming, a more stable edit can be made with the *insert mode,* wherein new shots are inserted in an already existing recording, without affecting the shots on either side of the insert. Insert editing requires the use of a blank videotape which then has a control track recorded on its entire length. The *control track* is a black video signal complete with sync pulses that will later serve as an electronic guide for the new information that is recorded onto the tape in the editing process.

To allow the operator to quickly locate the correct shot on the tape, *SMPTE time code* is often used. SMPTE (pronounced "sempty") stands for *Society of Motion Picture and Television Engineers,* the organization that established the time code. Time code records a unique number on the tape for each frame of the shot. As the length of the shot increases, seconds and minutes are added to the frame count. The time code numbers can be displayed by superimposing them on top of the picture seen on the video monitor. Time code allows the editor to quickly find the specific shot that he or she needs next in the assembly of the television program.

Semi-manual editing, known as *off-line editing,* is used primarily for news stories because it can be done very quickly. It is also used for low-cost productions that might be done for cable systems, corporate video, or for educational applications. Because the equipment for off-line editing is inexpensive, the systems are often used to create a *rough cut* of a television program. A director or producer can sit at the off-line edit controller for hours, experimenting with different shot sequences to tell a story in the most effective manner. Some off-line systems connected to an inexpensive PC can generate an edit list as the director experiments with shot selection. The final list, stored on disk, lists every shot in the program by tape number and time code address. Also included are audio cues and transitions needed to complete the program. The information on the disk can then be used to drive the VTR machines and the edit controller of a computer-based on-line editing system to put together a high-quality program.

The *on-line editing* system is usually used to control several source machines (in the AB-roll system) so that dissolves and fades between videotaped shots can be incorporated into the program. The system is used to control an audio mixer, a production switcher with a special-effects amplifier, and a character generator as well as the videotape players and recorders. The tape machines connected to this system are usually 1-inch broadcast-quality, open-reel recorders; high-quality ½-inch component, broadcast-quality recorders; or digital-broadcast cassette recorders. On-line systems are often used in postproduction of full-length television programs or in the production of broadcast commercials that require many shots and special effects in a 30-second spot. Almost every cable system, television station, instructional video system, and corporate production facility uses off-line videotape editing systems on a daily basis, but the more expensive on-line systems are usually found in large television stations in the bigger markets and in postproduction houses. The *"post house"* makes a business of producing, editing, and duplicating videotaped television programs for clients such as advertising agencies, television program syndicators, and corporate training or public relations departments.

Nonlinear systems are quickly being adopted as a timesaving tool for the video editor. Sometimes called *desktop video,* the ***nonlinear editing system*** uses a PC terminal to display control commands as well as audio and video signals. With keyboard commands the editor can cut-and-paste audio and video into a program. The system allows the editor to assemble the elements of the video program in much the same way as word processing software enables the writer to quickly edit text. Because the video portion of the program requires so much memory, the digitized television signal must be recorded on one of several external disk drives. The amount of external storage space determines the length of the video program that can be assembled. As the cost of adding disk drives decreases, more facilities will move to nonlinear editing systems.

Storage

Television storage equipment is used to hold a representation of the audio and video signal until it is needed for broadcast or display in the home via a videotape recorder or a *digital videodisc (DVD)* player. A television storage system can include photographic film, videotape, a videodisc, a DVD, or even the hard drive of a computer (see chapter 14). Future storage systems might incorporate the use of memory chips built into the "smart" television display device.

In the earliest days of television, programs were performed live in front of the television camera. High-quality entertainment and horrendous mistakes were transmitted together to the viewing audience. Film projectors were quickly adapted to match television's 30-frames-per-second (fps) standard so that more programming material, including B movies produced by the hundreds in Hollywood in the 1930s, would be available to feed the insatiable appetite of the television audience. The adoption of the film projector by the television station led to the first system of storing video pictures for later use. Electronic recording of the video signal became possible in 1956, when the Ampex Corporation marketed the quadrature video recorder, which used heavy reels of 2-inch-wide videotape. This format was the standard in professional video recording for almost twenty-five years, until the helical system was adopted for use by the broadcast industry.

Videotape Recording

Using the same theory that states that if a wire cuts through a magnetic field, an electric current will be created, a current flowing through a wire will also create a magnetic field. *Magnetic fields* are composed of invisible lines of force called ***flux.*** These lines of force also have ***polarity,*** or a north pole and a south pole. Ends of flux of the same polarity tend to repel each other; lines of force of the opposite polarity attract. Magnetic recording, either audio or video, uses these principles to convert electrical information into elements of magnetic polarity that can be stored for long periods of time and then retrieved and converted back into electric current.

The recording head of a VTR is a tiny electromagnet or a coil of wire wound around a soft metal, donut-shaped core. A slice taken out of the core creates a gap in which a magnetic field of varying intensity and polarity is created. Varying or modulating the current that creates the magnetic field yields changes in the field. If the magnetic field is brought close to a ferromagnetic material such as iron oxide, the magnetic polarity is remembered as patterns of residual magnetism created in the iron oxide.

To use this theory to devise a video storage medium, a ***bias current*** is applied to a recording head to create an electromagnetic field. The current creating the field is then modulated by the changing analog video signal, which in turn changes the intensity and polarity of the field. Magnetic particles embedded in the iron-oxide coating the videotape are aligned in corresponding patterns as the tape is brought in contact with the recording head. Because recording each new bit of information in the video signal requires a new spot on the tape, and because there is a tremendous amount of information necessary to create the signal, the tape must be moved past the recording head. In addition, the video head rotates on a drum at a high rate of speed. This is necessary to find adequate space for all of the information required to store the picture elements of the video signal. To do this the tape is wrapped around the head wheel so that the head transverses in a helical, or slanted, track across the tape (see figure 13.17). Each pass of the recording head in

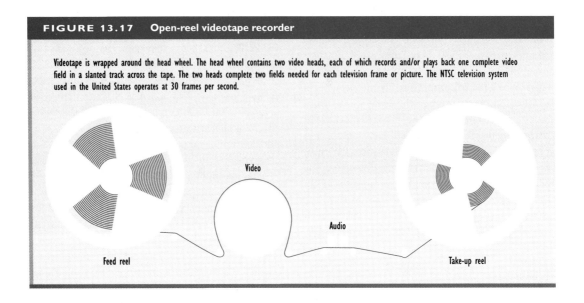

FIGURE 13.17 Open-reel videotape recorder

Videotape is wrapped around the head wheel. The head wheel contains two video heads, each of which records and/or plays back one complete video field in a slanted track across the tape. The two heads complete two fields needed for each television frame or picture. The NTSC television system used in the United States operates at 30 frames per second.

the *helical scan* across the tape lays down enough information to complete one frame (or one field, depending on the format) of the television picture. Additional information that is needed to create the control track, the time code data, and the audio signal is recorded longitudinally on the videotape by stationary recording heads.

The recorded information is played back by passing the playback head (similar to the recording head of a VTR) over the recorded magnetic fields. As the magnetic field is detected in the gap of the playback head, a slight current is created in the coil of wire around the head. Changes in the magnetic field that are recorded on the tape cause variations in the current that is generated in the head. This modulation of the current is a mirror image of the video signal that was originally recorded. The tiny signal created by the playback head is amplified and displayed on a video monitor or transmitted to the home receiver.

Videotape comes in a variety of formats to meet different needs. Most broadcast program production and most television station program playback are performed on reel-to-reel 1-inch tape (*1 inch* refers to the width of the tape). This format provides a stable, high-quality picture that is suitable for broadcast transmission. The machines can be used to create some special effects in the production process, such as slow motion. The disadvantages of this format are that both the VTRs and the tape stock are expensive and somewhat difficult to load and operate.

Some news, commercial, educational, corporate, and cable productions are still recorded on ¾-inch videocassettes. This format requires only small, lightweight, battery-powered equipment and is therefore ideal for use in the field. The *electronic news gathering (ENG)* revolution of the mid-1960s was fueled by the adoption of this format by local television stations to replace 16mm newsfilm.

Videotape formats that use ½-inch tape have been developed for several uses. The *Betacam* format has been adopted for ENG by most local television news departments. Although the tape is smaller and therefore has less storage space, the processing circuitry in the equipment allows for better picture quality than with the wider ¾-inch tape. The Betacam format is expensive, but the size advantage allows a camera and recorder to be docked together and carried into the field as one unit—a significant convenience factor in the daily rush to gather news stories. Some stations have used this small format in automated video playback machines that select short cassettes in a preprogrammed order for commercials, public service announcements, and station identifications that are aired during commercial and station breaks.

The ½-inch tape is also used extensively for home use. The *VHS* format is adequate for rented movies and for recording a program off the air for later viewing on the home screen, but it is not of sufficient stability or quality for use by a broadcast transmission system.

The development of the *S-VHS* videotape format has created a ½-inch format that, although having the advantages of small size and somewhat inexpensive equipment, exceeds the quality of the older ¾-inch format. Small production companies, corporate video departments, educational institutions, and cable television systems may adopt S-VHS for a low-cost production format.

The need for a smaller and more convenient handheld *camcorder*—a portable video camera with an attached or built-in VTR, forming a single unit—has led to the development of Hi-8, an 8mm tape format developed for producing home videos. Digital camcorders are fast replacing low-end equipment in both the consumer and the professional markets.

Digital Recording

As with other areas of electronic communication, the video-recording process is moving to digital technology, wherein the computer on/off codes replace the analog video signal. The resultant picture quality of the digital system is far superior to that of other videotape recording systems.

A popular digital format is the *Ampex D-2* system, which uses a 1-inch videocassette. The greatest advantage of the digital system is high quality, but the system offers other advantages, especially in the production process. For example, digital signals can be combined or layered over one another to create composite pictures of almost unlimited complexity. The D-2 format can also be used to *stillframe,* or freeze, individual pictures or to play scenes at a wide variety of speeds, from slow motion to fast-forward. Because of the expense of the system, its use is generally limited to specialized production tasks, but as the price of digital equipment drops, more applications will be found for digital videotape recording.

Summary

▶ The equipment that the broadcaster uses to create and to distribute television programs can be categorized in terms of five primary functions.

▶ Encoding equipment is necessary to change sound and light energy into electric energy—primarily the microphone and the television camera.

▶ Manipulation equipment is that which is required to shape the signal for the best technical quality and to select and change the signal for creative purposes: the video switcher, the processing amplifier, and the special-effects generator (SEG).

▶ Storage equipment holds information until it is needed for broadcast, cable cast, or home-video use and includes the videotape recorder (VTR), the videodisc, and computers servers.

▶ Transmission equipment includes the broadcast transmitter, coaxial cable, fiber optics, and satellite communications and is discussed in chapter 14.

▶ Display equipment includes the television receiver as well as advanced television (ATV) systems and high-definition television (HDTV). Future display technology may well include large, flat picture screens and three-dimensional picture recreation made possible with holography.

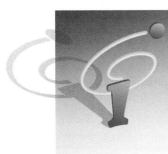

INFOTRAC COLLEGE EDITION EXERCISES

The technology needed to create radio and television programs has changed radically over the past several years. Broadcast standards and equipment have moved from analog to digital. The desktop computer has become the most powerful production tool for the broadcaster.

13.1 Chapter 13 examines audio and video technologies from old-format analog equipment still used by many broadcast stations to the newest digital standards mandated by the FCC.

   a. Using InfoTrac to access *Broadcasting & Cable*, locate articles on broadcast automation and the new equipment that makes it possible to run the station at a lower cost than hiring people.

   b. Using InfoTrac to access *Broadcasting & Cable*, locate articles about HDTV production and transmission equipment.

13.2 There are many new media technologies on the horizon, from digital television to the unlimited potential of the Internet.

   a. Using InfoTrac to access *Broadcasting & Cable,* find information on the use of the Internet to transmit streaming video for delivering television programs to the home.

   b. Will the traditional over-the-air radio or television station disappear with the adoption of the Internet to deliver audio and video information? Can you find articles in computer journals to support your position?

CHAPTER 13   AUDIO AND VIDEO SYSTEMS

WEB WATCH

Here is a list of a few URLs (Internet addresses) for some of the organizations or corporations discussed in this chapter. Please explore these Web sites and follow the links to learn more about the complex business of the electronic media. Add your descriptions and your own favorite sites at the end of the list. Please keep in mind that the dynamic nature of the Internet allows sites to come and go but also allows organizations to update information about themselves very quickly.

Address **Description**

http://www.panasonic.com/ _____

http://www.sony.com/ _____

http://www.jvc.com/ _____

http://www.harris.com/communications/ _____

http://www.systemassociates.com/ _____

http://www.broadcast.philips.com/fdesktop.asp _____

http://www.itelco-usa.com/ _____

http://www.avid.com/ _____

http://www.attc.org/ _____

http://www.hri.com/atl/system.html _____

Other favorite sites: _____

CHAPTER 14

Distribution Systems

Analog or digital, broadcast signals are governed by the laws of physics in their propagation and distribution.

The majority of the U.S. audience probably doesn't care how the signal gets from the radio or television station to the receiving set, so long as the programming is entertaining or informational, plentiful, and inexpensive. The technical communication systems should be transparent and not something that the user must think about. As electronic technology has developed over the past hundred years or so, it has become more sophisticated and much more user-friendly. With the advent of automatic frequency controls, fine tuning, and color correction, along with push-button tuning, the consumer has come to expect perfectly tuned, high-quality sound and picture by simply operating the on/off switch on the television set. The days of manually adjusting the receiver for optimal performance are long gone, but, unfortunately, with them went the basic knowledge of the natural factors that affect the reception of radio and television frequencies. People who now subscribe to cable television know that they must pay an extra fee each month, but beyond that they usually have little understanding of the technical differences between over-the-air broadcasting and coaxial-cable distribution.

Like most viewers, many people employed in broadcasting have also developed the attitude that the technology is magic that should be left strictly to the engineers. This mind-set is undesirable for a number of reasons. First, as communication equipment becomes more refined, compact, and reliable, little need exists for broadcast engineers at the local station to operate and repair equipment; consequently, engineers are no longer a ready resource of information about the technology. Broadcast managers without much technical background are now making decisions about new equipment purchases. Second, technical principles dictated by the natural laws surrounding the use of the electromagnetic spectrum directly affect policy considerations of the communication businesses, and the laws of physics affect economic potential in this business as in no other. Third, new business opportunities now being considered by the traditional broadcaster are better suited to one technology than another; therefore, knowledge of the limitations of the technology is necessary to implement those services. Fourth, an individual working in any industry should have some idea of how each aspect of that industry works. This includes knowing something about the technology that makes electronic communication and the subsequent radio and television businesses possible.

It is obviously important for the student of the electronic media to have an understanding of the scientific principles underlying electronic communications. This chapter provides basic information on the theory of electromagnetism and how that natural force is used in communications. In addition, the technical principles behind the operation of AM and FM radio, as well as television, cable, satellite systems, and fiber optics, in delivering the message to the home are presented. Although this chapter is not intended to prepare the student to become an electronic engineer, or even to pretend to be one, it should demystify the magic that has grown up around radio and television transmission.

The focus of this chapter, therefore, is on communication distribution systems—the technological means through which the message is sent from the radio or television station to the home or car receiver. That message can be delivered through one of three basic systems, which are shown in table 14.1.

TABLE 14.1	Electronic communication transmission and distribution technologies	
PHYSICAL	**WIRED**	**BROADCAST**
Mail services / xerography	Twisted pair / ISDN/DSL	Radio / television broadcast
Photographic film	Coaxial cable	Microwave / MMDS / ITFS
Audiotape / videotape / computer disk	Fiber-optic cable	Satellite communication / direct broadcast satellite
Compact disc / videodisc / CD-ROM		Cellular telephone / personal communication services
Computer memory chip		

Broadcast Communication Systems

The Electromagnetic Spectrum

The *electromagnetic spectrum* is a continuous range of frequencies of electromagnetic energy from *low frequency (LF)* of about 30 *Kilohertz (KHz)* (30,000 cycles per second) through the *medium frequencies (MF)*, where AM radio is located, to the *very high frequencies (VHF)* and *ultrahigh frequencies (UHF)* of FM radio and television (see figure 14.1). The upper end of the spectrum now used for radio wave transmission is in the frequency range of 30 to 300 *gigahertz (GHz)* and is used for satellite communication. Additional bands of frequencies exist above

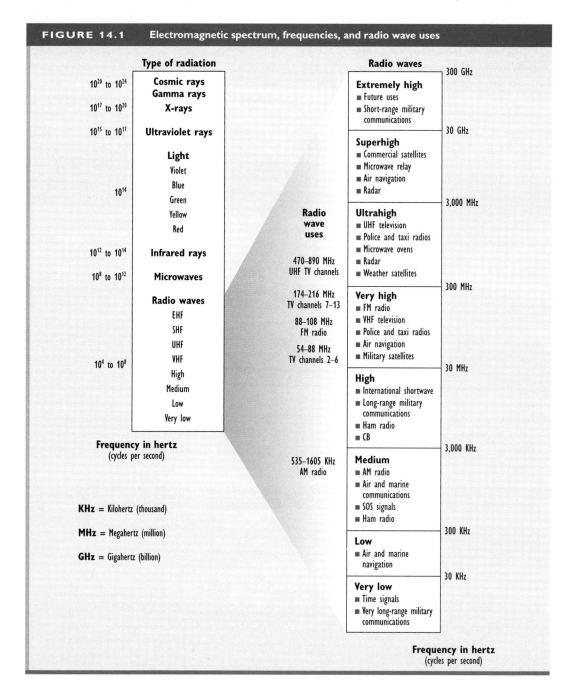

FIGURE 14.1 Electromagnetic spectrum, frequencies, and radio wave uses

those mentioned that are also used for radio transmission. For example, visible light frequencies are part of the electromagnetic spectrum, as are X-rays and gamma rays.

Electromagnetic energy occurs naturally. We are bombarded by light energy on a daily basis as well as by other forms of electromagnetic energy. Naturally generated electromagnetic energy in the lower frequencies includes that created by electrical storms and the static electric sparks caused by petting a cat's fur in a dry environment. The electromagnetic energy caused by an electrical storm can be detected by static on an AM radio and flashing of the picture on channels 2 or 3 on a television set.

Hertzian, or electromagnetic, waves have certain properties that make them useful in the communication process as well as properties that somewhat limit their use. The very fact that the waves occur naturally means that those generated for communication purposes often are subject to interference by electrical storms or sunspot activity. This natural interference can disrupt the communication process, as can artificial interference, such as that caused by the sparks of an electric motor or an automobile engine, or it can be generated by another radio transmitter. Because a particular frequency, or portion of the spectrum, can be used only once in a limited geographical location without interference, the spectrum is a limited natural resource.

Electromagnetic energy radiates outward in all directions from its source and travels at the speed of light. It operates most efficiently through the vacuum of space, allowing the signal to be carried in a wireless mode from the transmitter to the receiver both through space and through the earth's atmosphere. At the speed of 186,000 miles per second, the message can be received almost instantly anywhere around the globe.

The waves radiated from a source of electromagnetic energy dissipate as they travel away from the source. Theoretically, the waves continue to infinity, but practically their strength diminishes rapidly as they spread over greater territory. The diminishing wave strength, called *attenuation,* means that the signal will travel over only a limited geographical area before it becomes too weak to detect and use for communication purposes. Attenuation defines the coverage pattern, or contour, of the station's signal.

Different portions of the spectrum have different properties. For example, lower frequencies have longer wavelengths than do higher frequencies. The *wavelength* is the physical measure from one peak of the wave to the next. Each wave is composed of a complete oscillation, or cycle of energy; this distance is the wavelength. Wavelength is inversely related to *frequency,* or band designation: The shorter the wavelength, the higher the frequency.

Wavelength in the electromagnetic spectrum determines how the energy will react to obstacles in its path, as well as to other external forces. Low-frequency waves have a long wavelength and will bend around objects that they can't readily penetrate. Such waves will follow the contour of the earth and bend over hills and around buildings. Short waves generated at higher frequencies do not bend as easily; instead they radiate in nearly straight lines. Short waves will, however, bounce off objects that they cannot penetrate. Television signals, for example, reflect off buildings and mountains, and the even shorter waves used in satellite communications can be reflected, and therefore dissipated, by objects as small as raindrops.

Electromagnetic waves in the low and medium frequencies also react to soil conditions. Higher moisture levels in the soil enable signals with longer wavelengths to be transmitted farther than they would go in dry soil. This means that AM radio stations in valleys and near rivers send their signals greater distances for a given amount of power than can their counterparts in drier climates.

Longer wavelengths are also greatly affected by the ionosphere, a layer of gases situated between 60 and 600 miles above the earth. During the day it is heated by the sun and ionizes; but once the sun sets, the layer of gases cools. Long-wave radio signals are refracted back to earth by the thinner nonionized layer of gases. This causes the signal to travel much farther than during the day. Shortwave stations anticipate this and use the resulting skip pattern to send their signals much farther than normal (see figure 14.2). Skip is often a problem for the AM broadcast station, because the Federal Communications Commission (FCC) requires many of them to limit transmissions at night so they will not interfere with other stations broadcasting on the same frequency.

The wavelength also affects the amount of power required to send a signal a given distance: The shorter the wavelength, the more power is needed. For example, a television station operating on channel 2 has a carrier frequency of 54 to 60 MHz; it is limited

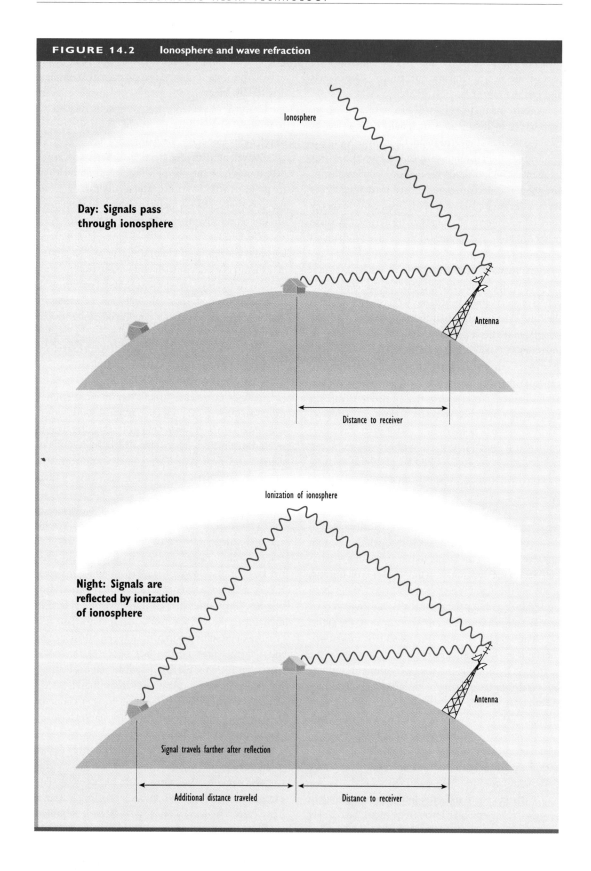

FIGURE 14.2 Ionosphere and wave refraction

to a maximum of 100,000 watts of power to send its omnidirectional signal about 60 miles with consistent quality. The station operating on channel 13 needs 316,000 watts of power to achieve about the same coverage area. Other factors affecting the station's coverage area include the height of the antenna above the average terrain and the electrical efficiency of the antenna—its *gain*.

Frequency Allocation
Because the different wavelengths of the spectrum react differently to various conditions, some frequencies are better suited to one communication service than others. Services such as television require a large amount of spectrum space to transmit all of the information required to compose the moving, color pictures that we see on-screen. Other services, such as satellite communication, require shorter wavelengths that will transmit through the ionosphere.

Each country has specific electronic communication needs but must share the natural electromagnetic spectrum with its neighbors, as the spectrum does not conform to borders. To maximize the use of this limited natural resource and not interfere with the transmission and reception of signals in other countries, the world's nations must cooperate in the use of electronic communication. We rely on the International Telecommunication Union (ITU) to sort out the utilization of the electromagnetic spectrum worldwide.

To assign frequencies for the expanding number of uses on the spectrum, the ITU has divided the world into three regions and allotted frequencies for different communication needs. Individual national governments may then issue licenses to operators to use portions of the spectrum in their respective countries.

The FCC licenses the use of the spectrum in the United States. As mentioned, AM radio is in the medium frequency and occupies space from 300 to 3,000 KHz. Each AM station uses 10 KHz of bandwidth (10,000 cycles of spectrum space) extending from 535 to 1,705 KHz. These frequencies constitute the standard broadcast band. The available bandwidth provides 117 channels for the use of AM radio stations. The high frequencies are between 3 and 30 MHz. Television channels occupy two different parts of the spectrum. Channels 2 through 13 are in the VHF portion of the spectrum, which runs from 30 to 300 MHz; channels 14 through 69 are in the UHF range in the 300 through 3,000 MHz portion of the spectrum.

Satellite news gathering truck serves as an uplink needed to transmit television signals on the Ku-band to a communications satellite located 22,300 miles above the equator.

Each television channel requires 6 MHz of bandwidth. There is a large space in the spectrum between channels 6 and 7 that is not used by television. A portion of this space is allocated to FM radio. One hundred FM channels, each requiring 200 KHz of the spectrum, are positioned between 88 and 108 MHz.

The multiple multipoint distribution system, what is called "wireless cable," uses frequencies at the top end of the UHF portion of the spectrum in the 2,500 MHz range. These television channels cannot be received by a television set that has a tuning range up to only 216 MHz; nor can the television receiver tune signals from a satellite dish *(TVRO, or television receive only)*. These signals, which are in the *superhigh frequency (SHF)* range of 3 to 30 GHz, must be converted downward to the range of the television receiver. Most satellite-delivered programming designed for use by cable subscribers is rebroadcast from the satellite to the earth station on *C-band*, which uses frequencies between 3 and 6 GHz. Another set of frequencies, called the *Ku-band*, extends from the 11 to 15 GHz range. These are used for, among other things, *satellite news gathering (SNG)* by local television stations and *direct broadcast satellite (DBS)* systems.

With the advent of high-definition television (HDTV), which requires massive amounts of bandwidth, and cellular telephone service, which may soon be requested by nearly all current telephone users, the FCC is under constant pressure to provide more licenses to use the electromagnetic spectrum. Efficient spectrum management will be a challenge in the twenty-first century. Luckily, as the demand for the spectrum has increased, new technology has made it possible to use higher frequencies effectively.

For example, for the VHF portion of the spectrum to be freed for the high demand for new cellular services, and for the spectrum to be made available for the projected HDTV services, the FCC mandated in the late 1990s that a new television service be implemented in the United States. The *Advanced Television Standards Committee (ATSC)* developed new standards for a digital television transmission system that would incorporate compression to make better use of the limited spectrum. Either a digital HDTV signal or up to four digital *advanced television (ATV)* channels could be squeezed into the normal 6 MHz bandwidth assigned to each television channel.

The FCC mandated that television broadcasters move transmission to the new digital standard on newly assigned UHF channels on a phased schedule, with network affiliates in large markets changing first and all stations broadcasting the ATV standard by the year 2003. All current analog and VHF *standard television (STV)* transmission is expected to be phased out by 2006. During the transition period, many television stations will be expected to broadcast both STV and ATV formats to meet the needs of the public as they buy new television receivers.

THE CARRIER WAVE

To use the electromagnetic spectrum for communication, several processes must be undertaken. First, a carrier wave must be created. A *carrier wave* consists of electromagnetic energy in the radio frequencies (see figure 14.3). The broadcast transmitter uses a quartz crystal, which vibrates at a specific frequency when shocked with electric current. This can be used to control the speed of oscillation of an alternating current. The crystal in the transmitter controls the frequency of the alternating current that creates the carrier wave. An alternating current changes direction,

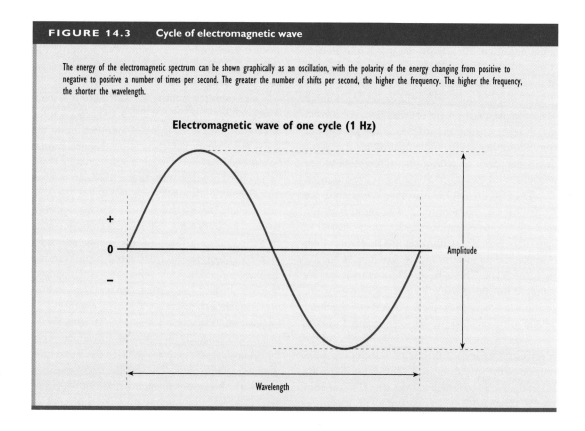

FIGURE 14.3 Cycle of electromagnetic wave

The energy of the electromagnetic spectrum can be shown graphically as an oscillation, with the polarity of the energy changing from positive to negative to positive a number of times per second. The greater the number of shifts per second, the higher the frequency. The higher the frequency, the shorter the wavelength.

or *polarity*, many times per second. For example, the electric current that powers the light bulbs in our homes changes its direction of flow 60 times per second. Alternating current needed to create radio waves must change direction much faster. An AM radio station operating at 620 KHz must create a signal with an alternating current that changes direction at the rate of 620,000 times per second. That means that the current must begin to flow in one direction, reach a peak, slow, stop, reverse direction, peak, slow, stop, and repeat the process 620,000 times during each second of operation.

Once the correct carrier frequency is achieved, the alternating current is amplified, then the high-power, high-frequency alternating current is sent through a coaxial cable or a waveguide to the transmitting antenna, which often sits atop a tower high above the transmitter. This alternating current radiated from the antenna—called *RF (radio frequency)*—constitutes the carrier wave of the station and is the frequency or channel on which the station operates. Information, or programming, can be modulated onto the carrier wave.

The Antenna

In its simplest form, an *antenna* is nothing more than a length of wire with an insulator at each end suspended above the ground. If that wire is cut to a length that is equal to one-half the wavelength of a radio frequency sent into it, electrons in the antenna will oscillate freely. As the electrons move back and forth on the antenna, an electromagnetic field is built up around it. As the expanding and contracting electromagnetic fields build and collapse around the antenna, the energy that is radiated into space is the radio wave that carries radio and television messages.

Radio frequency energy travels in a combination of direct waves, ground waves, and sky waves (see figure 14.4). *Direct waves* travel in a line of sight from the transmitting antenna to the receiving set. The signal transmitted via direct waves is often the most reliable and consistent, but it is limited in the distance that it will travel. Because of the curvature of the earth and the line-of-sight characteristics, high transmitting and receiving towers are needed to send the signal beyond the horizon. Services in the higher frequencies of the spectrum, such as television

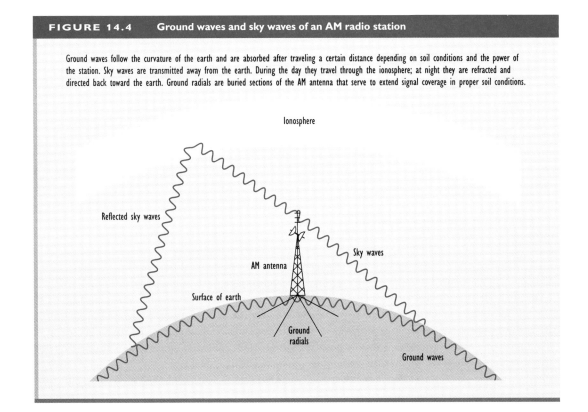

FIGURE 14.4 Ground waves and sky waves of an AM radio station

Ground waves follow the curvature of the earth and are absorbed after traveling a certain distance depending on soil conditions and the power of the station. Sky waves are transmitted away from the earth. During the day they travel through the ionosphere; at night they are refracted and directed back toward the earth. Ground radials are buried sections of the AM antenna that serve to extend signal coverage in proper soil conditions.

stations, FM radio stations, microwave relay stations, and satellite communication systems, all rely on direct waves for transmission.

Lower-frequency services, such as AM radio, rely in part on *ground waves*. Low frequencies and subsequent long wavelengths travel a few yards above the surface of the earth for hundreds or thousands of miles, depending on the frequency and power of the transmission. This ground signal follows the curvature of the earth.

Sky waves, the third type of radio wave, have somewhat more complex radiation patterns and are therefore unpredictable in their coverage area. As stated earlier, lower-frequency waves can be reflected off the ionosphere, which is composed of several layers of gases above the earth's breathable atmosphere. The widely spaced molecules of the gases can ionize when exposed to the bombardment of solar energy. As they ionize they allow the long waves of low-frequency transmissions to pass through. When night approaches, the sun no longer ionizes the atmosphere, and those layers become thinner; this changes the angle of refraction, and the waves are directed back to earth over a wide distance. The coverage area of an AM radio station expands greatly at night because of this phenomenon and accounts for our ability to pick up stations from distant cities late at night but not during the daytime hours.

The FCC, recognizing that lower-frequency signals can travel much farther at night, issues licenses to some AM stations that stipulate that those stations must *sign off,* or cease broadcasting, after sundown. Other stations must reduce their power at night so that their signal will not interfere with that of a station in a distant city operating on the same frequency. Still others are required to broadcast in a directional pattern at certain times.

The electromagnetic waves emanating from the antenna of a broadcast station can be controlled in a variety of ways, thus controlling the station's coverage area. Certainly, the easiest way to vary the coverage area is through the amount of electric power fed into the antenna. All other factors remaining the same, the more power fed into the antenna, the farther the coverage area will extend. Stations that are licensed to serve local populations are limited to low power, whereas stations designed to serve larger regions are authorized to operate at much higher power. Because higher frequencies require more power to send the signal a given distance than do lower ones, the FCC authorizes higher power for UHF television stations than for those operating in the VHF range. In addition, VHF channels 7 through 13 are given greater power than channels 2 through 6 so the coverage areas of both groups of frequencies are nearly equal. This is done for competitive purposes so that a station operating on channel 2 will not have the advantage of reaching a greater potential audience than one licensed to operate on channel 13. VHF TV frequencies will be phased out after 2006.

Another way to control a station's coverage pattern is through electrical *phasing* of an antenna array (several antennas connected to the same transmitter). If the waves from each element, or antenna, are peaking and reversing at exactly the same time, the waves are in phase and will reinforce each other to create a broader coverage pattern. If the waves are reversing in opposite directions at the same time, they are out of phase 180 degrees and will cancel each other out, thus reducing the coverage pattern to almost nothing. By controlling the phasing of an antenna array through delay lines, a distorted coverage pattern can be created that will reinforce and extend the coverage pattern on one side of the antenna and collapse it to almost nothing on the other side. This technique makes it possible for the broadcast station to direct most of its signal to a highly populated area and away from a sparsely settled one.

A third way to control the signal is through *polarization*—that is, through broadcasting the carrier wave in either a vertical pattern or a horizontal pattern. Waves in the electromagnetic spectrum can be transmitted by the antenna in either a horizontal or a vertical polarization in relationship to the horizon. Television signals are almost always broadcast with a horizontal polarization. FM radio used a circular polarization, which combines both vertical and horizontal signals needed to reach listeners in cars. For a signal to be detected from one of these services, the receiving antenna must be orientated the same way as the transmitting antenna—an antenna designed to receive the vertical signal will reject a horizontal pattern. Very little usable signal will be picked up if the receiving antenna is polarized 90 degrees from the transmitting antenna. This phenomenon is used to advantage in satellite communications, where the same frequencies can be used twice without interference if one antenna is horizontally polarized and the other set for vertical polarization.

The carrier wave transmitted and received by itself is of little value in the communication process. The content of the communication must be superimposed on the electromagnetic waves. This process is called *modulation*, which means a change or variation of the amplitude (AM) or frequency (FM) of the electromagnetic wave so that it carries sound or video information.

AM Radio

The second main function of the transmitter, after creating the carrier wave, is to modulate that wave so that the changes in current that represent the audio or video signal can be transmitted to the home receiver. The first radio broadcast systems were known as AM because the system of modulation used to superimpose the content on the carrier frequency was amplitude modulation.

In AM radio the transmitter creates a carrier wave of a set frequency and amplitude. A second electronic signal that varies in accordance to sound elements created by the radio microphone is then superimposed on the carrier frequency. This causes the amplitude of the carrier frequency to vary in relationship to the changes found in the audio signal.

Properties of AM

Many of the physical properties associated with AM radio that have vexed station managers, engineers, listeners, and the FCC over the years are associated with the long wavelengths of the medium frequency band and are discussed earlier in the chapter. These include ground conductivity and the fact that the ground wave will travel farther in moist soil than in dry; sky waves, which mean that the signal of an AM radio station will reflect back from the ionosphere to the earth in a different pattern at night than during the day, thus extending coverage patterns greatly when the sun sets; and gravity, which allows the longer waves at the lower frequencies to bend around natural and artificial obstacles such as mountains and buildings.

Other properties of AM radio come from the process of superimposing the signal on the carrier wave through amplitude modulation. For example, much of the static heard on AM radio is created by natural phenomena such as lightning. In addition, artificial static on AM radio is caused by electric motors, malfunctioning electrical spark systems in automobile engines, and electric power transmission lines. Each of these is a source of electromagnetic energy that can potentially block out the AM radio broadcast signal.

Only one radio or television station can broadcast on the same frequency in the same geographical area at the same time. If two stations attempt to transmit simultaneously on frequencies that are close together, interference called either *co-channel* or *adjacent channel* will result.

The FCC licenses stations to broadcast on specific frequencies with restricted power output. AM radio stations are licensed to operate as Class I to Class IV stations.

Class I, also called clear channel, stations usually operate with a maximum power of 50,000 watts and are designed to serve the remote rural areas as well as major cities. There is usually only one dominant station assigned to each clear channel. Because these stations are alone on a particular frequency, they can operate day and night at full power without interference. Class II stations are secondary but also operate on a clear channel frequency. There are actually five categories of Class II stations. Depending on the frequency used, these must reduce power, go directional, or in some cases leave the air at night so that they don't interfere with the signal of a Class I station. Class III stations are designed to serve a region, including a city and the rural area surrounding it. By limiting the power of the stations on these frequencies to a maximum of 5,000 watts, the FCC enables many stations to operate on the same frequency without interfering with one another's signals. Class IV stations are designed to serve local populations and are limited to a maximum power of 1,000 watts in the daytime and 250 watts after sundown. By operating at this low power, as many as 150 stations, each serving its own small town, can operate on the same frequency so long as the stations are separated geographically.

FM Radio

During the first three decades of radio broadcasting, audiences had to be satisfied with the fading signals, static, and low fidelity of the amplitude-modulated system. After World War II, the FCC authorized the present frequencies that are used for FM radio. The FM system, which utilizes an entirely new concept in modulating the audio information onto the carrier frequency—frequency modulation—did much to eliminate the constant static that had plagued AM. With the increased bandwidth allocated for FM broadcasting, a wider range of audio frequencies could be transmitted, more closely matching the sounds that the human ear could hear. In addition, the increased range of the spectrum that was allocated for the service allowed enough space for the transmission of stereo. In recent years FM has become the radio service of choice for most U.S. listeners because of this improved signal quality.

Properties of FM

The FCC has assigned the frequencies of 88 to 108 MHz for the use of FM broadcast radio. This assignment is in the very high frequency range of the electromagnetic spectrum, and therefore the signal with its shorter wavelength does not react to the physical environment in the same way as does the AM signal. In this portion of the spectrum, the signal travels from the transmitter to the receiver in a line of sight, using direct waves rather than following the curvature of the earth. This means that the transmission antenna must be high above the ground to reach a larger coverage area.

Because the FM service uses the VHF portion of the spectrum, the shorter waves are not reflected back by the ionization of the ionosphere. The signal travels into space and is not reflected back to earth at night. This means that the coverage area is the same at night as it is in the daytime. Therefore, the broadcaster can be assured of a reliable coverage pattern twenty-four hours a day and will not have to change power or direction of the signal at sunup and sundown. The audience is assured of a consistent signal.

Frequency Allocations

The FCC developed three classes of commercial FM station licenses that are based on the maximum *effective radiated power (ERP)* of the station. The effective radiated power is determined by a formula that multiplies the power of transmitter times the gain (or number of radiating elements) of the antenna times the height of the antenna above the average terrain.

FM stations are licensed as Class A, Class B, and Class C. All noncommercial, or public, stations are assigned frequencies in the first twenty channels (88 to 92 MHz), which have been reserved for that use. Class A stations are limited to an effective radiated power of 3 kilowatts (kW) and meant to serve local areas with a coverage area limited to about fifteen miles. These stations can be assigned to any of the three zones. Class B stations are limited to a maximum effective radiated power of 50 kW and generally cover an area of about thirty miles. Class C stations can operate with up to 100 kW of effective radiated power; the signal from these stations can cover an area of about sixty-five miles. To reduce the chance of co-channel interference, the FCC has also enacted rules concerning minimum mileage of separation among FM stations operating on the same channel.

FM Stereo and Subsidiary Services

Because the bandwidth is so great, it is relatively easy to transmit a stereo signal in FM radio. As does AM, the FM transmitter creates *sidebands,* which are frequencies above and below the center carrier frequency that can be used to carry additional information to the radio receiver. Almost all FM radio stations use a portion of their sidebands to carry the audio information for the two separate channels required for stereo music transmission. The technology to accomplish this service dates back almost fifty years and is called *multiplexing,* which means to combine and to send two signals over the same channel.

Whereas most FM stations use the sidebands for carriage of the stereo signal, some use their extra transmission capacity for other services. The FCC has licensed some FM stations for subsidiary-communication authorization (SCA) service, which allows the station to transmit a variety of communication services to specific receivers, using a portion of the FM sidebands. For example: Messages can be sent to pagers; music services, complete with proprietary commercials, can be transmitted to grocery stores; reading services for the blind can be sent into the home; or signals can even be sent to control peak electricity use in the home through the FM transmitter equipped for SCA service. As the need for wireless electronic communication services increases, the SCA service enables the broadcaster to develop new revenue streams to help offset the increasing costs of operating a radio station.

Broadcast Television

The transmission of moving pictures, color, and sound combined into a television signal over the electromagnetic spectrum, in theory, is no more difficult than the transmission of audio information. The challenge has been to squeeze all of the elements into a limited amount of spectrum space. For broadcast of STV signals, the United States uses the set of television standards established by the National Television System Committee (NTSC), which dictates the transmission of thirty individual pictures, each composed of 525 lines of information, each second. The vast amount of visual information, combined with synchronization signals, color information, audio, and, in many cases, stereo sound, demands maximum utilization of the 6 MHz of bandwidth that the FCC has assigned to each television channel. Because digital television (DTV) and high-definition television use a variety of standards developed by the Advanced Television Standards Committee, NTSC transmitters will be phased out by the year 2006.

To modulate the required information, two separate transmitters are used for each television channel (although they might be housed in the same enclosure and look like one device). One transmitter is used to superimpose the visual information on the carrier wave through amplitude modulation; the second, less-powerful transmitter uses frequency modulation to broadcast the audio signal. The output of the two transmitters is combined in a diplexer and sent to a common antenna, which is usually mounted on a tall tower to achieve maximum broadcast coverage.

Properties of Television Frequencies

As explained earlier in this chapter, current broadcast television frequencies are in two different portions of the electromagnetic spectrum: Channels 2 through 13 are in the VHF band, and channels 14 through 69 are in the UHF range. The frequencies in television are measured at the bottom end of each 6 MHz channel. Channel 2 starts at 54 MHz; channel 14 is at 470 MHz; and channel 69 is at 800 MHz. As the frequency increases, the spectrum takes on different characteristics. For example, the signals at the higher frequencies tend to travel in straight lines and become more resistant to effects of the earth or of the ionosphere. The portion of the spectrum used for television relies exclusively on direct waves. Sky and ground waves are not a factor at the shorter wavelengths in the VHF and UHF bands, therefore transmission between the television station and the home receiver must be line of sight to ensure a good picture. If the TV receiver is in a shadow of a large building or in a valley, chances are the signal will not be very good.

To overcome these problems, television antennas are located as high above average terrain as possible. (The FCC imposes rules about the height combined with power limitations.) In addition, the signal is horizontally polarized and directed toward the horizon so that the majority of the signal is sent toward potential television homes rather than allowed to radiate into space.

At the higher UHF frequencies, the signal pattern becomes even more erratic, and signal reflections from hills, buildings, and trees can make consistent reception difficult even in an otherwise strong reception area of the coverage pattern (see figure 14.5). All television frequencies are susceptible to *ghosting*, or double images, on the television screen caused by a reflected signal arriving at the set a millisecond after the direct wave, but UHF frequencies seem more prone to this type of distortion than are the VHF frequencies. These multiple reflections can make the picture almost unwatchable in some strong reception areas.

The coverage areas for television stations are measured as grade A and grade B contours, which are circles drawn on a map of the area around the transmitter (see figure 14.6). The grade A contour, or city grade, includes the area in which consistently good service can be expected at least 90 percent of the time. A good picture should be expected in the grade B contour at least 50 percent of the time. Digital signals perform differently, as they stop rather abruptly at the edge of the contour. The set either receives an ATV signal or not. The A and B contour will not exist after 2006.

Because of the varying power requirements of the different ranges of the spectrum, the higher power allocated to a station operating on channel 12 might create a coverage pattern similar to that of a station operating on channel 3 but with only one-third the power. A UHF station in the higher range of those frequencies might have a limited coverage pattern even though it is broadcasting with 5,000 kW of power. Low-band VHF stations have always had an

314 PART III ELECTRONIC MEDIA TECHNOLOGY

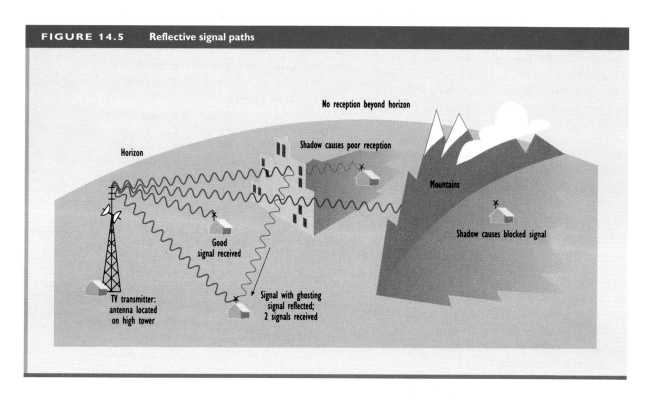

FIGURE 14.5 Reflective signal paths

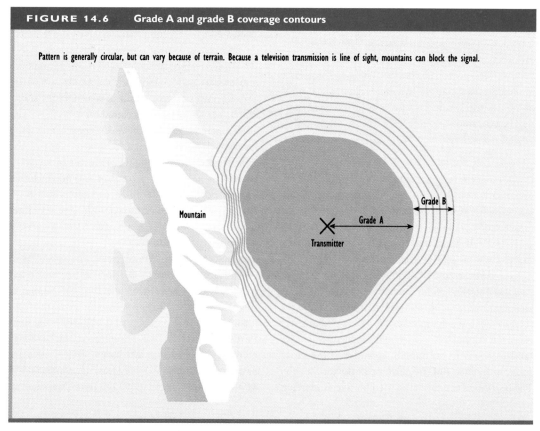

FIGURE 14.6 Grade A and grade B coverage contours

Pattern is generally circular, but can vary because of terrain. Because a television transmission is line of sight, mountains can block the signal.

advantage over other stations in both operating cost and coverage of the market. Because of the intricacies of signal propagation, requirements of high power, and difficulty of reception, UHF stations have been at a great competitive disadvantage until cable put them on par with the lower-frequency VHF stations, but the FCC has ordered that television broadcasting be digital in the future and use UHF frequencies. VHF channels will be reassigned to other services after the year 2006.

Stereo and Vertical-Blanking Services

Stereo, audio, and *SAP (second audio program)* use the sidebands of the audio portion of the television signal, but information can also be broadcast in unused spaces of the visual portion. The synchronization portion of the signal contains a *vertical blanking interval (VBI)*. Because no picture information is sent during that period, there is room to transmit a limited amount of information for purposes completely separate from the television program. This information normally includes coded signals that tell color television receivers the proper color phasing. In recent years the television networks have transmitted information that can be used to detect whether the affiliated television station has transmitted specific commercials.

Perhaps the best-known vertical blanking interval service is closed-captioning for the hearing impaired. Many network programs, as well as the programs of some local stations in larger markets, are transcribed into script form. The dialogue of the program is converted into text and stored electronically using a device similar to a stenograph. The resulting signal, sent in the vertical blanking interval, can be detected and displayed on the television screen as subtitles, so the hearing impaired can read the audio information while the subtitles are undetected by the majority of the audience.

Another VBI service was *teletext*. Multiple frames of such text information can be broadcast to the home during the broadcast of any television program. The teletext can include sports scores, stock market reports, weather information, news wire copy, or airline schedules. The conventional television receiver, equipped with a VBI decoder and a frame grabber, can select one of the hundreds of potential frames of information and display it on the screen indefinitely. To date, teletext services have not found a market, and most of the experiments with the service have been terminated. The potential remains, however, to transmit data to the home with the conventional television signal. A drawback is that the information broadcast is not interactive: The user can manipulate a certain amount of information that is displayed, but cannot interact directly with the computer generating the displayed information. Using compression, ATV will offer several television channels routinely, along with a variety of communication services, such as those previously described.

OTHER BROADCAST TECHNOLOGIES

Low-Power Television

First envisioned in the 1950s, *low-power television (LPTV)* became a reality in the early 1980s when the FCC began licensing stations to serve small communities and minority populations in larger markets. The transmission technology was based on an established television *translator* system that had become prominent in the west, providing small, isolated communities with a signal from a distant market. The translator is a low-power (usually 10 to 100 watts) television transmitter operating on a VHF or UHF frequency. The unmanned transmitter is usually located on a mountaintop above the community, where from that vantage point it can receive an off-air television signal from a conventional television station. The signal is demodulated, and the resultant audio and video signals are remodulated on a different frequency for retransmission to the community. Translators were often owned by television stations to extend their signals to significant communities beyond the normal coverage pattern, although some were established by community organizations as a precable attempt to bring a television signal to local viewers. A translator is licensed only to retransmit a television signal; it is not permitted to originate local programming.

Low-power television stations use the same type of transmitter as do translators, but they are established to serve a community with local programming. Because the technical standards are not as stringent, the LPTV station can broadcast a signal from ½-inch S-VHS videocassette recorders and use inexpensive cameras for production. The signal of the low-power station can be received on any conventional television set on either the UHF or VHF channels but, because of the power limitation, usually only within about ten miles of the transmitter.

After a slow start, caused in part by an overload of applications to the FCC for channels in the early 1980s, LPTV stations are now flourishing in number if not in profitability. Many large cities are served by LPTV stations that are programmed in languages other than English.

Microwave

We normally think of *microwave* communication systems as high-frequency, point-to-point services designed to transport signals from a studio to a transmitter site, or from an *electronic-news-gathering (ENG)* truck to the television studio. But other microwave services established to send information to the consumer have gained new importance in the past decade. These services are *multiple multipoint distribution systems (MMDSs)* and *instructional television fixed services (ITFS)*. Both use the same technology, but ITFS is designed for use by education for instructional television delivery, and the MMDS has become known as wireless cable. Wireless cable distributes traditional cable network signals to a local community by a closed-circuit broadcast system, thereby eliminating the need and expense to wire that community with coaxial cable.

Both ITFS and MMDS operate by modulating the audio and video signals on a low-power carrier wave that has frequency in the range of 2,500 MHz and above. The highest UHF television frequency the home receiver can tune is at 800 MHz, so the standard TV receiver cannot pick up the signals of ITFS and MMDS. To receive these services intended for educational or public audiences, the user needs a receive antenna, often connected to a microwave-style reflector dish in marginal signal areas, and a down-converter, needed to change the 2,500 MHz frequency signal to a low-band VHF channel that the conventional television receiver can display. Because extra equipment is needed, access to the signal can be controlled. Generally, wireless cable operators lease the antennas and down-converters to their customers, making it a pay service. MMDS channels were originally designed as a tariff service: The operator was granted the right by the FCC to lease the channel space to an end user at a fixed rate that was controlled by the FCC. Also in the same band of frequencies are *operational-fixed service (OFS)* channels, which provide a service to businesses for the transmission of video and data between offices. A bank might use OFS channels to communicate with all of its branches in one city, for example. MMDS operators are now providing fast wireless Internet service to homes in several large cities.

The ITFS system uses the frequencies from 2,500 to 2,690 MHz, and the FCC generally assigns frequencies to an operator in blocks of four channels. Normally, educational operators use ITFS channels for distance-learning purposes. For example, they are used to send instructional programs from a college or university to its branch campuses or to send business and engineering classes from a university to part-time students working for corporations in the area.

SATELLITES

Communication satellites have probably had the greatest impact of any new communication technology developed since the mid-sixties. In 1945 science fiction writer Arthur C. Clarke postulated that a satellite positioned in an orbit over the equator at 22,300 miles above the earth would travel at a rate of speed that would match the period of rotation of the earth and would therefore be geosynchronous. Because of its apparent stationary position, a geosynchronous satellite could be used as a communication platform from which radio transmissions would reach huge coverage areas from one transmitter. In fact, he surmised that the entire earth could be covered with

Inexpensive MMDS antenna used to receive television signals from wireless-cable operations.

reliable communication signals from three strategically positioned satellites. His theories became testable only after the launch of the first satellite.

Ted Turner's Superstation and HBO were being delivered to cable television systems nationwide by the mid-1970s. A decade later the conventional networks were delivering programming to their affiliates by satellite, and more than 150 cable services and networks were on "the bird." Syndicators were distributing television programs both internationally and domestically by satellite, and by the end of the 1980s much of the business data that turns the wheels of international commerce was transmitted by satellite. In addition, many of the nation's television stations were routinely transmitting their own live news segments back to their studios via SNG trucks.

Currently, more than thirty domestic communication satellites are in operation, transmitting or receiving on either the C-band or the Ku-band of frequencies (additional bands of frequencies, such as the *Ka-band,* may be used for satellite communication in the near future). As indicated earlier, the C-band frequencies range from 4 to 6 GHz (one gigahertz is equal to 1 billion hertz), and the Ku-band frequencies are in the 10 to 14 GHz range. Most of the satellites have transponders needed to transmit many television channels along with hundreds of audio and data channels. The transponder, which is powered by solar-charged batteries, receives a signal from a ground station, converts the signal to a different frequency, and rebroadcasts it back to the earth (see figure 14.7).

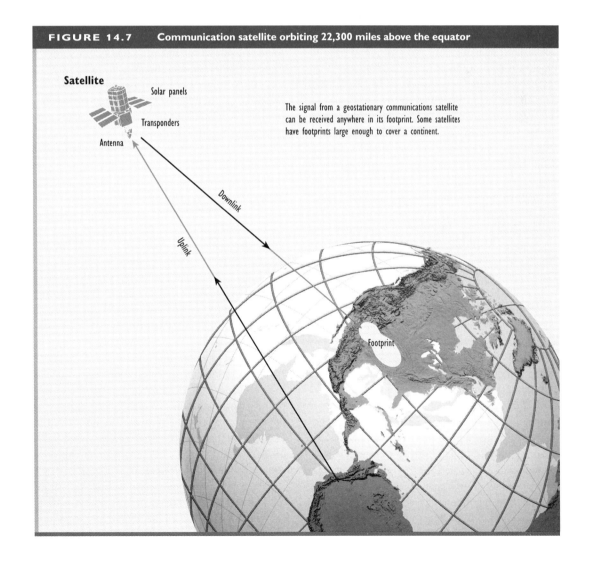

FIGURE 14.7 Communication satellite orbiting 22,300 miles above the equator

The signal from a geostationary communications satellite can be received anywhere in its footprint. Some satellites have footprints large enough to cover a continent.

There are several limitations to communication satellites. One is the power consumption required to transmit on a number of channels simultaneously. Power for the transponders is generated by solar arrays that recharge on-board battery systems for use when the satellite is in the shadow of the earth. Launch limitations of size and weight dictate the number of solar panels and batteries that can be included, therefore limiting the power consumption. Luckily, the efficiency of the electromagnetic waves traveling through space, coupled with high-gain directional antenna systems, make it possible for the extremely low-power signals to travel the first 22,280 miles with little attenuation. Most of the interference that the signal encounters is found in the few miles of atmosphere just above the earth's surface.

An additional limitation is the expected lifespan of the satellite. Some transponders, as well as power systems, are likely to fail during the ten- to fifteen-year satellite lifespan, but the factor that determines the life limitation is the propellant required to constantly maneuver the satellite to keep it in the proper orbit needed for a geostationary position. Once the pressurized propellant system is exhausted, the satellite is free to drift out of orbit and, therefore, out of alignment with the earth stations aimed toward it. Currently, as all satellites use the same bands of frequencies, the only way that the earth station can distinguish one satellite from another is through precise aiming of the dish along the azimuth and zenith. Satellites are parked in space about four degrees apart, which equals many miles of separation at the distance of 22,300 miles from the earth; however, it is a small arc of focus for the average dish.

Uplinks and Downlinks

The signal path from the ground station to the satellite is called the *uplink* (although that term can also apply to the hardware used), and the signal path from the transponder to the ground station is the *downlink*. A commercial ground station, called a *teleport*, is used to communicate with a number of satellites on several transponders simultaneously. The teleport consists of many transmitters feeding signals to antennas focused by large parabolic dishes. The teleport is near a major metropolitan area, because that is where most business is generated. Because the antennas must be shielded from extraneous *radio frequency (RF)* interference caused by terrestrial microwave paths, the dishes are often positioned in a valley so that the surrounding hills will shield them from the worst of the radio noise.

Another type of uplink is portable, owned by many television stations and used for satellite news gathering. Whereas the older C-band technology (4 to 6 GHz) required a dish of nearly forty feet in diameter that had to be transported on a trailer, the newer Ku-band antennas (10 to 14 GHz) are much smaller and, when complete with all of the associated equipment needed to originate a television program, fit neatly into a medium-size truck that is easily maneuverable in city streets yet self-supporting in terms of power when in the open country.

Many universities, television stations, and businesses operate small, stationary Ku-band uplinks. If the signal is to be carried on the C-band however, it is often received at a teleport and turned around—the signal is downlinked from the Ku-band transponder and uplinked to a different satellite on a C-band frequency so that the final customers can receive the signal on their C-band dishes.

The coverage pattern of the satellite, called the *footprint,* can be adjusted to cover different sections of the country. Some satellites are configured to create a second footprint, called a *sidelobe,* necessary to cover a secondary geographic area. The signal quality received from the satellite varies greatly depending on the latitude of the receiving dish. A dish in northern areas of the globe must be aimed lower to the horizon to focus on a satellite positioned over the equator, but the lower angle of the dish allows more ground noise to interfere with signal reception as well as requires the signal to transverse a greater depth of the atmosphere. To achieve an acceptable signal level, receivers in the higher latitudes must rely on a greater dish diameter to gather more signal.

The downlink signal is reflected by the parabolic dish to a central focal point. At that point the antenna, usually about one inch long, is connected directly to the *low-noise amplifier (LNA)*. The amplified signal is then transported via coaxial cable to the down-converter that strips the audio and video signal from the carrier. Additional circuitry is often included so that the audio and video signals can be remodulated on a low-band VHF channel for viewing on a conventional television receiver.

Although communication satellites were originally intended for point-to-point communication, some homeowners living in remote locations soon

TVRO (television receive-only) satellite dish is designed to receive programming transmitted from a satellite to a television station or a cable television system via C-band frequencies. Networks and syndicators deliver much of their programming by satellite communications.

Eighteen-inch DBS home antenna used to receive television signals directly from a satellite. In addition to the dish or antenna, the viewer must also invest in a converter needed to change the frequencies to those that the television set can use and must also pay a monthly fee for programming.

discovered that a dish could provide a television signal that they could not otherwise receive over the airwaves. In 1985 the FCC legalized the use of receive-only dishes for the public. These dishes became known as *TVROs (television receive-only)*.

The *satellite master-antenna television (SMATV) system* designates a TVRO that is connected to a private master television antenna system. These minicable television systems are often found in apartment complexes to provide television antenna service to the individual residents. The SMATV operator usually pays the copyright fees necessary to decode the signal and then distributes the satellite signals combined with antenna reception of the local television stations to each television set in the complex. Sometimes the price of the service is included in the rental fee of the apartment, and other times it is collected each month from the subscriber.

Direct broadcast satellite (DBS) sends the signal directly from the satellite to the individual home. To achieve an antenna size that will be acceptable to most homeowners, the satellite transponder must operate with a more concentrated footprint and with higher power.

After several years of experimentation and some false starts, several companies were formed to offer television services directly to the home via satellite. In 1994 DirecTV, Inc., was formed as a subsidiary of a major communication satellite operator—the Hughes Electronics Corporation. United States Satellite Broadcasting (USSB), Primestar, and Dish Network were also early DBS providers. Mergers and significant startup economics have limited the field to two major competitors. DirecTV, offering 210 channels as well as Spanish-language services, has more than 8 million subscribers; EchoStar Communications Corp. has about 4 million. The growth rate of the DBS services has been phenomenal, with about a 40 percent increase in subscribers in 1999 alone.

In late 1999 President Clinton signed into law an amendment to the Satellite Home Viewer Act that would allow DBS services to offer local network-affiliated stations to be redelivered via satellite in their own markets. This local-to-local service is seen as key to making DBS service competitive with cable by offering the local channels on a basis equal with the cable network channels. Currently, many dish owners must also install a television antenna to pick up local television stations.

The delivery of local stations, coupled with low equipment cost—$59 for a DirecTV dish and related equipment—is expected to spur on the growth of the DBS industry.

Wired Communication Systems

In addition to broadcast, electronic mass communication systems use copper, silver, aluminum, or glass wire for signal distribution. Telephone systems are the best-known wired communication systems.

Most telephone lines are two wire and voice grade, meaning that the signal they carry is susceptible to noise interference and is limited in frequency response to about 4,000 Hz. Although this limited frequency response is fine for conversation, even AM radio stations order higher-quality lines for remote transmissions. The two wires needed for signal transmission are twisted around one another rather than run parallel to each other. By twisting the wires, there is less chance of setting up inductive currents in the wires, and they tend to radiate less signal and block some outside electromagnetic interference.

Twisted-pair wires, usually in sets of four called *switched-56* (56,000 bits per second [bps] multiplied to an effective rate of 160,000 bps), are now used for the transmission of digital signals created by computers. Analog signals are frequency dependent; digital signals are time dependent. Therefore, the frequency limitations imposed by the twisted-pair wires are of less concern when the signal to be transmitted is digitized. By converting a video signal from analog to digital and using compression techniques, a full-motion picture can be sent over twisted-pair telephone lines even with their extremely limited frequency response.

Network Standards

In the past, the quality of the transmission line was measured in terms of bandwidth: Music transmitted over a wire required a greater bandwidth than voice for the receiver to pick up all of the subtle nuances. The enormous amount of data needed to create the full-motion, color television picture requires 6 MHz of bandwidth. Digital communication, however, depends on the speed that the information can be sent. *Baud rate* is a measure of the movement of the signal on a transmission line. A baud rate of 9,600 bps means that more information can be sent in a specific period of time than a baud rate of 1,200 bps. A greater bandwidth in analog communication equals a greater speed in digital communication.

The telephone companies' latest entry into the digital communication field is *Integrated Services Digital Network (ISDN)*—a set of standards that allows end-to-end digital telecommunication services. The ISDN systems would be capable of delivering all forms of communication services to business or residential users. The wire that delivers telephone signals into the home would also carry information that controls the environmental and security systems of the building. This single-wire technology is predicted to be the electronic superhighway of the twenty-first century.

Computer equipment manufacturers have long subscribed to *ASCII* (American Standard Code) and constructed equipment to the RS-232-C code to achieve some degree of compatibility. With the additional requirement that individual computers "talk" to one another, however, the architecture of the telecommunication networks had to be carefully designed. Equipment of similar architecture could communicate through bridges, but *gateways,* or access paths, had to be developed to allow different computer architectures to communicate.

The International Standards Organization (ISO) has been working on the development of a set of protocols that would allow a worldwide computer interface. Its OSI model (ISO/OSI) stands for *open-systems interconnect.* In the interim, the *Transport Control Protocol/Internet Protocol (TCP/IP)* allows users to communicate by computer terminal with other Internet subscribers around the world. The Internet, originally a U.S. federal government telecommunication network, provides an interconnection for personal computer users throughout the world.

Video Compression

The purpose of *video compression* is to squeeze the television picture and sound, which normally requires 6 MHz of bandwidth, into a telephone line that is limited to about 4 KHz. One of the first examples of video transmitted through twisted pair was the AT&T picture phone, introduced in the late 1960s but never developed or marketed. From that evolved the concept of slow-scan video, which can generate freeze-frame black-and-white still pictures that could be transmitted over conventional telephone lines. Slow-scan technology has limited use in teleconferencing and instruction but is perhaps of more value in security observation.

With the advent of higher-frequency telephone lines, digital technology, and video compression, near full-motion video can be transported over twisted-pair cable. Video to be sent is fed into a *codec,* or coder-decoder, to be compressed and converted into a digital signal. The key to compression is to transmit only the changing part of the signal and thus require considerably less time and bandwidth for transmission. This compression technique relies on both spatial and temporal redundancy: *Spatial redundancy* consists of the large areas of the television picture that have the same brightness and color information from one frame to the next, such as the sky or the colored backdrop behind a speaker. *Temporal redundancy* means that, because there is very little movement from one frame of the picture to the next, only that movement needs to be transmitted thirty times per second, not the entire picture.

Industry standards for video compression have yet to be fully developed, but if there is too much spatial compression, the picture lacks detail, and the color in the larger objects tends to bleed or blur into adjacent areas. Too much temporal compression causes the motion in the picture to become jerky. The goal is to develop a set of standards that gives a natural-appearing picture without requiring excessive bandwidth for transmission. Currently, compressed video is used for the two-way transmission of video for teleconferencing. In addition, the National Technological University uses digital compressed video to reach its student clients via satellite distribution using only 3 MHz of bandwidth to transmit full-motion color pictures of classroom instruction.

Digital compressed video will allow the further development of desktop video production and will make possible the delivery of video on demand to the home. This means that compression technology will allow the viewer to select a movie for viewing in the home from a catalog of thousands in a central library and have that program delivered over upgraded telephone circuits.

Streaming video is delivered over telephone lines to the home computer. The digital information is sent at the slow rate dictated by the quality of the telephone lines but is held and reassembled into a moving picture by the computer.

Although digital technology and video compression will allow the delivery of full-motion television signals through the switched telephone system in the near future, coaxial cable and fiber are now the wires of choice for the transmission of multiple television signals.

Cable Television Systems

Cable television systems began as **community antenna television (CATV)** in the late 1940s. The function of CATV was to intercept broadcast signals from distant television stations and deliver them into the homes of customers by placing a receiving antenna on a nearby hill, amplifying the signal, and sending it through coaxial cable or fiber-optic cable to the television set. Technically, cable television systems are composed of three parts. The head-end, the physical plant, and the home terminal. The **head-end** consists of the equipment used to pick up or, in some cases, to originate the signal. Customers are connected with coaxial cable.

Coaxial cable is constructed with a center *conductor,* which carries the high-frequency information; a *dielectric,* or insulator, which is often foam and is wrapped around the conductor; and a *shield,* made of braided copper or aluminum foil wrapped around the dielectric. The three layers are then covered with a rubber coating for protection from the elements.

This conductor configuration allows the cable to act like a pipeline for a portion of the electromagnetic spectrum. The high frequencies are contained around the center conductor and are not allowed to radiate into space. At the same time, the shield serves to create an interference-free environment for the signal. Perhaps the greatest advantage of the coaxial cable is that an almost unlimited range of frequencies can be transmitted through the wire at the same time.

Cable television systems take advantage of this capacity by transmitting between fifty and ninety television channels on one wire. The only limitation on the number of channels transmitted is in the amplifier used to boost the signal. If carriage of more than ninety channels is required, additional ones are carried on a second cable reusing the same frequencies.

The head-end receives signals from several sources. Normally, off-air signals from television stations are received by antennas that are set to the proper frequency. In addition, cable networks are picked up off communication satellites by TVROs that are dedicated to a single satellite. Also, some systems use point-to-point microwave links to bring in distant broadcast signals. About one-half of all cable systems use videocassette recorders to play back locally originated programming at the head-end. Other systems use twisted pair to deliver computer-generated text to message channels. The signals brought into the head-end building by any of these methods are processed and assigned a cable channel. Processing includes amplification as well as proper adjustment of the audio, video, and color levels. Signals received off-air and from satellite or microwave must be demodulated, removing the audio and video components from the original carrier frequencies.

The audio and video signals from all sources are remodulated onto a channel specific to that cable system. The information from a channel 3 television station received off-air might be carried on channel 27 on the cable system. By carefully controlling the frequencies of each of the cable modulators and limiting the audio and video levels, adjacent channels can be used to carry information without encountering interference. A misadjusted sound level on channel 3, however, might create interference problems on channel 4. Correctly balancing signal levels, although partially controlled by an *automatic-gain control (AGC)*, can be a nightmare for a technician working for a poorly designed system. Signal quality control is complicated by the fact that the lower channels travel farther with the same amount of amplification as do the higher ones. Amplifiers must therefore be adjusted to add more amplification to the lower frequencies and less to the higher ones.

Because the range of frequencies amplified by the cable system is continuous, most cable television systems use channels that are not tunable by older television receivers. Newer television sets that are cable-ready can tune the additional channels.

Low-band cable channels, starting at 54 MHz for channel 2, correspond with the over-the-air channels, but cable includes television channels in the spectrum space between VHF channels 6 and 7, which is devoted to other services, such as FM radio. Midband cable channels A through J are in this area on most cable systems now. Cable superband channels—7 through 13—are carried above the VHF high band. The number of channels that the system carries depends on the frequency range of the amplifiers. Systems operating with older 300 MHz amplifiers can carry about thirty-five channels, but those with 500 MHz systems often include up to sixty channels without resorting to a dual or second coaxial cable.

The output of all the channel modulators in the head-end building are combined and sent out on *trunk lines,* which carry the signal from the head-end to the various geographic areas of the community to be served. The trunk line, which consists of high-quality amplifiers spaced about 1,500 feet apart and large-diameter coaxial cable to minimize signal attenuation, is the first element of the physical plant. As the trunk passes streets or, more likely, alleys, signal is diverted by bridging amplifiers to *feeder lines* that are strung in back of the houses to be served. *Multitaps* are connected to the feeders about every four houses. These allow some of the signal to be diverted to the home by the *drop line,* which is attached to the television receiver. The feeder and drop lines can be constructed of less-expensive, smaller-diameter coaxial cable, because the signal is not required to travel far and the signal losses, therefore, are likely to be minimal. Cable installers and technicians are required to maintain the cable plant and to connect subscriber homes to the cable.

As the signal, whose strength is measured in *decibels (db),* travels through the physical plant, the resistance in the cable attenuates the current. Therefore, at intervals of about three times per mile, the signal must be reamplified. Each reamplification boosts the signal level back to the proper strength but also adds unwanted *noise,* or random electrical interference, to the signal. After about thirty re-amplifications, the noise, which is visible on the television screen in the form of "snow"—tiny spots of colored light over the picture—becomes objectionable. The physical length of any cable television system is therefore limited to about ten miles from the head-end. Systems that must serve a larger geographical area use several mini-head-ends called *hub sites.* Signals from the main head-end are delivered to the hub sites either by a high-quality coaxial cable called a *supertrunk* or through the airwaves on an *AML (amplitude-modulated-link)* system.

The home terminal of the cable television system consists of the coaxial cable drop line, which links the home to the physical plant, and usually a set-top converter, which gives older television receivers the ability to tune the midband and superband frequencies. In addition, the converter serves as a descrambler so that pay services can be received in the home. Some more-complex converters are addressable and have two-way capability, which allows the more technologically advanced cable systems to offer pay-per-view services to individual homes connected to the system.

Currently, data from addressable converters is sent *upstream*, or from the home to the head-end, at frequencies lower than the 54 MHz used for channel 2. Although the same coaxial cable can be used to carry this information upstream, a crossover filter is required at each amplifier to divert the upstream frequencies around the main amplifier and into a smaller amplifier that sends the signal in the proper direction.

Star Versus Tree Designs

The telephone system, which is designed for two-way switched service, has an architecture that resembles a star system, with rays of individual twisted-pair wires radiating in all directions to the homes from a central switching location (see figure 14.8a). Although a separate telephone wire to each home is expensive, it allows easy direct interconnection from one home to another. By interconnecting the exchange points, any home in the world with a telephone can be connected to any other home for direct two-way message

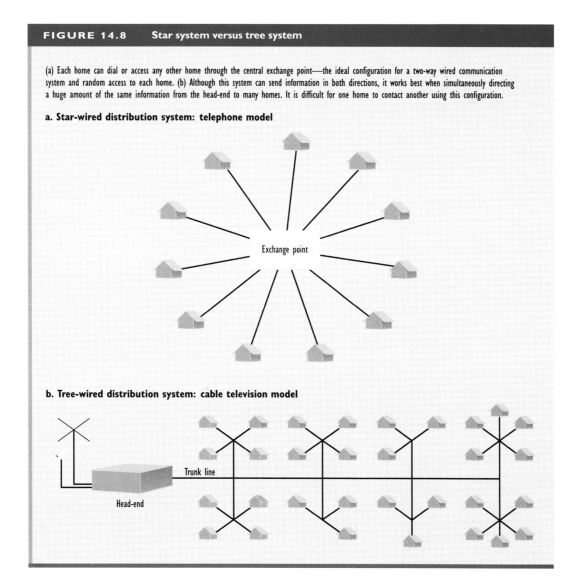

FIGURE 14.8   Star system versus tree system

(a) Each home can dial or access any other home through the central exchange point—the ideal configuration for a two-way wired communication system and random access to each home. (b) Although this system can send information in both directions, it works best when simultaneously directing a huge amount of the same information from the head-end to many homes. It is difficult for one home to contact another using this configuration.

a. Star-wired distribution system: telephone model

b. Tree-wired distribution system: cable television model

exchange. This star system configuration is an interpersonal communications model.

Cable television systems were originally designed on a mass communication model to deliver television signals one-way to many homes at the same time. The physical design that cable developed became known as a tree system: The large trunk lines carry the information to the feeder lines that branch up and down the alleys of the community (see figure 14.8b). Drop lines deliver the programming to the individual homes. Cable homes were never intended to electronically interact with one another for message transmission and can interact only with the head-end on a limited basis. Addressability, or the ability to send programming to only one home of the thousands that might be connected to the system, presents expensive technological problems for the cable system.

Fiber-optic Distribution Systems

Fiber-optic distribution systems, sometimes called *lightwave guides,* have proliferated since the first low-loss fibers were produced in 1970. The hair-thin glass strands now carry most of the long-distance telephone messages transmitted in the United States and carry data and telephone conversations across the ocean floors to link the continents. Television stations use fiber-optic links to send programming from the studio to a transmitter, and cable television systems are beginning to install fiber in place of the traditional coaxial cable trunk and feeder lines. Some experiments that are under way would use fiber to connect the home directly to cable or telephone distribution trunk lines.

The advantages of fiber-optics are considerable. The portion of the cable that carries the signal is constructed of a glass thread made from silicon, one of the most plentiful substances on earth, thereby making the cost of the cable low. In addition, the fiber cables are extremely thin, taking up a fraction of the space of the coaxial and twisted-pair wires that they replace. This becomes an important consideration as space in the conduits under our cities reaches a premium. The cable is also flexible and therefore easy to install. Probably the greatest advantage of the fiber system is that it can carry many times the information load of even the coaxial cable, thus making it the communication wire of choice with which to connect homes and businesses in the twenty-first century.

The fiber cable was originally developed as a method of channeling the output of early laser light communication experiments. The fiber consists of an extremely high grade of glass extruded as a fine wire. As the molten glass is drawn to the proper diameter, a cladding, or second coating of glass, is applied. The glass cladding acts as a mirror that reflects the light back into the fiber as the waves travel the twists and turns that the installed cable may take. The fiber, along with others, is packaged in a plastic sheath so that the twists and turns cannot be made so sharp that the glass threads break.

To turn the optical fiber into a communication system, both a light source and a light detector are needed. The light source can be either a light-emitting diode (LED) or a laser diode coupled directly to the end of the fiber. Either source creates high-frequency electromagnetic waves in the infrared spectrum. Visible light frequencies do not travel as far through the fiber and are used only to detect breaks in the glass. A photo diode is connected to the other end of the wire to detect light pulses and convert them into electric energy.

Both analog and digital signals can be sent through the fiber-optic system. Light-emitting diodes are used for digital systems, because they can be turned on and off quickly enough to create the binary code of 1's and 0's that is the language of the computer. AM or FM modulation techniques can also be used to vary the amplitude or the frequency of the laser light waves and send an analog signal through the fiber-optic wires.

At this time fiber-optic systems are used most frequently for point-to-point communication systems, because of the problems inherent in developing the many connections to the system needed for distribution to the home.

When the modulated light signal reaches a point of near attenuation, it must be amplified. In the fiber system, this means that the signal must be detected, converted to an electrical signal, amplified, and reconverted into light energy. The repeaters that are required in long-distance systems are expensive to install and maintain. Many more repeaters, as well as optical switches and couplers, are needed in a distribution system that takes the fiber into the home. It is less expensive at this point to convert the optical signal to an electrical one for distribution over the last mile to the home. This type of distribution is called a *hybrid fiber/coaxial (HFC) system.* Most new cable television systems are constructed as HFC systems.

One problem that has plagued fiber system contractors has been creating a good splice between two fibers. To achieve good light transfer across a joint, the ends of the glass must be polished and perpendicular to the path of the light. Then the two ends must be joined exactly, with no overlap or angle in the joint. Needless to say, this is difficult to accomplish when attempting to join two fibers only slightly thicker than a human hair in the cramped environment of underground wiring systems beneath a city street.

Because of the huge amount of spectrum space available in the light frequency range and the vast amounts of information that can be carried on fiber, however, this glass wire is expected to become the communication superhighway of the future. One fiber wire will be able to carry the combined loads of all the communication services now used by the typical U.S. home, with plenty of space left over for services not yet invented.

Physical Distribution

Another way to distribute the communication signal to the home is by recording it on some medium, such as magnetic tape or optical disc, and physically delivering it to the home. The advantages of the physical distribution system are economics of delivery, mass production of the information, and the ability of the user to play back the information many times. The major disadvantage, of course, is that the information cannot be sent to the user in real time; there is always a delay from the origin of the request to the time that the material can be used by the consumer. Although such a delay is of no consequence for many messages, it is intolerable for others.

Home Audio: The Phonograph

Although now difficult to find in common use, the phonograph provided a successful model of physical distribution of sound for more than a hundred years. The advantage of the phonograph record was that, for the first time, electronic communication signals could be permanently stored and mass produced at a low cost per unit. The tiny grooves of the phonograph record vary horizontally in relationship to the frequency and volume of the original audio source. As the needle follows the grooves, the vibration of the stylus in a magnetic field in the cartridge creates an analog electric current that is a representation of the varied current from the microphone that picked up the sound waves. The varied current is then amplified and used to drive loudspeakers that produce the sound we hear.

Magnetic Tape

In magnetic recording, the varying electric current is used to create a tiny magnetic field in a gap in a recording head. As the tape coated with magnetic material is pulled past the recording head, the field sets up patterns in the north/south magnetic orientation of the particles. Later, a playback head reads these magnetic variations as an electric current is created by the magnetic fields passing the head.

Although the magnetic-tape system made possible a higher-fidelity sound with less wear on the recording medium, information had to be recorded in a linear fashion, nullifying the economics of mass production that had been established with the phonograph record. The great advantage was that the magnetic system could be used immediately, because it didn't require a complex manufacturing process. The tape could simply be rewound and played back. This opened the door for the home recorder—as well as for the copyright problems associated with rerecording music owned by someone else. Those problems aside, broadcast stations now had an inexpensive and reliable production tool for the creation and storage of radio and later television programming.

Many formats of audio-recording systems have been developed. Major recording studios use reel-to-reel recorders with tape up to 2 inches in width that record twenty-four tracks of information simultaneously. Broadcast stations typically used open-reel, $1/4$-inch tape as well as $1/4$-inch cartridges and audiocassettes in the past. Because of its small size and convenience, the audiocassette has also become the format of choice for home audio systems. These analog recording systems are being replaced by the higher quality of *digital audiotape (DAT)*.

Optical Audio

With the ability to convert an analog signal into a digital one came the development of optical recording. Since its introduction in the 1980s, the CD player has all but taken the place of the traditional stereo phonograph record system. The **compact disc (CD)** is a $4^{11}/_{16}$-inch-diameter record made of a thin sheet of aluminum sandwiched between two sheets of plastic. Embedded in the aluminum are spiral tracks composed of a series of pits. The pits are read by the CD player as the binary digital code needed to re-create the information of the audio signal.

The CD player uses a low-power laser beam to read the pattern of pits on the disc by focusing the beam through the plastic coating to the reflective aluminum surface. A photo diode senses the light reflected from the disc. The smooth surface of the disc reflects the laser light, but the area occupied by the pit has a zero reflectance value: It appears to be a black hole, because the focal length of the laser beam stops short of the bottom of the pit. By spinning the disc at a high rate of speed and focusing the laser beam with a movable mirror to follow tracks of pits impressed into the surface, a digital code can be re-created.

There are several advantages to the CD player. First, its small size but high-density information-packing format allows a longer playing time than the traditional stereo long-playing record (LP) in one quarter of the size. About sixty spiral tracks can fit into the space taken by one groove on an LP, and there are 20,625 spiral revolutions on each compact disc. Second, the digital format does not degrade in quality over time, because there is virtually no mechanical wear to the disc. Third, the fidelity can be much higher because the CD has more space for information. In fact, if the recorded signal represented data, the entire text of a twenty-volume set of encyclopedias could be recorded on a single compact disc. Fourth, the system can be programmed to instantly access a specific audio cut from any segment of the disc, making cuing possible.

The CD has become the medium of choice for the home audio system as well as providing a high-quality format for most radio stations. In addition, this format in the form of *CD-ROM (compact disc–read-only memory)* is becoming an accepted medium for high-density information storage and use with personal computers.

Home Video

By the mid-1960s the home audio entertainment system, with its AM/FM tuner, stereo record changer, audiotape recorder, and multiple speakers was common in U.S. homes. It took another twenty years for the home video entertainment system to become popular, even though the television receiver was the most common appliance. The element that eventually made the home video entertainment system possible was the *videocassette recorder (VCR)*.

Sony introduced the first Betamax VCRs in the early 1970s. By the 1980s competing Japanese electronics companies developed the Video Home System, commonly known as *VHS*, which, although using the same width of tape, could record up to six hours of programming on one cassette.

Although the home recording systems were originally designed for use in making home movies, most were sold to record broadcast television programs off the air for time-shifting purposes. Operators of video-rental stores did not want the cost of stocking duplicate movie titles on both the Betamax and VHS formats. Research indicated that the VHS system, with its lower initial cost and its longer recording format, was more popular. After the Betamax tapes began to disappear from store shelves, the fate of Sony's home format was sealed. The VCR became primarily a movie playback machine, using the rented tape for programming. By the early 1990s, the VCR had become more popular than cable, with more than 80 percent of all television homes owning at least one machine.

Optical Video

In the early 1980s, two different formats of *videodisc* were introduced by different companies: the CAV (constant angle velocity) and the CLV (constant linear velocity). The two systems are similar in that they both can play back full-color video programs that are prerecorded on a 12-inch aluminum disc sandwiched between two sheets of plastic for protection. The aluminum surface has pits that contain the code for the re-creation of the electrical signal and, except for size, is similar to the audio CD.

The CAV system emerged as the format of choice for the videodisc, although not necessarily for the home video market. Its unique advantage of random access coupled with computer programming made it a natural for use in education and corporate video applications, because specific segments could be accessed for programmed instruction.

The CAV system, marketed by Magnavox, Pioneer, Phillips, Sony, and others, reads the information off the disc with a low-power laser beam in the same manner as the signal is read from a CD. Each spiral track provides enough information to complete one frame of the television picture.

Optical videodiscs can be mass-produced *read-only discs (ROD)*, write-once discs (WORM—write once, read many times), or erasable discs. Information that is to be encoded on the read-only videodisc is usually premastered on videotape. The signal from the tape is then converted to a digital format and used to modulate a laser beam that cuts holes in the surface of the metal of the glass-enclosed master disc. A duplicate of the master is made using a photochemical process that creates a metal "mother," or template, that can be used to create inexpensive copies of the master. This process is generally confined to a clean (or perfectly dust-free) room in a processing lab.

The *write-once disc* can be created in the same component–sized unit that plays back the disc. A small laser in the unit can be focused on the surface and modulated to melt a series of holes in the thin metal. These holes create the same zero reflectance as do the pits of the mass-produced disc. Write-once discs are generally used for archival storage of data.

Erasable discs are designed with two layers of recording material: A polymer is bonded to the thin aluminum material. When a write laser beam is directed to this surface, the heat generated is sufficient to soften and slightly distort the smooth surface, which then scatters the light beam from the read laser, thereby creating the same effect as the pits in the read-only disc. An air-space sandwich enclosed between the protective outer plastic coatings allows space for the material to distort. Reheating the medium with the laser in effect erases the old information and allows new material to be recorded. Because of the high quality obtainable with digital recording and the durability of the disc, the erasable videodisc system may become the recording format of the future for radio and television stations.

The *digital videodisc (DVD)* has become a major component of the home video entertainment system. The inexpensive mass-produced DVD has emerged as a significant medium for display of movies in the home, as its format allows information to be densely packed into a small amount of space. The DVD provides the high-quality movie along with trailers, outtakes, promotional music videos, interviews with the director, and additional languages with room to spare.

Combined with the home computer for control and the television set for viewing, the optical disc may become a common method of displaying other types of information in the home. We may find an inexpensive DVD packed as assembly instructions for do-it-yourself projects rather than the indecipherable printed instructions that are now universal.

DIGITIZING THE TECHNOLOGY

By converting the traditional analog communication signals to digital, several advantages can be realized. These include miniaturization of equipment with a subsequent reduction of power consumption, convenient storage of the information, ease of editing or manipulating the information, and no signal loss or degradation of the signal through the transmission or storage medium. Each reproduction, or generation, of the signal is an exact copy of the original with no loss of high-frequency response.

Digitization has led to the development of the compact disc, the video game, the palm-size video camcorder, the DVD, the smart TV with its picture-in-picture, and the handheld personal communication service (PCS)—or digital cellular telephone. The process of digitization involves sampling the analog audio or video signal at a specific time interval and converting the continuous analog waveform to a series of digital pulses. The more times the signal is sampled per second, the more closely it resembles the continuous original analog signal (see chapter 13).

The development of the technology needed to digitize the analog signal has brought about a marriage of information technology and the computer industry. Mass communication's encoding, storage, manipulation, transmission, and display technologies have all come to rely on circuits developed for computers. The chips now incorporated in these technologies allow the pictures and sounds emitted from our radio and television receivers to be of a constant high quality. Automatic control circuits have taken most of the guesswork out of adjusting receivers for the optimal quality. The next generation of computerized mass communication technology will integrate systems more completely so that information from the broadcast distribution media, the wired media, or the physical distribution systems can be accessed instantly to preselect and personalize entertainment and information of particular interest to the individual viewer.

Summary

▶ Mass communication messages can be delivered to the home by electromagnetic waves, wire, or cable or through the physical delivery of a recorded message.

▶ The broadcast portion of the electromagnetic spectrum is a continuous range of frequencies from about 30 KHz to 300 GHz.

▶ Signals can be added to an electromagnetic carrier wave by amplitude modulation, frequency modulation, or phase modulation.

▶ AM radio uses frequencies of 10 KHz in a bandwidth from 535 to 1,605 KHz.

▶ FM radio uses frequencies of 200 KHz in a bandwidth from 88 to 108 MHz.

▶ Television channels require a bandwidth of 6 MHz and are located in both the VHF and the UHF portions of the electromagnetic spectrum. VHF channels (2 through 13) use frequencies from 54 to 216 MHz. The UHF channels (14 through 69) use frequencies from 470 to 806 MHz.

▶ Higher frequencies in the electromagnetic spectrum have shorter wavelengths and travel in direct line-of-sight paths; longer wavelengths at the lower frequencies bend to follow the contours of the earth.

▶ Coaxial and fiber-optic cables are becoming popular delivery systems.

▶ Physical distribution systems include optical and magnetic recording formats.

InfoTrac College Edition Exercises

The normal speed at which communication technology advances is exponential, but every so often we make a quantum leap. The transistor allowed a leap of technology in the 1960s; digital transmission technology is providing a similar advancement into the new millennium.

14.1 Chapter 14 examines the various technologies that enable information and entertainment content to be delivered electronically to a mass audience. The standard AM broadcast station of the 1920s has evolved into a variety of electronic distribution systems. Great technological advances have been made during the eight years of broadcast history. Perhaps the technological innovation that holds the greatest potential for the broadcaster is the move to an all-digital world.

   a. Using InfoTrac to access *Broadcasting & Cable,* look for and report on articles about advanced television and new ATV standards.

   b. Find articles and prepare a report on the status of the FCC-mandated move of broadcast television from an analog standard to a digital one.

   c. Describe the status of high-definition television in the United States. What programs are currently broadcast in HDTV?

14.2 Cable television has been an important part of the media landscape for the past twenty years. More than two-thirds of American homes subscribe to cable, but the technology of coaxial cable is changing.

   a. Using InfoTrac to access *Broadcasting & Cable,* look for and report on articles about fiber distribution systems.

   b. Locate and report on articles on direct broadcast satellite systems that describe the challenges DBS presents to traditional cable television systems.

   c. Find and report on articles about the advantages of hard-wired communication systems. Do the same for wireless systems.

WEB WATCH

Here is a list of a few URLs (Internet addresses) for some of the organizations or corporations discussed in this chapter. Please explore these Web sites and follow the links to learn more about the complex business of the electronic media. Add your descriptions and your own favorite sites at the end of the list. Please keep in mind that the dynamic nature of the Internet allows sites to come and go but also allows organizations to update information about themselves very quickly.

Address	Description
http://www.phillips.com/telecom.htm	
http://www.jrselectronics.com/	
http://www.ictv.com/fast/	
http://www.gi.com/	
http://www.hns.com/	
http://www.omneon.com	
http://www.vxm.com/21r.52.html	
http://www.dvb.org/dvb_standards/dvb-dvbmmds.htm	
http://www.global-fiber-optics.com/	
http://www.dvdvideogroup.com/	
http://www.pioneer_ent.com/	
Other favorite sites:	

CHAPTER 15

Telecommunication is international in scope—satellite communication and the Internet know no political borders.

THE GLOBAL VILLAGE

In the early 1960s, media guru Marshall McLuhan predicted a world in which communication technology would allow citizens of geographically and politically diverse countries of the world to become virtual neighbors. He predicted that, through radio and television, we would all become neighbors of a "global village." Although the Internet has made this prediction practical, global electronic communication has been a reality for more than 150 years, beginning with the telegraph.

Telecommunication systems do not exist in a national cultural or economic vacuum. The electromagnetic energy broadcast by a radio or television station does not recognize international boundaries. Signals intended for the citizens of one country can often be received by people in other countries. Footprints created by direct broadcast satellites (DBSs) can cover large areas of the earth and serve people in many countries at the same time. In the recent past, some governments of primarily Second World (Communist) countries have attempted to control the electronic media crossing their borders, as well as tightly controlling the media within the countries themselves. Other nations have established international broadcasting systems to send ideological messages to people within closed borders.

Ideologies have sparked various international broadcast schemes, and economics have had an even greater effect on the distribution of broadcast content to other countries. One of the most important exports of the United States is television programming. Programming and movies sold to television networks and stations in other countries account for the second-highest volume of export dollars—after aerospace components and airplanes. Many television production companies, knowing that programming sold to networks in the United States will not turn a profit with the first broadcast, count on syndication and foreign distribution for their profits.

The television stations and networks in other countries that buy the American programming have a variety of ownership patterns and systems of control. Many countries established their internal broadcast systems on a public service model rather than as commercial businesses. Some countries view the broadcasting system as a method of educating the public—sometimes in the ideology of the government. Now, however, many countries are adopting a commercial model of broadcasting similar to that of the United States.

This chapter provides an overview of broadcasting systems in several other countries and compares them with those of the United States. In addition, the basics of international broadcasting are examined, and the global marketplace for television programming, with the resulting profits and problems, is also explored. One such problem is *cultural imperialism*—the transmission of the values of one society to another through the pictures and stories provided by the mass media. In addition, because the electromagnetic spectrum must be shared among all of the countries of the world, the chapter also touches on the international control systems that have been established to allocate frequencies on a global basis.

A Framework for Regulation

When broadcasting developed in the early part of the twentieth century, it was not limited to the United States. Several other countries developed the technology at about the same time. The control and use of the technology in some of those countries, however, took a very different direction from that in the United States. At the height of the cold war, mass communication researchers Fred Siebert, Theodore Peterson, and Wilber Schramm (1956) published a book of essays in which they detailed how the political philosophies of governments shaped the structure of the mass media systems that operated in their respective countries. They listed their four theories of the press as: Authoritarian, Libertarian, Soviet Communist, and Social Responsibility. Although the political structure of the world has changed significantly with the fall of the Berlin wall and the breakup of the Soviet Union, it is still interesting and relevant to examine the four theories of the press.

At one extreme end of a continuum of control is the *Authoritarian theory*, characterized by very tight control of the media by the government in power. Generally, this theory assumes that the public is not capable of understanding political problems and therefore should not be exposed to any ideas that are critical of the government. Under this form of control, the government often directly censors the content of the media. Over the years many governments have assumed dictatorial power over the people they rule. Certainly the regimes of Nazis and Fascists were dictatorships, as are the governments of several Central and South American countries and large parts of the Asian and African continents. But at the dawn of mass communication, the kings of England and France also exerted authoritarian power and control over the press. One of the main tenets of the Authoritarian theory is that of operating the press to support and keep the government in power.

At the other extreme of the continuum of control is the *Libertarian theory*. Many countries have adopted democratic forms of government and consequently some type of Libertarian media control. Under the Libertarian theory, the assumption is that the individual members of society have certain rights of "life, liberty, and the pursuit of happiness" and that the function of society and government is to advance the rights of the individual. Under such a system, the mass media are generally free to offer the public any and all information. In fact, one of the tenets of a democratic society is that there will be a "free marketplace of ideas." If the media freely provide information, the individual can pick and choose from all ideas concerning government and make the right choices as a voter. Under this theory the media are controlled by the "self-righting process of truth" and the courts in terms of libel, obscenity, copyright, and privacy issues. Under the Libertarian theory, the press serves as the "fourth estate"—as a check on the power of the government.

Although the *Soviet Communist theory* became moot with the breakup of the Soviet Union, many of its tenets are retained by the Communist governments of Cuba and the People's Republic of China. Under the Communist system, the media are owned and controlled by the state and generally provide information that aids the continued success of the state.

The *Social Responsibility theory* developed in the United States in the mid–twentieth century. Its basic tenet is that with the freedoms found in the Libertarian theory come certain responsibilities for the mass media: The press has an obligation to enlighten the public in a democratic society, to protect the rights of the individual by serving as a watchdog of government, and to raise conflict to the plane of discussion. Whereas the traditional Libertarian theory suggests that anyone with the economic means to access the public has the right to do so, the Social Responsibility theory postulates that anyone with anything to say has the right to address the public. This tenet becomes especially important in broadcasting, because the electromagnetic spectrum is a limited natural resource. Not everyone with the economic means to broadcast has the ability to control one of the limited channels that exist.

Much of the media regulation, both in the United States and in other countries, revolves around maximizing the use of the limited spectrum space without interfering with other broadcast voices. Every government, whether totalitarian or libertarian in philosophy, must develop rules and serve as a traffic cop for the use of the limited spectrum. The assignment of transmitter power, frequency, and hours of operation becomes especially important when operating a broadcast transmitter near the border with an adjacent country. The natural phenomenon of the electromagnetic spectrum does not conform to the political philosophy or geography of our world.

International Telecommunication Agreements

Because electromagnetic waves cannot be made to conform to the political boundaries of countries, international agreements are required so that the domestic broadcast signals of one country do not interfere with those of another. In addition, agreements on technical standards must be in place if international electronic communications are to take place—broadcast receivers in one country must be able to decode the signals transmitted from another country if communication between those countries is desired. The international use of the Internet is commonplace today because of agreements among countries on technical standards for electronic communications.

The need for such standards became apparent with the advent of electronic communications. Soon after the invention of the electric telegraph in 1837, telegraphy systems were being used to communicate information to facilitate the movements of trains, freight, and commerce in North America and Europe. A submarine telegraph cable connecting those two continents was finally completed in 1866. Wireless telegraphy between countries became possible early in the twentieth century.

To encourage international communications, the International Telegraph Union was formed at a telegraph convention in Paris in 1865. That organization was involved in establishing interconnection standards, developing methods for ensuring privacy of information, and overseeing tariffs or charges for the use of the international communication systems. When radio communication became possible, a similar organization—the International Radiotelegraphic Union (IRU)—was formed to discuss the problems of wireless communications. In one of the first radio conferences, the IRU established allocations of specific frequencies to be used for particular radio services. After the *Titanic* disaster in 1912, the IRU was able to break a monopoly that prevented Marconi company radio stations from communicating with non-Marconi-equipped ships at sea. After the Radio Act of 1912 was enacted in the United States, other countries adopted common emergency communication procedures so that ships in trouble could request help via radio. By 1932 the International Radiotelegraphic Union and the older International Telegraph Union merged to form the International Telecommunication Union (ITU).

Over the years the ITU has changed significantly to keep up with the increased complexity of telecommunication technology and issues. Now an agency of the United Nations, the ITU is based in Geneva, Switzerland. Regulations concerning spectrum usage are developed at World Radio Conferences, where representatives of the nearly 190 countries and three hundred public and private telecommunication organizations that are members meet to discuss international concerns. The Federal Communications Commission (FCC) serves as the U.S. representative on the ITU.

In a world of political conflict and competing ideologies, the ITU has had to deal with misinformation, propaganda broadcasts, copyright issues, frequency jamming, pirate stations, satellite parking spaces, and the physical cutting of communication cables.

In addition to the ITU, there are other international organizations that coordinate electronic communication utilization across political boundaries. Satellite communication has become an area of import in the past several decades. Intelsat (International Telecommunication Satellite) was established as a common carrier in the mid-1960s to develop and operate satellite links between countries for international data, telephone, and television transmission. In the early years, Intelsat held a monopoly on international satellite communication. It delivered international messages between *PTT*s (governmental *postal, telegraph,* and *telephone* services) in other countries and Comsat (Communications Satellite Corporation) in the United States. Now several additional private satellite corporations are operating globally, including PanAmSat, Eutelsat, Asiasat, Arabsat, and Intersputnik.

Another international organization actively involved in broadcasting across borders is the European Broadcasting Union (EBU). In the late twentieth century, the European Union was developed to enable fifteen nations in Europe to eliminate trade barriers and move to a single currency. As the European Union takes shape, transborder communications become important to strengthening the federation. The European Union has developed regulations for commercial broadcasting that include limits on advertising, programming content, and non-European programming. The EBU works to share programs among the European Union members.

Other regional international broadcast cooperatives exist in Africa, Asia, the Middle East, the Caribbean, and Central and South America. These organizations contribute to intercultural understanding through the sharing of broadcast programming.

INTERNATIONAL BROADCASTING

Every country in the world has developed its own national, or internal, electronic mass communication system for distribution of information and entertainment to its citizens. In addition, some governments broadcast signals to other countries—usually for political purposes. This is called *international broadcasting* because the intent is that the radio or television signals reach across international borders to listeners in other countries. The United States, Australia, China, Germany, Great Britain, and Russia are all major players in international, or external, broadcasting. The British Broadcasting Corporation

(BBC) operates one of the largest international broadcast services, World Service Radio. The United States offers another through its Voice of America.

Historically, most of the information broadcast internationally has been used for *propaganda* purposes—or to "reeducate" people in one country to the political objectives of another. World War I was probably the first large-scale international conflict in which the mass media were widely used for propaganda. As radio receivers were not yet common, posters, newspapers, and newsreel films were used extensively to shape the views of people in many of the warring countries about the evil goals of their enemies. Armies even used barrage balloons as platforms from which to drop leaflets urging the enemy troops to abandon the fight and surrender. Governments involved in the conflict quickly cut international telegraph cables and destroyed enemy wireless communication stations in hopes of creating chaos in military communications. The French even installed a jamming station atop the 1,000-foot-high Eiffel Tower to disrupt radio transmissions in Germany.

The Netherlands became one of the first countries after World War I to establish a regular international broadcasting service. The purpose was to reach people in its far-flung empire with items of news and entertainment from the homeland. The service was operated by the Phillips company, which is still a major player in telecommunication equipment design and manufacture.

Other early international broadcasters included the newly formed Soviet Union, which broadcast messages in celebration of the Bolshevik revolution. Other governments that began international broadcasting included Germany, France, Great Britain, and Japan—all of which had established empires of colonies around the world. By the mid-1930s Radio Luxembourg was broadcasting commercially sponsored entertainment programming to Western Europe, and Radio Vatican was reaching Catholics worldwide with religious programming.

The U.S. government did not take an early interest in international broadcasting, but commercial corporations such as Westinghouse, General Electric, and Radio Corporation of America (RCA) experimented with international transmission early on. Both the Columbia Broadcasting System (CBS) and the National Broadcasting Company (NBC) began

During World War II, when invading German forces occupied France, General Charles DeGaulle spoke to the French people from the studios of the BBC in England.

international broadcasts in the 1930s, with signals designed to reach people in Latin America and broadcast affiliates in Canada. One experiment in the United States was the superpower station WLW in Cincinnati, which broadcast a 500,000-watt signal that covered much of North America.

When the Fascists and the Nazis came to power in the 1930s, they joined the Soviets and the Japanese in waging a "radio war" against governments of neighboring countries. Their foreign-language broadcasts were designed to garner support from people in other countries for the Fascist or Nazi cause. By the late 1930s, the British government had given permission to the BBC to broadcast in foreign languages to counter the propaganda of Italy and Germany. The United States did not enter into the radio propaganda war until after World War II was well under way. The U.S. War Information Office was established in 1942, and the Voice of America (VOA) was born. The program, first broadcast on commercial shortwave stations, was to spread the word about the benefits of living in a democracy.

During World War II, international broadcasting was used by both sides in the conflict for a variety of purposes, of which propaganda certainly was one. Some messages were designed to give hope and comfort to citizens in occupied countries; others, such as those broadcast by "Tokyo Rose" in Japan, were used to foster fear and uncertainty in the ranks of U.S. troops fighting in the South Pacific. Both sides used international broadcasts as a method of sending coded messages to agents in other countries and to gather information about the enemy by monitoring their broadcasts.

Close on the heels of victory in World War II came the cold war. In 1948 Congress established a permanent international information agency, which provided funding to fight an information war against communism. By the early 1950s, the United States Information Agency was operating the Voice of America, which had become the official means for the U.S. government to reach citizens in other countries.

In addition to the VOA, Radio Free Europe and Radio Liberty were established after World War II to broadcast via shortwave to people behind what was then called the iron curtain.

Communism on the European continent and in the eastern hemisphere was considered a threat to the United States that had to be fought with both guns and words. Another Communist threat lay just ninety miles over the horizon from Miami, Florida, where Fidel Castro's Cuba became a Communist bloc country in America's backyard. After a near brush with World War III during the Cuban Missile Crisis and reports of attempts on Castro's life by the Central Intelligence Agency (CIA), the Voice of America established Radio Martí in 1983 to reach the Cuban people with positive messages about the United States. Television Martí was approved by Congress in 1990 to broadcast television pictures from a blimp (called an *aerostat*) moored over the everglades in Florida. These international services created much controversy among commercial broadcasters in the United States, who felt that the Cuban government's attempts to jam the international broadcasts would cause interference to commercial broadcast stations in the Southeast.

By the late 1980s, the U.S. Information Agency had adopted new technology to broadcast its messages of democracy abroad. The television service Worldnet was delivered by satellite to cable television and hotels throughout the world. Much of its original role of providing a news service from the United States has now been usurped by the ubiquitous Cable News Network (CNN).

International broadcasting can also be delivered for nonpolitical purposes. Some religious organizations use shortwave radio to broadcast their messages of faith around the world. As mentioned, Radio Vatican broadcasts a Catholic message. Radio Luxembourg has long broadcast a popular commercial entertainment service to Europeans looking for an upbeat alternative to public radio in their own countries, and Australia broadcasts native-language programs to islanders in the South Pacific. International broadcasting, or external broadcast services, is alive and well.

National Broadcast Systems

As we have seen, some countries operate external broadcast systems to reach people beyond their borders, but almost every country operates broadcast systems to serve the informational and entertainment needs of its own citizens. These systems are called **national systems**, *internal systems*, or *domestic systems*.

The purpose, structure, financing, and regulation of the national broadcast systems vary considerably from one country to the next, as there are many factors that determine the nature of the internal broadcast system of a particular country. These include the political ideology of the government, the economic base, the population density, the literacy rate, and the geography, to name but a few.

The domestic broadcast system of each country is different, but there are similarities too. For example, electronic communication systems that operate in First World or industrialized and capitalistic countries often have a different operating philosophy than those operating in Third World, or economically underdeveloped, countries. Second World countries, or those with Communist governments, often tightly control the messages that the internal media deliver to the citizenry.

In terms of government control of the internal broadcast media, most countries operated under one of two models. Broadcast systems in some countries were established to be owned and operated by the government—usually under a PTT department or a ministry of information. In other countries broadcast systems were established under a free-enterprise system, with the government licensing private owners to build and operate stations. Often these stations were commercial, with operating revenue derived from the sale of advertising time. Government-owned stations usually received their funding directly from the state treasury or from an annual license fee charged to each receiver owner. Such stations, often referred to as public stations, don't usually depend on advertising revenue for support. Interestingly, many countries that have relied on government-owned public broadcast systems in the past are now allowing privately owned commercial systems to operate as well.

The technology of radio broadcasting was developed in the early twentieth century. This coincided with the end of the Age of Imperialism, when many Western European countries had colonies around the world. It was said that the sun never sets on the British Empire. This was because the British had colonies in North America, South America, the Middle East, Africa, Asia, and Australia. The French, Dutch, Germans, Spanish, and Portuguese also had worldwide systems of colonies. International radio became a method of keeping in touch with the mother country, but equally important was that the domestic broadcast system model of the homeland was exported to the colonies. For example, the BBC became the model for India, South Africa, Nigeria, Australia, New Zealand, Jamaica, and parts of the Middle East. When the colonies, or members of the commonwealth, developed self-rule, the basic broadcast system remained in place.

Some internal broadcast television systems developed an American flavor when RCA sold studio and transmission equipment to countries in South America and the Middle East. American technicians needed to install and operate the equipment often became de facto consultants in the operation of the broadcast systems.

In the 1950s American programming began finding its way into foreign markets. Many countries found that it was far cheaper to buy high-quality American programming than it was to produce their own offerings. As was stated, American television programming has become one of the most profitable exports of the United States.

COMPARATIVE DOMESTIC BROADCAST SYSTEMS

It is impossible to describe in detail the internal broadcast systems of each of the many countries of the world. This section presents the basics of electronic media in several countries to provide an idea of the differences that exist. Because Canada and Mexico are our closest neighbors, their broadcast systems are briefly examined. In addition, broadcast systems of countries that have had an influence on the U.S. economy or culture are described. These include Great Britian, Germany, Turkey, Russia, Saudi Arabia, India, Japan, and the People's Republic of China.

CANADA

Although geographically very large, Canada is sparsely populated. Many of its major cities are located near the U.S. border and are within the coverage pattern of television stations licensed in the United States. In addition, about 75 percent of Canadian homes are wired for cable, which delivers not only U.S. television stations, but also U.S. cable networks.

Canada has a public broadcasting system, the Canadian Broadcast Corporation (CBC), that provides both radio and television services to the entire country. Programming is provided in both English and French in deference to the French-speaking people of Quebec. The CBC network is funded by the government, but it does sell some commercial time on the television service. It uses privately owned stations as outlets to broadcast its programming.

In addition to the publicly owned CBC, the major privately owned network in Canada is Canadian Television (CTV). There are also smaller regional networks such as Ontario TV. Most of the three hundred privately owned television stations are affiliated with a Canadian network. There are about seven hundred radio stations in Canada, as well as a domestic broadcast satellite system that serves indigenous populations in the vast northern spaces with educational programming.

The Canadian Radio and Television Commission (CRTC) oversees broadcasting in that country. One of its primary concerns has been cultural imperialism from the ubiquitous U.S. programming flowing over the border. As a result, Canadian stations are required to broadcast Canadian-produced programming about 60 percent of the time. There are also strict rules regarding violence in programming and many concerning children's programming. Consequently, Canada has a well-developed program production system and creates exceptional children's television shows. Some U.S. production companies are now shooting in Canada, where there is a pool of technical expertise and low production costs.

MEXICO

Mexico too has a broadcast system that includes both public and private broadcasting. The private system is a monopoly that has long dominated broadcasting in that country. Televisa, the primary broadcaster, includes three national networks based in Mexico City. They are under the ownership and control of one "well-connected" family. Televisa's channel 1 broadcasts *telenovelas,* Spanish-language serial dramas, much like our soap operas. They are very popular in Mexico, as well as in other countries, and are exported to both South and Central America. Some even find their way to the United States and are broadcast over the Spanish-language Telemundo and Univision networks.

Televisia's channel 2 provides audiences with foreign-produced programs. Most of these are imported from the United States, but, like in Canada, there is a limit on the amount of U.S. programming that can be broadcast on Mexican television.

Televisia's channel 3 broadcasts news and information programs. Televisa is a commercial broadcaster, and all of its channels are advertiser supported. The company has complete regulatory support from the Mexican government—in exchange for its support of the government through friendly news broadcasts.

There is also a public broadcast service in Mexico—the Mexican Television Cultural Network. Its few stations broadcast educational programming to the rural areas that are not served by private commercial broadcasting. It also operates a public television station in Mexico City.

GREAT BRITAIN

Great Britain includes the countries of England, Scotland, Wales, and Northern Ireland. Broadcasting began in the early 1920s as a public service in England. By 1927 the BBC had been established for the purpose of operating radio in the national interest and without government interference. The radio service was funded by an annual license fee charged to the homeowner for each receiver. Later additional radio services were added by the BBC, but private commercial radio was not yet available. By the 1950s many listeners were tuning in Radio Luxembourg for more-exciting programming choices. Even unlicensed **pirate radio** stations located on boats anchored offshore beyond

A reporter interviews an Atlas soccer player after their 2–0 win over Chivas. They played in front of a capacity crowd of 75,000 fans.

the 12-mile limit began broadcasting rock and roll, as well as commercials, to a younger audience.

By the late 1960s, the government had put the pirates out of business but recognized the need for more modern radio programming. The BBC now operates five national radio channels as well as local radio services. Radio 5 was first designed to reach a younger audience, but then it was switched to a talk format. Independent commercial radio stations have been available in Great Britain only since the mid-1990s, when, under the 1990 Broadcasting Act, national independent radio networks were allowed. The two national commercial radio networks now available are Classic FM and Virgin Radio.

The BBC has a long history in television. Some of the early experiments with television were conducted by Briton John Baird in the mid-1920s, and within a decade regular daily programs were telecast by the BBC. Television service was discontinued during the Second World War, but by the 1953 coronation of Queen Elizabeth, the BBC television service was popular with a sizable audience.

In 1954 the government authorized the first commercial television service—Independent Television—and the country was able to legally receive both public and commercial broadcasting for the first time. The Independent Television Authority divided the country into regions, and commercial operators were invited to apply for franchises to provide

Aziz Fadhil reading the BBC Word news in Swahili.

programming. Some of the programming corporations are familiar to U.S. viewers who watch British programming on the Public Broadcasting Service (PBS). The more familiar programming services include Thames TV, Carlton Television, Granada TV, and London Weekend Television.

Many citizens of Great Britain can now receive five terrestrial television services. BBC1, which started telecasting programs in 1936, now provides popular drama and comedy and is widely viewed. A second BBC channel, BBC2, provides programming similar to that found on public stations in the United States. The third channel is operated by the Independent Television Authority. Fifteen licenses are granted to provide the island nation with popular commercial television; each licensee serves a geographic region with television program production. The programs are then broadcast nationally over the third channel.

The fourth television service, channel 4, went on the air in 1982 as a second commercial television channel. Although programmed with high-quality shows, the channel 4 programming is affordable because, in addition to the sale of commercial time on channel 4, Independent Television is required to subsidize the channel out of its own advertising revenues. The mission of channel 4 stipulates that the channel provide service. In Wales, for example, it broadcasts a significant amount of programming in the Welsh language. The British people were offered a fifth choice of television programming in 1997 with the launch of channel 5.

A large portion of the British Isles is composed of hilly terrain—not at all conducive to easy television transmission and clear reception. Even so, cable television was not introduced until the late 1980s. Instead, many repeaters were used to bring the BBC and Independent Television signals to the small isolated valleys and rural areas. The cable systems that were built, some by U.S. telecommunication companies, have not been very successful.

Cable's lack of success might be due in part to direct broadcast satellite. Broadcast satellite services provide Great Britain with a wide variety of television signals. The Astra service, a DBS company based in Luxembourg, and Eutelsat provide television signals for customers in the United Kingdom, as well as much of western Europe. Recently, media magnate Rupert Murdoch bought a failing DBS service and revitalized it. His BSkyB direct broadcast service now provides television programming to Europe in much the same way as his STAR TV (Satellite Television Asia Region) service serves Asia.

In addition to DBS and terrestrial broadcasting, audiences in Great Britian own a large number of videocassette recorders (VCRs) and have access to teletext services. The teletext service requires that a decoder be attached to the television receiver, allowing sports scores, airline schedules, and stock market information broadcast by the television station to be selected and displayed on the home TV screen.

Germany

In 1989 East and West Germany were reunified after being governed by very different political systems for more than forty years. Although reunification placed an economic burden on Germany, it is a highly industrialized country whose population has a considerable amount of individual wealth and leisure time. Much of that time is spent in front of the television set, watching programming on videocassette or from satellite, public or privately owned broadcast stations, or cable television.

After the Second World War, broadcasting in Germany was decentralized so that government could not easily control the content of the media. A system was established with public broadcast stations serving regions. The stations were funded by a monthly fee on receivers and had complete control of the programming that they broadcast to their respective regions. Programming was expensive to produce, so a "network" structure was set up so that programming could be shared. Arbeitsgemeinschaft der Rundfunkanstalten Deutschlands (ARD), an association of public broadcast stations in Germany, collects programming produced by each regional station and distributes it as a national programming service. It provides both radio and television programming services to its regional broadcast stations.

A second "network" service, Zweite Deutsche Fernsehen (ZDF), provides television programming so that viewers can have a choice. Further choice was possible when ARD added 3rd Programs as a purely regional service. All three national services—ARD, ZDF, and 3rd Programs—are supported by license fees that viewers pay each month, but the programming services can also sell a regulated amount of commercial time.

In the past few years, privately owned commercial television stations have been permitted. Those, combined with DBS services, have allowed German viewers a large choice of programs compared with the three public channels that they have had in the past.

Turkey

Turkey has long been the crossroads of Europe and the Middle East, and there are many western influences in this predominately Muslim country. Although Turkey dates back a thousand years to the fall of the Byzantine Empire, the modern Republic of Turkey was established after World War I, at a time when radio was beginning to develop. Consequently, the first licenses were granted by the new government in 1926 to the Turkish Wireless Telephone Company to operate stations in the capital city of Ankara and its largest city—Istanbul. Within ten years broadcasting was taken away from the private company and became a state monopoly under the Directorate of Press Affairs.

In the early 1960s, a revised constitution established an autonomous agency—Turkish Radio and Television—to operate the public radio and television services. Turkish TV (TRT) began broadcasting in black-and-white in 1968 and switched to color in 1982. A second television channel was added in 1986, and a third was added in 1989. GAP TV was established to provide programs to the indigenous audiences in the eastern parts of the nation, and TV-4 was launched to provide educational programming. In the early 1990s, TRT-INT (TRT International) was established to target Turkish people who had emigrated to western Europe, Central Asia, and North Africa.

The government monopoly of broadcasting ended in the early 1990s, and the Supreme Council of Radio and Television (RTUK) was established to regulate the newly formed private radio and television stations. The first private television station—INTERSTAR—went on the air in 1990. Since then dozens of national, regional, and local stations have applied for licenses. Many of the Istanbul-based national stations offer programming to much of Europe and Asia via satellite.

Cable television has only recently been available in Turkey. Turkish Telecommunications Corporation offers subscribers TRT programs as well as many of the new private stations. In addition, U.S. cable networks such as CNN, CNBC, Nickelodeon, and Animal Planet are available on cable. Eurosport is a popular channel on cable, as is Number One, which is an MTV-like music service.

Multiple television channels have brought increased viewing in Turkey. Some estimate that more than 90 percent of the urban Turkish population watch television on a regular basis and that television is the most important source of news in the country. With the addition of the privately owned channels in the 1990s, broadcasting in Turkey has shifted from the BBC-based public service model to a U.S.-based commercial one.

Russia

Radio broadcasts began early in Russia. The Communists centralized radio under the direction of the party in 1917. By the mid-1920s cable radio was available in many of the large cities. Cable radio systems included a central broadcast receiver located in the center of a town or village, which was attached to a telephone wire that carried the signal to a speaker in the individual home. The speaker box often had a volume control and three buttons. Each button could select a different government-controlled program, but the speaker box did not have a tuning coil, so it could not pick up radio signals from outside the country. The cable radio system, funded by the state, provided an easy method of controlling what the listener could receive in terms of programming.

Cable radio programming provided during the time of the Soviet Union tended to have a propaganda message or be educational in nature. Entertainment programs were in short supply, and commercials and advertising did not exist, but because the cable radio equipment was cheap, sound quality was good, and license fees were reasonable, many homes used the service. Even today many cable radio systems in Russia remain.

During the Soviet years, all broadcasting was tightly controlled by the State Committee for Television and Radio—called Gostelradio. That organization made all the decisions concerning broadcasting in every Soviet state and satellite nation. The state-controlled Gostelradio organization was divided into three operating divisions: Soviet Central Television, Soviet Central Radio, and Radio Broadcasting for Foreign Countries.

Most people watched television on the Gostelradio First Channel, most of whose programming was heavily edited news about the wonders of the Soviet Union. There were also some movies, sports programs, and children's shows, but the philosophy of the channel was to provide mass political education for the people of the Soviet Union. The Gostelradio Second Channel was established to provide color television programs, but its content was often only reruns of First Channel programs. The Third Channel was a regional Moscow service, and the Fourth Channel contained educational and instructional programming.

Since the late 1980s and the fall of the Soviet Communist form of government, broadcasting—and indeed life in general—has changed dramatically in Russia. A precursor to that change was *glasnost,* or openness, a policy announced by Mikhail Gorbachev in the mid-1980s. After glasnost, radio and television news programs began reporting more news of the events of the world, not just of the Soviet Union. In 1990 a group of journalists from Soviet Central Television were allowed to establish the Russian Television and Radio Company (RTR) in Moscow. They first launched Russian Radio and then Russian Television, which began to program several hours a day on the Second Channel in competition with Gostelradio's Central Television.

After the collapse of the Soviet government, the management of broadcasting was transferred to the new Russian state government, and Gostelradio became RTO (State Radio and Television Company Ostankino). Both RTR and RTO provide programming to many state television stations operating in the rural areas.

In 1990 it became legal to operate a broadcast station independent of the state. Marathon TV was licensed in 1993 to broadcast in Moscow. At about the same time, Ted Turner entered into a joint venture to broadcast CNN with Russian subtitles and old MGM movies over channel 6.

Cable television became a reality in the early 1990s. It was easy to develop, because many of the vast apartment complexes in the urban areas were already wired to a master antenna system. The cable operator simply added videocassette recorders to play movies—often black-market copies pirated from other countries. Domestic television cable networks are now developing in Russia to provide additional programming.

The electronic mass media in Russia have undergone a tremendous change in a very short time. Broadcasting has moved from a strictly controlled instrument of government propaganda to a duel public and private system that is attempting to embrace capitalism. To date, both the economy and the governmental structure of the country are in turmoil. The future will determine if the Russian broadcast media will develop a strong economic base or will be supplanted by foreign videocassette and DBS programming.

SAUDI ARABIA

The kingdom of Saudi Arabia has long been a major oil supplier to the United States. Because of that, employees of major American oil companies began many years ago to take their western lifestyle to the desert sheikdom. Many of the western ways collided with the staunchly conservative Muslim beliefs, yet the ever increasing wealth generated by the oil revenues created a desire, at least among the wealthy ruling class, for the western lifestyle. Thus, in spite of religious opposition, both radio and television have been adopted in recent years.

Although a radio station was established in the early 1950s to counter Egyptian propaganda, domestic radio started in the 1960s, with the U.S. oil companies providing broadcast service to their employees. The kingdom became interested in broadcasting and contracted with RCA and the U.S. Army Corps of Engineers to construct radio stations for the dissemination of religious programming. In the mid-1960s, when television stations were constructed, the more conservative religious leaders protested strongly. The Koran does not permit graven images. Indeed, motion picture theaters are very rare in Muslim countries.

There are two national radio services in Saudi Arabia: the General Programming and the Holy Koran. The General Programming service provides news, children's shows, programs for housewives, and religious programs. The Holy Koran service provides religious discussions and readings from the Koran.

Television programming has been difficult for Saudi broadcasting. Most of the locally produced programming has been religious in nature. Some programming has been imported from Egypt (during times of peace), and a small amount has come from Britain and the United States. All programs are censored, and religion dictates what is acceptable content. At first women could not appear on TV; now they can, provided they dress and act conservatively. Dancing and gambling cannot be shown, women cannot be shown taking part in athletic games or sports, and all alcohol or references to it must be censored. Immoral scenes and scenes containing violence must also be deleted. Needless to say, much American television programming is considered inappropriate for Saudi television.

Many Saudi citizens have become wealthy and have been educated in western countries. When they return home, they want to take some of the western amenities back with them; because movie theaters are rare in Saudi Arabia, many have brought VCRs. The home video market in that country has become a huge business.

INDIA

Asia and the Indian subcontinent are awakening economically and are playing an increasing role in world politics. About 20 percent of the world's population lives in India—and that population is still growing. Most people are poor, illiterate, and farm to produce a small livelihood. Radio and television in village community centers provide many with all they know about the outside world.

Broadcasting in India started with radio clubs in the major cities in the early 1920s. The Indian Broadcasting Company was formed by a group of businessmen, but it floundered due to lack of audiences—it broadcast to the small group of Europeans living there. The British Colonial government took over the operation and requested assistance from the BBC. Naturally, broadcasting in India adopted the BBC model, with the licensing of receivers providing the funding to operate and program the system. The public radio system in India was named AIR (All India Radio).

When India gained its independence from Great Britain in 1947, AIR was transferred to the Department of Information and Broadcasting. Now a nationwide radio monopoly, AIR is funded directly by the government, but it also sells commercial advertising. Much of the programming is music, but AIR also broadcasts news and public affairs—it has one of the largest network news staffs in the world. Broadcasts are in English, Hindi, and about two dozen regional languages.

Doordarshan is the national television service in India, operating eighteen television stations and a network of more than five hundred transmitters. Its funding and structure are the same as those of AIR, but radio and television are operated independently of one another. Programming consists of educational material and news, along with sports and entertainment. Entertainment programming includes music, dance, soap operas, and religion. India also produces a large number of entertainment films, some of which find their way to the television screen.

Videocassettes and direct broadcast satellite have made tremendous inroads in India. Many VCRs are wired to rudimentary coaxial-cable systems that connect large apartment houses, and satellite dishes are seen nearly everywhere. STAR TV, owned by Rupert Murdoch, provides programming from the BBC, American broadcast networks, and MTV. Satellite-delivered ZEE TV provides Hindi-language programming. With videocassettes and satellite technology, Doordarshan no longer has a monopoly on what the Indian people watch.

Interestingly, one of the first uses of the broadcast satellite was in India. In 1975 Doordarshan, with NASA's cooperation, used the ATS-6 satellite for several months to broadcast health and agricultural information to rural Indian villages. In about twenty-five hundred villages, satellite dishes were constructed out of chickenwire to pick up the faint signals. Farmers would cluster around the village television receiver each evening to watch the instructional programs.

JAPAN

Although a small country geographically, Japan is densely populated and heavily industrialized and has a high gross national product. Its per-capita income is second only to that of the United States. The average Japanese wage earner works long hours and takes very little vacation time. Television, therefore, is a very important source of entertainment and relaxation, because its programming is available in the home twenty-four hours a day. Most Japanese homes have several television receivers, VCRs, and other electronic communication devices. Japan is the world leader in electronic communication hardware.

Nippon Hoso Kyokai (NHK), the Japanese national public network, began broadcasting in the mid-1920s and still exists today. After World War II, the General Headquarters of the Allied Powers and the Japanese government redefined radio law so that privately owned commercial stations could compete with NHK. One private company—Nippon Television (NTV)—was one of the first to offer television in Japan. NHK began television programming in 1953.

Currently, NHK operates two national public television networks, and there are five national commercial networks. These, with key stations based in Tokyo, include NTV, Tokyo Broadcasting System (TBS), Fuji TV, TV Asahi, and TV Toyko. Entertainment on the commercial networks is very popular, but most citizens turn to NHK for news.

Nippon Hoso Kyokai operates five television channels. Two are DBS services, and a third satellite-delivered service is high-definition television (HDTV). The two terrestrial services are G-TV (general programming) and E-TV (educational television), which provides instructional programming to the classroom. BS-1 (satellite) provides news and sports

Members of the British pop band the Spice Girls perform live before television cameramen in Tokyo, Japan.

from Japan and other countries; BS-2 (satellite) provides entertainment programming—some of which is imported from the United States.

NHK radio is the dominant sound medium in Japan, operating three national networks. Radio-1 airs news around the clock, Radio-2 offers educational programs, and Radio-3 is a music service. All NHK radio and television programs are funded by license fees.

With seven over-the-air television services and several DBS services (viewers can receive STAR TV from Hong Kong as well as WOWOW, a private-pay TV service), most homes do not subscribe to cable television.

Japan is a media-rich environment with the probability of even more commercial satellite-delivered signals in the future. The Japanese are interested in technology; they have HDTV presently and are moving into digital television in the near future. With multiple channels to feed, producing or finding enough programming is not easy; independent producers now provide most of the material that is seen on television. There is some discussion of privatizing NHK in the future.

People's Republic of China

In 1949, after years of civil war and occupation by the Japanese, a rebel army under the leadership of Mao Tse-tung established a Communist form of government—the People's Republic of China. The Nationalist Party leaders fled to the off-shore island of Formosa (now Taiwan). A government-in-exile was established that has become the Republic of China. With aid from the United States, Taiwan has developed into an industrialized democracy with a fairly high standard of living while, for many years, mainland China was caught up in the political turmoil of the Cultural Revolution. In recent years the "sleeping giant," with its population of more than 1 billion people, has begun to stir economically. In 1997 the wealthy capitalistic trade center of Hong Kong was returned to Chinese control after nearly a hundred years as a British colony.

Both Taipei (Taiwan) and Hong Kong are highly developed media centers. The electronic mass media in those two major cities are examined here, as is broadcasting in mainland China.

Radio broadcasting started in the early 1920s in Shanghai to serve the western business community in that trade center. When the Japanese invaded in 1937, broadcasting came to an end until after World War II, when the Communist Party took over and established the Central China Broadcasting Station in the capital city of Beijing. Television was first established in the late 1950s, and now both Beijing and Shanghai have four or five television channels, three of which are provided by Central China Television (CCTV). Local and regional stations exist throughout the country, with CCTV programming delivered by microwave and satellite (DBS dishes are illegal in the People's Republic of China).

The public broadcasting system is funded directly by the government, and no license fees are charged for the use of receivers. The Chinese broadcast media are operated as a branch of the government, and the mission of the stations is often simply to support government policies. Although most of the programming is produced locally, and its content is reviewed prior to broadcast, some BBC programs, as well as a number of U.S. programs, can be seen in China. Many Chinese leaders object to western influences but at the same time wish to further open the country to economic development.

Several commercial television stations have been established since 1990. Shanghai TV and Oriental TV sell commercial time, as do the public stations. The new commercial stations also broadcast U.S.-produced programs.

Hong Kong

Hong Kong, recently returned to Chinese control, is a special economic zone. It functioned as a capitalistic trade window for the People's Republic of China for years and is continuing in that role. The city is extremely wealthy, and its citizens have, in many cases, adopted a western lifestyle. Karaoke bars are popular, and their music is likely to have come directly from MTV.

Many U.S. media organizations have established Hong Kong offices to distribute programming in Asia. CNBC offers programming there as does the Cartoon Network. Murdoch's STAR TV is headquartered in Hong Kong, and its DBS service provides a broadcast footprint from Japan to the Middle East. Wharf TV provides cable television service to the thousands of high-rise apartment buildings that constitute the main living areas of the city.

Republic of China

The Republic of China, a country off the southeast coast of mainland China, comprises the island of Taiwan, the Pescadores, and other smaller islands. Taiwan has a heavy mass media concentration. More than eighty radio stations serve the island, and virtually every citizen owns a receiver. There are five islandwide television services, including a newly created public system. Nearly every home owns a television receiver, and signals are delivered throughout the mountainous terrain by a series of repeaters.

Taiwan Television Enterprise was established in the early 1960s. China Television Company began broadcasting a few years later, and Chinese Television System went on the air in 1971. All three systems operate from state-of-the-art studios in Taipei, and much of the programming is locally produced. It includes news as well as audience participation game shows and live daily soap operas. The sale of commercial time funds the stations. In 1997 Formosa Television began commercial broadcast operations from the south end of the island. The government-funded Public Television Service started its broadcast schedule in 1998.

Over-the-air broadcasting in Taiwan is now facing stiff competition from direct broadcast television and cable. Although cable television had existed for some time, the systems were operating illegally until the passage of the Cable Television Law in 1993. Now there are more than 140 cable television systems in operation, with many offering more than seventy channels. Cable penetration in Taiwan is 80 percent, which is the highest rate in the Asia Pacific region.

Worldwide Media Distribution

Broadcast stations have an insatiable appetite for programming. More channels demanded by the citizens of a country simply mean that more hours of programming must be found to meet their needs. Many countries produce their own programs because of language barriers and fears of cultural imperialism. Other countries turn to inexpensive syndicated programming to fill the broadcast hours. Great Britain, Egypt, Mexico, Brazil, Australia, and of course the United States produce a large volume of television product that is sold on the international market. Each year, buyers and sellers of television programs come together to conduct business at the European television programming fairs of MIP-TV and Monte Carlo. The outcomes of those deals often determine the profit margin for the production companies as well as what television programs a viewer in a Third World country might see. The television series *Dallas* gave much of the world a particular view of life in the United States, but the motivation for broadcasting that program in other countries was economics and not political ideology.

No matter what the motivation, the content of the television programs has an impact. Foreign visitors to the Grand Canyon often express disappointment in seeing the Native Americans riding in pickup trucks rather than on paint ponies.

Many media conglomerates have become multinational in their operations. News Corp. is an ideal example. Rupert Murdoch, owner of BSkyB in Great Britain and STAR TV in Hong Kong, also owns the Fox network and its affiliated television stations in the United States. Theoretically, most people on the globe are within reach of his television signals. CNN is another multinational operation, with news bureaus and programming outlets in most corners of the world. With the exception of a few African countries, CNN International distributes news on a worldwide basis to more than 130 countries and has news-gathering operations in many of them.

The traditional U.S. television networks—ABC, CBS, and NBC—all provide television news, sports, and entertainment programming on a contractual basis to broadcasters in other countries. NBC provides distribution of its cable television programming throughout the world; *The Tonight Show with Jay Leno* can be seen in London as well as in New York or Los Angeles. CBS distributes its *Evening News with Dan Rather* through the French company Canal Plus.

In addition to the sale of television programming on the international market, many U.S. media consultants have discovered lucrative foreign markets. As public broadcasting systems have been privatized and commercial advertiser–based stations have been established, broadcasters in many countries have turned to the U.S. news, management, and programming consultants to learn how to operate the media in a new capitalistic world. The news consultants who

Shooting the "Blue Light" Show at Moscow TV Center.

have shaped local television in America, such as McHugh-Hoffman and Frank N. Magid Associates, are now reinventing local television around the world. Thus ideas, as well as broadcast programs, are being exported by the media machine of the United States.

Other examples of multinational control of the media abound. Canadian companies, such as Rogers Cablesystems Ltd., have owned cable television systems in the United States for years. The Japanese own Sony Picture Studios in Hollywood, and several of the U.S. Baby Bell telephone operating companies made heavy investments in cable in Great Britain and Australia during the time that they could not operate cable systems in the United States.

Many of the world's economies have become interdependent. With that has come the need for people in one part of the world to have more information about their international neighbors and trading partners. The electronic mass media provide that information. With an increased flow of information among countries, perhaps we will develop a better understanding of other cultures and move closer to the global village envisioned by McLuhan so many years ago.

Summary

▶ Telecommunication systems do not exist in a vacuum. Electronic messages flow across international borders. Some messages are sent intentionally to politically influence people in other countries. This is called international, or external, broadcasting. Other messages unintentionally reach beyond a country's borders because of the physical laws governing the electromagnetic spectrum. Messages are also transmitted from one country to another because of the importation of television programming for entertainment purposes.

▶ Many countries are concerned about cultural imperialism—the transfer of the culture of one country to another via the media. Generally, the concern is that western, that is, American, values will overwhelm the historical cultures of other countries. Some countries have limited the amount of U.S. programming that the domestic broadcast system can import.

▶ Almost every country in the world has developed its own internal, or domestic, broadcasting system. The function and operation of systems vary depending on such factors as the political philosophy of the government, the literacy rate of the population, and the country's economic conditions and geography. Most domestic broadcast systems, however, fall into one of two models: either a public system funded by the government or by license fees, or a private system supported by the sale of commercial advertising.

▶ Domestic broadcast systems around the world underwent drastic changes at the end of the twentieth century. Many countries that had only public broadcasting systems in the past are now allowing private broadcasting systems to develop and compete with the government-supported public ones. The increase in the number of broadcast systems has meant that viewers have a greater choice of information and entertainment. The increased programming required must come from somewhere—usually a foreign market.

▶ Another trend in broadcasting in other countries is that the sale of commercial time to advertisers is becoming commonplace. Even countries with strong public systems funded by the government are turning to advertising to pay the ever increasing costs of running the stations. Audiences are accepting commercials, often as a desirable alternative to receiver license fees.

▶ Technology is also changing domestic broadcast systems. Many countries are turning to videocassette recorders (VCRs), cable television, and the direct broadcast satellite (DBS) to provide increased programming choices. Media companies in the United States are eager to tap the new worldwide markets for television program distribution. In the near future, digital broadcasting and the Internet may create additional opportunities for the production and sale of high-quality programming in the international marketplace.

INFOTRAC COLLEGE EDITION EXERCISES

Broadcasting and telecommunication is a worldwide phenomenon, but not all nations have followed the same political, economic, or regulatory path in designing and implementing electronic communication systems.

15.1 Chapter 15 provides a glimpse of the background and operation of electronic communication systems in several other countries. Other countries also have broadcasting systems with interesting histories and operating policies.

   a. Using InfoTrac to access *Journal of Broadcasting and Electronic Media*, look for articles about broadcasting in other countries, then write short synopses of three broadcast systems in other nations.

   b. Using InfoTrac to access computer or telecommunication journals such as *Worldwide Telecom, PC User, PC Week*, or *PC/Computing*, find articles on Internet use in other countries.

15.2 The broadcast system of a country generally reflects the political and cultural structure of the society in which it operates.

   a. Using InfoTrac to access publications such as *Russian Life, Scandinavian Studies, Journal of European Studies, Journal of Near East Studies, Journal of Southeast Asia Studies,* and *Asian Folklore Studies,* find and report on examples of differences in societies that may shape differences in the broadcast systems in those countries.

CHAPTER 15 THE GLOBAL VILLAGE 347

WEB WATCH

Here is a list of a few URLs (Internet addresses) for some of the organizations or corporations discussed in this chapter. Please explore these Web sites and follow the links to learn more about the complex business of the electronic media. Add your descriptions and your own favorite sites at the end of the list. Please keep in mind that the dynamic nature of the Internet allows sites to come and go but also allows organizations to update information about themselves very quickly.

Address	Description
http://www.voa.gov/	
http://www.itu.int/	
http://www.cbc.ca/	
http://www.metroradio.com.hk/index.php3	
http://www.twr.org.hk/	
http://www.hkba.org.hk/	
http://www.nhk.or.jp/index-e.html	
http://www.onr.com/user/fcantu	
http://www.rfi.fr/	
http://www.itn.co.uk/	
http://www.bbc.co.uk/	
http://www.powerfm.com.tr/start.htm	
Other favorite sites:	

CHAPTER 16

Effects and Influences of the Electronic Media

What electronic communication technology does to us, as a society, depends largely on what we, as individuals, do with it.

Effects and Influences of the Media

Our society depends on information processing as we enter the twenty-first century. The electronic media play a pivotal role in providing the information services that have become essential to business and leisure activities. We are influenced by the electronic media in virtually everything we do. From the moment we awake in the morning until we go to bed at night, we are utilizing some kind of electronic medium: We rely on radio and television for information and entertainment in the morning; we use the telephone at home, in the office, and in our vehicles to process information; we use computer terminals in our homes and offices throughout the day; and in the evening, when we are ready to relax, we return to radio or television for entertainment and information.

We are influenced by the electronic media in our daily decision making. Business decisions rely on information available from a variety of media sources. Consumer decisions are heavily influenced by media programming and advertising. Our tastes in music, art, and literature are also influenced by the media, as are political and news agendas.

Media professionals point to the positive influences of the media, but, of course, the electronic media also can have negative effects on society. Television, film, and music have been criticized for years for their alleged contributions to increasing violence among youth, advertisers have been criticized for exploiting children in television commercials, and journalists have been criticized for biases in reporting the news.

The mass media have also been blamed for a variety of society's ills, including the disintegration of the family, the decline of reading skills among youth, juvenile delinquency, and the decrease in physical activity in both children and adults.

Effects Theories

The effects of the mass media have been studied through a variety of research methods, the most common of which, as outlined in chapter 11, are historical-critical, experimental, content analysis, and survey. Although chapter 11 focuses primarily on the survey method used in audience research for radio and television programming, in the study of mass media effects, *experimental research* is often used. In this method, also known as *laboratory research,* an experimental group is tested for the effect of a particular variable while a control group that is not exposed to the variable is used for comparison. For example, a group of children views a television program portraying physical aggression while a second group views a program without aggression. The two groups are then observed in a play setting to determine if the variable, physical aggression, affects the behavior of the children in the experimental group.

The *content analysis* research method is used to examine the characteristics of radio and television programming. We might analyze prime-time television programs over a period of time to determine the presence of violence. In this method we might define *violence* as "acts of physical aggression by one character against another." We may choose to include verbal as well as physical aggression in our definition of violence. Through the analysis of programming content, we can determine if violence in prime-time television has increased or decreased over a period of time.

In the study of mass media effects, we generally look at the components involved in the mass communication process. A basic mass communication model (see Introduction, figure I.1) includes a source, a message, and a receiver. The message is delivered through a channel, which requires encoding and decoding processes. The message encounters physical and semantic noise in the process of traveling from the source to the receiver. Often we want to determine the effectiveness of our message, so most mass media research focuses on the effects of messages on members of the audience.

During the early history of radio and film, it was widely held that the mass media had a direct effect on audiences. This theory, known as the "magic bullet" or "hypodermic needle," proposed that all radio and film audiences would react to media stimuli as a group or in the same way. The theory did not account for differences in individuals, such as their family backgrounds, educational levels, and socioeconomic status. Further research about media effects suggested that these factors played major roles in audience behavior. In fact, the mass media appear to have more *indirect* than direct effects in that they work in conjunction with the influences of family, education, and peers. As a result, the media have different effects on different individuals. The different-effects theory was supported by Wilbur Schramm, who studied the effects of television on children in the early 1960s. He found that television viewing affected young viewers differently, depending on such variables as their relationships with their parents (Schramm 1963).

VIOLENCE THEORIES

Many studies of the effects of televised violence on children were conducted in the 1960s. An early study by Albert Bandura (1963) suggested a relationship between violent behavior on TV and aggressive behavior in children who viewed the behavior. Later studies sponsored by the U.S. government found this relationship as well, although it was difficult to predict which children would be affected and how.

Research on the effects of televised violence on individuals, especially children, has generated a variety of theories; the most prominent are catharsis, modeling, desensitization, and "mean world" syndrome.

According to the *catharsis theory*, televised violence serves as a release for average viewers. When the "good guy" takes out the "bad guy," viewers feel a healthy release of pent-up aggression. It is not socially acceptable to resort to physical violence themselves, so they release tension through televised violence.

The *modeling theory*, also known as *imitative behavior*, suggests that children learn to be aggressive by watching violent characters on TV. This effect was evident in the Bandura study. Again, the effects of televised violence tended to vary with different children. To some extent, children who already have aggressive tendencies may be more likely to identify with violent acts on TV. It appears that the research does not support a direct, causal relationship between televised violence and violent acts such as assault or murder. Nevertheless, the bulk of the research conducted does support an association between violence on TV and aggressive behavior in children.

The *desensitization theory* suggests that children become less sensitive, or even insensitive, to actual violence by watching violent acts on television. At some point, after watching countless murders and assaults on TV, some children have little or no reaction at all to violent acts in real life. They are not stimulated to commit violent acts, but neither do they feel shock. The theory suggests that children simply become passive dullards as a result of witnessing yet another murder on TV.

The *mean world syndrome*, which applies more to adults than to children, describes the reaction of TV viewers, particularly older people, to the world around them. Because they see so much violence on television, they feel that the real world in which they live is extremely violent. They have exaggerated fears of being attacked or killed on every street corner. Again, this effect has to be considered in the context of individual viewers and their environments. For many viewers, television provides a positive, even comforting, diversion from the routine of their daily lives. Still, televised violence continues to generate debate and concern among parents, consumer groups, and policymakers.

More-recent content studies of televised violence have reported conflicting conclusions. A 1996 study commissioned by the "Big Four" (ABC, CBS, NBC, and Fox) reported that the networks had made "modest improvement" in reducing violent programming. Yet another study, sponsored by the National Cable Television Association (NCTA) and released in 1998, reported that violence in Big Four prime-time network programming had actually increased. Violent programming also was up on independent television stations and on basic cable channels. Premium cable channels such as HBO were found to have the highest violence content—constituting more than 90 percent of programming. Researchers expressed concern that much of the violence portrayed was gratuitous or without consequences. They indicated that such portrayals could contribute to a desensitization to violence in general among child viewers.

The dramatic increase in random killings such as the high school shootings in Littleton, Colorado, prompted citizens and government officials to demand accountability for violent media programming.

TABLE 16.1 Network television program rating system

AGE-BASED RATING CATEGORIES (1997)		CONTENT-BASED CATEGORIES (1998)	
TVY	Suitable for all children	S	Sex
TVY7	Children older than age 7	V	Violence
TVG	General audience	D	Suggestive dialogue
TVPG	Parental guidance	L	Foul language
TV14	Older than age 14	FV	Fantasy violence
TVMA	Mature viewers		

After years of debate and thousands of academic studies, public pressure on Congress demanded consideration of violent and sexual content on television as the Telecommunications Act of 1996 was drafted. Title V of the act addresses obscenity and violence in the media. Perhaps the most controversial requirement of the act was the V-chip: Television receiver manufacturers were required to equip each new television set with an electronic chip that could be programmed by parents to block television programs containing sex and violence from coming into the home.

By itself, the chip would be completely ineffectual without a method of identifying programming content to be blocked, so when implementing the act the Federal Communications Commission (FCC) required the broadcast industry to develop a set of program ratings that would help parents identify programs that contained sex and violence. After outcries of censorship, the broadcast industry, realizing that parents overwhelmingly wanted a program rating system, developed ratings loosely modeled after those long used to classify theatrical motion pictures. The broadcast rating system, which went into effect in January 1997, displays a rating label in the upper-left corner of the television screen at the beginning of each dramatic program (news programs are exempt).

Sexual and violent content on cable television also came under the scrutiny of the Telecommunications Act. Adult programming such as is aired on the Playboy Channel, which had been simply scrambled in the homes of nonsubscribers, was now required to be completely covered. The Supreme Court upheld that requirement in early 1997. Obscenity on the Internet remains a problem, but regulation of the content is largely unenforceable because of the immensity of the system and the variety of regulatory philosophies that exist in the various countries throughout the world.

Media Stereotypes

Another major area of media criticism focuses on *stereotyping*, particularly in characterizations of women and ethnic minorities in television programming. There are countless examples of stereotypes in network television programs over the past sixty years. In early television situation comedies, women often were portrayed as housewives who stayed at home and baked cookies while Dad went out and earned a living. One episode of *Father Knows Best* showed Robert Young trying to teach his flustered wife how to drive the family car. She was portrayed as a befuddled fool who couldn't keep the gas and brake pedals straight.

Ethnic minorities, particularly African Americans, have been stereotyped in a variety of broadcast programs. A prime example can be found in the *Amos 'n Andy* network radio program that was so successful in the 1930s and 1940s. The African American characters were portrayed by white actors with exaggerated dialects. After successful runs on both NBC and CBS radio networks, the series crossed over to television in 1951. Obviously, the characters on television had to be played by African American actors, but the series received such criticism from the NAACP that it was canceled in 1953. The creators of the original radio program, Charles Correll and Freeman Gosden, maintained until their deaths that they had never intended to stereotype African Americans in the program.

Television programs have been analyzed for examples of stereotyping in a variety of content studies. Greenberg and Collette (1997) analyzed new characters introduced in fall network television seasons from 1966 to 1992. They found that new characters were primarily male as late as 1992, although the number of female characters did increase over the years. New-character gender was nearly balanced in 1980 and 1984, but there were never more women than men. Female characters tended to be younger than males. In regard to age in general, most new television characters were in the 20- to-35 age range. Fewer than 10 percent were over the age of 50, and only 2 percent were over 65.

Over the twenty-seven seasons that were analyzed, 88 percent of the new characters introduced were white, 10 percent were African American, and fewer than 1 percent were Hispanic, Asian, or Native American. The number of African American characters began to increase in 1984 and continued through 1992. Fox had the highest percentage of African American characters.

It is clear that throughout most of its history, women and ethnic minorities have been under-represented and stereotyped on network television. The concerns that arise from content studies relate to the effects of these portrayals on viewers. A number of studies indicate that African American teenagers watch more television than do white teenagers. Not surprisingly, teenagers tend to identify with role models who are of the same race. Some studies suggest that African Americans tend to view TV characters of all races as more realistic than do whites.

Among the most positive portrayals of African American characters were those on *The Cosby Show*. The program, introduced in 1984, portrayed the upper-middle-class family of an African American doctor and his wife, a lawyer. It remains one of the most successful situation comedies ever produced and, yet, has received both praise and criticism. Tucker (1997, 90) praised the series as "proving that a program about black Americans created and produced by a black American could be economically successful and popular with the mainstream broadcast television audience without resorting to traditional black stereotypes or deprecating humor." But there were others who criticized the program as an unrealistic portrayal of African American families. A 1988 *Television Quarterly* review referred to Cosby's character as "a kind of black Ozzie Nelson" and to the program as a "comfort zone for whites" (Tucker 1990, 101).

As the twentieth century drew to a close, it appeared that portrayals of women and minorities had improved. There were more female and minority role models on television and they were portrayed in more-positive images.

Children and Media Effects

There is little question that the electronic media, particularly television, play a significant role in the development and socialization of children. Even as infants, children are exposed to the media consumed by their parents. They begin watching television as preschoolers and are exposed to a variety of program

By the end of the 1990s, there were more positive women and minority characters in prime time television programming. Shown here is the cast of Twentieth Century Fox Television's *Chicago Hope*.

content. Many of today's young adults watched *Sesame Street*, a program generally considered to be a positive influence on children, regularly as preschoolers. But they also watched hours of violent cartoons and other "less desirable" programming.

Much of the research on children and the electronic media points to a strong modeling effect. Children learn from television what their roles in society might be and how to act toward other human beings. Of course, these lessons can be positive or negative, depending on the models portrayed in the programs viewed.

In the relationship between the child viewer and television, there are two factors that are very important to consider: attention and comprehension. Not surprisingly, younger children appear to have shorter attention spans than do older children. Jeffres (1997) suggests that very young children (two years or younger) do not "physically orient" themselves to the TV set as part of their play. They may be attracted by graphics and visual effects, but they are not as directly involved in program content as are older children. But as children mature, they become more aware of program content and structure and pay more attention to television. Watching television becomes a play activity in itself. It is at this point in a child's development that television begins to play a significant role.

As attention spans increase in children, so too do levels of comprehension. These factors are related in that as children comprehend more about a program's plot and character, they will pay more attention to it. Jeffres notes, however, that increased comprehension is not always necessary for a child to pay attention to a television program. He points to the difference between *Sesame Street* and *Mister Rogers' Neighborhood;* children tend to pay greater attention to the first program than to the second. This is evidently due to the graphics and production techniques used in *Sesame Street* that are not present in *Mister Rogers' Neighborhood*.

There is increasing concern among parents about the amount of time their children spend watching television and playing video games. In fact, television viewing is now being challenged by the computer and time spent on the Internet. The concern is not only that these activities replace physically active play, but that children may not be able to perceive the difference between reality and the fantasy of television programs.

Research on children's understanding of reality as portrayed on television suggests stages of development:

> In the first stage, children assume that the images they see on television are real objects that are physically present inside the small screen. In stage two, children do not believe that images they see on television act like real objects. Yet at this age children do not fully appreciate that televised images are pictorial representations of three-dimensional objects. Children in the third stage understand that television pictures are images of realities and that television represents external realities; however they do not comprehend that television images may misrepresent as well as represent external objects. Thus, children at this stage and the earlier stages tend to believe that television provides a fully accurate picture of external reality. It is not until the fourth stage (as other researchers have suggested) that children acquire a more complex understanding of the degree to which television does and does not portray life accurately or realistically (Jeffres 1997, 186–187).

In regard to age, other studies suggest that preschoolers and even children in the early elementary grades can still have problems distinguishing TV content from real life. By the time they become adolescents, however, most children perceive that television programs and actors are not real.

Although most research has focused on the *negative* effects of media on children, there are indications that there are positive, or *prosocial,* effects as well. Studies have shown that children learn social skills such as sharing, cooperation, and helping by watching television. Dominick (1990) points to a study of children viewing an episode of *The Waltons* that modeled cooperation. In a play setting following the program, the children were more willing to help one another than were children who had not seen the program. Wood (1983, 257) suggests that the media contribute to "social regularity . . . the inclination of the media—especially in reality content—to remind us of the rewards for social conformity and the punishments of deviant behavior."

The debate and concern over the effects of the electronic media on children will continue. Clearly, young television viewers are subject to both positive and negative media effects. But the degree of the effects appears to be significantly moderated by a child's demographics and environment. Factors such as age, intelligence, personality, and family relations play a major role in media effects.

Politics and Media Effects

The electronic media have become an integral part of the political process in the United States. Political campaigns for national offices (and even some local offices) virtually require television advertising and news coverage to be successful. There is a wealth of research available on the effects of the media on campaigns, voting, and candidate image.

Politicians recognized early in the history of radio and television that the media could be powerful factors in the election process. As noted in earlier chapters, legislation granting political candidates "equal opportunity" for access to radio and television stations was established in 1927. This section of the law, which is still in force, is based on the philosophy that the electronic media utilize public property—the electromagnetic spectrum. Station owners, as trustees of the public, cannot use their channels to favor one candidate over the other. This longstanding law, which does not apply to newspapers and other print media, is a testament by lawmakers to the potential power of radio and television.

The public has access to information on political candidates and issues through two basic avenues of the electronic media: news/public-affairs programming and political advertising. Radio and television news personnel serve as *gatekeepers* for the audience in that they decide what stories to cover in newscasts. This also provides an *agenda-setting* function by determining what issues are important enough to cover on a given day or week. These decisions by news personnel are not always in accord with the opinions of political candidates as to what are the important issues of the day. This was dramatically demonstrated in the prolonged media coverage of the sex scandals surrounding President Clinton during his second administration. There were many days when coverage of these scandals overshadowed other national and international events that some considered more important.

Before the advent of television, there was some concern that politicians might use the power of radio and film to manipulate the election process. This was based, in part, on the perceived "magic bullet" effects of Nazi propaganda films and radio programs like *War of the Worlds*. One of the landmark studies on media effects and politics was conducted in the 1940s by Paul Lazarsfeld and his colleagues (Lazarsfeld and Merton 1960). Based on their research of voter influences during the 1940 presidential election, they concluded that the media played a minimal role in voter behavior. As with other effects of the media, factors such as family, education, party affiliation, and peer groups exerted more influence on voting behavior than did the electronic media.

With the growth of television, media effects on the political process have become more pronounced. This is due in part to the increased use of the medium and in part to changes in the electorate. Jeffres (1997, 206) notes that "party affiliations and organizations have declined drastically; voters increasingly split their tickets between parties. Voters are more likely to base their decisions on issues and images rather than party affiliations." It is in the process of candidate image building that television seems to have the most pronounced effect on viewers. Because television is both a visual and a personal medium, it provides the viewer with a first impression of a candidate. This initial image is important to all political candidates, but it is critically important for those who are not well known to the voters.

The potential for television to play a major role in elections was reflected in the 1960 televised debates between John Kennedy and Richard Nixon. Surveys after the debates showed that most television viewers thought Kennedy won, whereas those who listened on radio favored Nixon. This may have been due to the perception that Kennedy was more dynamic and effective on camera than was Nixon, who appeared nervous and pale. Kennedy's more youthful, vibrant image on television evidently played a role in tilting a close election. Still, studies of the election process over time suggest that the two media effects that are the most common are *reinforcement* (strengthening of existing attitudes) and *crystallization* (sharpening of vaguely held attitudes). Television images can have a more pronounced effect on voters who have not made up their minds than on those who have already decided on a candidate.

Voters also receive information about and form images of political candidates through media advertising. In recent campaigns, televised political advertising has been criticized as mudslinging. Negative political commercials tend to focus on the opponent's shortcomings rather than on the candidate's own attributes. Negative advertising became so prominent by the 1988 presidential election that the television networks instituted what they called "adwatches" in their newscasts. Kaid (1996, 297) describes *adwatches* as "media critiques of candidate ads designed to inform the public about truthful or misleading advertising claims."

By the 1992 election, negative political commercials had become the focus of network news stories. In their content analysis of network television coverage of the 1988 and 1992 presidential elections, Kaid and her colleagues reported that the networks tended to focus more on negative politicals ads than on positive ones. They found that in the 1992 election, the networks focused more on Bush's negative ads than on Clinton's. Kaid suggests that this may have had an effect on voters: "The fact that Clinton's negative campaign went virtually unchallenged by media adwatches may be a factor in Clinton's decisive campaign victory" (Kaid 1996, 306).

Whether political candidates appear in television newscasts, political ads, or radio talk shows, the electronic media play an important role in campaigns and elections. Candidates are guaranteed by law the right to purchase advertising time on radio and television stations on an equal basis. But the electronic media represent but one of the many factors that influence voter behavior. Voters tend to selectively perceive media messages based on their political affiliations, peer influences, and socioeconomic backgrounds.

The electronic media played a crucial role in the 1992 election and the 1996 re-election of President Clinton.

Advertising and the Media

As noted earlier, advertising is the major source of revenue for the electronic media. Advertisers spend billions of dollars every year on radio and television ads that are designed to persuade the consumer to purchase products and services. Viewers and listeners are exposed to thousands of commercials during a year, but they give little thought to their cost or effect. Advertising must be effective in persuading consumers to buy, or advertisers would not spend so much money on commercials. It is obvious that advertising works, but *how* it works is less clear.

Advertising has its critics. They contend that it adds to the costs of products and services, that it creates artificial needs and desires in consumers, and that it exploits children. On the other hand, proponents argue that advertising is a necessary part of a free-enterprise system. It provides relevant product information that helps consumers make buying decisions, and it enhances competition among products to the benefit of the consumer. Some even argue that commercials have artistic value as creative expressions of the producers.

The effects of electronic media advertising are reflected in a variety of ways. There is a learning effect as consumers are exposed to information on products and services, and advertising can fulfill needs or desires of consumers for products and services. Of course, the effect that advertisers hope to achieve ultimately is the consumer's act of buying the product or service. The successful commercial will produce all three of these effects. The challenge is to gain the attention of the audience, hold that attention while delivering the message, and leave the audience with the motivation to buy. Jeffres (1997, 251) suggests that "ingredients of successful advertising appeal to consumers' desires for gratification or information; tastes or preferences based on self-images and goals; and a hierarchy of needs ranging from the physical (food and sex) to self-fulfillment."

Much of the concern about advertising effects revolves around ads targeted at children. Parents and critics charge that commercials in children's television programming promote products that are not healthy, such as sugared cereal, snack foods, and candy. They see these ads as exploitive, because children are more vulnerable to sales pitches than are adult viewers. Just as young children have trouble distinguishing between fantasy and reality in programs, they may also confuse commercial content with reality. Dominick (1990) suggests that children five to eight years old can identify commercials but they don't realize that the purpose of the commercials is to sell things. As children become older, they become more skeptical of commercials, particularly as they learn that products advertised often are not as they appear on television.

The potential for children to confuse TV ads with programs appears to be increased by program characters selling products. This practice, know as *host-selling*, was banned by the FCC in 1974 as an unfair advertising technique. In general, advertising is regulated by the Federal Trade Commission (FTC), but this FCC regulation was related to children's programming. The FCC also banned what it called *program-length commercials,* programs built around characters that were then sold as dolls or puppets. Some examples include *The Smurfs, Masters of the Universe,* and even *Sesame Street* characters. In 1985, when the FCC repealed commercial time limits on television, it effectively eliminated the ban on program-length commercials. However, Congress passed legislation in 1990 which brought back FCC commercial limits in programs designed specifically for children. Restrictions on host-selling in children's programs also have been maintained.

Summary

A wealth of research has been conducted on a wide variety of mass media effects, ranging from those of televised violence and sex to advertising and politics. If there is a common thread running through these studies, it is that media effects are moderated by other variables, such as peer and family influence, education, moral values, and socioeconomic status. Political commercials and programs are more likely to reinforce voter preferences than convert a voter from one party to another. The same is true for product ads: Commercials are effective, but they tend to influence a consumer's brand preference more than cause someone to adopt a new product. Still, mass media effects can be profound in some individuals under some circumstances, so criticism and research related to these effects continue.

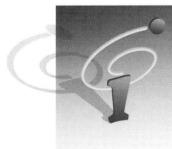

InfoTrac College Edition Exercises

Media studies suggest that effects on audience members are moderated by other variables, such as peer and family influence, education, moral values, and socioeconomic status.

16.1 You read in chapter 16 about stereotyping of characters in television programming. Use InfoTrac to locate the article by Mark R. Barner in *Journal of Broadcasting and Electronic Media* (Fall 1999) on stereotyping in children's educational television programs and answer the following questions.

   a. Barner's study found sex-role stereotyping in an overabundance of male behavior among main characters. What else makes these programs sex-role stereotypical?

   b. Barner predicted that female characters would be granted more-positive consequences for exhibiting "female" behavior in programs. Was this the case?

16.2 An aged-based rating system for network television programs was adopted in 1997, with content categories added in 1998. Use InfoTrac to locate the article by Sneegas and Plank in *Journal of Broadcasting and Electronic Media* (Fall 1998) on children's reactions to the rating system and answer the following questions.

   a. Were the children in the Sneegas and Plank study able to distinguish among the rating system categories?

   b. What differences, if any, where there between boys and girls in regard to parental rules for using the television ratings codes?

Web Watch

Here is a list of a few URLs (Internet addresses) for some of the organizations or corporations discussed in this chapter. Please explore these Web sites and follow the links to learn more about the complex business of the electronic media. Add your descriptions and your own favorite sites at the end of the list. Please keep in mind that the dynamic nature of the Internet allows sites to come and go but also allows organizations to update information about themselves very quickly.

Address	Description
http://www.nctvv.org/	
http://www.fcc.gov/vchip/	
http://www.familysafemedia.com/	
http://www.curse.free.com/	
http://www.tri-vision.ca	
http://www.tvguardian.com/	
http://www.tvguardian.net/	
http://www.lionlamb.org/	
http://www.sofcom.comau/TV/violence.html	
http://www.ithaca.edu/cretv/projects.html	
http://www.urich.edu/%7epsych/tvagslt.html	
Other favorite sites:	

Epilogue

You have now completed another step in your preparation for a professional position in the world of electronic media. It is our hope that the knowledge base you have gained from this text will lead to the development of the skills you need to excel in your chosen field. Creativity and good communication skills will be critically important in the media of the future. The challenge for media professionals in the twenty-first century will be to create, market, and deliver content to meet consumer demand. We hope this text plays a significant role in your preparation to meet that challenge.

Glossary

AB-roll editing   A videotape editing process that allows the picture on videotape A to be replaced by a picture on videotape B through a dissolve or other visual effect.

access programming   Programming produced for a cable channel by a local citizens' group, a community organization, an educational institution, or a government organization. The cable operator has only limited control over the content of the programming.

active listeners   The known radio listeners who have participated in station contests or called in to the station.

additive color   A process whereby the primary light colors of red, green, and blue can be added together to obtain all other colors, including what we see as white light.

addressable technology   The technology found in cable television or direct-broadcast satellite that allows each individual subscriber to be identified by an electronic code and programming to be delivered to that home only.

adjacencies   Spots sold in station breaks next to popular network programs. Because network programs usually attract larger audiences, adjacencies are often sold at higher rates.

administrative law   Law created by an administrative body, such as the Federal Communications Commission (FCC) or the Federal Trade Commission (FTC).

adult contemporary (AC)   A radio format incorporating contemporary music targeted to an adult audience ranging from 25 to 54 years of age.

advanced television (ATV)   A set of television standards developed in the late 1990s by the Advanced Television Standards Committee (ATSC) that allows a digital television signal to be broadcast to the home; it is now replacing those developed by the National Television System Committee (NTSC) for analog television.

afternoon drive time   A radio daypart designated as the hours between 3 P.M. and 7 P.M. when many listeners are in vehicles on the way home from work.

album-oriented rock (AOR)   A radio format incorporating songs other than hits from CDs or albums by popular artists.

AM (amplitude modulation)   A method of sending a broadcast message by encoding the information in constantly changing minute variations in the height of the wave of the electromagnetic signal.

analog   An electronic system wherein the values of the electronic signal are continuously changing to represent information carried.

anchor   An in-studio person who acts as the foundation of the on-air newscast presentation.

area of dominant influence (ADI)   A geographical area that reflects those counties in which the dominant share of television viewing is to home market stations.

ASCAP (American Society of Composers, Authors, and Publishers) An organization of authors that collects royalties from performances and distributes them to the member artists who hold copyrights.

aspect ratio The ratio of width to height of the video screen; most screens are 4 units wide by 3 units high.

assemble mode A type of video editing in which shots are added on videotape in a consecutive order without first recording on the videotape a control track of sync pulses necessary for the stability of the picture.

assignment editor (AE) The individual in the news department who monitors news events in the community and dispatches reporters and photographers to cover them.

audience flow The movement of viewers or listeners from one program to the next.

audience recycling The extent to which morning radio listeners tune in again in the afternoon.

audimeter An electronic device that records the time a television set is in use and the channel to which it is tuned.

auditorium testing A research method in which an audience sample watches television pilot programs or listens to new songs. Also known as *theater testing*.

avails Short for *spot availabilities*. Unsold periods of airtime for commercial spots. These are inventoried by the traffic department and available for sale to a commercial client.

average quarter-hour persons The average number of persons estimated to have listened to a radio station during any quarter-hour in a time period.

barter programming Revenue system wherein a portion of the commercial time in a program is presold by the program's producer to help defray production costs. Because of this arrangement, the cost for the station to license the program for broadcast is often lower, but the station has less commercial time in the program to sell.

baud rate The speed at which digital information can be transmitted, usually through modems over telephone lines.

beam splitter An optical device behind the lens of the television camera that separates light into the primary light colors and directs each color to the proper pickup device or chip.

Betacam A ½-inch-wide professional videotape format, often used in electronic newsgathering and for automated commercial and station break videotape playback.

bias current A small electric current that is applied to a recorder's audio or video head to align magnetic particles on the videotape prior to the recording process.

black In video terminology, the absence of picture information; to "fade to black" at the end of the video program.

block programming A scheduling strategy that groups together programs of a similar type to develop audience flow, such as a situation comedy followed by a second situation comedy followed by maybe a third and fourth. Audiences tuning into the first program will stay tuned for the subsequent ones.

broadband Cable or fiber-optic systems capable of carrying a wide range of the electromagnetic spectrum, thus allowing many channels of programming to be carried simultaneously.

call-out research Audience research method whereby radio listeners are called to survey music preferences.

camcorder A portable video camera with an attached or built-in videocassette recorder, forming a single handheld or shoulder-mounted unit.

carrier wave The portion of the electromagnetic spectrum used to transmit the specific channel to the receiver.

catharsis theory A media effects theory in which televised violence serves as a release of pent-up aggression for the average viewer.

cathode-ray tube Used as a video and computer display device in which streams of electrons scanning the rear of a glass tube light up photosensitive areas to create pictures.

CD-ROM Stands for *compact disc–read-only memory*. An optical storage device that allows digital information to be stored for use on a computer.

chain broadcasting An early term used to describe a broadcast network. Early radio stations were linked together by a "chain" of telephone wires so that several stations around the country could broadcast the same program at the same time, thus minimizing the costs of program production.

channel inertia The reluctance of the viewer to physically expend the energy to change the channel—to simply watch whatever comes on the screen next.

charge-coupled device (CCD) A digital chip composed of many pixels that hold an electric charge when exposed to light. The chip is used as a pickup device in a television camera to convert light energy to electric energy.

Chautauqua movement A time in the late-nineteenth and early-twentieth centuries, when educational and informational lecture and theatrical circuits were established to bring education and culture to the small towns of America in the form of tent show presentations.

checkerboarding A programming strategy that schedules different programs in the same time slot on different days of the week.

chroma key A video effect that replaces one primary color of the television picture with a picture from a second video source. The effect is often used to combine a weather map with a picture of the weathercaster pointing to different parts of the country.

chrominance The color information contained in the video signal.

churn The turnover of cable subscribers as they drop services or add them.

clock A diagram (resembling a clock) depicting an hour of radio programming, including music selections, announcer breaks, and commercials. Also called *hot clock* and *format wheel*.

closed-circuit television system (CCTV) A system in which the origination equipment is directly connected to the viewing equipment. The signal is not broadcast or cablecast and is limited in distribution.

clutter Too many spots in a commercial break. The fear is that the viewer will not remember the commercial message of any one spot if several are broadcast back-to-back.

coaxial cable A transmission wire used in cable or closed-circuit distribution that is shielded by an outer metal conductor that completely encircles an inner conductor. The two conductors are separated by a foam insulator.

coincidental method Means of gathering data by telephone in which respondents are asked to identify programs or stations they were watching or listening to when called.

color phase A timing of the color signal that, in effect, acts as a control to adjust the hue of the picture.

community antenna television (CATV) system The original name for what has become the cable television industry. Broadcast television signals were picked up by a large antenna system and delivered to a group of homes via a coaxial-cable distribution system.

compact disc (CD) A $4\frac{1}{2}$-inch plastic and aluminum disc that is used to record music for home use, taking the place of older versions of phonograph records.

compensation Money paid by a network to an affiliated station for the use of the station's airtime for the broadcast of the network programs, promotional announcements, and commercials.

concept testing The process of conducting audience research on a television program idea prior to putting that program or series into production.

conductor An element that is conducive to the flow of electrons. Usually a metal such as silver or copper.

consideration A legal synonym for money. It is used specifically in law relating to broadcast lottery, the material terms of a contest, and payola-plugola.

constitutional law Law that organizes government. It provides the patterns, direction, and the final interpretation of all other law. Each country has its own different constitutional framework.

contemporary-hit radio (CHR) Radio format that grew out of the Top 40 format and focuses on songs that are current hits on popular music charts.

content analysis A social science research method for the study of program or message content, such as aggression in television drama.

control board An electronic interface device that allows a radio broadcaster to control the source and volume of sounds feeding into an audiotape recorder or to the transmitter.

control track An electrical signal recorded on videotape that provides the synchronization information needed to control the speed of the transport during playback.

cooperative advertising A contractual arrangement between a wholesaler and a retailer to participate in a combined purchase of advertising time and/or space for the promotion of a product.

copyright A legal concept that gives the creator of a work (music, film, literature, television program, and so on) the right to profit from the sale of that work for presentation to an audience.

Copyright Royalty Tribunal (CRT) The body created for the purpose of collecting and distributing royalties paid by cable operators for use of broadcast station signals.

cost per point (CPP) The cost per rating point of reaching an audience of average quarter-hour persons that's equivalent to 1 percent of the population in a given demographic group.

cost per thousand (CPM) The cost of reaching one thousand persons or homes through commercial spots placed on radio or TV stations.

counterprogramming A scheduling strategy that puts a program of a different type in competition with programs that other stations are offering in the same time period. The practice offers an alternative to an audience.

cross-ownership One corporation owning two or more types of media outlets in the same market, such as a newspaper owning a radio station, or a television station owning a radio station. Most forms of cross-ownership are not allowed because of fears of restraint of trade.

cultural imperialism The transmission of the values of one society to another through the pictures and stories provided by the mass media.

cume The estimated number of different people who listened to a radio station for a minimum of five minutes in a quarter-hour within a reported daypart. Also known as *cumulative audience* or *reach*.

cut In video, an instantaneous change of the picture from one source to that of another. Also called *take*.

cyclical programming A program scheduling technique that repeats the program (or the format) throughout the day to save programming resources while reaching additional audiences that tune in at different times.

daypart A segment or division of the broadcast day on radio and television stations, used for programming and research purposes.

decibel (db) An electrical measure of sound intensity in audio, or signal strength in cable transmission.

delayed electronic feed (DEF) A network feed containing outtakes from network news or stories that never made the air; used by local stations within their own newscasts.

demodulator An electronic device that converts a transmitted signal into basic audio and video components.

demographics An audience research term referring to population categories such as age and gender. See also *psychographics*.

desensitization theory A media effects theory that suggests that children become less sensitive or even insensitive to violence by watching violent acts on television.

designated market area (DMA) A group of counties in which stations located in the metro area of a market or city achieve the largest audience share.

diary Written logs of listening or viewing patterns used by audience research companies to gather ratings data.

digital An electrical process that enables information to be coded into a series of on or off pulses of current.

digital audio workstation (DAW) A desktop computer loaded with an audio-editing and control software program that allows manipulation, amplification, and control of an audio signal.

digital videodisc (DVD) A home video playback format in which the picture and sound information is recorded digitally on a compact disc.

digital video effects The manipulation of the digital video signal to produce changes in the size, shape, or orientation of the picture, thus enabling the various attention-getting picture combinations and movements found in many television commercials.

direct broadcast satellite (DBS) A system of sending television programs directly from a communication satellite to the viewer's home television receiver through reception with a small 18-inch dish antenna and down-converter/tuner.

disc jockey (DJ) An individual who acts as host on a radio program. Also known as *on-air personality*.

dissolve A video effect in which the picture from one source is slowly replaced by the picture from another source.

distance learning The use of electronic media technology in education to deliver instruction to students outside the confines of the classroom.

downlink The electromagnetic signal path from the communication satellite to the earth.

duopoly The ownership of two or more broadcast stations of the same service (two FMs or two AMs) in one market. Also used to describe two broadcast systems, such as public and private, in one country.

edit controller A computerlike device with a keyboard that allows the selection of specific portions of a videotaped program and the precise control of the videotape recorder/players needed to assemble program segments on one machine into a completed program on a second machine.

efficiency of target audience An audience research formula that uses time-spent-listening estimates to determine how target audiences compare with total audiences.

electromagnetic spectrum A continuous range of frequencies of oscillation of energy as measured in Hertz (Hz), or cycles per second. Electric power is transmitted at low frequencies; radio and television at medium to ultrahigh frequencies; and visible light, X-rays, and gamma rays at the top end of the spectrum. Portions of the spectrum are also measured in wavelength, or distance, between the peaks of each cycle.

electronic news gathering (ENG) The process of using portable television cameras to take the news viewer on-location or directly to the scene of the news event through recorded videotape or live microwave transmission to the station.

evergreen A term used to describe a syndicated television series that remains popular with audiences and therefore receives good ratings over many years.

exclusive audience Persons who listened to only one radio station during a particular daypart.

exclusivity Refers to a program contract in which a station or network has purchased the rights to be the only broadcaster of that program in a specific market.

experimental research A research method in which an experimental group is tested for the effect of a particular variable while a control group that is not exposed to the variable is used for comparison. Also known as *laboratory research*.

fade An electronic process of slowly replacing a television picture with "black," or replacing "black" on the television screen with a television picture.

fiber-optic distribution systems Very small, pure, continuous threads spun out of glass that can transmit light energy over long distances.

field producer An individual who acts as a reporter's producer. Working directly with the assignment editor, conducting interviews and the initial research, he or she does everything a reporter does except appear on the air.

first-run program The first time that a television program or series is aired.

flight Each advertiser's schedule of commercial spots to be broadcast over a period of time. To reach a large audience and be effective, each commercial advertising spot must be broadcast many times; the flight is the sum total of all of those broadcasts.

flux Invisible lines of force which together compose magnetic fields.

FM (frequency modulation) A method of sending a broadcast message by encoding the information in constantly changing variations in the frequency of the transmission.

focus group A small group of people who are selected to give their individual opinions on a product or a television program for research purposes.

footprint The coverage area of a satellite broadcast signal.

freeze-frame An individual television picture with all motion stopped.

frequency A band designation (AM, FM, TV) assigned by the Federal Communications Commission to a city or broadcast station. Also used to describe the average number of times a person is exposed to a broadcast commercial schedule.

gatekeeper A person (or persons) who controls the information flow of any news item.

gateway A computer address and associated wiring that allows access to a computer network.

grid rate card A commercial airtime rate card listing not only different prices for different dayparts but also different prices for spots sold in the same daypart, taking into account the day-to-day demand for airtime. If demand is high, the advertiser pays the higher rate.

gross impressions The sum of the average quarter-hour persons for all commercial spots in a given schedule or run.

gross ratings points (GRP) The sum of all ratings points achieved for a particular commercial spot schedule or run.

hammocking A program scheduling strategy that places a new or low-rated television program between two high-rated ones. The audience flowing from one high-rated program to the next should give higher ratings to the program "hammocked" between the two successful ones.

head-end The physical area of the cable television plant that receives and processes the television signals before distributing them to the subscribers through co-axial or fiber-optic cable.

helical scan A technical method of recording information in a slanted track on videotape.

hertz (Hz) A name given (after Heinrich Hertz, 1857–1894) to describe the frequency per second of an alternating current or the frequency in cycles per second of an electromagnetic wave.

high-definition television (HDTV) A television broadcast standard that allows a high-resolution video signal (approximately twice the number of scanning lines of standard television), with an aspect ratio of 16 units wide by 9 units high, to be broadcast to the home.

historical-critical research Research method in which literature is examined to gather information on a particular subject.

homes using television (HUT) The percentage of all television households in a survey area with one or more sets in use during a specific time period. Also known as *households using television*.

horizontal programming A television programming strategy that schedules the same program at the same time Monday through Friday to build audience habit patterns. Also known as *stripping*.

hue The attribute of color that determines our perception of red, green, yellow, blue, and so forth.

hypoing Any unusual promotion or programming practice designated to distort the audience ratings results. Also called *hyping*.

infomercial A program-length video sales pitch, or commercial, for a product or service. Usually found on independent stations and late at night, when blocks of airtime cost less. These "programs" often pose as news or information, but demonstrate the advantages of buying a specific product.

insert mode A type of video editing in which new audio or video information is inserted into an already-existing videotape recording without affecting the shots on either side of the insert. Requires using a previously recorded control track as an electronic guide.

insertion equipment Automated videotape playback equipment that is used to add local commercials to a nationally distributed cable network such as ESPN or CNN.

instructional television fixed services (ITFS) A group of frequencies assigned by the Federal Communications Commission for use by educational organizations. Instructional programs can be "broadcast" in the 2,500 MHz frequency range at a maximum of 100 watts of power.

insulator A material with physical properties that inhibit the flow of electrons.

in-tab diaries The number of usable diaries actually tabulated to produce a market report or ratings book.

Integrated Services Digital Network (ISDN) A communication service proposed by telephone companies to provide a broadband wired communication system that could transport telephone, computer data, and video messages over wire.

international broadcasting Radio, television, and satellite transmission designed to be delivered across national borders to reach the population of another country.

Internet An informal but complex system of telephone wires, cables, and computer switches that allows personal computers worldwide to be connected in a network for information exchange.

KDKA The call letters of the first commercially licensed radio broadcast station in the United States, established by Westinghouse in Pittsburgh in 1920.

key An electronic cutting of one video image into another.

length of run The number of individual episodes that were produced for a series and are available for syndication. Generally, each episode can be broadcast a limited number of times during the license period.

libel Written defamation. Oral defamation is called *slander*. The law exists to protect the reputation of an individual and require accountability of the press.

license period The time frame in years within which a station has contractual right or approval to air a program in its local broadcast market.

line monitor A high-quality video display device that provides television station engineers and operators a picture of the broadcast signal.

listening locations Part of a radio market report designating places in which audiences listened to radio, such as at home or in a car.

local-origination (LO) programming Programming produced or provided locally by the cable operator. Often local sports coverage that can be produced easily and sold to advertisers in a local cable market.

low-power television (LPTV) A television station category licensed by the Federal Communications Commission to serve small geographical areas with a UHF television signal. The station is licensed to broadcast with limited power, so the coverage area is very small.

luminance The brightness portion of the video signal.

market audit A study of a broadcast/cable market that correlates retail or shopping activity with station audiences.

market study A study of a radio market conducted by program directors to determine format competition.

mean world syndrome A media effects theory that suggests that some TV viewers have exaggerated fears of the real world from watching excessive violence in the fantasy world of television. Also known as *cultivation*.

metro audience trends Data in radio market reports comparing shares, persons, and cume ratings for the five most recent metro survey periods.

metro market profile A section of a radio market report containing demographic and socioeconomic characteristics of the population.

metro survey area The core of the market—usually having the highest concentration of population.

microwave A high-frequency line-of-sight electromagnetic service used to transmit information from point to point.

midday Radio daypart designated as the listening hours between 10 A.M. and 3 P.M.

modeling theory A media effects theory that suggests that children learn to be aggressive by watching violent characters or models on TV. Also known as *imitative behavior*.

modem A shortened term for *modulator/demodulator*. An electronic device for sending and receiving computer data over telephone wires.

modulation The process of changing, or modulating, the amplitude (AM) or frequency (FM) of an electromagnetic wave so that it carries sound or video information.

modulator An electronic device for imposing audio and video signals onto a carrier frequency of the electromagnetic spectrum. Used in cable television to create the channels that carry the program to the home.

mood programming A program scheduling strategy that suggests that all programming of a station be of a similar type and directed toward a selected demographic group. Cable networks often use this strategy when selecting programming for MTV, ESPN, Comedy Central, and so forth. Also called *narrowcasting*.

morning drive time Radio daypart designated as the hours between 6 and 10 A.M., when listeners are likely to be in vehicles commuting to work.

multimedia The use of a computer to display interactive audio and video programs to educate and entertain.

multiple, multipoint distribution system (MMDS) A commercial, low-power, omnidirectional, broadcast distribution system that uses many channels in the 2,500 MHz range of frequencies. Because the systems are most often used to transmit cable networks such as CNN or ESPN to subscribers, this service is also known as *wireless cable*.

multisystem operator (MSO) A corporation that owns more than one cable television system. Many cable television owners find it economical to own and operate numerous individual cable television systems, each of which serves a different city.

music playlist A list compiled by a radio music director or program director of songs currently played on a station.

music sweep A series of songs played without interruption in a radio clock-hour; often designed to retain listeners from one quarter-hour to the next.

must-carry rule Federal Communications Commission rule requiring cable television systems to carry all local TV signals.

National Association of Broadcasters (NAB) An organization that lobbies before Congress and the Federal Communications Commission on behalf of broadcasters.

national system A broadcast system intended to serve the informational and entertainment needs of a country's own citizens. Also called *internal system* or *domestic system*.

National Television System Committee (NTSC) A nationwide engineering group that established the analog color television systems standard used in the United States.

network Linking chains of stations together by telephone wire or a transmitted electromagnetic signal for sharing programming or communicating information.

news director The individual who heads the news department within a station or corporate organization.

news insert A short, 60- to 90-second feature story produced outside of the news department. Such feature stories are often available from syndications. See also *video news release*.

news/talk format A radio format featuring consistent news programming, including international, national, local, sports, weather, and traffic mixed with talk segments.

nonbroadcast video The use of television production techniques and technology to create corporate and instructional training or motivational video messages.

nonlinear editing system A computer video-editing system that allows a television program to be assembled out of the final program order.

off-line editing Using lower-cost equipment in video editing to produce a "rough cut" of a television program.

off-network program A syndicated television program that had once been broadcast as part of a television network schedule. Also called *off-net program.*

on-line editing Using computerized video-editing equipment to assemble a television program.

oscillation The rhythmic variation in electric current, or the process of generating alternating current needed to create electromagnetic waves used for broadcasting.

overnight daypart A radio daypart designated as the listening hours between midnight and 6 A.M.

parsimony principle The repeated showing of a television program to maximize its use and the money spent to produce it.

passive listeners A random sample of possible or potential radio listeners drawn from the general population.

patent The legal means whereby the inventor of an apparatus is guaranteed exclusive right of ownership and revenue generation for a period of seventeen years. An inventor's or corporation's patent portfolio is a potential source of income for the patent holder, while allowing the public to benefit from the new device.

patent pools The process of an individual or a corporation sharing its patent rights with other corporations owning other patent rights so that collectively a superior device can be produced and sold to the public.

payola-plugola The unlawful practice of on-air personalities' accepting consideration for the promotion of specific music of products. See also *consideration*.

pay-per-view Marketing system whereby the cable subscriber pays individually for each program viewed.

PEG channels An acronym for the *public, educational,* and *government* access channels required of many cable television systems.

peoplemeter Electronic device that records television viewing patterns, including demographics, as viewers "punch in" to identify themselves. The *passive* peoplemeter would eliminate the need for viewer identity buttons by scanning the viewers' facial features.

perceptual survey In-depth telephone interviews of targeted listeners or viewers.

per-inquiry spot Commercial time sold by the broadcast station or cable system in exchange for payment for each unit of product sold in the station's coverage area.

persistence of vision A physiological effect of the human eye that allows the brain to retain a fleeting image after the eye no longer receives the light energy that makes up that image.

persons estimates Estimates of the number of persons listening to a radio station during a given time period (includes average quarter-hour and cume persons).

pickup pattern The physical area around the microphone within which the microphone can "hear" well, that is, has optimal sound pickup.

pilot A prototype television program used to audience-test a concept, plot, and stars to be used in a television series.

pirate radio An unlicensed broadcast station transmitting signals illegally.

pixel Short for *picture element*. An individual element in a CCD television camera pickup device that converts light focused at a specific location on the chip into electric energy. Also refers to a discrete point on a viewing screen.

point-of-information (POI) program A non-broadcast video demonstration message designed to be displayed in a retail outlet so that potential customers may learn how to benefit from the use of the product.

point-of-purchase (POP) program A non-broadcast video sales message designed to be displayed in a retail outlet where the product advertised on the program may be purchased.

polarity The positive or negative forces of an electric current.

polarization A method of creating a pattern of the electromagnetic waves in either a horizontal or a vertical orientation.

positioning The process of using advertising to create uniqueness of a product or service brand in the mind of the consumer.

postproduction The process of assembling the completed television program through editing, effects, and audio mixing.

POTS (plain old telephone system) An acronym used to describe the switched, voice-only telephone system that has been in use since the early 1900s, as compared with the newer, high-speed audio, video, and data links now available.

power programming A program scheduling strategy that places a program directly in competition with a program of the same type on another channel, such as news against news, or sitcom against sitcom. Also known as *challenge programming*.

precedent law Case law; often called tort, or discovery, law, it is found in the interpretations of law established by court rulings.

preview monitors Video screens that allow station personnel to view the output of video sources such as television cameras and videotape recorders before broadcasting the signal over the airwaves.

primary research Gathering information directly from subjects or respondents, as opposed to using information gathered by another source. See also *secondary research*.

privacy law Law established to ensure the rights of the individual versus the rights of the press.

proc amp Short for *processing amplifier*. Electronic equipment (often combined with a time base corrector) to allow the adjustment of video levels and color balance of the signal of a videotape recorder.

producer An individual at the network or television station who plans and often writes the television program or program segments.

product The term used in the television business to designate the television program.

product service information (PSI) program A video program designed for the purchaser of a product as an instructional device—to be used by the consumer in place of an instruction manual.

product shelf The term used in television to describe the channel available to be used for a program service.

program averages Audience estimates for television programs as listed in local market reports.

program title index An alphabetical list of television programs with station and broadcast time provided in local market reports.

propaganda The promotion of ideas or political doctrines through the use of the mass media.

proximity effect The emphasis of bass frequencies as a sound source moves closer to a microphone.

psychographics Audience research categories focusing on lifestyles or personalities. See also *demographics*.

PTT Short for governmentally operated *postal, telegraph,* and *telephone* communication services.

public broadcasting Noncommercial, educational broadcasting stations and networks, including National Public Radio (NPR) and the Public Broadcasting Service (PBS).

qualitative research Research based on observation of behavior or participation in a behavior or lifestyle.

quantitative research Research based on quantifiable data or information, such as numbers, that can be verified or replicated.

Radio Act of 1912 A law passed by Congress to regulate radio communication and the assignment and use of specific frequencies. It was primarily designed to control ship-to-shore communication for safety purposes, but became the basis for all governmental control of broadcast communications.

radio broadcast data system (RBDS) A radio signal incorporating data that can be coded by format for automatic tuning and digital display on receivers.

random sample A research sample in which every member of the population has an equal chance of being selected.

rating An estimate of the size of a radio or television audience, expressed as a percentage of the population or of the total number of TV households.

reach The total number of different people listening to or viewing a specific program or commercial.

recall method A means of gathering data by telephone in which the respondent is asked to recall radio listening or television viewing over the past 24 hours.

reliability The consistency or stability of the results of a research study; data that can be verified or replicated.

reporter The newsperson on the front line, on-location at the news event. It is his or her responsibility to direct the story's content and write, produce, and present the story.

rerun The repeat showing of a television program or series.

retransmission consent A Federal Communications Commission regulation whereby a cable television system must request permission from certain broadcast television stations to carry their signals, usually for consideration.

sample A set of respondents (viewers or listeners) selected for a study who are representative of the population from which they were selected.

sampling error A statistic that reflects the probability that an estimate is the same as the actual audience or percentage of the general population.

sampling frame A list of sampling units or respondents representing the population from which a sample was selected.

satellite master-antenna television (SMATV) system An antenna/satellite dish system that is used to receive television programming for private cable distribution to an apartment complex.

satellite news gathering (SNG) The use of a satellite uplink truck to send television news reports from a news event location back to the television station via a communication satellite.

saturation Describes the strength of a color or the degree to which the hue is diluted by white light.

second season The time (usually in late January) when the television networks schedule prime-time replacement series for programs that have been canceled because of low ratings.

secondary research Research based on data or information gathered by someone other than the primary researcher, such as research reports or articles. See also *primary research*.

semiconductor A silicon-based material used to construct electronic chips.

share trends Listing in local market reports of station shares over five ratings periods.

share Audience estimate expressed as a percentage of homes using television or persons listening to radio.

slander Oral defamation. Written defamation is called *libel*.

slivercasting Programming and schedules designed to reach a very narrowly defined demographic audience.

SMPTE time code A number that is assigned to each frame of a recorded videotape to aid in the editing process.

sound path The vehicle (or air, wires, and electronics) for carrying audio information from one point to another.

special-effects generator (SEG) Electronic equipment used to create video patterns and effects.

spinoff A new television program or series that is based on characters found in an older, successful series.

spot A short commercial advertisement broadcast on a radio or television station, or delivered to the home on a cable or direct-broadcast satellite channel. These are usually 10, 20, 30, or 60 seconds in length.

standard television (STV) A set of broadcast television standards developed by the National Television System Committee (NTSC) for analog broadcast television. It has been in use in the United States since the late 1940s, but is now being replaced by standards developed by the ATSC (Advanced Television Standards Committee) for digital television.

statutory law Laws passed by a legislative body, such as Congress or a state legislature. These laws are found in the acts of Congress and codes established to regulate.

stereotyping A fixed or conventional notion or conception, such as when television characters are given a narrow set of attributes that depict a group of people as having common characteristics that do not allow for individuality.

stop set A break in radio programming for commercials, announcements, DJ talk, and station promotion.

stripping A television programming strategy that schedules the same program at the same time Monday through Friday to build audience habit patterns. Also known as *horizontal programming*.

stunting The process of scheduling television specials and heavy promotion during rating sweeps periods to boost ratings.

superstation An independent broadcast television station that, although licensed by the Federal Communications Commission to serve a local market, broadcasts its signal nationally via satellite to cable television systems.

survey research A research method wherein sample respondents are surveyed through questionnaires, telephone calls, interviews, or electronic meters.

S-VHS A ½-inch videotape format used for educational and corporate video recording.

sync generator An electronic device for timing the scanning rate of all of the television equipment in the station as well as providing timing pulses to match the frame rate of a television receiver to the station equipment.

syndication The business of selling or leasing television programs and series to individual stations for broadcast in a local market.

talent In the context of news, it is any person who appears in an on-the-air newscast capacity.

target audience Segment or demographic of radio or TV audience for which a particular program, format, or commercial spot is intended.

TCP/IP (Transport Control Protocol/Internet Protocol) The programming language or code used by a personal computer to allow access to computer networks via the Internet.

telcos Short for *telephone companies*.

telecommunications The sending and receiving of meaningful messages over distance through the use of some electronic medium or device.

telegraph To write at a distance using a device that interrupts electric current sent through a wire. Meaningful messages can be sent by assigning a code of dots and dashes, or short and long interruptions in the current, to each letter of the alphabet.

telegraphy The process of using coded interruptions of current in a wire (or broadcast) to send a message over distance.

telephone surveys *See* survey research.

telephony The process of modulating voice or sound (frequency, volume, and timber) communications to send an interactive message using wire or wireless electronic technology.

teleprompter A device mounted in front of the lens of the television camera that displays the script to be read by the talent.

teletext Textual information broadcast during the vertical interval that can be displayed on the television screen as an additional service.

television households (TVHH) The number of households—the sampling unit for television viewing in the United States (or in a specific market area)—that actually contain a television receiver. This number includes nearly 98 percent of all of the homes in the United States.

television receive-only (TVRO) A name given to the communication satellite dish and related electronic equipment designed to pick up signals broadcast from a communication satellite.

tentpoling A television programming strategy that schedules a popular television program between two lower-rated programs to boost the ratings of the adjacent series.

tiers Groups of cable television channels that are packaged together and sold to the subscriber for one price. Tier 2, for example, would include more channels than a basic tier but would cost the subscriber more on a monthly basis.

time-base corrector (TBC) An electronic device used to synchronize the timing of a videotape recorder to other videotape recorders or to the electronic equipment of the television station.

time code A signal encoded on the videotape that provides a specific address for each electronic frame.

time period estimates Audience estimates for television programs as displayed in local market reports in 30-minute intervals.

time shifting A television programming strategy that schedules programs for playback at several times for the convenience of the audience. Also used to describe recording a television program at home for playback at a later time.

time spent listening An estimate of the length of time that radio audience members listen to a station during a specific period.

toll broadcasting An early name for commercial broadcasting in which the broadcaster charged a client a fee for the use of the broadcaster's facilities and airtime.

Top 40 format A radio programming style that features current music hits in a continuous rotation throughout the broadcast day.

trade-out An advertising arrangement wherein, in lieu of cash, the station provides commercial time to a local business in exchange for the product of that business. The product is often used by the station in its sales promotion efforts.

traffic A department in a station or network that schedules each second of the broadcast day with specific programs, public service announcements, station identification announcements, and commercials.

translator A low-powered television transmitter used to extend the signal of a broadcast station beyond its normal coverage area.

uplink The electromagnetic signal path from the earth to a communication satellite.

urban contemporary (UC) A radio format with roots in soul, disco, and rhythm and blues that focuses primarily on dance music.

vectorscope An electronic instrument used to display the color phasing information of the video signal.

vertical blanking interval (VBI) A period of time in the video signal in which the scanning beam of the kinescope picture tube retraces to the top of the tube to begin a new picture frame. Contains the sync pulses that keep the television receiver in step with the broadcast station. Can also be used to transmit additional information to the home via teletext.

VHS Short for *video home system*. A ½-inch videotape format commonly used for home videotape recorders.

video compression A technical process of fitting more digital video information into a specific bandwidth required for distribution. Once compressed, several programs can be broadcast in the same amount of electromagnetic space as was previously required for one channel.

videodisc A 12-inch plastic and aluminum digital electronic storage system used to record the video and audio of television programs. An older technology now largely replaced by the DVD.

video news release (VNR) A videotaped television news story prepared by a public relations organization for a corporation; it is sent to a broadcast television station in hopes that the corporation will receive free publicity if the story is used by the broadcast news department.

video on demand (VOD) A proposed electronic service that would instantaneously deliver a movie or television program to the consumer's home when the program is ordered.

video switcher An electronic device used to technically select a picture from one or more cameras or videotape recorders for inclusion in a television program or broadcast.

VSAT (Very Small Aperture Terminal) A tightly focused satellite communication system used for private businesses.

VU meter Stands for *volume-unit meter*. Measures volume units and shows visually the relative loudness of amplified sound.

waveform monitor An electronic instrument used to provide a graphic representation of the video picture so that it can be properly adjusted for brightness and contrast.

wavelength A measure of the distance from peak to peak of the cycles of an electromagnetic wave. The higher the frequency in Hertz, the shorter the wave length.

windowing Releasing a program in different distribution channels at different times to maximize revenue. For example, a movie might be scheduled first for theatrical distribution, then for pay-per-view, followed by home video, then network television distribution, and finally for sale to cable networks and local television stations.

wipe A video transition in which the picture from one video source moves from right to left (or left to right) to replace a picture from a second source.

Wireless Ship Act of 1910 The first major law passed by Congress concerning radio telegraphy. It required merchant ships carrying fifty or more passengers to have a wireless telegraph, or radio, to send distress messages if needed.

References and Suggested Readings

Chapter 1

Allen, F. L. 1959. *The big change.* New York: Harper & Row.

Archer, G. L. 1938. *History of radio.* New York: American Historical Society.

Barnouw, E. 1966. *A tower in Babel: A history of broadcasting in the United States to 1933.* New York: Oxford University Press.

Barnouw, E. 1975. *Tube of plenty: The evolution of American television.* New York: Oxford University Press.

Barton, B. 1926. Which knew not Joseph. H. D. Lindgren, ed., *Modern speeches.* New York: F. S. Crofts & Co.: 406–412. [Speech was given in 1923.]

Baudino, J. E., and J. M. Kittross. 1977. Broadcasting's oldest stations: An examination of four claimants. *Journal of Broadcasting* (Winter) 21(1): 61–83.

Bilby, K. 1986. *The general: David Sarnoff and the rise of the communications industry.* New York: Harper & Row.

Borman, E. G. 1958. This is Huey P. Long talking. *Journal of Broadcasting* (Spring) 2: 111–122.

Briggs, A. 1961. *The history of broadcasting in the United Kingdom,* vol. 1. New York: Oxford University Press.

Cantril, H., and G. W. Allport. 1941. *The psychology of radio.* New York: Peter Smith. [For a broadcast copy of the speech, listen to Phonoarchive tape 4116, National Archives, Milo Ryan Phonoarchive Collection.]

Carnegie, A. 1920. *Autobiography of Andrew Carnegie.* New York: Houghton Mifflin.

Congressional Record. 1926. Vol. 67, pt. 5:5488.

Dill, C. C. 1927. Traffic cop for the air. *The American Review of Reviews* (February) 75: 191–195.

Douglas, G. 1987. *The early days of radio broadcasting.* Jefferson City, NC: McFarland.

Everson, G. (n.d.). Autobiography (Philo T. Farnsworth Papers). Tempe: Arizona State University Library, Special Collections.

Everson, G. 1949. *The story of television.* New York: W. W. Norton.

Fang, I. E. 1977. *Those radio commentators.* Ames: Iowa State University Press.

Federal Communications Commission. (1941, May). Report on Chain Broadcasting. Order 37, Docket 5060. Washington, D.C.: Federal Communications Commission, 5–20.

Reginald A. Fessenden to S. M. Kinterner. 29 January 1932. Clark Manuscript Collection, Smithsonian Institution, Division of Electricity and Nuclear Energy, CWC 135–246A.

Girauld, C. 1949. The press radio war: 1933–1935. *Public Opinion Quarterly* 13: 263.

Godfrey, D. G. 1982. Canadian Marconi: CFCF, The forgotten first. *Canadian Journal of Communication* 8(4): 56–69.

Godfrey, D. G. 1991. CBS World News Roundup: Setting the stage for the next half century. *American Journalism* 7: 164–172.

Greene, T. P., ed. 1955. *American imperialism in 1898.* Boston: D. C. Heath.

Harding, W. G. 1921. Business and government. W. G. Harding, ed., *Our common country.* Indianapolis: Bobbs-Merrill Publishers: 20–23.

Heffner, R. 1952. *Documentary history of the United States.* Bloomington: Indiana University Press.

Hofer, S. F. 1979. Philo Farnsworth: Television's pioneer. *Journal of Broadcasting* 23(2): 153.

Hofstadter, R. 1955. *Social Darwinism in American thought.* Boston: Beacon Press.

Hoover, H. 1924. *The reminiscences of Herbert Clark Hoover.* New York: Oral History Project, Columbia University Oral Research Office.

How chain broadcasting is accomplished. 1929. *Radio Broadcast* (June): 65–67.

How much it costs to broadcast. 1926. *Radio Broadcast* (September): 367–371.

Inglis, A. F. 1990. *Behind the tube: A history of broadcasting technology and business.* Boston: Focal Press.

Josephson, M. 1934. *The robber barons: The American capitalists.* New York: Harcourt, Brace.

Landry, R. J. 1938. Edward R. Murrow. *Scribner's* (December) 104: 7–11.

Lessing, L. 1969. *Man of high fidelity: Edwin Howard Armstrong.* New York: Bantam Books.

Lichty, L. W., and M. C. Topping. 1975. *American broadcasting: A source book on the history of radio and television.* New York: Hastings House Publishers.

Lovece, F. 1989. Is it TV's 50th birthday or not? *Channels* (June) 9(6): 9.

Marconi, D. 1962. *My father Marconi.* New York: McGraw-Hill.

McChesney, R. W. 1990. The battle for U.S. airwaves, 1928–1935. *Journal of Communication* (Autumn) 40(4): 29–33.

Moffett, E. A. 1986. Hometown radio in 1942: The role of the local station during the first year of total war. *American Journalism* 3(2): 87–98.

Moreau, L. R. 1989. The feminine touch in telecommunications. *A.W.A. Review* 4: 70–83.

Murrow, E. R. 1958. *We take you back* (CBS Radio News Documentary, Phonoarchive Tape 4065). Washington, D.C.: National Archives. March 13.

Newspaper versus news broadcasts. 1938. *Fortune* (April): 104–109.

Paley, W. S. 1963. Foreword. Milo Ryan, ed., *History in sound* (p. v). Seattle: University of Washington Press.

Paley, W. S. 1979. *As it happened: A memoir.* New York: Doubleday.

Parrington, V. L. 1927–1930. *Main currents in American thought,* vols. 1–3. New York: Harcourt, Brace.

Phillips, T. A. 1929. An estimate of set sales. *Radio Broadcast* (September): 270–272.

Pratte, A. 1993–94. Going along for the ride on the prosperity bandwagon: Not war between the editors and radio: 1923–1941. *Journal of Radio Studies* 2: 123–139.

Roosevelt, F. D. (1940, December 29). Arsenal of democracy (radio address). E. Wrage & B. Baskerville, eds., Contemporary forum: 246–254. Seattle: University of Washington Press.

Ropp, T. 1985. Philo Farnsworth: Forgotten father of television. *Media History Digest* (Summer) 5(2): 42–58.

Shirer, W. L. 1984. *Twentieth century journey: The nightmare years, 1930–1940.* Boston: Little, Brown.

Sivowitch, E. N. 1970–1971. A technological survey of broadcasting's pre-history, 1876–1920. *Journal of Broadcasting* (Winter) 15(1): 1–15.

Smith, R. F. 1959–1960. Oldest station in the nation. *Journal of Broadcasting* (Winter) 4(1): 40–55.

Smith, R. R. 1965. Origins of radio network news commentary. *Journal of Broadcasting* (Spring) 9: 113–122.

Smith, S. B. 1990. *In all his glory: The life of William S. Paley.* New York: Simon & Schuster.

Spalding, J. W. 1963–1964. 1928: Radio becomes a mass advertising medium. *Journal of Broadcasting* (Winter) 8(1): 31–44.

Sterling, C. H., and J. M. Kittross. 1990. *Stay tuned: A concise history of American broadcasting.* 2d ed. Belmont, CA: Wadsworth.

Strong, J. 1885. *Our country: Its possible future and present crisis.* New York: American Home Missionary Society.

Summers, R. E., and H. B. Summers. 1968. *Broadcasting and the public.* Belmont, CA: Wadsworth.

Sussking, C. P. 1962. Popov and the beginnings of radio telegraphy. *Proceedings of the Institute of Radio Engineers* (October): 2036–2037.

Third National Radio Conference. 1924. *Proceedings and recommendations for radio regulation.* Washington, D.C.: U.S. Government Printing Office.

White, P. W. 1947. *News on the air.* New York: Harcourt, Brace.

Wrage, E., and B. Baskerville. 1960. *American forum.* Seattle: University of Washington Press.

Chapter 2

Baldwin, T. F., and D. S. McVoy. 1988. *Cable communication.* 2d ed. Englewood Cliffs, NJ: Prentice Hall.

Barnouw, E. 1966. *A tower in Babel: A history of broadcasting in the United States to 1933.* New York: Oxford University Press.

Barnouw, E. 1968. *The golden web: A history of broadcasting in the United States 1933–53.* New York: Oxford University Press.

Barnouw, E. 1970. *The image empire: A history of broadcasting in the United States from 1953.* New York: Oxford University Press.

Bilby, K. 1986. *The general: David Sarnoff and the rise of the communications industry.* New York: Harper & Row.

Blakely, R. J. 1979. *To serve the public interest: Educational broadcasting in the United States.* Syracuse, NY: Syracuse University Press.

Brooks, T., and E. Marsh. 1979. *The complete directory of prime time network TV shows, 1946–present.* New York: Ballantine Books.

Bunzel, R. 1991. Filling the magic box. *Broadcasting* (December 9): 27–34.

CCET (Carnegie Commission on Educational Television). 1967. *Public television: A program for action.* New York: Harper & Row.

Fabe, M. 1979. *TV game shows.* Garden City, NY: Doubleday.

Federal Communications Commission (FCC). *Sixth report and order* (Federal Register 3905, 3908). Washington, D.C.: GPO, 1952, April 14.

Friendly, F. W. 1968. *Due to circumstances beyond our control.* New York: Random House.

Kahn, F. J., ed. 1968. *Documents of American broadcasting.* New York: Appleton-Century-Crofts.

Land, H. W., and Associates. 1967. *The hidden medium: A status report on educational radio in the United States.* Washington, D.C.: National Association of Educational Broadcasters.

Leigh, F. A. 1984. College radio: Effects of Class D rules change. *Feedback* 26(2): 18–21.

McDonald, J. F. 1979. *Don't touch that dial: Radio programming in American life from 1920 to 1960.* Chicago: Nelson Hall.

National Association of Broadcasters (NAB). 1975. *The television code.* Washington, D.C.: NAB.

Rose, B. G., ed. 1985. *TV genres: A handbook and reference guide.* Westport, CT: Greenwood Press.

Routt, E., McGrath, J. B., and F. Weiss. 1978. *The radio format conundrum.* New York: Hastings House.

Rules changes relating to noncommercial educational FM broadcast stations (Federal Register, 3412–3416). (1979, January 16). Washington, D.C.: U.S. Government Printing Office.

Special edition: The first sixty years. 1991. *Broadcasting* (December): 56, 60.

Winship, M. 1988. *Television.* New York: Random House.

Chapter 3

Black, U. D. 1987. *Data communications and distributed networks.* 2d ed. Englewood Cliffs, NJ: Prentice-Hall.

Brooks, J. 1975. *Telephone: The first hundred years.* San Francisco: Harper & Row.

Calhoun, G. 1988. *Digital cellular radio.* Norwood, MA: Artech House.

Clarke, A. C. 1958. *Voice across the sea.* New York: Harper & Brothers.

Coe, L. 1993. *The telegraph: A history of Morse's invention and its predecessors in the United States.* Jefferson, NC: McFarland.

Coll, S. 1986. *The deal of the century: The breakup of AT&T.* New York: Atheneum.

Everett, J., ed. 1992. *VSATs: Very small aperture terminals.* London: Peter Peregrinus.

Fischer, C. S. 1992. *America calling: A social history of the telephone to 1940.* Los Angeles: University of California Press.

Harlow, A. F. 1971. *Old wires and new waves.* New York: Arno Press and *New York Times.*

Hordeski, M. F. 1989. *Communications networks.* Blue Ridge Summit, PA: Tab Books.

Kahaner, L. 1986. *On the line: The men of MCI—who took on AT&T, risked everything and won!* New York: Warner Books.

Kraus, C. R., and A. W. Duerig. 1988. *The rape of Ma Bell: The criminal wrecking of the best telephone system in the world.* Secaucus, NJ: Lyle Stuart.

Lipartito, K. 1989. *The Bell system and regional business: The telephone in the south, 1877–1920.* Baltimore: Johns Hopkins University Press.

Martin, D. H. 1991. *Communication satellites: 1958–1992.* El Segundo, CA: Aerospace Corporation.

Reid, J. D. 1974. *The telegraph in America.* New York: Arno Press.

Smith, W. F., ed. 1969. *The story of the Pony Express.* San Francisco: Pony Express History and Art Gallery.

Temin, P., with L. Galambos. 1987. *The fall of the Bell system: A study in prices and politics.* New York: Cambridge.

Williams, A. 1928. *Telegraphy & telephony.* New York: Thomas Nelson & Sons.

Chapter 4

FCC regulations, summaries, and updates are published by Pike and Fischer, Inc., Bethesda, MD. They are also available in the *Broadcast and Cable Yearbook,* published annually by R. R. Bowker, New Providence, NJ. Specific laws are part of the Code of Federal Regulations, published by the Government Printing Office (GPO) and available in many libraries.

Allen, F. L. 1931. *Only yesterday: An informal history of the nineteen-twenties.* New York: Harper & Row.

Bensman, M. R. 1970. The Zenith-WJAZ case and the chaos of 1926–27. *Journal of Broadcasting* (Fall) 14: 423–440.

Bittner, J. R. 1982. *Broadcast law and regulation.* Englewood Cliffs, NJ: Prentice-Hall.

Congressional Record. 1926. 67 (11): 12615.

Congressional Record. 1927. 68 (3): 3027.

Coolidge, C. 1968. *Foundations of the republic: Speeches and addresses.* Freeport, NY: Books for Libraries Press.

Ernst, M. L. 1926. Who shall control the air? *The Nation* (April 21) 122 (3172): 44.

Garvey, D. E. 1976. Secretary Hoover and the quest for broadcast regulation. *Journalism History* (Autumn) 3: 66.

Godfrey, D. G. 1977. The 1927 Radio Act: People and politics. *Journalism History* (Autumn) 4(3): 74–78.

Godfrey, D. G. 1979. Senator Dill and the 1927 Radio Act. *Journal of Broadcasting* (Fall) 23(4): 477–489.

Godfrey, D. G., and V. E. Limburg. 1990. Rogue elephant of radio legislation: Senator William E. Borah. *Journalism Quarterly* (Spring) 67(1): 214–224.

Guback, T. H. 1968. Political broadcasting and the public policy. *Journal of Broadcasting* (Summer) 12(8): 191–211.

Harding, W. D. 1921. *Our common country.* Indianapolis: Bobbs-Merrill.

Holt, D. 1967. The origin of public interest in broadcasting. *Educational Broadcasting Review* (October) 1(1): 15.

Hoover, H. 1952. *The memoirs of Herbert Hoover: The Cabinet and the presidency, 1920–1933.* New York: Macmillan.

Jansky, C. M. 1957. The contributions of Herbert Hoover to broadcasting. *Journal of Broadcasting* (Summer) 1: 241–249.

Johnson, N. 1976. *How to talk back to your television.* Boston: Little, Brown.

Kahn, F. J. 1984. *Documents of American broadcasting.* New York: Appleton-Century-Crofts.

McChesney, R. W. 1988. Franklin Roosevelt, his administration, and the Communication Act of 1934. *American Journalism* 5(4): 204–229.

National Association of Broadcasters. 1988. *Political broadcast catechism.* Washington, D.C.: National Association of Broadcasters.

National Association of Broadcasters. 1990. *Broadcast regulation: A review of 1989 and a preview of 1990.* Washington, D.C.: National Association of Broadcasters.

Pember, D. R. 1990. *Mass media law.* Dubuque, IA: Wm. C. Brown.

Political spellbinding by radio. 1924. *Popular Mechanics Magazine* (December) 42(6): 881.

Sarno, E. F., Jr. 1969. The national radio conferences. *Journal of Broadcasting* (Spring) 13(2): 189–202.

Chapter 5

FCC regulations, summaries, and updates are published by Pike and Fischer, Inc., Bethesda, MD. They are also available in the *Broadcast and Cable Yearbook,* published annually by R. R. Bowker, New Providence, NJ. Specific laws are part of the Code of Federal Regulations, published by the Government Printing Office and available in many libraries.

Bensman, M. R. 1986. *Broadcast regulation: Selected cases and decisions.* Lanham, MD: University Press of America.

Creech, K. G. 1993. *Electronic media law and regulation.* Boston: Focal Press.

Ferris, C. 1995. *Cable television law: A video communications practice guide.* Albany, NY: Matthew Bender.

REFERENCES AND SUGGESTED READINGS

Gillmor, D. M., and J. A. Barron. 1990. *Mass communication law.* St. Paul: West Publications.

Kahn, F. J. 1984. *Documents of American broadcasting.* New York: Appleton-Century-Crofts.

Miller, P. 1990. *Media law for producers.* White Plains, NY: Knowledge Publications.

National Association of Broadcasters. 1995. *Broadcast regulation.* Washington, D.C.: National Association of Broadcasters.

Pember, D. R. 1993. *Mass media law.* Madison, WI: Wm. G. Brown Communications.

Chapter 6

Baldwin, T. F., and D. S. McVoy. 1988. *Cable communication.* 2d ed. Englewood Cliffs, NJ: Prentice Hall.

Broadcasting and Cable Yearbook 1998. 1998. New Providence, NJ: R. R. Bowker.

Cable network subscriber counts. 1990. *Cable World* (June 14): 98.

Capogrosso, E. 1986. *The IDC labor guide, 1987.* Burbank, CA: IDC Services.

Fratrik, M. R., ed. 1986. *The small market television manager's guide.* Washington, D.C.: The National Association of Broadcasters.

Higgins, J. Top 25 MSOs. 2000. *Broadcasting & Cable* (May 24): 34.

O'Donnell, L. B., C. Hausman, and P. Benoit. 1989. *Radio station operations: Management and employee perspectives.* Belmont, CA.: Wadsworth Publishing.

Pringle, P. K., Starr, M. F., and McCavitt, W. E. 1991. *Electronic media management.* 2d ed. Boston: Focal Press.

Sherman, B. L. 1995. *Telecommunications management: The broadcast and cable industries.* 2d ed. New York: McGraw Hill.

Top 25 television groups. 2000. *Broadcasting & Cable* (January 24): 72.

Chapter 7

Moriarty, S. E. 1991. *Creative advertising: Theory and practice.* 2d ed. Englewood Cliffs, NJ: Prentice-Hall.

Orlik, P. B. 1994. *Broadcast/cable copywriting.* 5th ed. Needham Heights, MA: Allyn & Bacon.

Warner, C., and J. Buchman. 1991. *Broadcast and cable selling.* 2d ed. Belmont, CA: Wadsworth.

Zeigler, S. K., and H. H. Howard. 1984. *Broadcast advertising.* 2d ed. Columbus, OH: Grid.

Chapter 8

Carroll, R. L., and D. M. Davis. 1993. *Electronic media programming: Strategies and decision making.* New York: McGraw-Hill.

Cohen, E. 1991. *How America found out about the Gulf War.* (August) Coral Springs, FL: Birch Scarborough Research.

Gatski, J. 1993. Broadcasters get full dose of RDS. *Radio World* (May 26).

Hilliard, R. L. 1985. *Radio broadcasting.* 3d ed. New York: Longman.

Keith, M. C. 1987. *Radio programming: Consultancy and formatics.* Boston: Focal Press.

McFarland, D. T. 1990. *Contemporary radio programming strategies.* Hillsdale, NJ: Lawrence Erlbaum.

Ochs, E. 1993. Radio research: Beyond the numbers. *Gavin* (May 28): 24–27.

Radio & Records. 1991. *Ratings report and directory*, vol. 1. Los Angeles: Radio & Records, Inc.

Routt, E., J. B. McGrath, and F. Weiss. 1978. *The radio format conundrum.* New York: Hastings House.

Whetmore, E. J. 1981. *The magic medium: An introduction to radio in America.* Belmont, CA: Wadsworth.

Chapter 9

Carroll, R. L., and D. M. Davis. 1993. *Electronic media programming: Strategies and decision making.* New York: McGraw-Hill.

Clift, C. E., J. C. Richie, and A. M. Greer. 1993. *Broadcast and cable programming: The current perspective.* 2d ed. Dubuque, IA: Kendall/Hunt.

Eastman, S. T. 1993. *Broadcast/cable programming: Strategies and practices.* 4th ed. Belmont, CA: Wadsworth.

Lichty, L. W., and J. M. Ripley. 1969. *American broadcasting: introduction and analysis: Readings.* Madison, WI: College Printing and Publishing.

Mair, G. 1988. *Inside HBO.* New York: Dodd Mead.

Nielsen Media Research National Audience Demographics Report. 1996. (December). Reprinted in *Broadcasting & Cable Yearbook*, New Providence, NJ: R. R. Bowker (1998).

Whittemore, H. 1990. *CNN: The Inside Story.* Boston: Little, Brown.

Chapter 10

Paparazzi Ruining Lives, Actors Testify. *The Honolulu Advertiser*, Friday, May 22, 1998, A-16.

Bluem, A. W. 1965. *Documentary in American television.* New York: Hastings House Publishers.

Chancellor, J. 1983. Speech delivered at the annual meeting of the Radio and Television News Directors Association.

Chester, G. 1946. The press radio war: 1933–1935. *Public Opinion Quarterly*: 164–172.

Communicator. Radio-Television News Directors Association's monthly periodical.

Fang, I. 1980. *Television news, radio news.* St. Paul: Rada Press.

Foote, J. S., and M. E. Steele. 1986. Degree of conformity in lead stories in early evening network TV newscasts. *Journalism Quarterly* 63: 19–23.

Galician, M. L. 1986. Perceptions of good news and bad news on television. *Journalism Quarterly* 63: 622–616.

Kammer, Jerry. News Stars Create Backlash, Journalist Says. *The Arizona Republic.* Saturday, January 24, 1998, B-2.

Murrow, E. R. 1958. A broadcaster talks to his colleagues. Speech delivered at the annual meeting of the Radio and Television News Directors Association.

Paley, W. S. 1963. Foreword. Milo Ryan, ed., *History in sound* (p. v). Seattle: University of Washington Press.

Reynolds, O. T., and L. Thonssen. 1961. The reporter as orator: Edward R. Murrow. Loren Reid, ed., *American Address: Studies in Honor of Albert Craig Baird.* Columbia: University of Missouri Press.

Ries, Al and J. Trout. 1986. *Positioning: the battle for your mind.* New York: McGraw-Hill.

Sanders, T. 1992. *Lead (first) stories as a predictor for negative perceptions of television news.* Paper presented at the Broadcast Education Association's Annual Convention.

Schramm, W. 1949. The nature of news. *Journalism Quarterly*: 259.

Schramm, W. 1957. *Responsibility in mass communication.* New York: Harper & Brothers Publishing.

Shirer, William L. 1976; reprint 1990. *Twentieth century journey: a memoir of a life and times,* vol. 2 "Twentieth century: The nightmare years." New York: Simon and Schuster.

Shook, F., and D. Lattimore. 1982. *The broadcast news process.* Englewood, CO: Morton.

Van Der Werf, Martin. Bush Comfortable With Iraq Mission. *The Arizona Republic.* Wednesday, May 6, 1998, A-16.

White, D. M. 1964. The "gatekeeper:" A case study in the selection of news. In L. A. Dexter and D. M. White, eds, *People, society, and mass communication.* New York: Free Press.

White, P. W. 1947. *News on the air.* New York: Harcourt Brace.

Whitemore, H. 1990. *CNN: The inside story.* Boston: Little, Brown.

Chapter 11

Arbitron Company. 1991. *Description of methodology.* Laurel, MD: Arbitron.

Carroll, R. L., and D. M. Davis. 1993. *Electronic media programming: strategies and decision making.* New York: McGraw-Hill.

Chadwick, B. A., H. M. Bahr, and S. L. Albrecht. 1984. *Social science research methods.* Englewood Cliffs, NJ: Prentice Hall.

Dominick, J. R., and J. E. Fletcher. 1985. *Broadcasting research methods.* Boston: Allyn & Bacon.

Nielsen Media Research. 1990. *Nielsen station index reference supplement.* New York: Nielsen.

Chapter 12

The Annenberg/CPB Project. 1991. *Pathways to success: Using technologies to reach distant learners.* Washington, D.C.: Corporation for Public Broadcasting.

The Annenberg/CPB Project. 1992. *Going the distance: A handbook for developing distance degree programs.* Washington, D.C.: Corporation for Public Broadcasting.

Budd, J. F., Jr. 1983. *Corporate video in focus: A management guide to private TV.* Englewood Cliffs, NJ: Prentice-Hall.

Gayeski, D. M. 1983. *Corporate and instructional video: Design and production.* Englewood Cliffs, NJ: Prentice-Hall.

Greenberg, J. S., and J. M. Biedenbach, eds. 1987. *Compendium on uses of television in engineering education.* Washington, D.C.: American Society for Engineering Education.

Hausman, C. 1991. *Institutional video: Planning, budgeting, production, and evaluation.* Belmont, CA: Wadsworth.

Lewis, R., and D. Spencer. 1986. *What is open learning?* London: Council for Educational Technology.

Richardson, A. R., ed. 1992. *Corporate and organizational video.* New York: McGraw-Hill.

Stokes, J. T. 1988. *The business of nonbroadcast television: Corporate and institutional video budgets, facilities and applications.* White Plains, NY: Knowledge Industry Publications.

Wood, D. N., and D. G. Wylie. 1977. *Educational telecommunications.* Belmont, CA: Wadsworth.

Chapter 13

Alkin, G. 1981. *Sound recording and reproduction.* Boston: Focal Press.

Alkin, G. 1989. *Sound techniques for video and TV.* Boston: Focal Press.

Alten, S. 1999. *Audio in media.* 5th ed. Belmont, CA: Wadsworth Publishing.

Bartlett, B. 1987. *Introduction to professional recording techniques.* Indianapolis: H. W. Sams.

Benson, B., ed. 1988. *Audio engineering handbook.* New York: McGraw-Hill.

Burroughs, L. 1974. *Microphones: Design and application.* Plainview, NY: Sagamore.

Clifford, M. 1986. *Microphones.* 3d ed. Blue Ridge Summit, PA: Tab.

Everest, F. A. 1986. *Acoustic techniques for home and studio.* 2d ed. Blue Ridge Summit, PA: Tab.

Fielding, K. 1990. *Introduction to television production.* New York: Longman.

Hanson, J. 1987. *Understanding video: Applications, impact, and theory.* Newbury Park, CA: Sage.

Hasling, J. 1980. *Fundamentals of radio broadcasting.* New York: McGraw-Hill.

Huber, D. M. 1987. *Audio production techniques for video.* Indianapolis: H. W. Sams.

Kindem, G. 1987. *The moving image: Production principles and practices.* Glenview, IL: Scott, Foresman.

Mathias, H., and R. Patterson. 1985. *Electronic cinematography: Achieving photographic control over the video image.* Belmont, CA: Wadsworth.

McLeish, R. 1988. *The technique of radio production.* 2d ed. Boston: Focal Press.

Oringel, R. 1989. *Audio control handbook.* 6th ed. Boston: Focal Press.

Runstein, R., and D. Huber. 1989. *Modern recording techniques.* 3d ed. Indianapolis: H. W. Sams.

Smith, F. L. 1990. *Perspectives on radio and television.* 3d ed. New York: Harper & Row.

Sweeney, D. 1986. *Demystifying compact discs: A guide to digital audio.* Blue Ridge Summit, PA: Tab.

Watkinson, J. 1988. The art of digital audio. Boston: Focal Press.

Yoakam, R. D., and C. F. Cremer. 1985. *ENG: Television news and the new technology.* New York: Random House.

Zettl, H. 1990. *Sight, sound, motion: Applied media aesthetics.* 2d ed. Belmont, CA: Wadsworth.

Zettl, H. 2000. *Television production handbook.* 7th ed. Belmont, CA: Wadsworth.

Chapter 14

Baldwin, T. F., and D. S. McVoy. 1988. *Cable communication.* 2d ed. Englewood Cliffs, NJ: Prentice-Hall.

Cheo, P. K. 1985. *Fiber optics: Devices and systems.* Englewood Cliffs, NJ: Prentice-Hall.

Dayton, R. L. 1991. *Telecommunications: The transmission of information.* New York: McGraw-Hill.

Gross, L. S. 1990. *The new television technologies.* 3d ed. Dubuque, IA: Wm. C. Brown.

Isailovic, J. 1987. *Videodisc systems: Theory and applications.* Englewood Cliffs, NJ: Prentice Hall.

Mirabito, M. M., and B. L. Morgenstern. 1990. *The new communications technologies.* Boston: Focal Press.

Palais, J. C. 1992. *Fiber optic communications.* 3d ed. Englewood Cliffs, NJ: Prentice-Hall.

Pohlmann, K. C., ed. 1991. *Advanced digital audio.* Carmel, IN: Howard W. Sams.

Schrader, R. L. 1985. *Electronic communication.* New York: McGraw-Hill.

Withers, D. J. 1991. *Radio spectrum management.* London: Peter Peregrinus.

Chapter 15

Fortner, R. S. 1993. *International communication: history, conflict, and control of the global metropolis.* Belmont, CA: Wadsworth, Inc.

Gross, L. S. 1995. *The international world of electronic media.* New York: McGraw-Hill, Inc.

Martin, L. J., and A. G. Chaudhary. 1983. *Comparative mass media systems.* New York: Longman Inc.

McLuhan, M. and Fiore, Q. 1967. *The medium is the message: An inventory of effects.* New York: Bantam Books.

Paulu, B. 1981. *Television and radio in the United Kingdom.* Minneapolis, MN: The University of Minnesota Press.

Siebert, F. S., T. Peterson, and W. Schramm. 1956. *The four theories of the press.* Urbana, Ill: The University of Illinois Press.

Chapter 16

Bandura, Albert, Dorthea Ross, and Shiela A. Ross. Imitation of film-mediated aggressive models, *Journal of Abnormal and Social Psychology.* 1963. 66(1): 3–11.

Broadcasting & Cable. 1996. 126(44). Cahners, Washington, D.C.

Broadcasting & Cable. 1998, 128(17). Cahners, Washington, D.C.

Beadle, M. E. 1998. "Correll, Charles and Freeman Gosden." In *Historical dictionary of American radio,* D. G. Godfrey and F. A. Leigh, eds. Westport, CT: Greenwood.

DeFleur, M. L., and E. E. Dennis. 1981. *Understanding mass communication.* Boston: Houghton Mifflin.

Dominick, J. R. 1990. *The dynamics of mass communication.* 3d ed. New York: McGraw-Hill.

Greenberg, B. S., and L. Collette. 1997. "The changing faces on tv: A demographic analysis of network television's new seasons, 1966–1992." *Journal of Broadcasting & Electronic Media* (Winter) 41(1): 1–13.

Jeffres, L. W. 1997. *Mass media effects.* 2d ed. Prospect Heights, IL: Waveland Press.

Kaid, L. L., J. C. Tedesco, and L. M. McKinnon. 1996. "Presidential ads as nightly news: A content analysis of 1988 and 1992 televised adwatches." *Journal of Broadcasting & Electronic Media,* 40(3): 297–308.

Lazarsfeld, Paul F. and Robert K. Merton, Mass communication, popular taste and organized social action," *Mass Communications,* Wilbur Schramm, ed. Urbana: University of Illinois Press, 1960.

Schramm, Wilbur. Communication development and the development process, *Communications and Political Development,* Lucian W. Pye, ed. Princeton: Princeton University Press, 1963.

Tucker, L. R. 1997. "Was the revolution televised?: Professional criticism about the Cosby Show and the essentialization of black cultural expression." *Journal of Broadcasting & Electronic Media* (Winter) 41(1): 90–107.

Wood, D. N. 1983. *Mass media and the individual.* St. Paul, MN: West Publishing Co.

Index

A. C. Nielsen Media Research Company, 142, 151, 234, 244–246
AB-roll editing, 295
ABC, 44, 133, 134, 176, 177
ABC radio networks, 176
Above-the-line personnel, 142
AC, 169
Access programming, 202, 203
Account executives, 155–157
Acid rock, 43
Action for Children's Television (ACT), 110
Active listeners, 239
Additive color, 289
Addressable technology, 163
ADI, 235
Adjacencies, 155
Adjacent channel, 311
Administrative laws, 99
Adult alternative, 171, 172
Adult contemporary (AC), 169
Adult learning, 263–268
Adult standards, 171
Adult western, 193
Advanced television (ATV), 290, 308
Advanced Television Standards Committee (ATSC), 308
Advertisers, 151–157
Advertising, 355, 356
Advertising agencies, 157, 158
Affiliate contracts, 135
Afternoon drive time, 168
Album-oriented rock (AOR), 43, 170
All in the Family, 55
All-news stations, 172
Allen, Frederick Lewis, 19, 79
AlliedSignal Aerospace, 259
Alternative-music formats, 171

AM band expansion, 113
AM radio, 311
AM stereo, 113
Ambush journalism, 228
American Society of Composers, Authors and Publishers (ASCAP), 26
Ameritech, 71
Ampex D-2 system, 299
Amplitude, 273
Analog-tape recording process, 281
Anchors, 214, 224, 228
Antenna, 309
AOR (album-oriented rock), 43, 170
AP, 215
Arbitron, 234–236, 242–244
Arbitron Radio Market Reports, 242–244
Archer, Gleason L., 17
Area of dominant influence (ADI), 235
Armstrong, Edwin Howard, 33, 34
Arnaz, Desi, 44
ARPAnet, 73
ASCAP, 26
ASCII, 320
Aspect ratio, 290
Assemble mode, 296
Assignment editor (AE), 212
Associated Press (AP), 215
AT&T, 7, 23–25, 63, 65–71, 74, 82, 128
AT&T, breakup of, 69–71
ATSC, 308
Attenuation, 305
ATV, 290, 308
Audience, 158–161
Audience analysis and marketing, 153, 232–253
criticisms of, 250, 251
data analysis, 239–241

data gathering, 237–239
early research companies, 234
market area, 235, 236
research reports, 242–246
research sample, 236, 237
sales/marketing applications, 246–250
types of research, 233, 234
Audience demographics, 150
Audience flow, 188, 189
Audience promotion, 160
Audience recycling, 246
Audience research. *See* Audience analysis and marketing
Audio and video systems, 270–301
analog vs. digital transmission, 281, 282
audio signal storage equipment, 279, 280
computer-generated video, 293
decoding technology, 287–291
digital broadcasting, 283, 284
digital recording, 299
display technologies, 287–291
edit controllers, 296, 297
encoding technology, 291–293
monitoring systems, 294, 295
processing amplifiers, 295
signal manipulation, 293–297
sound, 272, 273
sound path, 274–285
storage, 297–299
television receiver, 287–289
tools of the trade, 271
video camera, 291–293
video production switcher, 293, 294
videotape recording, 297–299
vision, 285–287

INDEX

Audio console, 278, 279
Audio signal storage equipment, 279, 280
Auditorium testing, 239
Authoritarian theory, 331
Autry, Gene, 46
Avails, 153, 196
Average quarter-hour persons, 240
Awards, 144

Baird, John, 67
Balanced programming, 85, 86
Ball, Lucille, 44, 45
Bandura, Albert, 350
Barter, 196, 197
Barter programming, 156
Barton, Bruce, 18
Bates v. State Bar of Arizona, 110
Beam splitter, 291
Beautiful music, 43, 171
Bell, Alexander Graham, 14, 63, 64, 65
Bell Atlantic, 71
Bell Labs, 67, 68
Bell South, 71
Below-the-line personnel, 142
Bennett, James Gordon, 62
Berle, Milton, 44
Betacam, 298
Bias current, 297
Bidirectional microphone, 277
Bilby, Kenneth, 17, 25
Bill C-58, 110–111
Billboards, 160
Biltmore Agreement, 31
Biondi, Frank, 200
Birch Radio, 235
Black, 293
Blease, Cole, 81
Block programming, 189
Blue Book, 88
Bluem, A. William, 209
Borah, William E., 81, 82
Boutique advertising agencies, 158
Brand image, 152
Briggs, Asa, 21
Broadcast communication system, 303–320. *See also* Distribution systems
Broadcast console, 278, 279
Broadcast frequencies, 120

Broadcast license, 120
Broadcast station, 119–121, 125
Broadcast television programming, 182–205
   access channels, 202, 203
   audiences, 185
   cable programming, 197–203
   dayparts, 186
   local origination, 202
   network programming, 192, 193
   pay networks, 199
   program sources, 187, 188
   program types, 187
   programming costs, 196, 197
   programming department, 183–185
   scheduling strategies, 188–192
   superstations, 199
   syndication, 193–196
Broadcasting's prehistory, 14–16
Brown, Charles, 70
Bryan, William Jennings, 81
Bumpers, 110
Bus, 293
Bush, George, 227
Business infrastructure, 144
Business year, 155

Cable Communications Policy Act of 1984, 54
Cable economics, 161–163
Cable News Network (CNN), 201, 218, 219
Cable programming, 197–203
Cable Services Bureau, 101
Cable Television Consumer Protection Act of 1992, 54
Cable television systems, 321–324
Cable testing, 239
Cable TV, 52–54
Caesar, Sid, 44
Call-out research, 239
Call-out surveys, 159
Camcorder, 299
Camera, 291–293
Camera control units, 295
Canada, 110–111, 336
Candlestick-style telephone, 65
Carcione, Joe, 220
Cardioid pickup pattern, 277, 278
Carlin, George, 108

Carnegie Commission on Educational Television, 50
Carrier wave, 308–311
Carterfone decision, 69
Cash, 196
Cash-plus-barter syndication, 197
Cassidy, Hopalang, 46
Catharsis theory, 350
Cathode-ray tube (CRT), 288
CATV systems, 52–54, 127, 128, 321
CAV system, 326
CBS, 26, 27, 134
CCDs, 292
CCTV systems, 264–266
CD, 325
CD player, 325, 326
CD-ROM, 326
Cease-and-desist orders, 103
Celler, Emanuel, 23
Cellular telephone, 74, 75
Censorship, 80–82, 91
CFCF, 21
Chain broadcasting, 122
Challenge programming, 191
Chancellor, John, 226
Channel inertia, 189
Character generator, 293
Charge-coupled devices (CCDs), 292
Charlie McCarthy, 107
Chautauqua movement, 21
Checkerboarding, 190
Children, 352, 353, 356
Children's television advertising, 110
China, 343
CHR, 169
Chroma key, 294
Chrominance, 289
Churn, 198
Cigarette advertising, 112
Clarke, Arthur C., 68
Class A stations, 312
Class B stations, 312
Class C stations, 312
Class D stations, 49, 50
Class I stations, 311
Class II stations, 311
Class III stations, 311
Class IV stations, 311
Classic rock, 170
Clinton, Bill, 93, 319

Clock, 168
Closed-captioning, 315
Closed-circuit television (CCTV) systems, 264–266
Clutter, 155
CLV system, 326
CNN, 201, 218, 219
Co-channel, 311
Coates, George A., 26
Coaxial cable, 321
Coca, Imogene, 44
Coincidental method, 238
Cold calls, 156
Columbia Phonograph Corporation, 26, 27
Combo operation, 279
Commercial and economic structure, 148–165
   advertisers, 151–157
   advertising agencies, 157, 158
   audience, 158–161
   audience demographics, 150
   cable economics, 161–163
   competition, 150
   coverage area, 150, 151
   market size, 150
   programming, 159
   promotion, 160, 161
   revenue sources, 163, 164
   sales department, 152–157
   selling tools, 154–157
   station representative companies, 158
Commercial television networks, 123
Committee on Children's TV v. General Foods, 110
Common Carrier Bureau, 101
Communication, 5
Communication distribution systems. *See* Distribution systems
Communication Satellite Act of 1962, 68
Communication satellites, 316–320
Communications Act of 1934, 86–88. *See also* Section 312-7A; Section 315
Compact disc (CD), 325
Compensation, 188
Competition, 150
Compliance and Information Bureau, 101
Computer-generated video, 293
Comsat, 68

Concept testing, 192
Condenser microphones, 275, 276
Conductors, 286
Constitutional law, 99
Construction permit (CP), 85
Consultants, 140–142, 222, 223, 239
Contemporary-hit radio (CHR), 169
Content analysis, 233, 349
Content providers, 138–140
Control track, 296
Conus Communications, Inc., 217
Convergence of technologies, 7, 8
Converging systems (telcos/cable), 71, 72
Cook, Fred J., 90
Coolidge, Calvin, 24, 79
Cooperative advertising programs, 112
Copy, 5
Copyright, 196
Copyright Act of 1976, 103, 112
Copyright Royalty Tribunal (CRT), 112
Corporate monopoly, 82, 109
Corporate stations, 20
Corporate underwriting, 157
Corporate videos, 255–263
   documentation, 260, 261
   future of, 263
   information, 259
   motivation, 260
   news releases, 262
   organizational structure of department, 262, 263
   orientation, 258
   product demonstrations, 261, 262
   public relations, 262
   sales information, 261
   training, 258, 259
Corporation for Public Broadcasting (CPB), 50, 51
Correll, Charles, 351
Cost per point (CPP), 154, 246
Cost per thousand (CPM), 154, 247
Coughlin, Charles, 29
Counterprogramming, 151, 191
Country, 170
Coverage area, 150, 151
Cowett, Ed, 70
CP, 85
CPB, 50, 51
CPBS-UIB, 27
CPM, 154, 247

CPP, 154, 246
Criminal Code, 104
Cross-ownership, 125
Cross-promotion, 160
Crossley, Archibald, 234
Crowley, Steve, 220
CRT, 113, 288
Crystallization, 354
CSRs, 132
Cultural imperialism, 331
Cume, 155
Cume persons, 240, 248
Customer service representatives (CSRs), 132
Cyclical programming, 192

DAB, 178, 284
Darrow, Clarence, 81
DAT machines, 281, 282
Data analysis, 239–241
Data communication networks, 72, 73
Data gathering, 237–239
Davis, Elmer, 32
DAW, 285
Dayparts
   radio, 167, 168
   TV, 186
Daytime, 186
DBSs, 75, 319, 320
de Forest, Lee, 15, 16, 66
de Gaulle, Charles, 334
Deceptive advertising, 109
Decoding technology, 287–291
Delayed electronic feed (DEF), 216
Delivery, 196
Demodulation, 283
Demodulator, 72
Demographic composition, 120
Demographics, 119, 234
Desensitization theory, 350
Designated market area (DMA), 235
DIALOG, 217
Diaries, 237, 238
Different-effects theory, 349
Digital audio broadcasting (DAB), 178, 284
Digital audio workstation (DAW), 285
Digital audiotape (DAT) recorders, 281, 282
Digital broadcasting, 283, 284
Digital recording, 299

Digital-tape recording process, 281
Digital technology, 8
Digital video effects, 294
Digital videodisc (DVD), 297, 327
Digitization, 327
Dill, Clarence C., 79, 80, 82, 107
Dingell, John, 106
Direct broadcast satellite (DBS), 75, 319, 320
Direct waves, 309
Disc jockey (DJ), 39–41, 167
Disco, 43
Discovery law, 99
Disneyland, 46
Display technologies, 287–291
Dissolve, 294
Distance learning, 263–268
Distribution systems, 302–329
    AM radio, 311
    broadcast communication systems, 304–320
    broadcast television, 313–315
    cable television systems, 321–324
    carrier wave, 308–311
    digitization, 327
    fiber-optic systems, 324, 325
    FM radio, 311, 312
    home audio, 325
    home video, 326
    lower-power television, 315
    microwave, 316
    optical audio, 325, 326
    optical video, 326, 327
    physical distribution, 325–327
    satellites, 316–320
    vertical-blanking services, 315
    video compression, 320, 321
    wired communication systems, 320–325
DJ, 39–41, 167
DMA, 235
DMA ratings, 244
Dolan, Chuck, 200
Domestic systems, 335
Donovan, William, 84
Double-billing, 112
Downlink, 318
Drawbaugh, Daniel, 64
Drive time, 168
Duke, Red, 220

DuMont Network, 44–46
Duopoly, 177
DVD, 297, 327

Early Bird, 68
Early-early morning, 186
Early fringe, 186
Early morning, 186
Easy listening, 170, 171
EBU, 333
Economics. *See* Commercial and economic structure
Edell, Dean, 220
Edison, Thomas, 63, 65, 286
Edit controllers, 296, 297
Editors, 212
Education-access channels, 203
Educational TV, 49–52
EEO programs, 109
Effective radiated power (ERP), 312
Effects/influence of the media, 348–357
    advertising, 355, 356
    children, 352, 353, 356
    effects theories, 349–351
    media stereotypes, 351, 352
    politics, 354, 355
    violence theories, 350, 351
Effects theories, 349–351
Efficiency of target audience, 246
Eisner, Michael, 133
Electromagnetic spectrum, 120, 285, 304
Electromagnetic waves, 305
Electron flow, 287
Electron gun, 288
Electronic manipulation equipment, 271
Electronic Media Rating Council, 144
Electronic meters, 238
Electronic news gathering (ENG) technology, 217, 218, 298
Elevator music, 43, 171
Elk Hills oil scandal, 79
Emmys, 144
Encoding equipment/technology, 271, 291–293
ENG technology, 217, 218, 298
Engineering societies, 144
Enhanced underwriting, 157
Equal employment opportunity (EEO) programs, 109

Erasable disc, 327
Ernst, M. L., 82
ERP, 312
ETV stations, 50
European Broadcasting Union (EBU), 333
European phony war, 30
Everson, George, 34
Exclusive audience, 244
Exclusivity, 196
Experimental research, 233, 349

f-stop, 291
Fade, 294
Faders, 278
Fairness doctrine, 89–91, 106, 107
False advertising, 109
Family viewing standard, 55
Fantasy era, 193
Faraway Hill, 45
Farnsworth, Philo T., 34, 35
FCC, 99–103
FCC rule-making process, 102
FCC rules/regulations, 104–112
Federal Communications Commission (FCC), 99–103
Federal Radio Commission (FRC), 85
Federal Trade Commission (FTC), 109
Fessenden, Reginald A., 15
Fiber-optic distribution systems, 324, 325
Field, Cyrus W., 61, 62
Field producers, 214
Financial consultants, 141
Financial interest and syndication (FinSyn) rules, 54, 135
First Amendment, 103
First network broadcast, 23, 24
First-run programs, 48, 140, 188, 195
First situation comedy, 45
First soap opera, 45
First televised presidential debate, 47
Flight, 153
Flux, 297
FM radio, 33, 34, 41–43, 311, 312
Focus groups, 159, 239
Footprint, 318
Forfeitures, 103
Format funding, 284
Format radio, 40–42

Format wheel, 168
Fowler, Mark, 93
Fox, Michael J., 228
Fox Broadcasting Corporation, 134
Franchising wars, 54
Frank N. Magid Associates, 239
Fraudulent billing, 112
FRC, 85
Freedom of Information Act, 104
Freeze-frame, 295
Frequency, 248, 273
Frequency discount, 155
Freud, Sigmund, 18
Fringe time, 140
FTC, 109
Fuchs, Michael, 200
Future media business, 10
Futures file, 212

Gain, 307
Game show scandal, 46, 47
Game shows, 193
Gatekeepers, 212
Geosynchronous satellite, 68
Germany, 338, 339
Ghosting, 313
Global village, 330–347
    Canada, 336
    China, 343
    Germany, 338, 339
    Great Britain, 337, 338
    Hong Kong, 343
    India, 341, 342
    international broadcasting, 333–335
    international telecommunication agreements, 332, 333
    Japan, 342
    Mexico, 337
    national broadcasting systems, 335, 336
    Russia, 339, 340
    Saudi Arabia, 340, 341
    Taiwan, 343, 344
    Turkey, 339
    worldwide media distribution, 344, 345
Good music, 170
Gosden, Freeman, 351
Gould, Jay, 62
Government-access channels, 203

Grade A contour, 313
Grade B contour, 313
Grand Ole Opry, 39
Gray, Elisha, 64
Great Britain, 337, 338
Great Debates Law, 92
Great Lakes opinion, 86, 88
Greene, Harold H., 70, 71
Grid rate card, 155
Gross impressions, 247
Gross ratings points (GRP), 246
Ground waves, 310
Group-owned stations, 124
GRP, 246
Guilds, 142, 143

Halberstam, David, 228
Hammocking, 190
Hanks, Tom, 228
Harding, Warren G., 18, 19, 79
Hargis, Billy J., 90
HBO, 55, 136, 199, 200
HDTV, 113–114, 290
Helical scan, 298
Heliograph, 59
Hennock, Freida, 44
Herrold, Charles D., 21
Hertz (Hz), 273
Hertz, Heinrich, 14, 273
Hertzian waves, 305
HFC systems, 324
"Hidden Medium: Educational Radio, The," 50
High-definition television (HDTV), 113–114, 290
Historical-critical research, 233
History in Sound, 31
Hollings, Earnest F., 106
Hollywood, 46
Home audio, 325
Home Box Office (HBO), 55, 136, 199, 200
Home video, 326
Homes using television (HUT), 241
Hong Kong, 343
Hooper, C. E., 234
Hoover, Herbert, 22, 28, 67, 80, 81, 83, 84
Horizontal programming, 190
Host-selling, 356

Hot clock, 168
Hubbard, Gardiner G., 64, 65
Hubbard, Mabel, 64
Hue, 289
HUT, 241
Hybrid fiber/coaxial (HFC) system, 324
Hypodermic needle, 349
Hypoing, 109

I Love Lucy, 44, 45
Idiot sitcom, 193
Imitative behavior, 350
In-flow, 189
In-house research, 239
In-tab diaries, 240
Indecency, 107, 108, 176, 177
India, 341, 342
Industrial age, 13
Industrial video, 255. *See also* Corporate videos
Influence of the media. *See* Effects/influence of the media
Infomercials, 156, 157, 262
Information programming. *See* News and information programming
Information storage equipment, 271
Inner workings. *See* Audio and video systems
Insert mode, 296
Instructional television, 263–268
Instructional television fixed service (ITFS), 266, 267, 316
Insulators, 286
Integrated Services Digital Network (ISDN), 73, 320
Intelsat, 333
Internal systems, 335
International advertising, 110–111
International aspects. *See* Global village
International broadcasting, 333–335
International Bureau, 101
International Morse Code, 60
International Radiotelegraphic Union (IRU), 333
International telecommunication agreements, 332, 333
International Telecommunication Union (ITU), 333
Internet, 3, 73
Inventory, 155
IRU, 333
ISDN, 73, 320

ISO/OSI, 320
Issue programming, 55
ITFS, 266, 267, 316
ITT, 63
ITU, 333

Japan, 342
Jawboning, 102
Jazz Singer, The, 67
Johnson, Nicholas, 109
Judson, Arthur, 26

Kahn, Frank, 104
Kaltenborn, H. V., 32, 207
KDKA, 21
Kennedy, John, 47, 92, 354
Key, 294
Kraft Television Theatre, 44
Ku-band, 307

La Follette, Robert, 81
La Palina Boy, 27
Laboratory research, 349
LANs, 73
Late fringe, 186
Lavaliers, 277
Lawyers and advertising, 110
Lazarsfeld, Paul, 354
LCD screen, 290
Lease access, 202
Leblanc, Maurice, 34
Legislation. *See* Regulations
Length of run, 196
Lessing, Lawrence, 33
Letters, 103
Levin, Jerry, 200
Libel and slander laws, 103
Libertarian theory, 332
License period, 196
Lichty, Lawrence, 187
Life and Legend of Wyatt Earp, 46
Light color wheel, 289
Lightwave guides, 324
Limbaugh, Rush, 174
Line monitor, 295
Liquid-crystal display (LCD), 284, 290
Listening locations, 244
LNA, 318
Local area networks (LANs), 73

Local news, 210, 211
Local-origination (LO) programming, 202
Local owners, 125, 126
Lone Ranger, 27
Long, Huey P., 29
LO programming, 202
Lotteries, 114
Low-noise amplifier (LNA), 318
Low-power television (LPTV), 315
LPTV, 315
Luminance, 289
Lux Radio Theatre, 39

M*A*S*H, 48, 49
Macdonald, Torbert, 93
Mackay, John W., 62
Magic bullet, 349, 354
Magnetic recording, 297, 325
Major audience appeals, 187
Management consultants, 141
Marconi, Guglielmo, 14, 15
Maritime radio, 16
Market area, 235, 236
Market audits, 250
Market size, 150
Market study, 173
Martin, Ann, 213
Mary Kay and Johnny, 45
Mass communication process, 5
Mass Media Bureau, 101
Maxwell, James Clerk, 14
MBS, 27, 87, 175
McCarthy, Joseph, 208
McLendon, Gordon, 40, 41
McGowan, William, 69, 70
MCI, 69
McLuhan, Marshall, 331
Mean world syndrome, 350
Meblin, Dave, 220
Media Audit, 235
Media barons, 109
Media conglomerates, 125, 133, 134
Media effects. *See* Effects/influence of the media
Media functions, 6, 7
Media megamergers, 130, 131
Media organization and ownership, 118–147
    affiliate contracts, 135

    ancillary businesses in broadcasting, 132–144
    broadcast station, 119–121
    business infrastructure, 144
    cable system organizational structure, 131, 132
    consultants, 140–142
    content providers, 138–140
    cross-ownership, 125
    group owners, 124
    local owners, 125, 126
    mergers, 130, 131
    MSOs, 127–130
    multinational aspects, 131
    music licensing, 143
    network/affiliate relationships, 134, 135
    networks, 133–136
    O&O stations, 122, 123
    ownership limits, 109, 121, 122, 177
    professional associations, 143, 144
    satellite-delivered cable networks, 136–138
    station organization, 126, 127
    unions/guilds, 142, 143
    vertical integration, 129, 130
Media stereotypes, 351, 352
Metro audience trends, 242
Metro market profile, 242
Metro survey area (MSA), 235
Mexico, 337
Microphones, 275–278
Microwave communication, 316
Millennium coverage, 189
Miller test, 108
Miller v. California, 108
Minow, Newton, 48, 102
Mixing, 278
MMDS, 316
Mobile telephone service, 74
Modeling theory, 350
Modem, 72
Modified Final Judgment, 70
Modulation, 311
Modulator, 72
Monitoring systems, 294, 295
Mood programming, 191
Morgan, J. P., 66
Morning drive time, 168
Morse, Samuel F. B., 14, 59–61

Morse code, 60, 62
Moving-coil microphone, 275, 276
Moving-coil speaker, 283
MSA, 235
MSNBC, 73
MSO, 127–130
MTV, 43
Multimedia, 8–10
Multiple multipoint distribution system (MMDS), 316
Multiple ownership, 109
Multiplexing, 312
Multisystem owner (MSO), 127–130
Murdoch, Rupert, 4, 122, 134, 344
Murrow, Edward R., 29, 32, 207, 208, 223, 224
Music formats, 43
Music licensing, 143
Music playlist, 174
Music sweep, 170
Music syndication services, 175
Must-carry rule, 53, 113, 198
Mutual Broadcasting System (MBS), 27, 87, 175

NAB, 55, 143
NAEB, 50
Narrowcasting, 191
National Association for Better Broadcasting v. FCC, 110
National Association of Broadcasters (NAB), 55, 143
National Association of Educational Broadcasters (NAEB), 50
National Association of Television Programming Executives (NATPE), 143
National broadcasting systems, 335, 336
National Cable Television Association (NCTA), 143
National Educational Radio (NER), 50, 51
National Educational Television (NET), 50
NATPE, 143
NBC, 25, 134
NBC Blue, 25
NBC Supernet, 73
NCE television stations, 49, 50
NCTA, 143
NER, 50, 51
NET, 50

Network/affiliate relationships, 134, 135
Network Cable MediaWatch service, 235
Network contract, 26
Network feeds, 216
Network news, 207–210
Network programming, 192, 193
Network rerun programs, 48
Network television program rating system, 351
Networks, 133–136
New Deal, 28, 29
News and information programming, 206–231
    ambush journalism, 228
    anchors, 214, 224, 228
    CNN, 218, 219
    consultants, 222, 223
    content treatment, 221
    lead story, 227
    leads (in and out), 222
    local news, 210, 211
    network feeds, 216
    network news, 207–210
    news bureaus, 216, 217
    news inserts, 220
    newsroom, 212–215
    programming strategies, 220–224
    promotions, 221, 222
    satellite news gathering, 217, 218
    scooping, 227
    specialized wire/computer services, 216
    technological innovation, 211, 212
    trends, 224–229
    video news releases, 219, 220
    wire services, 215, 216
News bureaus, 216, 217
News consultants, 142, 222, 223
News director, 212
News doctor, 228
News inserts, 220
News magazines, 193
News releases, 262
News star, 228
News talent, 223, 224
News/talk format, 172, 173
Newsroom, 212–215
Niche formats, 169, 171, 175
Nielsen Media Research Company, 142, 151, 234, 244–246

Nielsen Station Index, 234, 244–246
Nielsen Television Index, 234, 246
1927 Radio Act, 83–86
1960 Program Policy Statement, 88, 89
Nipkow, Paul, 34
Nixon, Richard, 47, 92, 354
Nobel, Edward J., 87
Nonbroadcast television, 255. *See also* Corporate videos
Noncommercial educational broadcasting, 49–52
Nondestructive edits, 285
Nonlinear editing system, 297
NSFNet, 73
NSI Plus, 234
Nynex, 71

O&O stations, 122, 123
Obscenity, 107–108
Off-line editing, 296
Off-network programs, 48, 140, 188, 195
Office of Administrative Law Judges, 100
Office of Engineering and Technology, 99
Office of Legislative and Intergovernmental Affairs, 100
Office of Plans and Policy, 99
Office of Public Affairs, 99
Office of the General Counsel, 100
Office of the Managing Director, 99
OFS channels, 316
Omnidirectional microphones, 276, 277
On-line editing, 296
Open-reel videotape recorder, 298
Operational regulations. *See also* Regulations, 98–117
    children's television advertising, 110
    equal employment opportunity, 109–110
    fairness doctrine, 106–107
    false/deceptive advertising, 109
    FCC, 99–103
    FCC rules/regulations, 104–112
    fraudulent billing, 112
    hypoing, 109
    indecency/obscenity, 107, 108
    international advertising, 110–111
    multiple ownership, 109

news distortion, 108
payola/plugola, 41, 111–112
professional advertising, 110
public files, 111
recordings, 111
section 312-7A, 104–105
section 315, 91, 92, 104–106
sponsored programs, 111
station identification, 111
statutory/related applied law, 103, 104
tobacco advertising, 112
Operational-fixed service (OFS) channels, 316
Optical audio, 325, 326
Optical video, 326, 327
Organization. See Media organization and ownership
Organizational video, 255. See also Corporate videos
Oscillation, 272
OSI model (ISO/OSI), 320
Our Country: Its Possible Future and Present Crisis, 16
Out-flow, 189
Overdubbing, 279
Overnight, 186
Overnight daypart, 168
Owned and operated (O&O) stations, 122, 123
Ownership. See Media organization and ownership
Ownership limits, 109, 121, 122, 177

Pacific Bell, 71
Pacifica case, 108
Paley, William S., 26, 27, 31, 44, 133, 134, 207
Paparazzi, 228
Parrington, V. L., 13
Parsimony principle, 192
Participation advertising, 47
Passive listeners, 239
Passive peoplemeter, 238
Pastore, John, 92, 93
Pay-per-view (PPV), 72, 163, 199
Payola, 41, 111–112
PBS, 51, 52, 188, 267
PC, 72, 73
PEG channels, 202
Penetration, 144

People's Republic of China, 343, 344
Peoplemeter, 238
Per-inquiry spots, 157
Perceptual surveys, 239
Persistence of vision, 285
Personal computer (PC), 72, 73
Personal digital assistant, 3
Personal interviews, 238
Persons estimates, 240
Peterson, Theodore, 331
Phasing, 310
Phonograph, 325
Physical distribution, 303, 325–327
Pickup pattern, 275
Pilot, 192
Playback amplifier, 280
Plugola, 111–112
Pocket peoplemeter, 238
POI programs, 261
Point-of-information (POI) programs, 261
Point-of-purchase (POP) programs, 261
Polarization, 310
Political campaign law, 91, 92
Politics, 354, 355
Pony Express, 59, 60
POP programs, 261
Positioning, 152
Post house, 296
Postal Telegraph, 62, 63
Postproduction, 295
Potentiometer, 278
Power programming, 191
Precedent case, 99
Precedent law, 99
Premium channels, 199
Presentation or display equipment, 271
Press-radio war, 31
Preview monitors, 295
Primary research, 233
Prime time, 186
Prime-time access rule (PTAR), 48, 135
Prime-time soaps, 193
Privacy laws, 104
Private television, 255. See also Corporate videos
Proc amp, 295
Processing amplifiers, 295

Producer, 212
Product, 139, 149, 186
Product demonstration, 261, 262
Product service information (PSI), 261
Production console, 279
Production switchers, 293, 294
Profane material, 107, 108
Professional advertising, 110
Professional associations, 143, 144
Program-appeal research, 186, 187
Program averages, 244
Program directors, 173, 174
Program-length commercials, 356
Program policy statement of 1960, 88, 89
Program title index, 244
Programming, 159
   news, 206–231. See also News and information programming
   radio, 166–181. See also Radio programming
   TV, 182–205. See also Broadcast television programming
Programming and news consultants, 142
Progressive rock, 43
Projection television systems, 290
Promise-versus-performance regulation, 89
Promos, 153
Promotion(s), 160, 161, 196, 221, 222
Propaganda, 334
Proximity effect, 275
PSAs, 153
PSI, 261
Psychographics, 155
PTAR, 48, 135
Public-access channel, 202
Public broadcasting, 50–52
Public Broadcasting Act, 51
Public Broadcasting Service (PBS), 51, 52, 186, 267
Public files, 111
Public interest standard, 86
Public relations, 160, 161, 262
Public service announcements (PSAs), 153

Qualitative research, 233
Quantitative research, 233
Quarter-hour persons, 240

Radio
    AM, 311
    AM band expansion/stereo, 113
    censorship, 80–82
    FM, 33, 34, 41–43, 311, 312
    history, 12–35, 39–43
    maritime, 16
    music formats, 43
    payola, 41, 111–112
    production and distribution, 139
    programming, 166–181. *See also* Radio programming
    public interest standard, 86
    research reports, 242–244
    Top 40 format, 41
Radio Act of 1912, 17, 18
Radio Act of 1927, 83–86
Radio and Television News Directors Association (RTNDA), 143
Radio broadcast data system (RBDS), 178
Radio clock, 168, 169
Radio dayparts, 167, 168
Radio frequency (RF), 309
Radio Martí, 335
Radio pioneers, 13
Radio programming, 166–181
    clock, 168, 169
    consolidation, 177, 178
    dayparts, 167, 168
    indecency, 176, 177
    information (news/talk format), 173
    local programming, 173, 174
    music formats, 169–172
    new technology, 178
    sources of programming, 174, 175
    syndication, 175
Radio syndication, 175
Radio wave, 282
Raised eyebrow, 85, 88
Random sample, 236
Rate cards, 155, 248
Rating(s), 234, 240, 241
Ratings books, 242–244
Ratings numbers, 154
RBDS, 178
RBOCs, 7, 71
RCA, 19, 20, 24, 25
Reach, 240
Read-only videodisc, 327

Reagan, Ronald, 70
Reality programs, 187
Reality shows, 136
Recall method, 238
Receiver, 287–289
Record amplifier, 280
Recorders, 279, 280
Recordings, 111
Red Lion decision, 89, 90
Reflective signal paths, 314
Regional Bell operating companies (RBOCs), 7, 71
Regulations
    cable/new technology, 113–115
    Communication Act of 1934, 86–88
    FCC, 99–103
    1927 Radio Act, 83–86
    operational, 98–117. *See also* Operating regulations
    section 312-7A, 104–105
    section 315, 91, 92, 104–106
    Telecommunications Act of 1996, 93–95
Reinforcement, 354
Reis, Philipp, 63
Reiser, Paul, 228
Relevance era, 193
Religious stations, 171
Report on Chain Broadcasting, 87
Reporter, 214
Reruns, 48, 140, 196
Research, 233. *See also* Audience analysis and marketing
Research reports, 242–246
Research sample, 236, 237
Retransmission consent, 113, 198
Reuters, 216
Revenue sources, 163, 164
Review Board, 100
RF (radio frequency), 309
Ribbon microphone, 275, 276
Ripley, Joseph, 187
Rockefeller, John D., 13
ROD, 327
Roosevelt, Franklin D., 28, 29
ROS, 153, 155
Rosenburgh, Leo H., 21
Rough cut, 296
RS-232-C code, 320
RTNDA, 143

Run of schedule (ROS), 153, 155
Runyon, Randy, 33
Russia, 339, 340

S-VHS, 299
Sales department, 152–157
Sales/marketing applications, 246–250
Sales service, 153
Sample, 236, 237
Sampling frame, 236
Sarnoff, David, 17, 22, 25, 26, 33–35, 133
Sarnoff, Robert, 92
Satellite-delivered cable networks, 136–138
Satellite dishes, 128, 136, 319
Satellite master-antenna television (SMATV), 319
Satellite news gathering (SNG), 217, 218
Satellites, 316–320
Saturation, 289
Saudi Arabia, 340, 341
Scanning, 288
Scarborough Report, 235, 249
Scheduling strategies, 188–192
Schlessinger, Laura, 174
Schramm, Wilbur, 207, 331, 349
Scooping, 227
Scopes monkey trial, 80, 81
Second season, 192
Secondary research, 233
Section 312-7A, 104–105
Section 315, 91, 92, 104–106
SEG, 294
Selling tools, 154–157
Semaphore, 59
Semiconductors, 286
Servo-zoom mechanisms, 291
"Seven Dirty Words," 108
Share, 240, 241
SHF, 307
Shirer, William L., 207
Shock jocks, 176
Short-term and conditional renewals, 103
Sidebands, 312
Sidelobe, 318
Siebert, Fred, 331
Siepnamm, Charles A., 88
Sign off, 310

Signal distribution, 282
Signal generation, 275
Signal manipulation, 293–297
Signal processing, 278
Signal reception, 282
Signal storage, 279
Signal transmission equipment, 271
Silent-witness program, 108
Sixth Amendment, 103
Sixth Report and Order, 44, 50
60 Minutes, 47
$64,000 Question, The, 47, 156
Sky waves, 310
Sliders, 278
Slivercasting, 191
SMATV, 319
Smith, Howard K., 47
SMPTE time code, 296
SNG, 217, 218
Social responsibility theory, 332
Socially relevant issues, 55
Sound, 272
Sound bites, 5
Sound path, 274–285
Sources of media law, 99
Southwestern Bell, 71
Soviet Communist theory, 332
Spanish-language stations, 171
Spatial redundancy, 321
SPC, 51
Speaker, 283
Special-effects generator (SEG), 294
Specialized and foreign wire services, 216
Spinoff, 193
Sponsored programs, 111
Spots, 150, 246
Standard television (STV), 308
Stanton, Frank, 209
Star-wired distribution system, 323
Station identification, 111, 153, 168
Station Program Cooperative (SPC), 51
Station promotional announcements (promos), 153
Station representative companies, 158
Statutory laws, 99
Steinmetz, Charles P., 16
Stereotyping, 351, 352
Stern, Howard, 108, 174, 176
Stillframe, 299

Stop set, 168
Storz, Robert H., 40
Storz, Todd, 40, 41
Streisand, Barbra, 228
Stripping, 190, 191
Stunting, 191
STV, 308
Subscription fees, 164
Subscription television stations, 163
Summers, Harrison, 187
Superhigh frequency (SHF), 307
Superstations, 199
Survey research, 233
Sync generator, 288
Syndicated exclusivity rule, 53
Syndicated programs, 48
Syndication
    radio, 175
    TV, 48, 139, 140, 193–196

Tabloid journalism, 108
Tabloid photographers, 228
Taiwan, 343, 344
Talent, 214, 223, 224
Talk-radio programs, 174
Tape dropout, 281
TCP/IP computer protocol, 73, 320
Teapot Dome oil scandal, 79
Technical consultants, 141
Telcos, 72
Telecommunication, 59
Telecommunication networks, 73
Telecommunications Act of 1996, 54, 72, 93–95, 99, 113
Telegraph, 13, 14, 59–63
Telegraph Hill, 59
Telegraphy, 14
Telephone, 63–71
    AT&T (Ma Bell), 67–70
    breakup of AT&T, 69–71
    converging telephone and cable, 71, 72
    early competition, 66
    expanding business, 65, 66
    invention, 63–65
    satellite communications, 68
Telephone bypass, 72
Telephone directory, 236
Telephone interviewing, 238
Telephony, 14

Teleport, 318
Teleprinter, 62
Teletext, 315
Television
    cable TV, 52–54
    children's advertising, 110
    commercial networks, 123
    educational TV, 49–52
    Emmys, 144
    freeze of new applications, 44
    game show scandal, 46, 47
    golden age, 44–46
    growth/transition, 43
    HDTV, 113–114
    history, 34, 35, 43–52
    Hollywood, move to, 46
    issue programming, 55
    LPTV, 315
    O&O stations, 123
    programming, 182–205.
        See also Broadcast television programming
    PTAR, 48
    public broadcasting, 50–52
    rating system, 351
    research reports, 244–246
    syndication, 48, 139, 140, 193–196
    UHF channels, 313
    VHF channels, 313
Television dayparts, 186
Television household (TVHH), 237
Television Martí, 335
Television production companies, 139
Television receiver, 287–289
Television storage equipment, 297–299
Telstar, 68
Temporal redundancy, 321
Tentpoling, 190
Terry, E. M., 21
Texaco Star Theater, 44, 156
Theater testing, 239
Thomas, Lowell, 207
Thomas, Norman, 81
Time-base correctors, 295
Time period estimates, 244
Time shifting, 238
Time spent listening (TSL), 244, 246
Tisch, Laurence A., 134
Titanic, 16, 17
Tobacco advertising, 112

Tokyo Rose, 334
Toll broadcasting, 21–23
Tools of the trade, 271. *See also* Audio and video systems
Top 40 format, 40, 41, 169
Tower Antennas, 128
Trade-outs, 157
Traffic, 153
Transduction, 282
Translator, 315
Tree-wired distribution system, 323
TSL, 244, 246
Turkey, 339
Turner, Ted, 136, 199, 201, 218, 317
TV. *See* Television
TV camera, 291, 292
TVHH, 237
TVRO dishes, 128, 136, 319
Twentieth Century: The Nightmare Years (Shirer), 207
Twenty-One, 47
Twisted-pair wires, 320

U.S. Neutrality Acts, 30
U.S. v. AT&T, 70
US West, 71
UC, 170
UHF channels, 313
UIB, 26
Underwriting announcements, 157
Unidirectional microphone, 277, 278
Unions, 142, 143
United Independent Broadcasters, Inc. (UIB), 26
United Press International (UPI), 215
Uplink, 318
Urban contemporary (UC), 170

Vail, Alfred, 61
Vail, Theodore N., 65, 66
Van Deerlin, Lionel, 93
VBI, 315
VCR, 326
Vectorscope, 295
Vertical-blanking signal (VBI), 315
Vertical integration, 129, 130
VHF channels, 313
VHS, 298, 299
Video camera, 291–293
Video compression, 320, 321
Video control room, 271
Video DJ, 43
Video news release (VNR), 219, 220, 262
Video on demand (VOD), 72
Video production switcher, 293, 294
Video switcher, 293
Videocassette recorder (VCR), 326
Videodisc, 326, 327
Videotape recording, 297–299
Viewer paying, 164
Violence theories, 350, 351
Vision, 285
Vitaphone, 67
VJ, 43
VNR, 219, 220, 262
VOA, 334, 335
VOD, 72
Voice of America (VOA), 334, 335
Voice-over, 279
VSAT, 75
VU meters, 278

War of the Worlds, 29, 30
Watson, Thomas, 64
Wave refraction, 305, 306

Waveform monitor, 295
Wavelength, 305
Wavelengths, 17, 273
WEAF, 21–23
Welles, Orson, 29, 30
West, Mae, 107
Westar, 63
Western Electric, 67, 68
Western Union, 62, 63
Westwood One Radio Networks, 174, 175
Wheatstone, Sir Charles, 59
White, Paul, 207
White, Wallace H., Jr., 80, 82–84
White Radio Bill, 83
Who Wants to Be a Millionaire, 193
Wilson, Woodrow, 66
Windowing, 195
Wire services, 215, 216
Wired communication systems, 303, 320–325
Wireless cable, 307
Wireless telecommunication networks, 74, 75
Wireless Telecommunications Bureau, 101
Worldwide media distribution, 344, 345
WORM, 327
Write-once disc, 327
Writer's Guild v. FCC, 55
WTVH, 107

Your Hit Parade, 39
Your Show of Shows, 44

Zapple, Nicholas, 106
Zapple doctrine, 106
Zoom lens, 291
Zworykin, Vladimir K., 34, 35